AQA
A-level

Biology

Pauline Lowrie
Mark Smith

Mike Bailey
Bill Indge
Martin Rowland

Approval message from AQA

This textbook has been approved by AQA for use with our qualification. This means that we have checked that it broadly covers the specification and we are satisfied with the overall quality. Full details of our approval process can be found on our website.

We approve textbooks because we know how important it is for teachers and students to have the right resources to support their teaching and learning. However, the publisher is ultimately responsible for the editorial control and quality of this book.

Please note that when teaching the *AQA A-level Biology* course, you must refer to AQA's specification as your definitive source of information. While this book has been written to match the specification, it does not provide complete coverage of every aspect of the course.

A wide range of other useful resources can be found on the relevant subject pages of our website: www.aqa.org.uk.

HODDER
EDUCATION
AN HACHETTE UK COMPANY

Orders: please contact Hachette UK Distribution, Hely Hutchinson Centre, Milton Road, Didcot, Oxfordshire, OX11 7HH. Telephone: +44 (0)1235 827827. Email education@hachette.co.uk. Lines are open from 9 a.m. to 5 p.m., Monday to Friday. You can also order through our website: www.hoddereducation.co.uk

AQA A-level Biology Student Book 1 published in 2015

AQA A-level Biology Student Book 2 published in 2015

This combined edition published in 2019 by
Hodder Education,
An Hachette UK Company
Carmelite House
50 Victoria Embankment
London EC4Y 0DZ

www.hoddereducation.co.uk

Impression number   10 9 8 7 6 5

Year   2023 2022

Cover photo

Typeset in 11/13 pt ITC Berkeley Oldstyle Std by Aptara, Inc.

Printed by CPI Group (UK) Ltd, Croydon, CR0 4YY

A catalogue record for this title is available from the British Library.

ISBN: 978 1 5104 6978 5

MIX
Paper | Supporting
responsible forestry
FSC™ C104740
FSC
www.fsc.org

Contents

Go online for answers and extended glossaries:
www.hoddereducation.co.uk/AQABiology

Get the most from this book

Welcome to the **AQA A-level Biology Student's Book**. This book covers all content for the AQA AS Biology specification and the AQA A-level Biology specification.

The following features have been included to help you get the most from this book.

Prior knowledge

This is a short list of topics that you should be familiar with before starting a chapter. The questions will help to test your understanding.

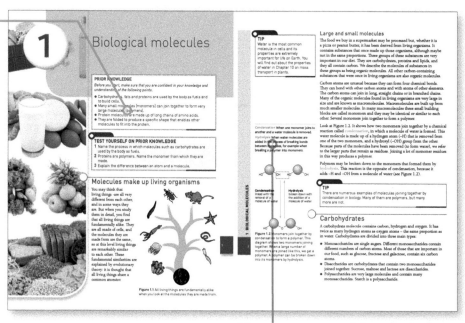

Key terms and formulae

These are highlighted in the text and definitions are given in the margin to help you pick out and learn these important concepts.

Tips

These highlight important facts, common misconceptions and signpost you towards other relevant topics.

Test yourself questions

These short questions, found throughout each chapter, are useful for checking your understanding as you progress through a topic.

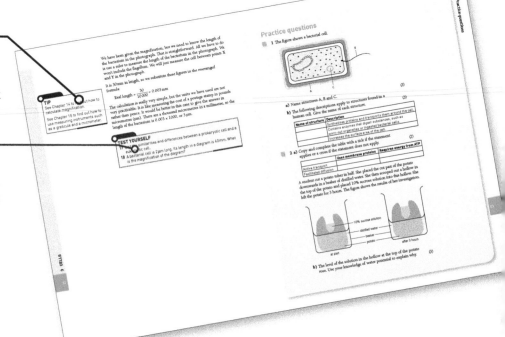

Activities and Required practicals

These practical-based activities will help consolidate your learning and test your practical skills. AQA's required practicals are clearly highlighted.

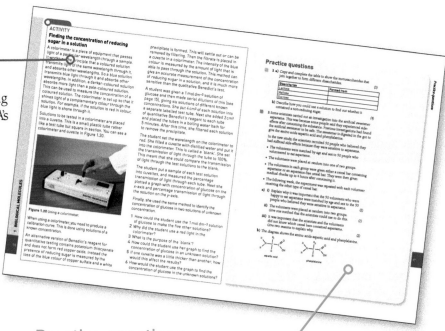

Practice questions

You will find Practice questions at the end of every chapter. These follow the style of the different types of questions you might see in your examination, including multiple-choice questions, and are colour coded to highlight the level of difficulty. Test your understanding even further, with Maths questions and Stretch and challenge questions.

- Green – Basic questions that everyone should be able to answer without difficulty.
- Orange – Questions that are a regular feature of exams and that all competent candidates should be able to handle.
- Purple – More demanding questions which the best candidates should be able to do.
- Stretch and challenge – Questions for the most able candidates to test their full understanding and sometimes their ability to use ideas in a novel situation.

Extension

Throughout the book you will also find Extension boxes, which contain extra material to deepen your understanding of a topic.

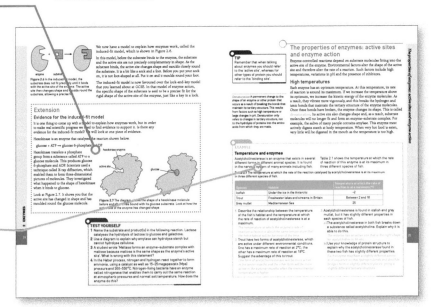

Examples

Examples of questions and calculations feature full workings and sample answers.

Dedicated chapters for developing your **Maths** and **Practical skills** and **Preparing for your exam** can be found at the back of this book.

Acknowledgements

The Publisher would like to thank the following for permission to reproduce copyright material:

p.1 © evgenia sh – Fotolia; **p.3** © Yi Liu – Fotolia; **p.5** © Photo Insolite Realite/Science Photo Library; **p.16** © Martyn F. Chillmaid/Science Photo Library; **p.19** © fotografiche. eu – Fotolia; **p.28** © Patrik Stedrak – Fotolia; **p.36** © Sinclair Stammers/Science Photo Library; **p.37** *t* © Dr Jeremy Burgess/ Science Photo Library, *b* © Science Photo Library; **p.39** *t* © CNRI/ Science Photo Library, *b* © Susumu Nishinaga/Science Photo Library; **p.40** © A.R. Cavaliere, Dept of Biology, Gettysburg College, Gettysburg; **p.42** *t* © Biophoto Associates/Science Photo Library, *b* © Sinclair Stammers/Science Photo Library; **p.50** © Dr Gopal Murti/Science Photo Library; **p.51** © Moredun Animal Health Ltd/Science Photo Library; **p.56** *l* © Adrian T Sumner/Science Photo Library, *r* © A. Barrington Brown/Science Photo Library; **p.57** © Science Photo Library; **p.62** © Adrian T Sumner/Science Photo Library; **p.73** © M.I. Walker/Science Photo Library; **p.74** © REUTERS/Daniel LeClair; **p.78** © Power And Syred/Science Photo Library; **p.80** © M.I. Walker/Science Photo Library; **p.82** © CNRI/Science Photo Library; **p.87** © yuuuu – Fotolia; **p.88** © Peter Menzel/Science Photo Library; **p.89** © Steve Gschmeissner/Science Photo Library; **p.90** © Dr Klaus Boller/Science Photo Library; **p.108** © Photographee. eu – Fotolia; **p.110** © Pascal Goetgheluck/Science Photo Library; **p.112** © Microfield Scientific Ltd/Science Photo Library; **p.115** © eAlisa – Fotolia; **p.116** © Gordon Museum, GKT Medical School, King's College London; **p.119** *l* © James Steveson/Science Photo Library, *r* © CNRI/Science Photo Library; **p.124** *both* © James Steveson/Science Photo Library; **p.130** © Eye Of Science/ Science Photo Library; **p.131** © Power And Syred/Science Photo Library; **p.139** © Elena Schweitzer – Fotolia; **p.140** © Henri Bureau/Sygma/Corbis/VCG via Getty Images; **p.142** © Elena Schweitzer – Fotolia; **p.154** © Susumu Nishinaga/Science Photo Library; **p.156** © CNRI/Science Photo Library; **p.159** *t* © Mauro Fermariello/Science Photo Library, *b* © Andy Crump, TDR, WHO/Science Photo Library; **p.160** © D. Phillips/Science Photo Library; **p.165** © John Radcliffe Hospital/Science Photo Library; **p.167** © Dr Vim Jesudason; **p.168** *all* © Dr Vim Jesudason; **p. 169** *all* © Dr Vim Jesudason; **p.181** *l* © Garry DeLong – Fotolia, *m* © Steve Gschmeissner/Science Photo Library, *r* © Dr Keith Wheeler/Science Photo Library; **p.184** © enskanto – Fotolia; **p.185** © Dr Keith Wheeler/Science Photo Library; **p.189** © Brookhaven National Laboratory/Science Photo Library; **p.192** © hecke71 – Fotolia; **p.193** © L. Willatt, East Anglian Regional Genetics Service/Science Photo Library; **p.196** © Alila Medical Images/Alamy; **p.198** *l* © Piotr Filbrandt – Fotolia, *r* © daibui – Fotolia; **p.202** © CDC/Gilda L. Jones, Courtesy Public Health Image Library; **p.210** © nickos68 – Fotolia; **p.211** *l* © Fotokon – Fotolia, *r* © fuujin – Fotolia; **p.213** *l* © ihervas – Fotolia, *r* © plrang – Fotolia; **p.214** *t* © birdiegal – Fotolia, *b* © Erni – Fotolia; **p.215** *l* © helmutvogler – Fotolia, *m* © Rich Lindie – Fotolia, *r* © nickos68 – Fotolia; **p.220** © Kajornyot – Fotolia; **p.225** © Jim Laws/Alamy; **p.231** © Mark Smith; **p.232** *t* © Mark Smith, *b* © andrewmroland – Fotolia; **p.235** © david crosbie – Fotolia; **p.239** © Yi Liu – Fotolia.com; **p.242** © Dr. Jeremy Burgess/ Science Photo Library; **p.243** © Mark Smith; **p.251** © tsach – Fotolia.com; **p.256** © Bill Indge; **p.258** © Bill Indge; **p.262** © birdiegal – Fotolia.com; **p.267** © Sylvie Bouchard – Fotolia. com; **p.268** © Gary Braasch/Getty Images; **p.272** © Bill Indge;

p.275 © Dr Jeremy Burgess/Science Photo Library; **p.279** © SOLLUB – Fotolia.com; **p.281** © blickwinkel/Alamy; **p.286** © tilzit – Fotolia.com; **p.287** © Erni – Fotolia.com; **p.290** © Piotr Naskrecki/Getty Images; **p.293** © Graham Jordan/Science Photo Library; **p.299** © Image Source/Alamy; **p.301** © CNRI/ Science Photo Library; **p.307** © Ed Reschke/Getty Images; **p.308** © Science Pictures Limited/Science Photo Library; **p.316** © Dr. Don Fawcett/Getty Images; **p.318** © Manfred Kage/Science Photo Library; **p.324** © jirapong – Fotolia.com; **p.325** *l* © Luke Horsten/Getty Images, *r* © Woodfall/Photoshot; **p.327** *l* © Lehtikuva OY/REX, *r* © Mark Dadswell/Getty Images Sport; **p.330** *l* © Steve Gschmeissner/Science Photo Library, *r* © Biology Media/Science Photo Library; **p.331** © Don Fawcett/Science Photo Library; **p.338** © Dr. Fred Hossler/Visuals Unlimited, Inc.; **p.340** © Ed Reschke – Oxford Scientific/Getty Images; **p.343** © Steve Gschmeissner – Science Photo Library/Getty Images; **p.360** © Carolina Biological Supply, Co/Visuals Unlimited, Inc.; **p.363** © Biosphoto Superstock; **p.368** Used with permission, courtesy of the Wisconsin Fast Plants Program, College of Agricultural and Life Sciences: University of Wisconsin-Madison: www.fastplants. org; **p.387** © Chris Mattison/FLPA; **p.388** © giedriius – Fotolia. com; **p.389** *l* © Piotr Filbrandt – Fotolia.com, *r* © daibui – Fotolia.com; **p.395** © Eye Of Science/Science Photo Library; **p.397** *tl* © kjekol – Fotolia.com, *tr* © JohanSwanepoel – Fotolia. com, *bl* © zuzule – Fotolia.com, *br* © Reddogs – Fotolia. com; **p.398** © Steve Hopkin/Ardea; **p.400** © Gail Johnson – Fotolia. com; **p.404** © Soru Epotok – Fotolia.com; **p.405** © Mark Smith; **p.410** © arolina66 – Fotolia.com; **p.416** *l* © pisotckii – Fotolia. com, *r* © Andrea Wilhelm – Fotolia.com; **p.417** © Nazzu – Fotollia.com; **p.419** *both* © Bill Indge; **p.420** © alekswolff – Fotolia.com; **p.422** © videnko – Fotolia.com; **p.426** © DenisNata – Fotolia.com; **p.427** © Getty Images; **p.435** © romaneau – Fotolia.com; **p.436** © Mark Thomas/Science Photo Library; **p.444** © MBI/Alamy; **p.456** © Maximilian Stock Ltd/Science Photo Library; **p.457** *t* © Naturepix/Alamy, *b* © Nick Cobbing/ Rex Features; **p.463** © Tek Image – Science Photo Library/Getty Images; **p.464** © David Parker/Science Photo Library; **p.478** © Martin Shields/Alamy; **p.485** *t* © Biology Media/Science Photo Library, *b* © Dr. Gladden Willis – Visuals Unlimited/Getty Images; **p.492** © josephotographie.de – Fotolia.com; **p.500** © Mark Smith; **p.502** © Mark Smith **p.503** *tl* © Wikipedia: http:// commons.wikimedia.org/wiki/Category:Colorimeters#/media/ File:Chlorine_colorimeter.JPG, *tm* © Joao Inacio/Getty Images, *tr* © Filip Ristevski - Fotolia, *b* © Martyn F. Chillmaid/Science Photo Library; **p.505** *l* © Alan John Lander Phillips – E+/Getty Images, *r* © Steve Gschmeissner/Science Photo Library; **p.506** © Serpil_ Borlu – iStock via Thinkstock; **p.507** *t* © Dr Keith Wheeler/ Science Photo Library, *b* © Dr. Don Fawcett/Getty Images; **p.508** *both* © Mark Smith; **p.509** *t* © Pauline Calder, *b* Courtesy of Dominikmatus via Wikipedia (http://creativecommons.org/ licenses/by-sa/2.5/); **p.510** *t* © Mark Smith; **p.511** *t* © Mark Smith, *b* © Alexander Gospodinov – Fotolia; **p.512** © Mark Smith; **p.513** © Dr Brian Beal; **p.514** *both* © Mark Smith; **p.517** © milanmarkovic78 – Fotolia.com.

t = top, *b* = bottom, *l* = left, *r* = right, *m* = middle

Every effort has been made to trace all copyright holders, but if any have been inadvertently overlooked, the Publisher will be pleased to make the necessary arrangements at the first opportunity.

1

Biological molecules

TEST YOURSELF ON PRIOR KNOWLEDGE

1 Name the process in which molecules such as carbohydrates are used by the body as fuels.
2 Proteins are polymers. Name the monomer from which they are made.
3 Explain the difference between an atom and a molecule.

Molecules make up living organisms

You may think that living things -are all very different from each other, and in some ways they are. But when you study them in detail, you find that all living things are fundamentally alike. They are all made of cells, and the molecules they are made from are the same, so at this level living things are remarkably similar to each other. These fundamental similarities are explained by evolutionary theory: it is thought that all living things share a common ancestor.

Figure 1.1 All living things are fundamentally alike when you look at the molecules they are made from.

1

Large and small molecules

The food we buy in a supermarket may be processed but, whether it is a pizza or peanut butter, it has been derived from living organisms. It contains substances that once made up those organisms, although maybe not in the same proportions. Three groups of these substances are very important in our diet. They are carbohydrates, proteins and lipids, and they all contain carbon. We describe the molecules of substances in these groups as being organic molecules. All other carbon-containing substances that were once in living organisms are also organic molecules.

Carbon atoms are unusual because they can form four chemical bonds. They can bond with other carbon atoms and with atoms of other elements. The carbon atoms can join in long, straight chains or in branched chains. Many of the organic molecules found in living organisms are very large in size and are known as macromolecules. Macromolecules are built up from much smaller molecules. In many macromolecules these small building blocks are called monomers and they may be identical or similar to each other. Several monomers join together to form a polymer.

Look at Figure 1.2. It shows how two monomers join together by a chemical reaction called condensation, in which a molecule of water is formed. This water molecule is made up of a hydrogen atom (–H) that is removed from one of the two monomers, and a hydroxyl (–OH) group from the other. Because parts of the molecules have been removed (to form water), we refer to the larger parts that remain as residues. Joining a lot of monomer residues in this way produces a polymer.

Polymers may be broken down to the monomers that formed them by hydrolysis. This reaction is the opposite of condensation, because it adds –H and –OH from a molecule of water (see Figure 1.2).

Condensation When one monomer joins to another and a water molecule is removed.

Hydrolysis When water molecules are added in the process of breaking bonds between molecules, for example when breaking a polymer into monomers.

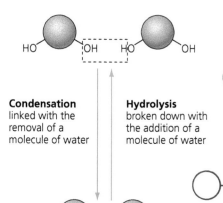

Condensation
linked with the removal of a molecule of water

Hydrolysis
broken down with the addition of a molecule of water

Figure 1.2 Monomers join together by condensation to form a polymer. This diagram shows two monomers joining together. When a large number of monomers are joined like this, we get a polymer. A polymer can be broken down into its monomers by hydrolysis.

Carbohydrates

A carbohydrate molecule contains carbon, hydrogen and oxygen. It has twice as many hydrogen atoms as oxygen atoms – the same proportion as in water. Carbohydrates are divided into three main types:

- Monosaccharides are single sugars. Different monosaccharides contain different numbers of carbon atoms. Most of those that are important in our food, such as glucose, fructose and galactose, contain six carbon atoms.
- Disaccharides are carbohydrates that contain two monosaccharides joined together. Sucrose, maltose and lactose are disaccharides.
- Polysaccharides are very large molecules and contain many monosaccharides. Starch is a polysaccharide.

Figure 1.3 Most of the carbohydrate that we eat comes from plants. This crop is sugar cane, a plant that stores sucrose in its stem. The carbohydrate stored by other food plants such as potatoes and cereals is starch.

Glucose and other sugars

Glucose is a monosaccharide, so it is a single sugar. Its molecular formula is $C_6H_{12}O_6$. This formula simply tells us how many atoms of each element there are in each glucose molecule.

Now look at the structural formulae shown in Figure 1.4. They show a molecule of α-glucose and a molecule of β-glucose. Count each type of atom in diagram (a). There are 6 carbon atoms, 12 hydrogen atoms and 6 oxygen atoms, equal to the numbers of different atoms shown by the molecular formula, $C_6H_{12}O_6$. This diagram also shows you how the atoms are arranged.

All glucose molecules have the same formula, $C_6H_{12}O_6$. However, there are two different kinds of glucose. This is because the atoms in the glucose molecule can be arranged in different ways, called isomers. Figure 1.4 shows the arrangement of the atoms in the two different kinds of glucose.

(a)
(b)

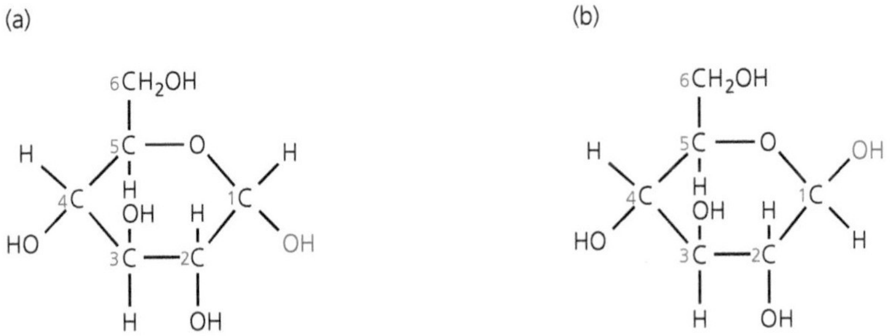

Figure 1.4 (a) An α-glucose molecule; and (b) a β-glucose molecule.

Look at the way that the –H and –OH groups are bonded to the carbon atom on the right-hand side (1C) in β-glucose. Now look at the –H and –OH groups bonded to the carbon atom on the left-hand side (C4). Notice that they are bonded the opposite way round. Compare this with the diagram of α-glucose. Here, both –H groups are above the carbon atoms, and both –OH groups are below the carbon atoms.

Galactose and fructose are also monosaccharides and have exactly the same molecular formula as α-glucose. However, the atoms that make up these molecules are arranged in different ways. This means that, although all three substances are sugars, they have slightly different structures. This gives them slightly different properties.

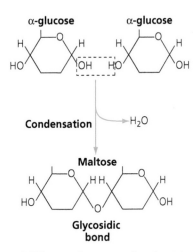

Figure 1.5 Two α-glucose molecules join together by condensation to give a molecule of the disaccharide maltose.

Monosaccharides such as α-glucose are the monomers that join together to make many other carbohydrates. Two α-glucose molecules join by condensation to form a molecule of the disaccharide maltose. The bond forms between carbon 1 of one α-glucose molecule and carbon 4 of the other; such a bond is called a glycosidic bond (see Figure 1.5).

Other disaccharides form in a similar way. Lactose, for example, is the sugar found in milk. It is formed in a condensation reaction between a molecule of α-glucose and a molecule of another monosaccharide, galactose. Sucrose is formed from α-glucose and fructose.

Glycosidic bond A chemical bond formed as the result of condensation between two monosaccharides.

Reduction is gain of electrons or hydrogen.

When sugars such as α-glucose are boiled with Benedict's solution, an orange precipitate is formed because Cu(II) ions in the Benedict's solution are reduced to orange Cu(I) ions. This reaction occurs because of the way the chemical groups are arranged in such sugars. These sugars are therefore called reducing sugars. Fructose, maltose and galactose are also reducing sugars.

Extension

Structure of amylose and amylopectin

Figure 1.6 shows the structure of starch. You can see that amylose is a long chain of α-glucose molecules. They are linked by 1,4-glycosidic bonds. This chain is coiled into a spiral and its coils are held in place by chemical bonds called hydrogen bonds. Amylopectin is also a polymer of α-glucose but its molecules are branched due to 1,6-glycosidic bonds.

In amylose, the α-glucose molecules are linked by 1,4-glycosidic bonds. Notice that the −CH₂OH

side-chains all stick out on the same side. This arrangement causes the chains of α-glucose molecules to coil into spirals as shown in Figure 1.6. Amylopectin molecules have branches because some of the α-glucose molecules form bonds between carbon atoms 1 and 6 instead of 1 and 4. This enables starch molecules to fold up compactly.

Amylose
consists of a long chain of α-glucose residues

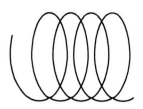

The chain is coiled into a spiral. Hydrogen bonds hold this spiral in shape.

Amylopectin
consists of branched chains of α-glucose residues

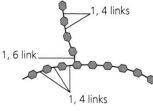

Figure 1.6 Starch consists of amylose and amylopectin. Starch from different plants contains different amounts of these two substances.

TIP
You can remember that oxidation is loss of electrons or hydrogen and reduction is gain of electrons or hydrogen by using the mnemonic OILRIG: Oxidation Is Loss, Reduction Is Gain.

Sucrose does not give an orange precipitate with Benedict's solution; it is a non-reducing sugar. However, when boiled with dilute acid, sucrose is hydrolysed to monosaccharides. The sucrose molecules are hydrolysed into α-glucose and fructose, both reducing sugars. Then it will give a positive test with Benedict's solution. (See page 14 for details of qualitative tests.)

Starch

Starch, a substance found in plants, is one of the most important fuels in the human diet. It makes up about 30% of what we eat. Starch is a mixture of two substances, amylose and amylopectin. Both these substances are polymers made from a large number of α-glucose molecules joined together by condensation reactions. In the biochemical test for starch, you add a drop of iodine solution. A blue-black colour indicates the presence of starch.

Storage molecules

Starch for storage

We use the starch from plants as a fuel. For many plants, starch is a storage compound, both for short-term storage overnight when photosynthesis cannot occur, and for long-term storage, for example in seeds and in the

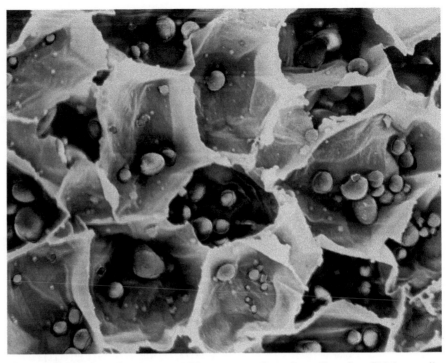

Figure 1.7 Starch molecules can fold up compactly and can therefore fit into small storage organelles, such as the starch grains in potato tuber cells, shown here. The starch grains are shown in green.

organs such as bulbs and tubers that survive through the winter. It is particularly suited for storage because it is insoluble and so does not diffuse out of cells easily or have any effects on water potential and thus osmosis.

As a storage compound it is important that starch can be easily synthesised and broken down. Plants have enzymes that can rapidly carry out these processes.

We have a digestive enzyme called amylase that hydrolyses the starch in our diet to maltose. This can then be hydrolysed into glucose, which is needed to provide a source of fuel for respiration.

Glycogen for storage

Animals such as humans do not rebuild excess glucose into starch for storage. Instead, we make it into a polysaccharide similar to starch called glycogen.

Like amylopectin, glycogen also consists of α-glucose chains with both 1,4- and 1,6-glycosidic bonds, but the 1,6 bonds are much more frequent, so the molecules are much more branched (see Figure 1.8, overleaf). This makes glycogen molecules even more compact than starch molecules. In humans, some glycogen is stored in the muscles as a readily accessible store of glucose close to the site where the rate of respiration is regularly raised very rapidly. The liver stores larger reserves of glycogen and continually breaks it down to maintain a stable blood glucose concentration.

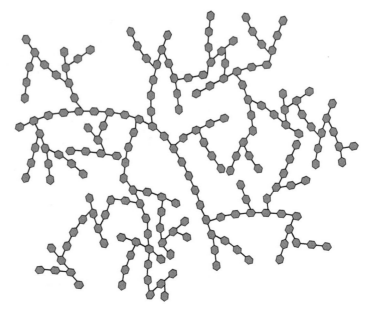

Figure 1.8 A glycogen molecule.

Cellulose for strength

The main substance in a plant cell wall is the carbohydrate cellulose. Like starch, cellulose is a polysaccharide and is a polymer of glucose. The monomer in cellulose is β-glucose.

In cellulose, the β-glucose molecules join together in chains by condensation. As when starch chains are made from α-glucose molecules, glycosidic bonds are formed. But in the cellulose chains, every other β-glucose is 'upside-down', so the –CH$_2$OH side-chains stick out alternately on opposite sides, as you can see in Figure 1.9. This 'alternate' bonding makes the cellulose molecules very straight. They are also very long. They line up parallel with each other and become linked together by many hydrogen bonds. Although each hydrogen bond is weak, many together lead to strong binding between the molecules.

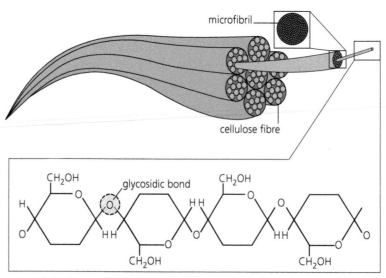

Long chain of 1,4 linked β-glucose residues. Hydrogen bonds link these chains together to form microfibrils.

Hydrogen bond A chemical bond important in the three-dimensional structure of biological molecules. Hydrogen bonds require relatively little energy to break.

Figure 1.9 Cellulose is a polymer of β-glucose molecules joined by glycosidic bonds. Its molecules are long and straight and form fibres that are very strong. Cellulose gives cell walls their strength and resistance to being stretched.

Small bundles of cellulose molecules make very thin fibres, called microfibrils. These microfibrils are remarkably strong. They have the ability to withstand stretching as steel fibres of the same diameter. Groups of microfibrils are joined together to make thicker, stronger fibres, just as a piece of string is made from many thinner strands. In cell walls, these fibres are criss-crossed, making the walls resistant to stretching in any direction.

Cellulose is structurally so well suited to its functions of supporting cells and limiting water intake that it is found throughout the plant kingdom. It is probably the most abundant carbohydrate. Surprisingly, neither humans nor any other mammal are able to make an enzyme that can digest cellulose. There are bacteria and fungi that do make such an enzyme, and these play an important role in recycling the constituents of cellulose. This is fortunate, since otherwise the world would have disappeared under cellulose long ago. Mammals such as cattle and rabbits, whose diet consists largely of plants, carry bacteria in their guts that hydrolyse cellulose, so they can make use of the energy in the large quantities of cellulose in their food. Humans, however, have no means of extracting the energy stored in cellulose.

TEST YOURSELF

1 Explain, using a diagram, how α-glucose molecules join together by condensation.
2 List the differences between the structures of cellulose molecules and starch molecules.
3 Starch is insoluble and does not affect osmosis in cells in which it is stored. Explain how these properties make starch a good storage compound.
4 What features of glycogen make it useful as a storage molecule in muscle tissue?
5 The molecular formula of galactose is $C_6H_{12}O_6$. What is the molecular formula of a molecule of lactose?
6 Starch molecules from different plants may differ from each other. Give two ways in which they might be different.

Lipids

The term 'lipids' covers a group of substances that includes fats and oils (triglycerides), steroids and sterols, and waxes. Two groups of lipids are especially significant. These are triglycerides and phospholipids. You will learn in Chapter 3 that the cell-surface membrane is made up of lipids and proteins.

Triglycerides

The commonest lipids found in living organisms are triglycerides. Most of the triglycerides found in animals are known as fats. They are solid at a temperature of about 20 °C. A triglyceride is made up of a molecule of glycerol and three fatty acid molecules. The basic structures of these molecules are shown in Figure 1.10, overleaf.

As Figure 1.10 shows, there are two kinds of fatty acids. Saturated fatty acids have only single bonds between the carbon atoms. Unsaturated fatty acids have at least one double bond between carbon atoms. In general,

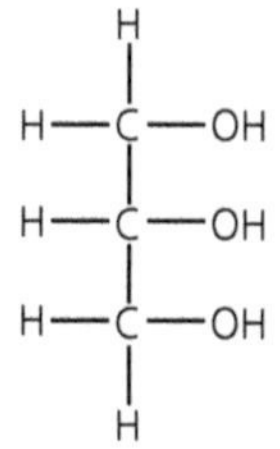

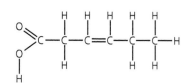

(a) Glycerol is a type of alcohol. It has three –OH groups, each of which can undergo a condensation reaction with a fatty acid.

(b) This is the simplest formula for a fatty acid molecule. The letter R represents a hydrocarbon chain consisting of carbon and hydrogen atoms.

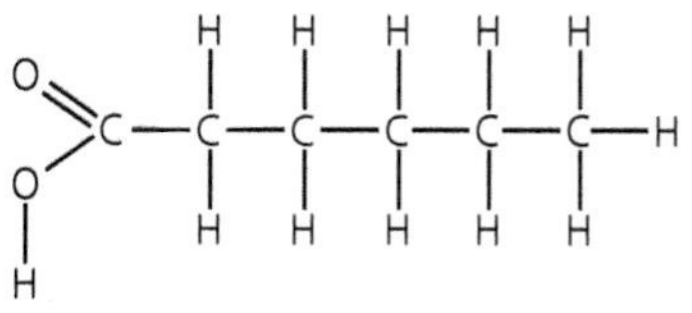

(c) In saturated fatty acids, each of the carbon atoms in this chain, with the exception of the last, has two hydrogen atoms joined to it. The bonds between the carbon atoms are single bonds.

(d) In unsaturated fatty acids, there are one or more double bonds between the carbon atoms in the chain. Because of this, some carbon atoms will be joined only to a single hydrogen atom.

Figure 1.10 The basic structure of a molecule of (a) glycerol and (b) fatty acid; (c) shows the structure of a saturated fatty acid and (d) shows the structure of an unsaturated fatty acid.

saturated fatty acids have higher melting points than unsaturated fatty acids. Fats that are solid at room temperature, such as lard or butter, tend to have more saturated fatty acids in them, while oils that are liquid at room temperature, such as sunflower or olive oil, have more unsaturated fatty acids.

Glycerol is a type of alcohol. Look at Figure 1.10 (a). You will see that there are three –OH groups in glycerol. These groups allow the molecule to join with three fatty acids to produce a triglyceride. Figure 1.10 (b) is the simplest possible way of showing the structure of a fatty acid molecule. The letter R represents a chain of hydrogen and carbon atoms. In the fatty acids found in animal cells there are often 14 to 16 carbon atoms in this chain.

When a triglyceride is formed, a molecule of water is removed as each of the three fatty acids joins to the glycerol. You may remember that this type of chemical reaction is called condensation (see page 2). The formation of a triglyceride from glycerol and fatty acids is shown in Figure 1.11. The bond formed between the glycerol and the fatty acid is called an ester bond.

You can use the emulsion test to test for lipids such as triglycerides. Crush a little of the test material and mix it thoroughly with ethanol. Pour the resulting solution into water in a test tube. A white emulsion shows that a lipid is present.

- Draw a diagram to show a glycerol molecule.
- Draw three fatty acid molecules 'the wrong way round' next to it.

$$
\begin{array}{c}
\text{H} \\
| \\
\text{H—C—O\!H \quad HO\!OC.R} \\
| \\
\text{H—C—O\!H \quad HO\!OC.R} \\
| \\
\text{H—C—O\!H \quad HO\!OC.R} \\
| \\
\text{H}
\end{array}
$$

glycerol   fatty acids

Figure 1.11 This diagram is a simple way of showing how a molecule of glycerol joins with three fatty acid molecules to form a triglyceride.

- Remove three molecules of water, taking the H from the glycerol and the –OH from the fatty acids.

$$
\begin{array}{c}
\text{H} \\
| \\
\text{H—C—OOC.R} \\
| \\
\text{H—C—OOC.R} \quad + \quad 3H_2O \\
| \\
\text{H—C—OOC.R} \\
| \\
\text{H}
\end{array}
$$

- Close everything up to show the completed triglyceride.

Phospholipids

A phospholipid has a very similar structure to a triglyceride, but as you can see from Figure 1.12, it contains a phosphate group instead of one of the fatty acids. It is quite a good idea to think of a phospholipid as having a 'head' consisting of glycerol and phosphate and a 'tail' containing the long chains of hydrogen and carbon atoms in the two fatty acids. The presence of the phosphate group means that the 'head' is attracted to water. It is therefore described as being **hydrophilic** or 'water loving'. The hydrocarbon tails do not mix with water, so this end of the molecule is described as **hydrophobic** or 'water hating'.

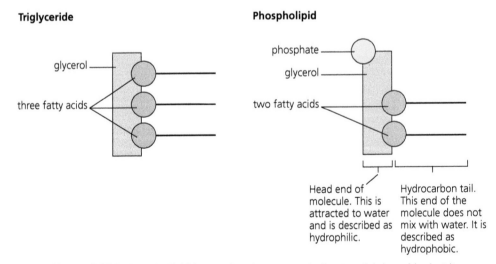

Figure 1.12 A phospholipid has a structure very similar to a triglyceride, but it contains a phosphate group instead of one of the fatty acids.

When phospholipids are mixed with water, they arrange themselves in a double layer with their hydrophobic tails pointing inwards and their hydrophilic heads pointing outwards. This double layer is called a phospholipid bilayer and forms the basis of membranes in and around cells.

9

7 Triglycerides are not polymers. Explain why.

8 Carbohydrates and triglycerides are both made of carbon, hydrogen and oxygen atoms. Explain how the proportions of these atoms are different in carbohydrates and triglycerides.

9 How is a triglyceride different from a phospholipid?

ACTIVITY

Fatty acids in milk

Milk contains triglycerides. Scientists investigated whether the fatty acids in human breast milk depend on the food that the mother eats. The scientists collected samples of breast milk from two groups of women. The women in one group were vegans and only ate food obtained from plants. Those in the other group, the control group, ate food obtained from both animals and plants. Table 1.1 shows the concentrations of different fatty acids in the milk samples.

Table 1.1 The concentrations of different fatty acids in vegan and control group milk samples.

Fatty acid	Number of double bonds in hydrocarbon chain	Number of carbon atoms in hydrocarbon chain	Concentration of fatty acid in breast milk sample/mg g^{-1}	
			Vegan group	Control group
Lauric	0	12	39	33
Myristic	0	14	68	80
Palmitic	0	16	166	276
Stearic	0	18	52	108
Palmitoleic	1	16	12	36
Oleic	1	18	313	353
Linoleic	2	18	317	69
Linolenic	3	18	15	8

1 The first four fatty acids in the table are saturated fatty acids. Explain why they are described as saturated.

2 Construct a table to show all of the following:
- the total concentration of saturated fatty acids in milk from the vegan group
- the total concentration of unsaturated fatty acids in milk from the vegan group
- the total concentration of saturated fatty acids in milk from the control group
- the total concentration of unsaturated fatty acids in milk from the control group.

3 Use an example from the table to explain what is meant by a polyunsaturated fatty acid.

4 Describe the difference between the total concentration of polyunsaturated fatty acids in milk produced by the vegan group and by the control group. Suggest an explanation for this difference.

Proteins

Earlier in this chapter, we saw that starch is a polymer made up of a single type of monomer, α-glucose. Whether these α-glucose monomers are joined to form straight chains or branched chains, they still form starch. Different types of starch are very similar.

Proteins are different. The basic building blocks of proteins are amino acids. There are 20 different amino acids found in almost all living organisms, which is indirect evidence for evolution. These amino acids can be joined in a range of different orders. In any living organism, there are a huge number of different proteins and they have many different functions.

If we take a single tissue, such as blood, we can get some idea of just how varied and important are the roles of proteins. Human blood is red because it contains haemoglobin. This is an iron-containing protein that plays an extremely important part in transporting oxygen from the lungs to respiring cells. When you cut yourself, blood soon clots. This is because another protein, fibrin, forms a mesh of threads over the surface of the wound, trapping red blood cells and forming a scab. Blood also contains enzymes, which are proteins. The antibodies produced by white blood cells are also proteins, and are important in protecting the body against disease.

The biuret reaction enables us to test for a protein. Sodium hydroxide solution is added to a test sample, and then a few drops of dilute copper sulfate solution. If there is a protein present, the solution will turn mauve.

Amino acids: the building blocks of proteins

All 20 amino acids have the same general structure. Look at Figure 1.13. Notice that there is a central carbon atom called the α-carbon and that it is attached to four groups of atoms. There is an amine group (–NH₂). This is the group that gives the molecule its name. Then we have a carboxyl group (the acid; –COOH) and a hydrogen atom (–H). These three features are exactly the same in all 20 amino acids. The fourth group, called the R-group, differs from one amino acid to another. As well as showing the general structure of an amino acid, Figure 1.13 also shows the structures of three particular amino acids found in proteins. In each of these three amino acids (and in the other 17), it is only the R-group that is different.

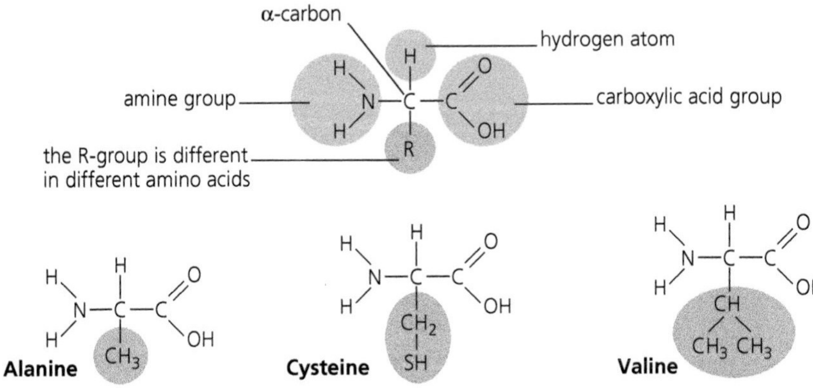

Peptide bond A chemical bond formed between two amino acids as a result of condensation.

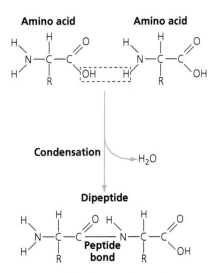

Figure 1.14 Joining amino acids.

α-carbon
amine group
hydrogen atom
carboxylic acid group
the R-group is different in different amino acids

Alanine **Cysteine** **Valine**

Figure 1.13 The structure of amino acids.

Amino acids join together by condensation reactions. Look at Figure 1.14. You can see that a hydrogen atom is removed from the amino group of one amino acid. This combines with an –OH group removed from the carboxylic acid of the other amino acid, forming a molecule of water. The bond formed between the two amino acid residues is called a peptide bond. Joining two amino acids together produces a dipeptide. When many amino acids are joined in this way, they form an unbranched chain called a polypeptide. Polypeptides can be broken down again by hydrolysis into the amino acids from which they are made.

11

Polypeptides and proteins

Primary structure The sequence of amino acids in a polypeptide.

A protein consists of one or more polypeptide chains folded into a complex three-dimensional shape. Different proteins have different shapes, determined by the order in which the amino acids are arranged in the polypeptide chains. The sequence of amino acids in a polypeptide chain is the primary structure of a protein (see Figure 1.15).

Figure 1.15 This diagram shows the primary structure of an enzyme called ribonuclease. The names of the amino acids that make up this protein have been abbreviated. Ribonuclease has 124 amino acids. Some proteins, such as antibodies, are much larger and contain many more amino acids.

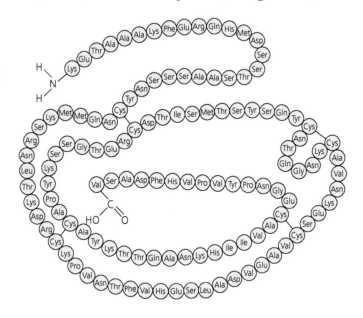

We shall see in Chapter 4 that genes carry the genetic information that enables cells to make polypeptides and ensures that the sequence of amino acids is the same in all molecules of a particular polypeptide. Changing a single one of these amino acids may be enough to cause a change in the shape of the protein and prevent it from carrying out its normal function.

Secondary structure The shape the polypeptide chain folds into, such as an alpha helix or a beta pleated sheet.

Parts of a polypeptide chain are twisted and folded. This is the secondary structure of a protein.

Extension

Secondary structure: α-helix and β-pleated sheet

Sometimes the chain, or part of it, coils to produce a spiral or α-helix. Other parts of the polypeptide may form a β-pleated sheet; this occurs where two or more parts of the chain run parallel to each other and are linked to each other by hydrogen bonds.

The sequence of amino acids in the polypeptide decides whether an α-helix or a β-pleated sheet is formed. Some sequences are more likely to form an α-helix, while others form a β-pleated sheet, as in Figure 1.16.

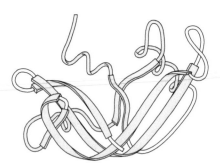

Figure 1.16 Here is another diagram of a ribonuclease molecule, this time showing its secondary structure. The three spiral yellow parts of the polypeptide chain are where it is coiled into an α-helix. The flat blue sections show where the chain is folded to form a β-pleated sheet.

Tertiary structure Gives a protein the characteristic complex, three-dimensional shape that is closely related to its function.

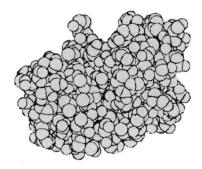

Figure 1.17 The model in this diagram shows the tertiary structure of a ribonuclease molecule. (The shapes represent atoms.) This is the way the whole polypeptide is folded.

> **TIP**
> Enzymes are proteins. You will learn more about them in Chapter 2.

(a)

(b)

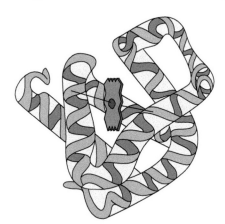

Figure 1.18 (a) Fibrous and (b) globular proteins.

The twisted and folded chain may fold up further to give the whole polypeptide molecule a globular shape. The complex folding of the whole molecule is the tertiary structure of the protein (Figure 1.17).

As with the secondary structure, the tertiary structure is also determined by the sequence of amino acids in the polypeptide chain. All molecules of a particular protein have the same sequence of amino acids, so under the same conditions they will all fold in the same way to produce molecules with the same tertiary structure.

Tertiary structure is extremely important and is very closely related to the function of the protein. Different types of bond form between different amino acids within the protein and the types of bond help to maintain the shape of the protein. These bonds include the following:

● Hydrogen bonds (see page 6), which form between the R-groups of a variety of amino acids. These bonds are not strong. They are easily broken, but there are many of them.
● Ionic bonds, which form between an amino acid with a positive charge and an amino acid with a negative charge, if they are close enough to each other. These are not strong bonds and are easily broken.
● Disulfide bridges, which form between amino acids that contain sulfur in their R-groups. These are quite strong covalent bonds, less easily broken than hydrogen bonds or ionic bonds.

There are two categories of proteins, differing in their tertiary structure. Fibrous proteins are typically long and thin, and they are insoluble. They often have structural functions, such as keratin in hair or collagen that makes up a lot of connective tissue in our bodies. Globular proteins are more spherical in shape. They are soluble and have biochemical functions, such as enzymes or myoglobin, a pigment that stores oxygen in muscle tissue.

Some proteins have more than one polypeptide chain. We describe a protein that is made up from two or more polypeptide chains as having a quaternary structure. The polypeptide chains are held together by the same sorts of chemical bond that maintain the tertiary structure. The ribonuclease molecule shown in Figure 1.17 does not have a quaternary structure because it consists of only one polypeptide chain. The red pigment in our blood, haemoglobin, is a protein that does have a quaternary structure. A molecule of human haemoglobin has four polypeptide chains.

> **TEST YOURSELF**
> **10** Polypeptides can be made up from 20 different amino acids. A tripeptide is a polypeptide consisting of three amino acids. How many different tripeptides is it possible to make?
> **11** Give one way in which the formation of a peptide bond is similar to the formation of a glycosidic bond.
> **12** Egg white contains a protein. Which one (or more) of the following occurs when egg white is heated in a water bath containing water at 100 °C?
>   **A** Glycosidic bonds are broken.
>   **B** The protein is killed by the heat.
>   **C** The bonds holding the tertiary structure are broken.
>   **D** The protein is hydrolysed.

Qualitative tests for substances in food

There are some tests that can be carried out to find out which substances are present in samples of food and other substances. These tests are summarised in Table 1.2.

TIP
Qualitative tests detect the presence of a substance but do not show exactly how much is present.

Table 1.2 Tests for food substances.

Substance	Test	Brief details of test	Positive result
Protein	Biuret test	• Add sodium hydroxide to the test sample. • Add a few drops of dilute copper sulfate solution.	Solution turns mauve
Carbohydrates Reducing sugars	Benedict's test	• Heat test sample with Benedict's reagent.	Orange-red precipitate is formed
Non-reducing sugars		• Check that there is no reducing sugar present by heating part of the sample with Benedict's solution. • Hydrolyse rest of sample by heating with dilute hydrochloric acid. • Neutralise by adding sodium hydrogencarbonate. • Test sample with Benedict's solution.	Orange-red precipitate is formed
Starch	Iodine test	• Add iodine solution.	Turns blue-black
Lipid	Emulsion test	• Dissolve the test sample by shaking with ethanol. • Pour the resulting solution into water in a test tube.	A white emulsion is formed

Care should be taken with the chemicals and methods described above. Consult the CLEAPSS Hazcards for each chemical, and ensure sufficient safety precautions are taken with chemicals and hot water.

ACTIVITY

A student was given three tubes. The table shows the contents of each tube.

Tube	Contents of tube
A	Protein and protease (an enzyme that digests protein) that have been left together at room temperature for an hour.
B	Sucrose, starch, lipid
C	Glucose, protein

The student carried out tests for food substances on all three tubes. The table below shows the results she obtained.

1 Copy the table and complete the left-hand column to indicate which tube gave which results.

Tube	Benedict's test for reducing sugars	Benedict's test for non-reducing sugars	Iodine test for starch	Emulsion test for lipids	Biuret test for protein
	Orange-red precipitate formed	Test not carried out	Stayed yellowy-brown	Stayed clear	Mauve
	Stayed blue	Stayed blue	Stayed yellowy-brown	Stayed clear	Mauve
	Stayed blue	Orange-red precipitate formed	Blue-black colour	Milky-white emulsion formed	Stayed blue

2 Why did the student decide not to carry out the non-reducing sugar test on one of the tubes?

3 Explain why two tubes gave a positive result for the biuret test.

4 Suggest how the student could use qualitative tests to distinguish between three solutions containing different concentrations of glucose.

ACTIVITY

A student decided to investigate the sensitivity of the iodine test for starch. She was given a 1% starch solution. She used this to make serial dilutions as shown in Figure 1.19.

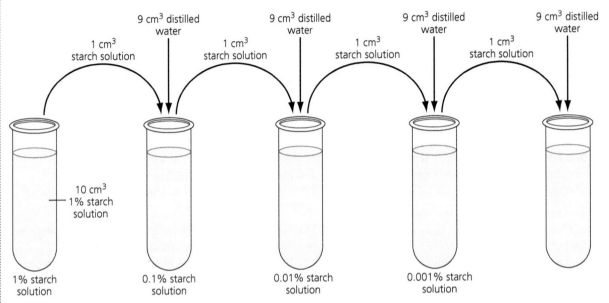

Figure 1.19 Making serial dilutions of starch.

After this she added 0.5 cm³ of iodine solution to each starch solution and looked for a blue-black colouration.

1 How would a technician make up a 1% solution of starch?
2 Give the concentration of the final solution in the diagram.
3 The blue-black colouration that occurs when iodine interacts with starch is caused by the iodine molecules becoming trapped in the amylose helix. Use this information to explain why the sensitivity of the iodine test can be different depending on the type of starch present.

TIP
You do not need to be able to recall the details of this practical activity.

Finding the concentration of reducing sugar in a solution

A colorimeter is a piece of equipment that passes light of a particular wavelength through a sample. It works on the principle that a coloured solution transmits light of the same wavelength through it, and absorbs other wavelengths. So a blue solution transmits blue light through it and absorbs other wavelengths. In addition, a darker-coloured solution absorbs more light than a pale-coloured solution. This can be used to measure the concentration of a coloured solution. The colorimeter is set up so that it shines light of a complementary colour through the solution. For example, if the solution is red in colour, blue light is shone through it.

Solutions to be tested in a colorimeter are placed into a cuvette. This is a small plastic tube rather like a test tube but square in section. You can see a colorimeter and cuvette in Figure 1.20.

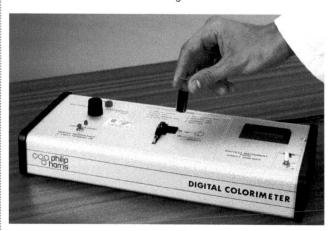

Figure 1.20 Using a colorimeter.

When using a colorimeter, you need to produce a calibration curve. This is done using solutions of a known concentration.

An alternative version of Benedict's reagent for quantitative testing contains potassium thiocyanate and does not form red copper oxide. Instead the presence of reducing sugar is measured by the loss of the blue colour of copper sulfate and a white precipitate is formed. This will settle out or can be removed by filtering. Then the filtrate is placed in a cuvette in a colorimeter. The intensity of the blue colour is measured by the amount of light that is able to pass through the solution. This method can give an accurate measurement of the concentration of reducing sugar in a solution, and it is much more sensitive than the qualitative Benedict's test.

A student was given a $1 \, mol \, dm^{-3}$ solution of glucose and then made serial dilutions of this (see page 15), giving six solutions of different known concentrations. She put $4 \, cm^3$ of each solution into a separate labelled test tube. Next she added $2 \, cm^3$ of quantitative Benedict's reagent to each tube and placed the tubes in a boiling water bath for 5 minutes. After this time, she filtered each solution to remove the precipitate.

The student set the wavelength on the colorimeter to red. She filled a cuvette with distilled water and put it into the colorimeter. This is called a 'blank'. She set the transmission of light through the tube to 100%. This meant that she could compare the transmission of light through the test solutions to the blank.

The student put a sample of each test solution into cuvettes, and measured the percentage transmission of light through each tube. Next she plotted a graph with concentration of glucose on the x-axis and percentage transmission of light through the solution on the y-axis.

Finally, she used the same method to identify the concentration of glucose in two solutions of unknown concentration.

1 How could the student use the $1 \, mol \, dm^{-3}$ solution of glucose to make the five other solutions?
2 Why did the student use a red light in the colorimeter?
3 What is the purpose of the 'blank'?
4 How could the student use her graph to find the concentration of glucose in an unknown solution?
5 If one cuvette was a little thicker than another, how would this affect the results?
6 How would the student use the graph to find the concentration of glucose in the unknown solutions?

Practice questions

1 a) Copy and complete the table to show the monosaccharides that join together to form different disaccharides. *(3)*

Disaccharide	Formed from
Lactose	
Maltose	
Sucrose	

b) Describe how you could test a solution to find out whether it contained a non-reducing sugar. *(4)*

2 Some scientists carried out an investigation into the artificial sweetener aspartame. This was because some people said they experienced side-effects after consuming the substance. Previous investigations had found the artificial sweetener to be safe. The sweetener is digested in the gut to give the amino acids aspartic acid and phenylalanine.

In the new study, the scientists recruited 50 people who believed they had suffered side-effects because they were sensitive to aspartame.

- The volunteers were matched by age and sex to 50 people who volunteered to eat aspartame.

- The volunteers were placed at random into one of two groups.

- The volunteers in each group were given either a cereal bar containing aspartame or an aspartame-free cereal bar. They were then given medical checks up to 4 hours after consuming it.

- The following week, the experiment was repeated with each volunteer receiving the other type of cereal bar.

a) i) Explain why it was important that the 50 volunteers who were happy to eat aspartame were matched by age and sex to the 50 people who believed they were sensitive to aspartame. *(2)*

ii) The volunteers were placed at random into two groups. Give one method that the scientists could use to do this. *(1)*

iii) It was important that the scientists and the volunteers did not know which cereal bars contained aspartame. Give two reasons to explain why. *(2)*

b) The diagram shows the amino acids aspartic acid and phenylalanine.

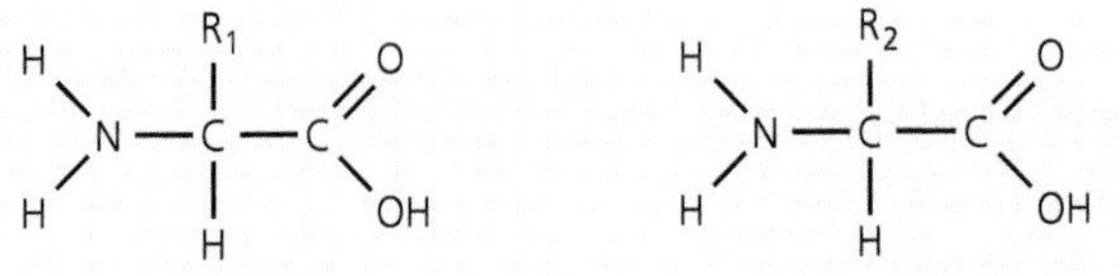

aspartic acid   phenylalanine

Aspartame is made by joining these two amino acids together.

i) Draw a molecule of aspartame. *(2)*

ii) Name the reaction that occurs when these two amino acids are joined together. *(1)*

3 The diagram shows two fatty acids.

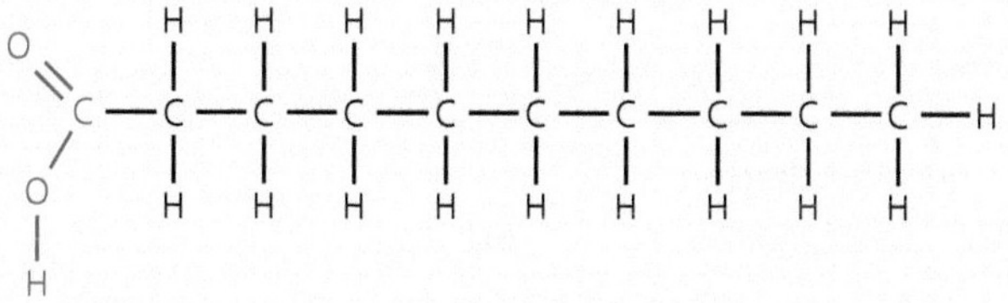

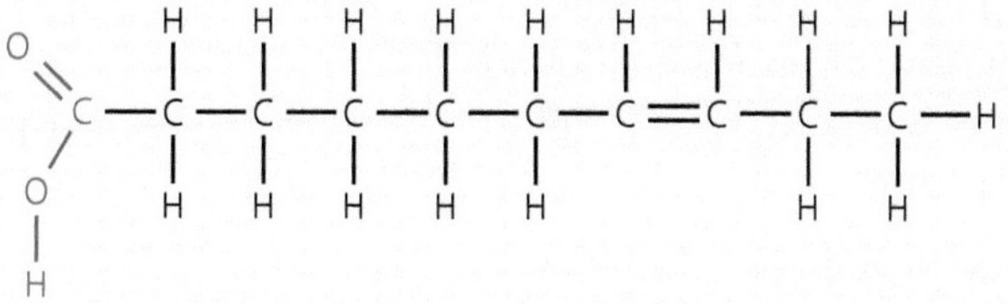

a) i) Which of these fatty acids is saturated? Explain your answer. *(1)*

ii) Name the reaction involved when three fatty acids combine with a glycerol molecule to form a triglyceride. *(1)*

b) Describe a test you could perform to show that a mixture contains lipids. *(2)*

c) Give one similarity and one difference between the structure of a phospholipid and the structure of a triglyceride. *(2)*

4 a) Copy and complete the table below with a tick if the statement is true and a cross if it is not true. *(3)*

Statement	Proteins	Polysaccharides	Lipids
Molecule is a polymer			
Contains amino acids			
Contains nitrogen atoms			

b) i) A protein has a tertiary structure but not a quaternary structure. How many polypeptides does it contain? *(1)*

ii) Name two kinds of bond that hold a protein in its tertiary structure. *(2)*

Stretch and challenge

5 Research the differences between D- and L-isomers of molecules. You should find that the **amino acids** in most living organisms are the **L-isomers**, and the **monosaccharides** in most living organisms are the **D-isomers**. Suggest why this is indirect evidence for evolution.

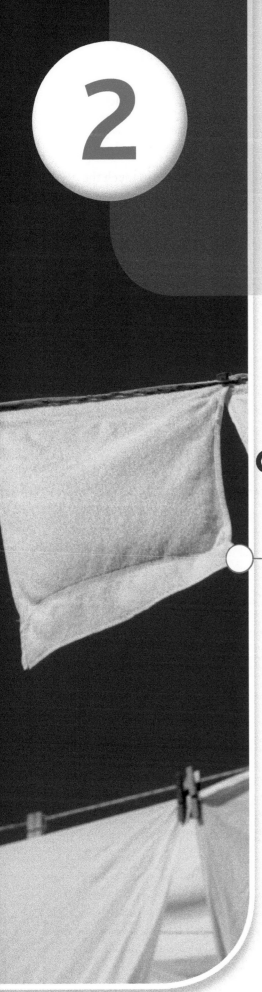

2

Enzymes

TEST YOURSELF ON PRIOR KNOWLEDGE
1 Name two factors that affect an enzyme's activity.
2 Suggest why changing the shape of an enzyme stops it from working.

Introduction

There are many chemical reactions taking place in living organisms, mainly inside the cells. The sum of all these reactions is known as **metabolism**. Some of these reactions hydrolyse larger molecules into smaller ones, while others join smaller molecules together to make bigger ones. There are enzymes in biological washing powder, which help to clean clothes at lower temperatures.

Molecules known as enzymes are required for metabolic reactions to take place. Enzymes are **biological catalysts**. They speed up the rate of chemical reactions, enabling reactions that take place in living organisms to occur fast enough for life to continue. As they are catalysts, and so remain unchanged during reactions, they can work many times.

Enzymes are made of protein and, as you will see, they have specific shapes that enable them to function. You learned in Chapter 1 (page 13) that globular proteins have a complex tertiary structure, and this is very important in explaining how enzymes work.

Enzyme A protein that speeds up the rate of a chemical reaction in a living organism. An enzyme acts as catalyst for specific chemical reactions, converting a specific set of reactants (called substrates) into specific products.

How do enzymes work?

Have you ever found an old newspaper, one that was perhaps months or even years old? If you have, you probably noticed that it had turned a yellow-brown colour. What had happened? The paper had reacted with oxygen in the air, but it is a very, very slow reaction.

We can easily speed up this reaction. All we need to do is to touch the corner of the paper with a lighted match. The paper bursts into flame, reacting with oxygen in the air much more quickly. This reaction involves combustion. The newspaper is fuel and contains chemical potential energy. In order to release this energy in the chemical reaction with oxygen, we must supply some energy at the start. This is where the match comes in. It provides the activation energy necessary to start the reaction.

We sometimes compare what happens here with what would happen if you had a large rock at the top of a steep hill. There is a lot of potential energy in this situation but, under normal conditions, the rock will just sit there. However, give it a push and it will roll all the way down to the bottom of the hill. In other words, supply activation energy at the start, and the rock will give up a lot of its potential energy. Look at Figure 2.1. This is a graph showing, in a different way, the idea of the energy changes that take place as a chemical reaction progresses.

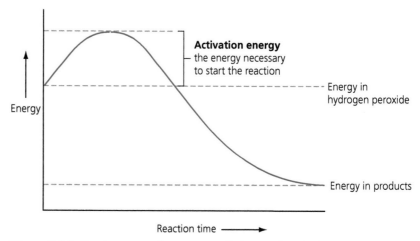

Figure 2.1 Activation energy is necessary before a chemical reaction will take place.

Hydrogen peroxide is a substance produced by reactions in many living cells. It is harmful, so it is removed. It breaks down very slowly to give the products water and oxygen:

$$2H_2O_2 \rightarrow 2H_2O + O_2$$

The products of this reaction, water and oxygen, are not only harmless, they are extremely useful to the organism.

We can pour some hydrogen peroxide into a test tube and it will break down slowly. We can make the reaction go faster by heating the hydrogen peroxide. Clearly, inside the body we cannot use a match or a Bunsen burner to heat our cells to increase the rate at which the hydrogen peroxide they produce is broken down. This is where enzymes come in.

Cells produce an enzyme called catalase. Its **substrate**, the substance on which an enzyme acts, is hydrogen peroxide. Catalase lowers the activation energy needed to start the breakdown reaction of hydrogen peroxide. As a consequence, the hydrogen peroxide breaks down rapidly at the relatively low temperatures found inside living cells. Figure 2.2 shows this as a graph. Note that there is the same chemical potential energy at the start of the two reactions and there is the same amount in the products of both reactions. The enzyme has simply lowered the activation energy necessary to start the reaction.

Figure 2.2 Adding an enzyme lowers the activation energy necessary to start a chemical reaction.

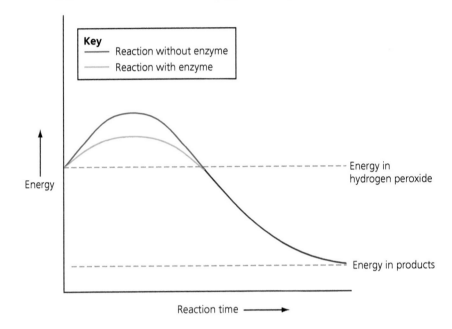

Extension

Enzyme shape and enzyme function: ribonuclease

A skill that a biologist must have is to be able to interpret unfamiliar data. Here we will look at some research carried out by Christian Anfinsen in the 1960s. It was very important research because it demonstrated the relationship between the structure of an enzyme and its function.

Anfinsen investigated the enzyme ribonuclease. This enzyme hydrolyses ribonucleic acid (RNA). There are different forms of ribonuclease. Look at the diagram in Figure 2.3, which shows the structure of one form of this enzyme.

The numbers on the diagram show positions along the polypeptide chain. For example, number 26 is the twenty-sixth amino acid along the chain. The large shaded dots represent the amino acid cysteine. This amino acid contains sulfur. Chemical bonds called disulfide bonds, or disulfide bridges (see Chapter 1, page 13), form between sulfur-containing

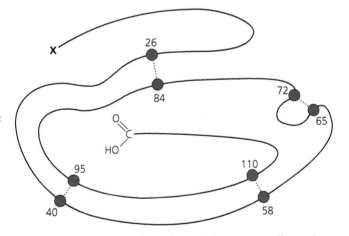

Figure 2.3 A molecule of one form of the enzyme ribonuclease.

21

amino acids. They are quite strong bonds and help to hold the tertiary structure of a protein together. In Figure 2.3 on the previous page, the amino acids at positions 26 and 84 are cysteine. The dotted line between them is a disulfide bridge.

When we look at unfamiliar data, it is very important to take the time to think about it carefully and make sure that we understand it. We should be able to answer some basic questions about Figure 2.3.

1 What chemical group is shown by the letter X on the diagram?

This is an amino (–NH$_2$) group. Look back at Figure 1.14 on page 11. You can see that the dipeptide has a –COOH group at one end and an –NH$_2$ group at the other. The same is true of a polypeptide. There is always a –COOH group at one end and an –NH$_2$ group at the other. Since the –COOH group is shown on the diagram, position X must be where there is an –NH$_2$ group.

2 Figure 2.3 shows one form of the enzyme ribonuclease. How do you think other forms of ribonuclease will be different?

Different forms of ribonuclease will have different amino acids in different positions. In other words, each form will have a slightly different primary structure. However, we would not expect them to differ much, because they are all forms of the same enzyme, ribonuclease.

3 All molecules of this form of ribonuclease have the same tertiary structure. Use information from the diagram to explain why.

This should be quite simple, if you have understood the diagram. All molecules of this form of ribonuclease will have the same sequence of amino acids. The cysteine molecules will therefore always be in the same positions and the disulfide bridges will form in the same place.

Hopefully, you have understood what the diagram in Figure 2.3 tells you. We will now look at the steps in Anfinsen's investigation.

- Anfinsen started by measuring the activity of untreated ribonuclease.
- He then treated the ribonuclease with mercaptoethanol. This substance broke the disulfide bridges in the ribonuclease molecules. He measured the activity of the treated ribonuclease.
- Finally, he removed the mercaptoethanol. As a result, the disulfide bonds re-formed. His results are shown in Figure 2.4.

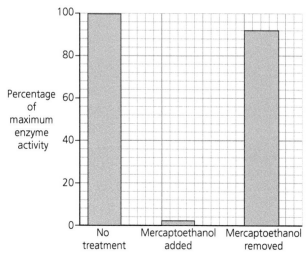

Figure 2.4 The effects of different experimental treatments on the rate of reaction of ribonuclease.

Again, the data shown in this bar chart are probably unfamiliar. We need to take the time to make sure that we understand the graph. We will start by looking at the axes. The x-axis, the horizontal one, shows the three different treatments. The y-axis, the vertical one, shows the enzyme activity. This is given as the percentage of maximum enzyme activity, so a value of 100 represents the fastest that the enzyme could possibly react.

To make sure that we really understand the information in the graph, we should ask the following, for example.

4 What does the second bar show?

When the ribonuclease is treated with mercaptoethanol, the enzyme is not very reactive. It shows only about 2% of its maximum activity.

5 Why did mercaptoethanol have such an effect on the activity of ribonuclease?

The tertiary structure of ribonuclease is held together by disulfide bridges. These have been broken, so the polypeptide chain loses its shape. We say that it is denatured. This means that the site into which the substrate molecules fit also loses its shape. As a result, the substrate will no longer bind to the enzyme. Not surprisingly, the activity of the enzyme falls.

6 What happened when the mercaptoethanol was removed from the ribonuclease?

The bonds re-formed and the tertiary structure of the enzyme was restored. The enzyme was functional again.

Enzymes, substrates and products

If we mix an enzyme solution with biuret reagent, the solution goes violet in colour. This is the test for proteins, and we can use it to show that enzymes are proteins. Now look at Figure 2.5. It shows some of the biochemical reactions that take place in a typical cell.

Each of the 520 dots is a particular substance and the lines connecting these dots are biochemical reactions. What is really important to understand is that each of these reactions is controlled by a different enzyme, so a single cell contains hundreds of different enzymes. These enzymes differ quite a lot in size. Some are rather small molecules and some are relatively enormous, but all of them are proteins.

In addition, each enzyme has a unique shape, or tertiary structure. Somewhere on the surface of the enzyme, a group of amino acids forms a pocket. This pocket is the active site of the enzyme. When an enzyme catalyses a particular chemical reaction, a substrate molecule collides with the active site and binds with it to form an unstable intermediate substance called an enzyme–substrate complex. This complex then breaks down to the product molecules. The enzyme molecule is not used up in the reaction. It is now free to bind with another substrate molecule.

An enzyme–substrate complex forms and breaks down rapidly, forming a complex that lowers the activation energy necessary to trigger the reaction. Scientists suggest different ways in which the activation energy might be lowered. The enzyme–substrate complex might bring together substrate molecules in positions that allow the reaction to take place more easily. Or it might put the substrate molecule under stress so that bonds break more readily.

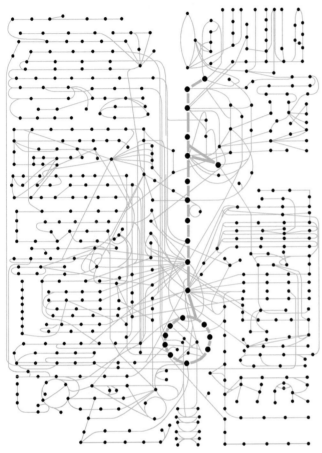

Figure 2.5 Believe it or not, this is a simplified diagram of some of the biochemical reactions that take place in a single cell. The dots represent different substances and the reactions are shown as lines linking the dots. Each one of the reactions is controlled by a different enzyme.

Active site Part of an enzyme molecule, not part of the substrate. The active site is the part of the enzyme molecule into which the substrate fits during a biochemical reaction. Active site and substrate therefore have complementary shapes; they do not have the same shape.

Enzymes and models

Scientists use models to help them explain their observations. We have known for a long time that enzymes are specific, meaning that each enzyme catalyses just one type of reaction with one type of substrate. Amylase, for example, is an enzyme that hydrolyses starch. It breaks down starch to maltose. If we mix amylase and protein, however, nothing will happen, even though the breakdown of proteins to amino acids also involves hydrolysis. Amylase only hydrolyses its specific substrate, starch. It won't hydrolyse any other substance.

Induced-fit model

We now know a lot about protein structure. For example, new techniques reveal that proteins are not rigid structures and we have found out that various parts of an enzyme molecule move in response to a change in its environment. Some of these movements are small. Others are quite large and happen when the substrate binds to the enzyme.

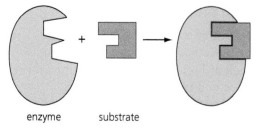

enzyme   substrate

Figure 2.6 In the induced-fit model, the substrate does not fit precisely until it binds with the active site of the enzyme. The active site then changes shape and moulds round the substrate, allowing a precise fit.

We now have a model to explain how enzymes work, called the induced-fit model, which is shown in Figure 2.6.

In this model, before the substrate binds to the enzyme, the substrate and the active site are not precisely complementary in shape. As the substrate binds, the active site changes shape and moulds closely round the substrate. It is a bit like a sock and a foot. Before you put your sock on, it is not foot-shaped at all. Put it on and it moulds round your foot.

The induced-fit model is now favoured over the lock-and-key model that you learned about at GCSE. In that model of enzyme action, the specific shape of the substrate is said to be a precise fit for the rigid shape of the active site of the enzyme, just like a key in a lock.

Extension

Evidence for the induced-fit model

It is one thing to come up with a model to explain how enzymes work, but in order to make real scientific progress we need to find evidence to support it. Is there any evidence for the induced-fit model? We will look at one piece of evidence.

Hexokinase is an enzyme that catalyses the reaction shown below.

glucose + ATP → glucose 6-phosphate + ADP

Hexokinase transfers a phosphate group from a substance called ATP to a glucose molecule. This produces glucose 6-phosphate and ADP. Scientists used a technique called X-ray diffraction, which enabled them to form three-dimensional pictures of molecules. They investigated what happened to the shape of hexokinase when it binds to glucose.

Look at Figure 2.7. It shows you that the active site has changed in shape and has moulded round the glucose molecule.

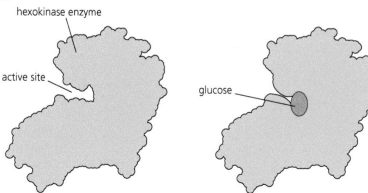

Figure 2.7 The diagram shows the shape of a hexokinase molecule before and after it has bound with its glucose substrate. Look at how the active site of the enzyme has changed shape.

TEST YOURSELF

1 Name the substrate and product(s) in the following reaction. Lactase catalyses the hydrolysis of lactose to glucose and galactose.
2 Use a diagram to explain why amylase can hydrolyse starch but cannot hydrolyse cellulose.
3 A student wrote 'Maltase forms an enzyme–substrate complex with maltose because maltose is the same shape as the enzyme's active site'. What is wrong with this statement?
4 In the Haber process, nitrogen and hydrogen react together to form ammonia, using a catalyst as well as 15–25 megapascals (Mpa) pressure and 300–550°C. Nitrogen-fixing bacteria have an enzyme called nitrogenase that enables them to carry out the same reaction at atmospheric pressures and normal soil temperature. How does the enzyme do this?

Denaturation A permanent change to the shape of an enzyme or other protein that occurs as a result of breaking the bonds that maintain its tertiary structure. This results from factors such as high temperature or large changes in pH. Denaturation only refers to changes in tertiary structure, not to the hydrolysis of proteins into the amino acids from which they are made.

The properties of enzymes: active sites and enzyme action

Enzyme-controlled reactions depend on substrate molecules fitting into the active site of the enzyme. Environmental factors alter the shape of the active site and therefore alter the rate of a reaction. Such factors include high temperatures, variations in pH and the presence of inhibitors.

High temperatures

Each enzyme has an optimum temperature. At this temperature, its rate of reaction is around its maximum. If we increase the temperature above its optimum, we increase the kinetic energy of the enzyme molecules. As a result, they vibrate more vigorously, and this breaks the hydrogen and ionic bonds that maintain the tertiary structure of the enzyme molecules. Once these bonds have broken, the enzyme changes its shape. This is called denaturation. Its active site also changes shape and, as a result, substrate molecules will no longer fit and form an enzyme–substrate complex. For example, the saliva of many people contains amylase. This enzyme most actively digests starch at body temperature. When very hot food is eaten, very little will be digested in the mouth as the temperature is too high.

EXAMPLE

Temperature and enzymes

Acetylcholinesterase is an enzyme that exists in several different forms in different animal species. It is found in the nervous system of many animals including fish.

Table 2.1 shows the temperature at which the rate of reaction of this enzyme is at its maximum in three different species of fish.

Table 2.1 The temperature at which the rate of the reaction catalysed by acetylcholinesterase is at its maximum in three different species of fish.

Species	Habitat	Temperature at which the rate of reaction is at a maximum/°C
Icefish	Under the ice in the Antarctic	–2
Trout	Freshwater lakes and streams in Britain	Between 2 and 18
Grey mullet	Mediterranean Sea	25

1 Describe the relationship between the temperature of the fish's habitat and the temperature at which the rate of reaction of acetylcholinesterase is at a maximum.
The temperature at which the enzyme's rate of reaction is fastest is about the same temperature that the fish's habitat will be for most of the time.

2 Trout have two forms of acetylcholinesterase, which are active under different environmental conditions. One has a maximum rate of reaction at 2°C; the other has a maximum rate of reaction at 18°C. Suggest the advantage of this to trout.
One form of acetylcholinesterase will be more active in the winter months and the other form will be more active in the summer months when the environmental temperature is higher.

3 Acetylcholinesterase is found in icefish and grey mullet, but it has slightly different properties in each species of fish.
a) The acetylcholinesterase in both fish breaks down a substance called acetylcholine. Explain why it is able to do this.
The enzyme has an active site that is the right shape for acetylcholine to fit into.

b) Use your knowledge of protein structure to explain why the acetylcholinesterase found in these two fish has slightly different properties.
Each enzyme may have a slightly different amino acid sequence so that hydrogen bonds and disulfide bridges form in slightly different places, giving a slightly different tertiary structure.

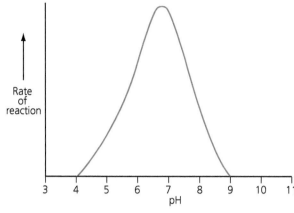

Figure 2.8 Enzymes work efficiently over a narrow pH range. This graph shows the effect of pH on a typical enzyme from a human cell.

pH

The pH of a solution is a measure of its hydrogen ion concentration. The higher the concentration of hydrogen ions (H^+), the lower the pH, and the more acid the solution. pH is a logarithmic scale, so as the pH scale falls one point from 7 to 6, the concentration of hydrogen ions increases 10 times.

Look at the graph in Figure 2.8. It shows that the rate of a reaction rises to a peak at pH 6.5. It then falls sharply. The peak value of 6.5 is the optimum pH for this enzyme.

Changing the pH above or below the optimum of an enzyme affects the rate at which the enzyme works. The change in pH alters the concentration of hydrogen ions (H^+) or hydroxyl ions (OH^-) in the surrounding solution. If the change is small, the main effect is to alter charges on the amino acids that make up the active site of the enzyme. As a result, substrate molecules no longer bind. A large change in pH breaks the hydrogen and ionic bonds that maintain the tertiary structure of the enzyme. The result is that the enzyme is denatured.

Inhibitors

Inhibitors are substances that slow down the rate of enzyme-controlled reactions. Competitive inhibitors are molecules that are very similar in shape to the substrate of the enzyme, as shown in Figure 2.9. They are described as competitive inhibitors because they compete with the substrate for the active site. They fit into the active site and block it. This prevents substrate molecules from entering and stops an enzyme–substrate complex being formed.

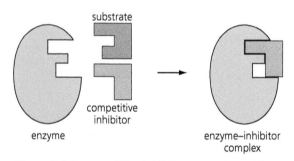

Figure 2.9 A competitive inhibitor competes with substrate molecules for the active site of the enzyme.

Figure 2.10 shows how non-competitive inhibitors work. Notice that a non-competitive inhibitor doesn't fit into the active site of the enzyme and block it in the way that a competitive inhibitor does. Instead, it binds somewhere else on the enzyme. This causes the enzyme, including the active site, to change shape and, as a result, substrate molecules no longer fit and so no enzyme–substrate complex is formed.

The higher the concentration of a competitive inhibitor, the greater the degree of inhibition. In other words, the effect of a competitive inhibitor can be reduced by adding more substrate. On the other hand, changing the concentration of substrate has no effect on the degree of inhibition shown by a non-competitive inhibitor.

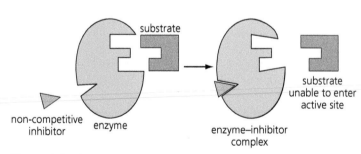

Figure 2.10 Non-competitive inhibitors bind to the enzyme somewhere other than the active site. This causes the active site to change shape.

TEST YOURSELF

5 Uric acid is a substance that is made in the body. Gout is a painful condition caused when too much uric acid is produced and crystals form in the joints. Gout can be controlled by a drug called allopurinol. Allopurinol is a competitive inhibitor. Suggest how allopurinol controls gout.

6 Ethylene glycol is a substance found in antifreeze. Sometimes it is consumed accidentally. Ethylene glycol itself is not poisonous but it is broken down in the body by the enzyme alcohol dehydrogenase to substances that are toxic. One way for doctors to treat ethylene glycol poisoning is to give the person a drink containing alcohol. Suggest how this works to prevent the poisoning becoming fatal.

7 Figure 2.11 shows the effect of a competitive or non-competitive inhibitor on an enzyme-controlled reaction. Explain why adding more substrate reduces the effect of a competitive inhibitor, but adding more substrate does not reduce the effect of a non-competitive inhibitor.

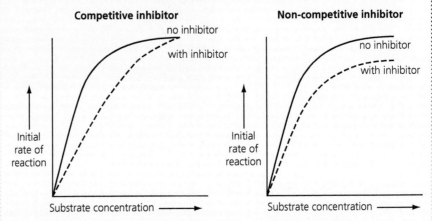

Figure 2.11 The effect of a competitive or non-competitive inhibitor on an enzyme-controlled reaction.

8 Copy the graphs shown in Figure 2.11 and add a line to each to show the effect of adding a greater amount of inhibitor.

> **TIP**
> Make sure that you understand the effects of different enzyme/inhibitor concentrations on rate in competitive and non-competitive inhibition.

The properties of enzymes: collisions come first

An enzyme and its substrate must come together before an enzyme-controlled reaction can take place. They must collide with each other with enough energy to break existing chemical bonds and form new ones. The greater the number of successful collisions in a given period of time, the faster will be the rate of reaction. Increasing the temperature and increasing the concentration of the substrate or of the enzyme can increase the probability that a successful collision will take place and that an enzyme–substrate complex will form.

Temperature

We have already seen that high temperatures denature enzymes and stop them from working. At temperatures below the optimum, an increase has a different effect. An increase in temperature increases the kinetic energy of the enzyme and substrate molecules. As a result, they move faster. This increases the probability that enzyme and substrate molecules will collide with each other. In most enzyme-controlled reactions, a rise of 10°C more or less doubles the rate of reaction, provided that temperature stays within an acceptable range around the optimum.

Figure 2.12 The tuatara is a New Zealand reptile. Like other reptiles, the tuatara cannot regulate its body temperature independently of the temperature of its environment (it is an ectotherm). The tuatara is active and its enzymes can digest food at temperatures as low as 6°C. With an increase in temperature, its enzymes work faster and it digests its food faster.

An increase in temperature therefore increases the rate of reaction until the temperature reaches an optimum value, and then the rate of denaturation also increases. Around 10°C above the optimum temperature, although the initial rate of reaction is very fast, it soon falls as the enzyme quickly becomes denatured. This is summarised in Figure 2.13.

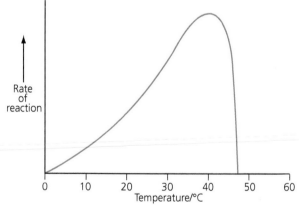

Figure 2.13 The effect of temperature on the rate of reaction of an enzyme-controlled reaction.

Rates of reaction

The rate of a reaction is how fast it is going over a given time. Catalase catalyses the reaction in which hydrogen peroxide is broken down to water and oxygen. One way of following the progress of this reaction is to measure the volume of oxygen produced.

Extension

In reality, denaturation occurs below the optimum; the optimum is the balance between slowing due to denaturation and increasing due to more collisions.

This is because there are two rates of two reactions: the rate of denaturation and the rate of enzyme–substrate reaction. Both are temperature dependent.

EXAMPLE

Using catalase to measure rates of reaction

1 Suppose 25 cm³ of oxygen was produced in 20 seconds. What would be the rate of this reaction over this 20 second period?

   The rate would be the volume of oxygen produced divided by the time the reaction took. In other words, 25/20 or 1.2529cm³ per second.

2 When we carry out this reaction in a tube in a laboratory, its rate changes. As time progresses, it slows down until it comes to a complete stop. Explain why the rate of the reaction slows down. Use your knowledge of the way in which enzymes work.

   The number of hydrogen peroxide molecules will get fewer and fewer as the substrate is broken down. The number of molecules of catalase, however, does not change. This means that there will be fewer collisions between substrate molecules and the active sites of the enzyme molecules. Fewer collisions mean a slower rate of reaction.

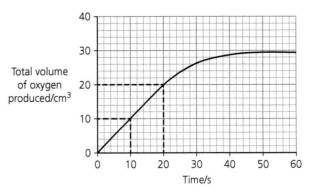

Figure 2.14 Graph showing the breakdown of hydrogen peroxide by catalase.

The graph in Figure 2.14 shows a progress curve for a reaction in which catalase is involved in breaking down hydrogen peroxide.

3 Calculate the rate of the reaction between 10 and 20 seconds. Explain how you arrived at your answer.

   Using a graph to calculate the rate of a reaction is really no different from the calculation that you have just carried out. In this case, you read off the total volume of oxygen produced between 10 and 20 seconds. This is 20 cm³ – 10 cm³. 10 cm³ of oxygen is therefore produced in 10 seconds, so the rate is 10/10 or 1 cm³ per second.

Suppose, however, that you want to look at what is happening a little later on when the rate is changing.

4 Calculate the rate of the reaction at 30 seconds.

   You cannot divide the volume produced by 30. This will give you the mean rate over the whole 40 second period. You want the rate at a particular time, 30 seconds. Figure 2.15 shows what you should do.

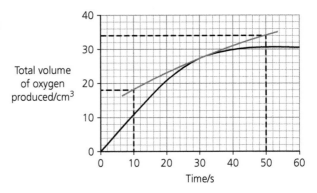

1. Draw a tangent to the curve at the point at which you are interested. Here it is 30 seconds.
2. Use the tangent to read the volume of oxygen produced. Here it is 34 – 18 or 16 cm³.
3. Use the tangent to read off the time in which this volume is produced. Here it is 50 – 10 or 40 s.
4. Calculate the rate by dividing the volume by the time. Here it is 16/40 or 0.4 cm³ s⁻¹.

Figure 2.15 Advice on answering questions: calculate the rate of the reaction at 30 seconds. Note that a tangent is a straight line that touches but doesn't cross a curved line.

Investigation into the effect of a named variable on the rate of an enzyme-controlled reaction

Note: This is just one example of how you might tackle this required practical.

A student decided to investigate the effect of pH on a protein-digesting enzyme. She set up several boiling tubes containing a mixture of $2\,cm^3$ of a protease enzyme and $5\,cm^3$ of a buffer solution as follows. In each tube she placed a glass capillary tube containing solidified egg white (a protein). She measured the length of egg white in each capillary tube before putting it in each tube. She left all the tubes, stoppered, in an incubator at 30°C for 12 hours. After this time, she removed the capillary tubes of egg white from the boiling tubes and measured the length of egg white remaining. The results are shown in Table 2.2.

SAFETY

Avoid direct contact with the enzymes in general, and particularly proteases. Enzymes can cause allergic reactions, and sometimes sensitisation. Wear eye protection, and wash off splashes from skin. People known to be sensitive to enzymes should avoid the activity, or wear gloves (they should be removed carefully to avoid contamination). Plant or fungal proteins should be used if possible, and care taken when measuring protein cylinders, to avoid dislodging the material.

Table 2.2

pH	Initial length of egg white in tube/mm	Final length of egg white in tube/mm
4.8	54	47
5.6	50	45
5.8	52	42
6.2	54	40
6.6	52	32
6.8	53	27
7.2	52	17
7.6	48	7
8.2	47	15
8.6	52	23
9.0	54	37
9.6	53	49

1 Calculate the percentage of egg white that has been digested in each tube.
2 Plot a graph showing the dependent variable on the *y*-axis against the independent variable on the *x*-axis.
3 Why was it important to calculate the percentage of egg white digested, rather than length of egg white in the tube?
4 Suggest a suitable control for this investigation. Explain why this is needed.
5 Why was it necessary to put stoppers in the tubes before incubating them?
6 The student used a buffer solution to create the correct pH in each tube. Explain why she used a buffer, rather than adding acid or alkali.
7 The student decided that the optimum pH for this enzyme was 7.6. Do you agree with this? Give reasons for your answer.

Limiting factors and substrate concentration

The shape of the curve in Figure 2.16 is one that you will often come across. It is an example that has nothing to do with biology. Supporters are going to a football match. To get into the ground, they have to pass through turnstiles. The graph shows the rate at which the supporters get into the ground plotted against the number of people outside who are trying to get in.

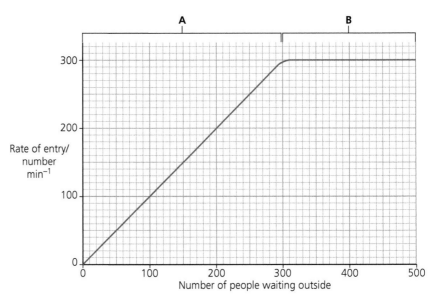

Figure 2.16 The rate of entry into a football ground plotted against the number of people outside.

We have divided this curve into two parts. Look at part A first. It shows that the rate of entry into the ground is directly proportional to the number of people outside. Three hours before the game starts, very few people are trying to get in. They can go straight to a turnstile and walk through. As more and more people arrive, the rate of entry into the ground increases.

There comes a point, however, when there are so many people outside that all the turnstiles are working as fast as possible and queues start to build up. The rate of entry to the ground cannot get any faster. We are now on the part of the curve labelled B. We say that over part A of the curve, the number of people outside is the limiting factor as it limits the rate of entry to the ground. The curve levels out in part B. It does not matter how much faster supporters arrive at the ground, the rate of entry stays the same. Something else is acting as the limiting factor. It is probably the number of turnstiles.

We will now look at a biological example that is based on the same principles. Figure 2.17 shows what happens to the rate of reaction when the concentration of the substrate is increased but enzyme concentration is kept constant. In this reaction, both the temperature and the pH are at their optimum values.

Figure 2.17 The effect of substrate concentration on the rate of an enzyme-controlled reaction.

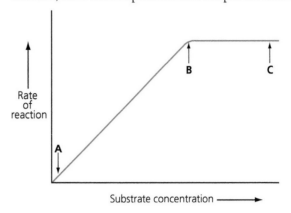

The effect of substrate concentration on the rate of an enzyme-controlled reaction

1 Look at the part of the curve between point A and B on Figure 2.17. What limits the rate of reaction over this part of the curve?
   The answer is substrate concentration.
2 What is the evidence from the graph for this answer?
   As we increase the concentration of the substrate, the rate of reaction also increases.
3 What causes the rate of reaction to increase over this part of the curve?
   The more substrate molecules there are, the greater is the probability that one of these molecules will collide
   successfully with the active site of an enzyme and a reaction will take place.
4 Look at the part of the curve between B and C. How does increasing the substrate concentration after point B affect the rate of reaction?
   The rate of reaction stays the same.
5 What caused the rate of reaction to stay the same over this part of the curve?
   At any one time, all the enzyme active sites are occupied. The enzyme cannot work any faster. The only way that the rate of reaction can be increased is to increase the number of enzyme molecules.

TIP

Look out for curves of the shape shown in Figures 2.16 and 2.17. They are very common in biology and their explanation relies on the same principles every time. We will call what we plot on the *x*-axis *X*, and what we plot on the *y*-axis *Y*.

- The first part of the curve rises. *Y* is limited by *X* because an increase in *X* produces an increase in *Y*.
- The *y*-axis value on the second part of the curve stays constant. On this part of the curve, an increase in *X* has no effect on *Y*. Something other than *X* is limiting *Y*.

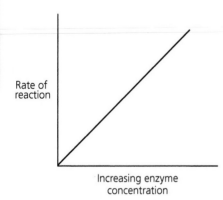

Figure 2.18 The effect of enzyme concentration on reaction rate.

The effect of enzyme concentration on reaction rate

Increasing the enzyme concentration will increase the rate of an enzyme-controlled reaction, provided that there is enough substrate present. This is because adding more enzymes increases the number of active sites for the substrate molecules to collide with. This is shown in Figure 2.18.

Of course, the linear increase in rate of reaction only occurs until substrate concentration becomes limiting.

TEST YOURSELF

9 As its body temperature increases from 15°C to 30°C, the rate at which a crocodile digests its food increases. What causes this increase?

10 An enzyme works most quickly at a pH of 6.4. A change in pH from 6.4 to 6.8 has a large effect on the rate of the reaction controlled by this enzyme. Use your knowledge of pH to explain why a small change in pH has a large effect on the rate of the reaction.

11 Sketch some axes with the x-axis labelled 'enzyme concentration' and the y-axis labelled 'rate of reaction' and draw on a line showing the rate of reaction in the presence of excess substrate.

12 Sketch some axes with the x-axis labelled 'enzyme concentration' and the y-axis labelled 'rate of reaction' and add a line to show the rate of reaction when the substrate concentration becomes a limiting factor.

13 The graph in Figure 2.19 shows the concentration of product when a concentration A of an enzyme is added to excess substrate. Sketch a line to show the concentration of product when the concentration of enzyme added is 2A but excess substrate is present.

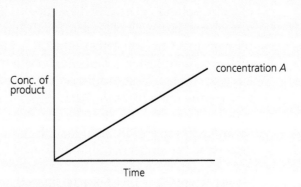

Figure 2.19 Concentration of product when an enzyme is added to excess substrate.

14 Enzymes used for student practicals are usually stored in the prep room refrigerator. Explain why.

ACTIVITY

Making a better oral rehydration solution

This is an investigation that could be carried out in a laboratory. How would you go about it?

Oral rehydration solutions (ORS) can be used by doctors to save the lives of children suffering from diarrhoeal diseases, often following natural disasters such as floods or typhoons. While children are suffering from diarrhoea, they are not absorbing the nutrients they need. As a result, they become weaker and weaker and less able to fight off the effects of the next attack of diarrhoea. Scientists wanted to develop a very effective ORS, one that not only prevents dehydration but also provides the valuable nutrients that a developing child needs.

The scientists investigated an ORS based on the starch in one type of rice flour. They found that it rehydrated diarrhoea sufferers and also helped to overcome malnutrition. However, at high concentrations, a rice-flour ORS is so thick that children cannot swallow it or take it from a feeding bottle. The scientists decided to use the enzyme amylase to digest the starch. This reduces its viscosity (thickness) so that a patient can drink it from a cup or a feeding bottle. They investigated the effect of different concentrations of amylase on reducing the viscosity of rice-flour solution.

TIP
See Chapter 15, on practical skills, to get ideas about investigating a scientific question.

Practice questions

1 A student carried out an investigation to find the effect of temperature on the rate of digestion of starch by amylase. She pipetted 5 cm³ of 1% starch solution into each of three test tubes. She then added six drops of iodine solution to each of these tubes. The student took another three tubes. Into each of these tubes she pipetted 2 cm³ of amylase solution and 1 cm³ of a buffer solution at pH 7. She placed one test tube containing starch solution and one test tube containing amylase into each of three water baths, at 10°C, 20°C and 35°C. She left the tubes in the water baths for 10 minutes. After this, she poured the tube containing amylase into the tube containing the starch solution, mixed the contents thoroughly and found the time taken for the blue colour to disappear. The student's results are shown in the table.

Temperature/°C	Time taken for blue colour to disappear/s
10	180
20	125
35	75

a) i) Describe how you could make 20 cm³ of 1% starch solution. (2)

   ii) Explain why the student added a buffer solution to the tubes containing amylase. (2)

   iii) The student left the tubes in the water baths for 10 minutes before mixing them together. Explain why. (2)

b) Suggest a suitable control for this investigation, explaining why it is necessary. (2)

c) The student recorded the time taken for all the starch to be digested. Suggest how she could use these results to calculate the rate of reaction. Explain your answer. (2)

2 The diagram shows a biochemical pathway that takes place inside a cell. It also shows the molecular structures of two of the substances in the pathway, glutamic acid and glucosamine 6-phosphate.

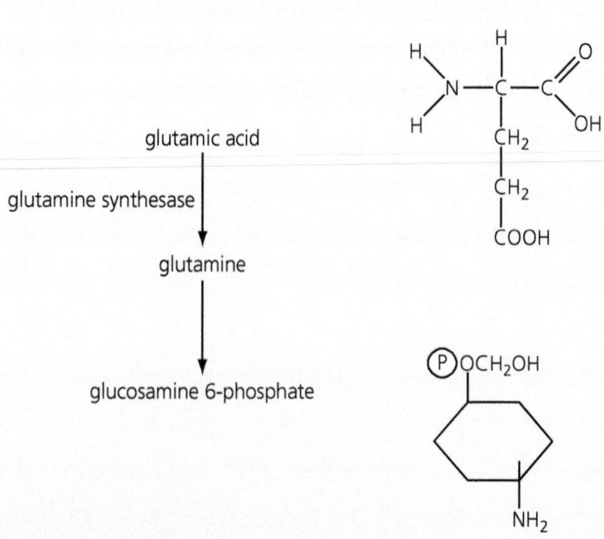

a) What type of substance is glutamic acid? Use the diagram to give the reason for your answer. (3)

b) Glucosamine 6-phosphate inhibits the enzyme glutamine synthetase. It is a non-competitive inhibitor.

   i) What information in the diagram suggests that glucosamine 6-phosphate is not a competitive inhibitor? Explain your answer. (2)

   ii) Explain how glucosamine 6-phosphate inhibits glutamate synthetase. (3)

c) Suggest a possible advantage to an organism of each of the following.

   i) The product formed in a biochemical pathway inhibits one of the enzymes in the pathway. (2)

   ii) The enzyme that is inhibited is at the start of the pathway. (2)

   iii) The product is usually a non-competitive inhibitor. (2)

3 Gelatine is a protein. When a gelatine solution cools, it sets to form a jelly. Fresh pineapple juice contains an enzyme that digests protein. A student investigated the effect of pineapple juice on the setting of jelly. He set up three different tubes of gelatine and recorded which had set after 3 hours. The contents of each tube and his results are shown in the table.

Tube	Contents of tube	Jelly set
A	$6\,cm^3$ gelatine + $2\,cm^3$ pineapple juice + $2\,cm^3$ water	No
B	$6\,cm^3$ gelatine + $2\,cm^3$ pineapple juice + $2\,cm^3$ hydrochloric acid	Yes
C	$6\,cm^3$ gelatine + $2\,cm^3$ boiled pineapple juice + $2\,cm^3$ water	Yes

a) Explain why $2\,cm^3$ of water was added to tubes A and C. (2)

b) Explain the results of tube A and tube B. (4)

c) What was the purpose of tube C ? (3)

Stretch and challenge

4 How are the properties of enzymes found in organisms that live in extreme environments, such as very cold or very hot places, different from those of enzymes found in most other organisms? How does the molecular structure of enzymes in organisms that live in hot environments allow them to resist denaturation?

5 How are enzymes used in industry? Focus on the range of uses and also the benefits of using enzymes rather than alternative chemical processes. (For example, in food manufacture; in producing 'stone-washed' jeans; in the production of paper; in cleaning materials, etc.)

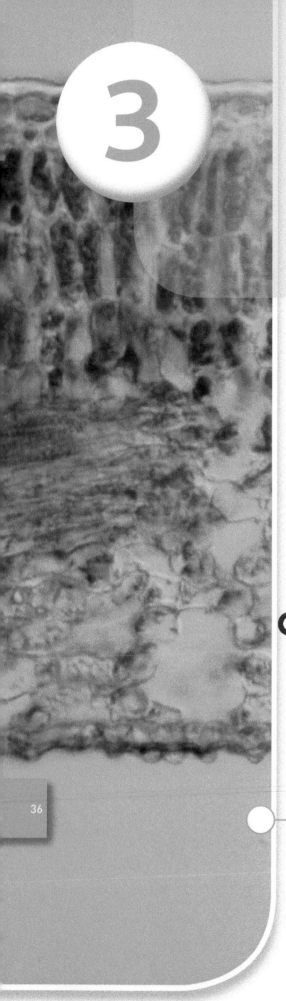

3 Cells

PRIOR KNOWLEDGE

Before you start, make sure that you are confident in your knowledge and understanding of the following points.

- Most human and animal cells have the following parts:
  - a nucleus, which controls the activities of the cell
  - cytoplasm, in which most of the chemical reactions take place
  - a cell membrane, which controls the passage of substances into and out of the cell
  - mitochondria, which are where where respiration occurs
  - ribosomes, which are where protein synthesis occurs.
- Plant and algal cells also have a cell wall made of cellulose, which strengthens the cell. Plant cells often have:
  - chloroplasts, which absorb light to make sugar
  - a permanent vacuole filled with cell sap.
- A bacterial cell consists of cytoplasm and a membrane surrounded by a cell wall; the genes are not in a distinct nucleus.
- Dissolved substances can move into and out of cells by diffusion.
- Diffusion is the spreading of the particles of a gas, or of any substance in solution, resulting in a net movement from a region where they are of a higher concentration to a region with a lower concentration. The greater the difference in concentration, the faster the rate of diffusion.

TEST YOURSELF ON PRIOR KNOWLEDGE

1 There are more mitochondria in a cell from an insect's wing muscles than in a cell from the lining of the insect's gut. Explain the advantage of this.
2 Give two structural differences between an animal cell and a plant cell.
3 A cell that produces enzymes contains a lot of ribosomes. Explain the advantage of this.
4 What features of cellulose make it useful for building plant cell walls?

Introduction

It was only when the microscope had been invented that scientists could start to study the detailed structure of living things and see that organisms are made up of cells. The Englishman Robert Hooke (1635–1703) was a brilliant self-taught scientist who made his own microscope. He made a microscope much better than anyone had ever made before, so he could see structures that had never been seen before. You can see his microscope in Figure 3.1.

Figure 3.1 Robert Hooke's microscope in *Micrographia* (1665).

Robert Hooke used his microscope to look at all kinds of things. He made a drawing of cork as seen under his microscope. He noticed that the cork was made up of empty spaces with walls around them. He called them **cells** after the Latin word *cellus*, meaning 'little room'. He calculated the number of cells in a cubic inch to be 1 259 712 000. He did not see all the smaller structures within the cell, because he was looking at the empty spaces between the cell walls in the cork, but he was the first person to realize that cells are the building blocks of living things.

We now know that two of the kinds of cell in living things are eukaryotic cells and prokaryotic cells. Eukaryotic cells are cells that contain a nucleus, such as are found in plants, animals and fungi. Prokaryotic cells include bacterial cells, and these do not contain a nucleus.

Eukaryotic cell A cell containing a nucleus and other membrane-bound organelles.

Prokaryotic cell A cell that does not contain a membrane-bound nucleus or any membrane-bound organelles.

Studying eukaryotic cells

The cells that line the small intestine of a human are called **epithelial** cells. Figure 3.2 shows epithelial cells from the human small intestine. This photograph was taken through an optical microscope.

A

Figure 3.2 Epithelial cells from the human small intestine. This photograph has been magnified 3300 times.

Look at it carefully and you will see that each cell contains a large structure, which is the **nucleus**. We call animal and plant cells eukaryotic cells because they possess a nucleus. The word eukaryote means 'true nucleus'.

You will also see that the boundary of the cell, where it lines the lumen of the intestine, shows up as a rather fuzzy thick line. If we magnify this a bit more, it does not help a lot. All we get is a slightly thicker fuzzy line. Magnification on its own is not enough. What we need is greater resolution. Magnification is making things larger. Resolution involves distinguishing between objects that are close together. To resolve objects that are close together, we need a microscope that will produce a sharper image.

Light waves limit the resolution of an optical microscope. Using light, it is impossible to resolve two objects that are closer than half the wavelength of the light by which they are viewed. The wavelength of visible light is between 500 and 650 nanometres (nm), so it would be impossible to design an optical microscope using visible light that would distinguish between objects closer than half of this value. That is good enough for a lot of purposes. It is certainly fine for looking at cells from animals and plants. But it won't let you see the very small structures inside a cell. For that we need an electron microscope.

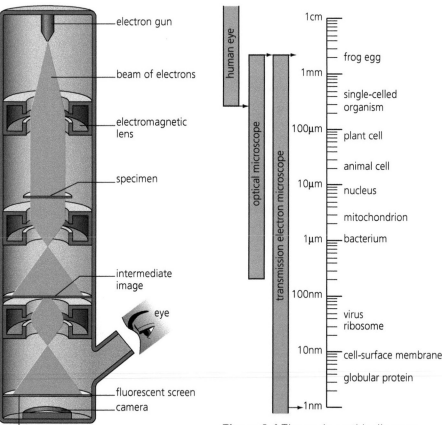

Figure 3.4 The scale on this diagram goes from 1 nm to 1 cm. These values have been plotted on a log scale because this is the best way of representing such a large range of measurements. The diagram shows that a human eye is able to see large single-celled organisms. With an optical microscope we can see things as small as bacteria. With modern transmission electron microscopes we can see large molecules.

Figure 3.3 The diagram shows the main features of a transmission electron microscope.

The transmission electron microscope

If the wavelength of light limits the resolution of an optical microscope, then one solution is to use a beam of electrons instead. Electrons have very much smaller wavelengths than light, so a beam of electrons should be able to resolve two objects that are very close together. That is the way an electron microscope works, as you can see in Figure 3.3.

We cut a very thin section through the tissue that we are going to examine (thin enough to let electrons through). This section, the specimen, is preserved and stained with the salts of heavy metals, such as uranium and lead. Then it is put inside a sealed chamber in the microscope. The air is sucked out of the chamber and this produces a vacuum. Finally, a series of magnetic lenses focuses a beam of electrons through the specimen and produces an image on a screen. Electrons pass more easily through some parts of the section than others. The electrons pass less easily through parts that are stained with heavy metals. This produces contrast between different parts of the specimen.

Magnification This tells you how many times bigger the image is than real life.

Resolution The ability to see two structures very close together as separate structures.

3 CELLS

38

The big advantage of using a transmission electron microscope is its resolving power. It lets us see the structure of a cell in much more detail than we could ever hope to see with an optical microscope. Look at Figure 3.4. It gives you a clear idea of just what can be seen with a human eye, a good quality optical microscope and a transmission electron microscope.

The scanning electron microscope

A **scanning electron microscope** works in a slightly different way from a transmission electron microscope. In a scanning electron microscope, the electron beam bounces off the surface of the object. It is particularly useful for looking at three-dimensional structures such as viruses.

A high resolving power means that a transmission electron microscope has a useful magnification of up to 100 000 times.

Electron microscopes have their limitations

You might think that a transmission electron microscope with a magnification of 100 000 times is the perfect instrument to investigate cell structure. However, such a microscope has limitations. Because there is a vacuum inside, all the water must be removed from the specimen. This means that you cannot use a transmission electron microscope to look at living cells. They must be dead. Also, the lengthy treatment required to prepare specimens means that artefacts can be introduced, which look like real structures but are actually the results of preserving and staining.

ACTIVITY

Sections through different planes

There are also problems in interpreting what you see. Some of these arise because a very thin slice has to be cut through the specimen. Look at Figure 3.5a. It shows a red blood cell. The shape of a red blood cell is often described as a biconcave disc because it is thinner in the centre and thicker at the edge.

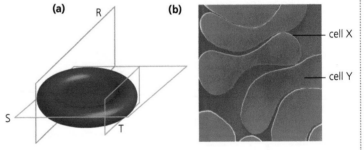

Figure 3.5 (a) A drawing of a red blood cell. Sections have been cut through it in different planes. (b) A photograph of sections through human red blood cells seen with a transmission electron microscope. A photograph like this is called an electron micrograph.

1 Make simple drawings to show what the cut surface of the cell in Figure 3.5a would look like if it were cut through each of the planes shown in the figure.

2 Look at Figure 3.5b. It shows very thin sections of red blood cells. Through which of the three planes shown in diagram (a) is the cell labelled X cut?

3 The cell labelled Y has a bent shape. This bent shape resulted when the cell was sliced. Suggest what caused this bent shape.

When the beam of electrons in a transmission electron microscope strikes a specimen, some of the electrons pass straight through and some are scattered by dense parts of the specimen. The parts of the specimen that scatter the electrons appear dark coloured on an electron micrograph.

4 Red blood cells show as a uniform dark colour on an electron micrograph. Explain why.

5 A human red blood cell measures 7 μm in diameter. Calculate the magnification of the electron micrograph in Figure 3.5b (see page 51 for how to do this).

Figure 3.6 shows a scanning electron micrograph of red blood cells.

6 What additional information can you get about red blood cells from Figure 3.6?

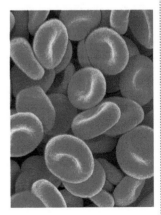

Figure 3.6 A scanning electron micrograph of red blood cells.

The ultrastructure of eukaryotic cells

Figure 3.2 shows some epithelial cells from the human small intestine as they appear when looked at with an optical microscope. Now look at Figure 3.7. This shows part of one of the same cells, but it has been taken with a transmission electron microscope, so you can see much more detail.

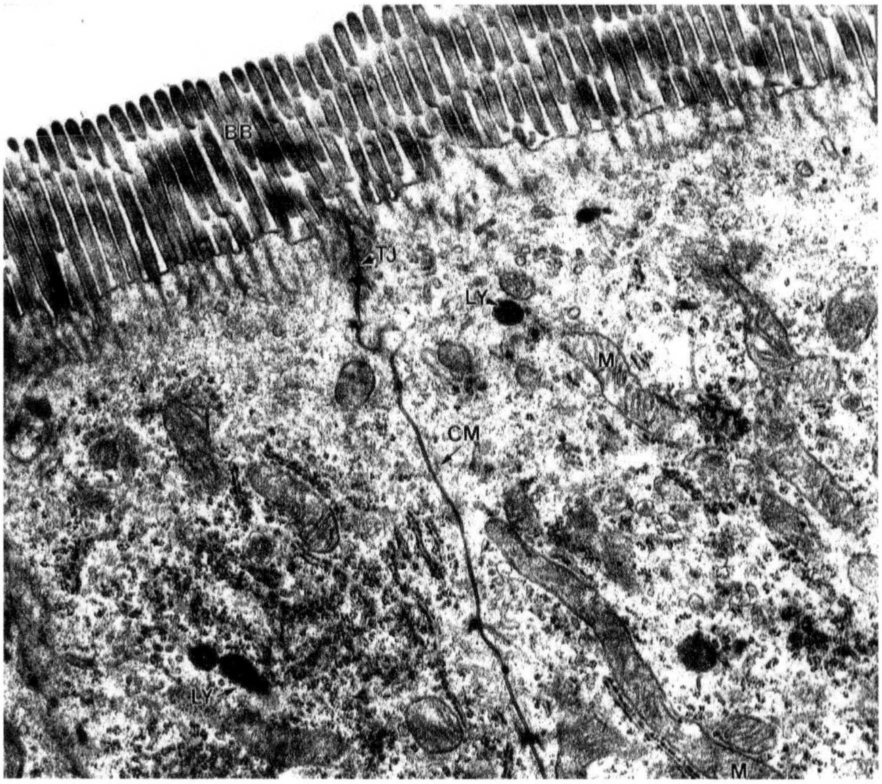

Figure 3.7 An electron micrograph of an epithelial cell from the small intestine.

Look again at the boundary of the cell, where it lines the lumen of the intestine. It shows that the fuzzy thick line that you saw in Figure 3.2 is made up of tiny finger-like folds in the membrane called **microvilli**. The nucleus can be seen clearly as well. The rest of the cell is made up of cytoplasm in which there are many tiny structures called **organelles**. These organelles have particular functions in the cell. Table 3.1 summarises the functions of the organelles found in an epithelial cell from the small intestine.

Because of the high resolution of a transmission electron microscope, the organelles in the cell can be seen clearly. You could not see most of these organelles with an optical microscope.

Separating cell organelles

Understanding the structure of an organelle is not the same as understanding its function. To find out about function, biologists need a pure sample containing lots of the organelle that they want to investigate. We separate cell organelles from each other using the process of **cell fractionation**. In this process, a suitable sample of tissue is broken up and then centrifuged at different speeds. Figure 3.8 is a flow chart that summarises the main steps in this process.

The cells in a tissue are broken open in a **homogeniser**. This is a machine rather like a kitchen blender. The tissue is suspended in a buffer solution which keeps the pH constant. This solution is kept cold and has the same water potential as the tissue.

↓

The homogenised mixture is filtered. This removes large pieces of tissue that have not been broken up.

↓

The filtrate is now put in a **centrifuge** and spun at low speed. Large organelles such as nuclei fall to the bottom of the centrifuge tube where they form a **pellet**. They can be resuspended in a fresh solution if they are wanted.

↓

The liquid or **supernatant** is now spun in the centrifuge again. This time, smaller organelles such as mitochondria separate out into a pellet.

↓

Figure 3.8 The organelles in a sample of tissue can be separated from each other by the process of cell fractionation in a centrifuge. This flow chart shows the main steps in the process.

Table 3.1 The main organelles found in an epithelial cell from the small intestine.

Organelle	Main features	Main function
Cell-surface membrane	The membrane found around the outside of a cell. It is made up of lipids and proteins.	Controls the passage of substances into and out of the cell.
Nucleus	The largest organelle in the cell. It is surrounded by a nuclear envelope consisting of two membrane layers. There are many holes in the envelope called nuclear pores.	Contains the DNA, which holds the genetic information necessary for controlling the cell.
Mitochondrion	A sausage-shaped organelle. It is surrounded by two membrane layers. The inner one is folded and forms structures called cristae.	Produces ATP during respiration. The molecule ATP is the source of energy for the cell's activities.
Lysosome	An organelle containing digestive enzymes called lysozymes. These enzymes are separated from the rest of the cell by the membrane that surrounds the lysosome.	Digests unwanted material in the cell.
Ribosome	A very small organelle, not surrounded by a membrane.	Assembles protein molecules from amino acids.
Rough endoplasmic reticulum	Endoplasmic reticulum is made of membranes that form a series of tubes in the cytoplasm of the cell. The membranes of rough endoplasmic reticulum are covered with ribosomes.	Synthesises and transports proteins around the cell.
Smooth endoplasmic reticulum	Similar to rough endoplasmic reticulum, but the membranes do not have ribosomes.	Synthesises lipids.
Golgi apparatus	A stack of flattened sacs, each surrounded by a membrane. Vesicles are continually pinched off from the ends of these sacs.	Packages and processes molecules such as proteins for use in other parts of the cell, or for export to outside the cell. Forms lysosomes.

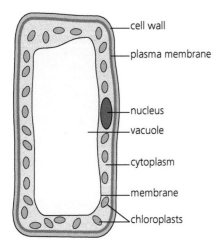

Figure 3.9 A simple diagram showing a single cell from the palisade layer of a leaf.

Plant cells

Figure 3.9 shows a single cell from the palisade layer of a leaf. You will notice that, like an animal cell, it is eukaryotic.

Cells in flowering plants have all the organelles that are found in animal cells. However, they have some fundamental differences. A key difference between plants and animals is that plants photosynthesise. This process requires light. To make glucose by photosynthesis, plants have **chlorophyll** to absorb light. Since chlorophyll is only useful in areas exposed to light, it is not present in all plant cells. For example, in underground roots chlorophyll would have no function. In most plants the leaves are where the majority of photosynthesis takes place, in the palisade mesophyll and spongy mesophyll. You can see this in Figure 3.10, overleaf.

Figure 3.10 A photomicrograph of a vertical section across a leaf, as seen with an optical microscope (x 32).

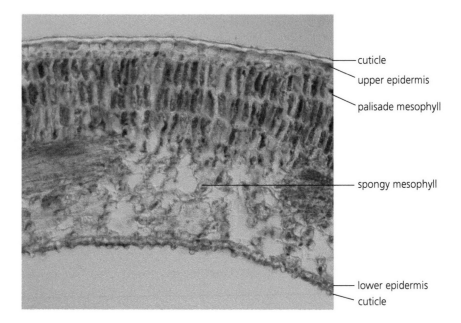

cuticle
upper epidermis
palisade mesophyll

spongy mesophyll

lower epidermis
cuticle

As you can see in Figure 3.11, the chlorophyll is not dispersed throughout the cells but is contained in separate organelles called chloroplasts. When viewed with an electron microscope, a chloroplast can be seen to have a complex structure that adapts it for photosynthesis.

Figure 3.11 An electron micrograph of a chloroplast (x 16 000).

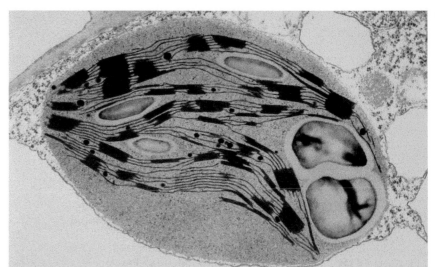

The chlorophyll molecules are embedded in the membranes of disc-shaped structures. In some areas, the disc-shaped structures are stacked up, rather like piles of coins. These stacks show up under low magnification in an electron microscope as dark, grainy patches. There are usually about 50 dark areas in a chloroplast, linked together by a network of the membranes. These structures hold the chlorophyll molecules in positions where the maximum amount of the light that falls on the chloroplast reaches them. The chlorophyll molecules use the light energy to split water into hydrogen ions and oxygen.

The surrounding **stroma** contains enzymes. These enzymes catalyse a series of reactions that use the hydrogen ions, electrons and carbon dioxide to make glucose. The chloroplast is surrounded by a double membrane, which enables control and localisation of substrates and ensures that the enzymes are held inside the chloroplast close to the chlorophyll while allowing free movement of small molecules such as carbon dioxide and water. The structure of the chloroplast ensures that this complex process happens efficiently in a small space with optimum use of available light energy. Excess glucose is converted to insoluble starch and stored temporarily in starch grains in the chloroplast.

Plant cell walls

You may have noticed that one major difference between animal and plant cells is that an animal cell has no wall, whereas the palisade cell has a wall outside the cell-surface membrane. This wall is fairly thin. It is not rigid like a brick wall, but it does give the cell strength, as it resists being stretched. It is the strength of the cell wall that stops a leaf cell from expanding and taking in so much water that it bursts.

The leaves of most flowering plants stick out from the stems. But, as you know if you have ever forgotten to water houseplants, when a plant is short of water the leaves flop down – they wilt. For the leaves to stay flat, facing the light, they need water that is taken up through the roots. It enters a leaf cell by osmosis and fills the vacuole. The water pushes against the wall and makes the cell firm, just as air pumped into a tyre makes the tyre hard. The cell wall is strong enough to prevent the cell from bursting. Animal cells, however, have no cell wall. If, for example, a red blood cell took in too much water, its cell-surface membrane would burst. Animals have systems that control the water content of the blood and tissue fluid surrounding cells and stop this happening.

All life on Earth exists as cells, which all have common features. This is indirect evidence for evolution. However, not all organisms consist of eukaryotic cells. Prokaryotic cells are described in Chapter 6.

TEST YOURSELF

4 Compare the cells shown in figures 3.7 and 3.9. Describe three ways in which the structures of the cells are similar and three ways in which they differ.

5 Calculate the actual length of the chloroplast in Figure 3.11 in micrometres.

6 Explain how a rising glucose concentration inside a chloroplast might result in damage to the chloroplast.

TIP
Cell-surface membranes are found in both prokaryotic and eukaryotic cells. Their structure is the same in both types of cell.

The cell-surface membrane

You saw in Table 3.1 that the **cell-surface membrane** is made up of lipids and proteins. The commonest lipids found in living organisms are triglycerides. You learned about these in Chapter 1 (pages 7–9). Most of the lipids found in a cell membrane are phospholipids. You will remember that a phospholipid has a 'head' consisting of glycerol and phosphate and a 'tail' containing the long hydrocarbon chains of the fatty acids.

This means that when phospholipids are mixed with water they arrange themselves in a double layer with their hydrophobic tails pointing inwards and their hydrophilic heads pointing outwards. This double layer is called a **phospholipid bilayer** and forms the basis of membranes in and around cells.

The fluid mosaic model

A cell-surface membrane is only about 7 nm thick, so we cannot see all the details of its structure, even with an electron microscope. Because of this, biologists have produced a model to explain its properties. This is called the **fluid mosaic model**. The model was given this name because it describes how the molecules of the different substances that make up the membrane are arranged in a mosaic. Not all of these molecules stay in one place.

They move around, so we also describe the bilayer as being fluid. All cell membranes have this structure, not just the cell-surface membrane.

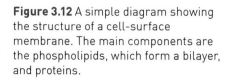

Figure 3.12 A simple diagram showing the structure of a cell-surface membrane. The main components are the phospholipids, which form a bilayer, and proteins.

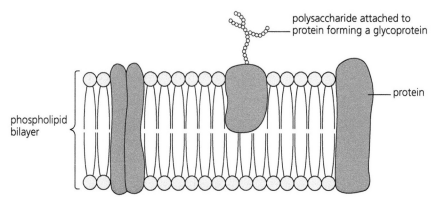

Look at Figure 3.12. This is a very simple diagram showing a section through a cell-surface membrane. It illustrates how the phospholipids and proteins are arranged. The membrane is based on a phospholipid bilayer. Very small non-polar molecules, and molecules that dissolve in lipids, can pass easily through this bilayer. Water-soluble substances, however, must pass through channels and carriers in the protein molecules that span the membrane. The phospholipid layer therefore forms a very important barrier. Since molecules of some substances are unable to pass through it directly, passage into or out of the cell is controlled by the protein molecules in the membrane.

Some of the proteins move freely in the phospholipid bilayer of the cell-surface membrane. Others are attached to both the cell-surface membrane and structures in the cytoplasm of the cell.

Membrane proteins have a variety of different functions, as follows.

- They may act as enzymes. Enzymes that digest disaccharides are found in the cell-surface membranes of the epithelial cells that line the small intestine.
- They may act as channels through the membrane to allow specific ions or molecules through.
- They act as carrier proteins and play an important part in transporting substances into and out of the cell.
- They act as receptors for hormones. A hormone will only act on a cell that has the right protein receptors in its cell-surface membrane or cytoplasm.
- They act as molecules that are important in cell recognition, and may act as antigens.

Carbohydrates are attached to lipids and proteins on the outside of the cell-surface membrane, forming glycolipids and glycoproteins. They are important in allowing cells to recognise one another. Cholesterol molecules are also found in the cell-surface membrane of animal cells, where they add strength and prevent movement of other molecules in the membrane.

Diffusion, osmosis and active transport

Diffusion

Diffusion is the random movement of the ions or atoms or molecules that make up a substance from where they are at a high concentration to where they are at a lower concentration. In other words, particles of the substance diffuse down a **concentration gradient**. Think about what happens if you put a drop of ink into a beaker of water. The ink molecules will gradually spread through the water. They will have diffused from where they were in a high concentration in the original drop to where they are in a lower concentration in the surrounding water.

These particles are moving at random. You can see this quite easily if you look at a tiny drop of toothpaste mixed with water under the microscope. The toothpaste particles move around, twisting and turning. They are moving because the molecules of water in which they are suspended are moving at random and are bumping into them. The kinetic energy that the molecules possess results in this movement.

In a solid, the particles are packed closely together and can only vibrate; in a liquid, the molecules are free to move, but are close together, so bump into each other and change direction; and in a gas, the molecules travel much further before colliding with each other.

Diffusion is also one of the ways in which substances pass into and out of cells. During aerobic respiration, for example, cells produce carbon dioxide. So there is a higher concentration of carbon dioxide inside a cell than outside. Like oxygen, carbon dioxide is a small non-polar molecule and will diffuse from where it is in a high concentration inside a cell through the cell-surface membrane to where it is in a low concentration outside the cell. Surfaces through which diffusion takes place are called **exchange surfaces**.

The **rate of diffusion** is the amount diffused through the surface divided by the time taken. This depends on a number of factors. These include the following.

- Temperature: molecules move faster at higher temperatures, so the higher the temperature, the faster the rate of diffusion.
- Surface area: the greater the surface area of the exchange surface, the faster the total rate of diffusion
- The difference in concentration on either side of the exchange surface. The greater this difference, the faster the rate of diffusion. In the intestine, the blood is continually transporting the products of digestion away from the intestine wall. This ensures a greater concentration gradient and a faster rate of diffusion.
- A thin exchange surface. Diffusion is only efficient over very short distances. Exchange surfaces such as the epithelium of the intestine are just one cell thick.

TEST YOURSELF

7 Fish have gills. They use these gills to obtain oxygen from the water by diffusion. Suggest three features of fish gills that are adaptations for efficient diffusion.

8 Which type of microscope would be best to view:
  a) a single-celled organism swimming through some pond water?
  b) the surface of a bacterial cell?
  c) the internal structure of a chloroplast?

9 Small, lipid-soluble substances enter cells most quickly. Explain why.

10 In cell fractionation, why is a buffer solution used?

Facilitated diffusion

Large, water-soluble molecules such as those of glucose cannot pass directly through the phospholipid bilayer of a cell-surface membrane. They need to be taken across by carrier proteins in the membrane. These carrier proteins have a binding site on their surface, which has a specific shape. A glucose carrier, for example, has a binding site into which only glucose molecules will fit. In addition, different sorts of cells have different carrier proteins. This explains why a particular cell will take up some substances but not others.

Diffusing molecules bind to a carrier protein. The protein changes shape and takes the molecules through the membrane.

Channel proteins help the diffusion of ions. Some ion channels have gates that open and close.

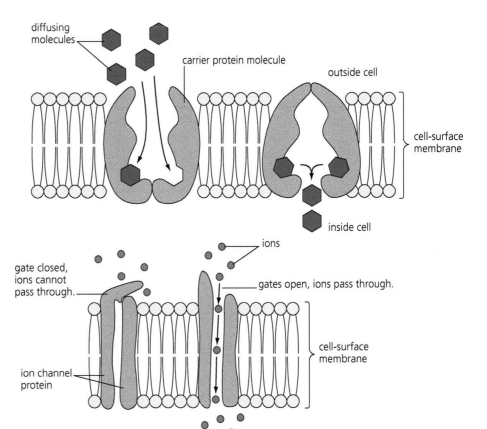

Figure 3.13 In facilitated diffusion, carrier proteins in the cell membrane assist in the transport of substances into the cell.

Look at Figure 3.13. You can see that carrier proteins change shape when they bind to diffusing molecules, such as glucose molecules. When this happens, the carrier proteins carry the molecules through the membrane and into the cell. This process, in which a protein carrier transfers a molecule that would not otherwise pass through the membrane, is called **facilitated diffusion**. Another kind of facilitated diffusion uses channel proteins. These are proteins that have a water-filled centre which water-soluble molecules and ions can diffuse through. As with simple diffusion, facilitated diffusion relies on the kinetic energy of the molecules. It is also described as a passive process because it does not require hydrolysis of ATP from respiration.

REQUIRED PRACTICAL 4

Investigation into the effect of a named variable on the permeability of cell-surface membranes

Note: This is just one example of how you might tackle this required practical.

Beetroot cells contain pigments called **betalains** that give the tissue its dark purple-red colour. The pigment is contained in the cell vacuole.

A student decided to investigate the permeability of beetroot membranes. She used a cork borer to cut several 'cores' of beetroot tissue. She cut these cores into slices 2mm thick and placed them in a beaker of cold water for 2 hours.

The student then took 21 test tubes. She used a graduated pipette to place exactly $5\,cm^3$ of distilled water in each tube. She labelled the tubes with the temperature of the water bath she was going to place them in. She put three tubes in each water bath. The temperatures of the water baths were 0, 25, 40, 50, 65, 85 and 95°C. She left the tubes in the water baths for 5 minutes.

SAFETY

Take care when using hot water – a water temperature greater than 50°C can cause scalding damage to the skin.

After this, she removed five beetroot discs from the beaker. She handled them carefully using forceps and gently blotted them dry using filter paper. Then she put the discs in the first tube and made a note of the time. She repeated this for each tube.

After exactly 45 minutes, the student shook each tube carefully and gently removed the beetroot discs. She placed a white tile behind each tube and made a note of its colour.

Next, the student set the colorimeter to a blue-green filter (530 nm) and inserted a cuvette containing distilled water. She set the absorbance to 100%.

Then she inserted a sample of the solution from each tube in turn into the colorimeter and measured the percentage absorbance. She obtained the following results.

Temperature/°C	Colour	Colorimeter reading/% absorption of light			
		Tube 1	Tube 2	Tube 3	Mean
0	Clear and colourless	0.0	0.1	0.0	0.0
25	Very pale pink	3.9	4.1	4.3	4.1
40	Very pale pink	19.3	20.4	19.9	19.9
50	Pale pink	46.2	44.3	43.5	44.7
65	Pink	76.4	78.3	75.3	76.7
85	Dark pink	99.6	98.8	98.2	98.9
95	Red	100.0	100.0	.100.0	100.0

1 Why were the beetroot discs placed in a beaker of cold distilled water for 2 hours before the investigation?

2 Why did the student handle the discs carefully using forceps, and why were they carefully blotted dry before being added to the tubes of water?

3 Why were the tubes of water left in the water bath for 5 minutes before the discs were added?

4 Why did the student use a red-green filter in the colorimeter?

5 Why did the student put a cuvette containing distilled water into the colorimeter before putting the test solutions in?

6 Plot a graph of these results. Use this graph to describe the effect of temperature on the amount of pigment released from beetroot tissue.

7 Use your knowledge of the structure of cell membranes to explain this effect.

8 Suggest at least two limitations of the student's method. Explain how these might cause small inaccuracies in the results obtained, and suggest how the effects of these limitations could be reduced.

Water potential and osmosis

Look at Figure 3.14, overleaf. It shows water molecules surrounded by a membrane. These water molecules are in constant motion. As they move around randomly, some of them will hit the membrane. The collision of

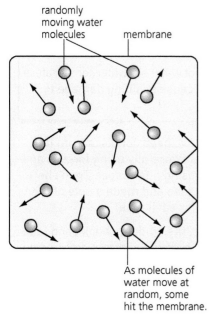

randomly
moving water
molecules

membrane

As molecules of
water move at
random, some
hit the membrane.

Figure 3.14 Water molecules move at random. Some will hit the surrounding membrane and create a pressure on it. This pressure is the water potential.

Osmosis The net movement of water molecules from a solution with a higher water potential to a solution with a lower water potential through a selectively permeable membrane.

the molecules with the membrane creates a pressure on it. This pressure is known as the **water potential** and is measured in units of pressure, usually kilopascals (kPa).

Obviously, the more water molecules that are present and able to move about freely, the greater the water potential. The greatest number of water molecules that it is possible to have in a given volume is in pure water, because nothing else is present. Pure water therefore has the highest water potential. It is given a value of zero. All other solutions will have a value less than this. They will have a negative water potential.

Now look at Figure 3.15. This shows a cell surrounded by distilled water. The cell-surface membrane separates the cytoplasm of the cell from the surrounding water. It is **selectively permeable**. This means that it allows small molecules such as water to pass through but not larger molecules. The cytoplasm of the cell contains many soluble molecules and ions. They attract water molecules, which form a 'shell' round them. These water molecules can no longer move around freely in the cytoplasm. Therefore, there is a much higher concentration of free water molecules in the water surrounding the cell than there is in the cell's cytoplasm.

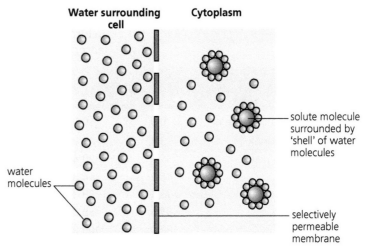

Water surrounding cell   **Cytoplasm**

solute molecule
surrounded by
'shell' of water
molecules

water
molecules

selectively
permeable
membrane

Figure 3.15 In this diagram, the concentration of water molecules that are able to move freely is higher outside the cell than inside the cell. As a result, water will move into the cell by osmosis.

The water potential is higher outside the cell than inside it, and therefore there is net movement of water molecules from the distilled water into the cell. Water molecules will also show net diffusion across a membrane from any solution with a higher water potential to a solution with a lower water potential. This is osmosis. We can, therefore define osmosis in terms of water potential.

If you put an animal cell into a solution with the same water potential as the cell, there will be no net movement of water by osmosis into or out of the cell so it will stay the same.

If you put an animal cell into a solution with a lower water potential than the cell, there will be net movement of water out of the cell and it will shrink in size. The same thing will happen to a plant cell, except that the cell membrane and its contents will shrink away from the cell wall. The gap between the cell membrane and the cell wall will be filled with external solution.

If an animal cell is placed in a solution with a higher water potential than the cell, it will take in water by osmosis and swell up. Eventually it will burst. A plant cell placed in such a solution will also take in water by osmosis, and will swell up to become firm, but it will not burst because the cell wall acts as a protective 'cage' around it, stopping it increasing in volume excessively.

REQUIRED PRACTICAL 3

Production of a dilution series of a solute to produce a calibration curve with which to identify the water potential of plant tissue

Note: This is just one example of how you might tackle this required practical.

TIP
See Figure 1.19 on page 15, for how to make up a dilution series.

A student was given a 1.0 mol dm^{-3} solution of sodium chloride and a beaker of distilled water. He used this to put 20 cm^3 of five different sodium chloride solutions into five Petri dishes as follows: 1.0, 0.75, 0.5, 0.25 and 0.0 mol dm^{-3} sodium chloride.

1 Copy and complete the table below to show how the student made up these solutions, called a dilution series.

Dish	Volume of 1.0 mol dm^{-3} sodium chloride solution/cm^3	Volume of distilled water/cm^3
1 1.0 mol dm^{-3} sodium chloride		
2 0.75 mol dm^{-3} sodium chloride		
3 0.50 mol dm^{-3} sodium chloride		
4 0.25 mol dm^{-3} sodium chloride		
5 0.0 mol dm^{-3} sodium chloride		

Next, the student used a cork borer to cut several 'cores' of potato. He then cut the cores into discs, with each disc about 2 mm thick. He weighed the discs in batches of five, made a note of the mass, and then placed the discs into one of the dishes. This was repeated for the other four dishes. He left the discs in the dishes for 2 hours.

SAFETY
Take care when using knives and cork borers. Your teacher will demonstrate to you the correct, safe technique when cutting cores and discs of potato.

After this time, the student carefully removed the discs from each dish using forceps. He carefully blotted them dry using filter paper, and re-weighed them in the same batches of five. He recorded the results on a table. He calculated the percentage change in mass of the discs.

2 Copy and complete this table to process the data appropriately.

Dish	Initial mass/g	Final mass/g	
1 1.0 mol dm^{-3} sodium chloride	3.83	2.87	
2 0.75 mol dm^{-3} sodium chloride	4.07	3.62	
3 0.50 mol dm^{-3} sodium chloride	4.02	2.85	
4 0.25 mol dm^{-3} sodium chloride	3.99	3.84	
5 0.0 mol dm^{-3} sodium chloride	3.96	4.73	

3 It was important that the student handled the discs carefully when removing them from the dish and blotting them dry with filter paper. Explain why.

4 The mass of potato discs placed in each dish was not exactly the same. Does this affect the validity of the student's results? Explain your answer.

5 Plot a suitable graph of the processed data to find the concentration of sodium chloride solution that has the same water potential as the potato cells.

6 Explain the changes in mass of the potato tissue. Use the terms 'osmosis' and 'water potential' in your answer.

7 Identify at least two limitations of this investigation, and suggest how they could be overcome.

Active transport

Most cells are able to take up substances that are present in lower concentrations outside the cells than inside. Plant cells, for example, contain mineral ions that are present in very small concentrations in the surrounding soil. **Active transport** is a process by which a cell takes up a substance *against* a concentration gradient.

As with facilitated diffusion, protein carrier molecules are involved, and they transport the substance across the membrane. The difference is, however, that active transport requires external energy. This energy comes from molecules of the substance ATP produced during respiration. Cells in which a lot of active transport takes place, such as the epithelial cells lining the small intestine, have large numbers of mitochondria, which produce the necessary ATP.

> **TEST YOURSELF**
> 11 Give one difference between simple diffusion and facilitated diffusion.
> 12 Give one similarity and two differences between facilitated diffusion and active transport.
> 13 Suggest how channel proteins are specific for certain molecules or ions.
> 14 List the similarities and differences between simple diffusion and osmosis.
> 15 Sometimes, in a hospital, a patient who has lost a lot of blood is given saline (sterile solution of sodium chloride) to compensate for the loss of blood. Why is it important that the saline solution have the same water potential as the blood cells?
> 16 In cell fractionation, why is the buffer used (a) of the same water potential as the cell and (b) ice-cold?

Prokaryotes and their structure

Bacteria are very small, much smaller than the human cells that some of them infect. Small size is a feature of the cells of all bacteria, and although most features of eukaryotic cells also apply to prokaryotic cells, they do differ from eukaryotic cells in a number of other ways. One is that a bacterial cell does not have a nucleus. It does contain DNA, but this DNA is only present as a circular molecule in the cytoplasm of the cell and not in a nucleus. It is not associated with proteins to form chromosomes.

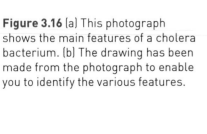

Figure 3.16 (a) This photograph shows the main features of a cholera bacterium. (b) The drawing has been made from the photograph to enable you to identify the various features.

(a)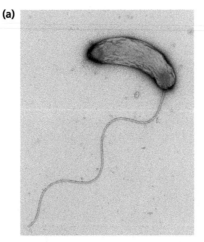

(b)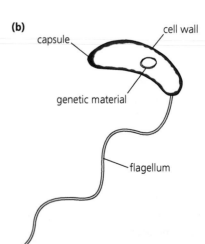

capsule

cell wall

genetic material

flagellum

We describe bacterial cells as **prokaryotes** and bacteria as **prokaryotic organisms**. The word prokaryote means 'before the nucleus'. Another feature of bacteria is that some of the DNA they contain, the genetic material, is found in tiny loops called **plasmids**.

A bacterium is surrounded by a cell-surface membrane. (This membrane is also called the plasma membrane.) Outside this membrane is a **cell wall**, but, unlike plant cell walls, it is not made from cellulose. It contains a type of glycoprotein called **murein**. Outside the cell wall there may be another layer. This is the protective **capsule**. This is only present in some bacteria. It helps to stop the bacterium drying out, or from being attacked by white blood cells. It can also store toxins produced by the bacterium. Some bacteria have at least one long whip-like **flagellum**, which helps the cell to move. Flagella are found only in some species of bacteria. Bacteria do not have any membrane-bound organelles, such as endoplasmic reticulum or mitochondria. However, they do have ribosomes that carry out protein synthesis. These ribosomes are smaller than ribosomes in eukaryotes.

Look now at Figure 3.17. This shows a diagram of a bacterium with its main features.

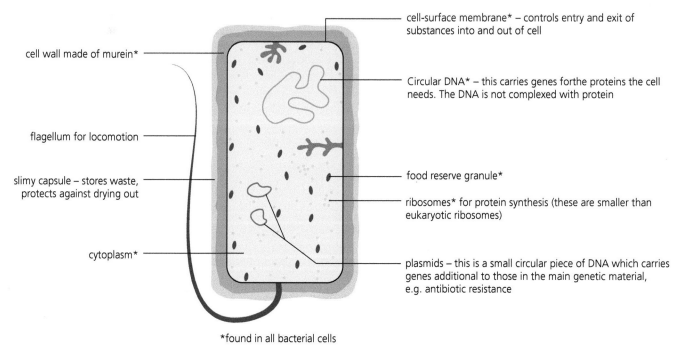

cell wall made of murein*

flagellum for locomotion

slimy capsule – stores waste, protects against drying out

cytoplasm*

cell-surface membrane* – controls entry and exit of substances into and out of cell

Circular DNA* – this carries genes forthe proteins the cell needs. The DNA is not complexed with protein

food reserve granule*

ribosomes* for protein synthesis (these are smaller than eukaryotic ribosomes)

plasmids – this is a small circular piece of DNA which carries genes additional to those in the main genetic material, e.g. antibiotic resistance

*found in all bacterial cells

Figure 3.17 The main features of a bacterium. Bacteria are tiny, generally ranging in size from 0.1 to 5 µm in length.

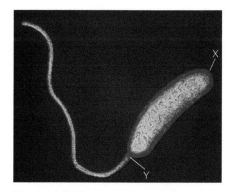

Figure 3.18 This photograph shows a cholera bacterium (× 10 000).

How to calculate size from an image

Look at Figure 3.18. It shows a single cholera bacterium magnified 10 000 times. We will use this information to work out its actual size.

The magnification of an object in a photograph is its length in the photograph divided by its real length. We can write this as a simple formula:

$$\text{Magnification} = \frac{\text{size of image}}{\text{size of real object}}$$

If we rearrange this formula, we can calculate the real length of the cholera bacterium in the photograph:

$$\text{Size of real object} = \frac{\text{size of image}}{\text{magnification}}$$

We have been given the magnification, but we need to know the length of the bacterium in the photograph. That is straightforward. All we have to do is use a ruler to measure the length of the bacterium in the photograph. We won't include the flagellum. We will just measure the cell between points **X** and **Y** in the photograph.

It is 30 mm in length, so we substitute these figures in the rearranged formula:

$$\text{Real length} = \frac{30}{10\,000} = 0.003\,\text{mm}$$

The calculation is really very simple, but the units we have used are not very practicable. It is like measuring the cost of a postage stamp in pounds rather than pence. It would be better in this case to give the answer in micrometres (μm). There are a thousand micrometres in a millimetre, so the length of the bacterium is 0.003 × 1000, or 3 μm.

TIP

See Chapter 14 to find out how to calculate magnification.

See Chapter 15 to find out how to use measuring instruments such as a graticule and a micrometer.

TEST YOURSELF

17 List the similarities and differences between a prokaryotic cell and a eukaryotic cell.

18 A bacterial cell is 2 μm long. Its length in a diagram is 40 mm. What is the magnification of the diagram?

Practice questions

1 The figure shows a bacterial cell.

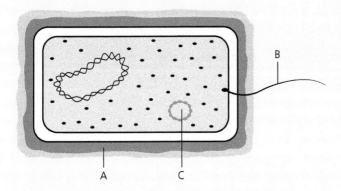

a) Name structures A, B and C. (3)

b) The following descriptions apply to structures found in a
human cell. Give the name of each structure. (3)

Name of structure	Description
	Synthesises proteins and transports them around the cell.
	Contains enzymes that digest substances, such as worn-out organelles or ingested bacterial cells.
	Increases the surface area of the cell.

2 a) Copy and complete the table with a tick if the statement
applies or a cross if the statement does not apply. (2)

	Uses membrane proteins	Requires energy from ATP
Active transport		
Facilitated diffusion		

A student cut a potato tuber in half. She placed the cut part of the potato
downwards in a beaker of distilled water. She then scooped out a hollow in
the top of the potato and placed 10% sucrose solution into this hollow. She
left the potato for 3 hours. The figure shows the results of her investigation.

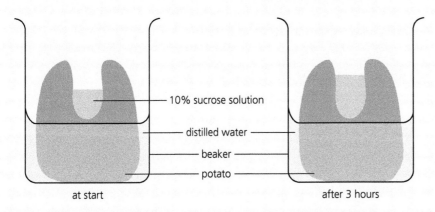

b) The level of the solution in the hollow at the top of the potato
rose. Use your knowledge of water potential to explain why. (3)

3 A student cut 30 discs of potato tissue. Each disc was identical in size and shape. She weighed groups of five discs, then put the discs into a dish containing distilled water. This was repeated five more times, with each group of discs being placed in a different concentration of sodium chloride solution. All the dishes were left in a refrigerator for 24 hours. After this, the student carefully dried the potato discs and re-weighed them. She calculated the ratio of the starting mass : final mass for each group of discs. For ease of plotting, a ratio of starting mass to final mass of 0.95 : 1 is plotted as 0.95. The results are shown on the graph.

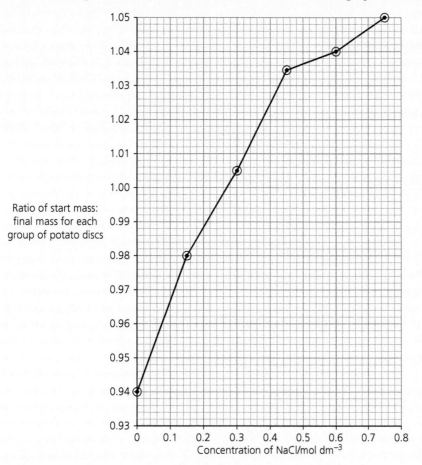

a) Explain the results obtained with a $0.75 \, mol \, dm^{-3}$ sodium chloride solution. (3)

b) The student calculated the ratio of the start mass : final mass for each group of discs, rather than the change in mass. Explain why. (2)

c) Use the graph to suggest the concentration of sodium chloride that has the same water potential as the potato cells. (1)

d) Explain why:

i) the discs were used in groups of five (2)

ii) It was important that the potato discs were identical in size and shape. (2)

4 a) The plasma membrane is described as fluid mosaic. Explain why. (2)

The graph shows the rate of diffusion of two different molecules into a cell. (2)

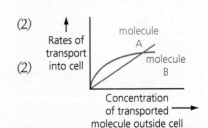

b) i) Molecule A enters the cell by simple diffusion, but molecule B enters the cell by facilitated diffusion. Explain the evidence from the graph to show that molecule B enters the cell by facilitated diffusion. (2)

The table gives some information about molecules A and B.

Molecule	Description
	Small, lipid-soluble
	Water-soluble ion

ii) Identify molecule A and then copy and complete the table above. Explain your answer. (2)

Beetroot is a vegetable. Its root cells contain a red pigment. In an investigation, discs of tissue were cut from beetroot tissue. All the discs were similar in size and shape. They were rinsed thoroughly, then each disc was placed separately into a test tube containing 10 cm³ of distilled water at various temperatures. After 1 minute the disc was removed from the water and a colorimeter was used to measure the percentage transmission of light through the solution. The lower the transmission, the more red pigment had leaked out of the beetroot disc into the water. The investigation was repeated twice, so that three readings were obtained for each temperature. The results of the investigation are recorded in the table.

Temperature /°C	Colour of tube after disc removed	Colorimeter reading/% transmission of light			
		Tube 1	Tube 2	Tube 3	Mean
0	Clear	100	98	99	99
22	Very pale pink	94	95	96	95
42	Very pale pink	80	77	79	78
63	Pink	27	29	31	29
87	Dark pink	1	1	1	1
93	Red	0	0	0	0

c) i) Describe these results. (2)

ii) Use your knowledge of the structure of cell membranes to explain these results. (2)

Stretch and challenge

5 Describe the detailed structure of chloroplasts and mitochondria. Evaluate the hypothesis that chloroplasts and mitochondria have developed from prokaryotic cells that were engulfed by primitive cells many millions of years ago.

6 Describe the structure of cells of the archaebacteria. Compare and contrast them with the 'standard' prokaryotic cells of the eubacteria.

DNA and protein synthesis

TEST YOURSELF ON PRIOR KNOWLEDGE
1 Put the following in order of size, starting with the smallest: gene, nucleus, chromosome.
2 Sketch the shape of a DNA molecule.
3 Mature human red blood cells have no nucleus. Explain the advantage of this.
4 Immature red blood cells are called reticulocytes, and these cells do have a nucleus. Explain why.

Introduction

The two men in Figure 4.1 are James Watson and Francis Crick. They worked out the structure of **DNA (deoxyribonucleic acid)**. The photo shows them standing by the model of DNA that they produced in 1953.

In 1952, James Watson had completed his PhD in the USA and was carrying out research at the University of Cambridge in the UK. Here he met Francis Crick, who had yet to finish the research for his PhD in physics and was becoming quite bored with it. The two men were fascinated by DNA and spent many hours discussing its possible structure.

Figure 4.1 James Watson and Francis Crick first proposed the structure of DNA that you will learn about in this chapter. Here, the two researchers are posing in front of the DNA model they built in 1953. Crick is pointing at the model.

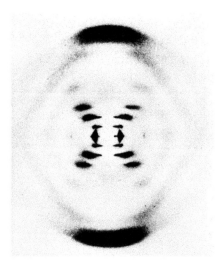

Figure 4.2 Rosalind Franklin produced this photograph from X-rays diffracted through a DNA molecule on to a photographic plate, and it told her that DNA is a double helix. Unfortunately for her, Watson and Crick saw her evidence and used it before she did.

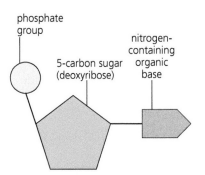

Figure 4.3 A single DNA nucleotide is made from a molecule of a five-carbon sugar (a pentose) called deoxyribose, a phosphate group and an organic base. Nucleotides are the monomers from which nucleic acids are made.

phosphate group

5-carbon sugar (deoxyribose)

nitrogen-containing organic base

Nucleotides The subunits from which DNA is made. They are formed from a pentose sugar, a phosphate group and a nitrogen-containing organic base.

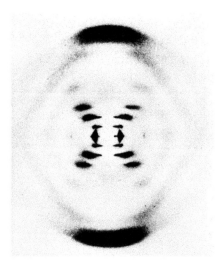

What other scientists had found out

They knew from experiments performed in the previous 20 years that DNA is the hereditary material of organisms: a substance that carries information for characteristics from one generation to the next. They knew that DNA was one type of the substances found in cells called **nucleic acids**. They also knew that it contained chemical groups of atoms called deoxyribose and phosphate, and four types of a group called organic bases. These bases are adenine, thymine, cytosine and guanine. Watson and Crick also knew Chargaff's rule: that the number of adenine bases in a molecule of DNA is always the same as the number of thymine bases, and similarly for the pair of bases cytosine and guanine. Maurice Wilkins and Rosalind Franklin working at Kings College in London had evidence that DNA was a helical structure (Figure 4.2).

What Watson and Crick did

Using solely this information, molecular models and some very inspired guesswork, Watson and Crick came up with the structure of DNA. They published a 900-word report of their proposed structure in the journal *Nature* in April 1953 and won a Nobel Prize in Physiology or Medicine in 1962 for their discovery. Reflecting their reliance on the work of others, their Nobel Prize was shared with Wilkins. Unfortunately, by that time, Rosalind Franklin had died of cancer; otherwise she would also have shared the Nobel Prize.

DNA structure

Nucleotides

The structure of DNA and the way that it carries information is the same in all organisms, which is indirect evidence for evolution. Nucleic acids are polymers made up of repeated subunits. In this case, the subunit is called a nucleotide. Figure 4.3 shows the structure of a single nucleotide from DNA. It has three components, as follows.

- First there is a **pentose** (five-carbon sugar): DNA gets part of its name from the sugar in each of its nucleotides. The sugar is called **deoxyribose** (which is the sugar ribose with one oxygen atom missing).
- Second is a **phosphate group**. This has a negative charge, which makes DNA a highly charged molecule. This negative charge enables us to separate fragments of DNA by a technique called electrophoresis.
- Third is a **nitrogen-containing organic base**. Each nucleotide contains one of four bases. Two are **adenine** (A) and **guanine** (G) (bases known as purines). The other two are **thymine** (T) and **cytosine** (C) (bases known as pyrimidines). Each base contains some nitrogen atoms in its structure.

> **TIP**
> You will learn about the technique of electrophoresis in Chapter 24.

Polynucleotide strands

Figure 4.4 shows how two nucleotides join together by a condensation reaction (see Chapter 1, page 2). You can see that two deoxyribose groups (also called residues) become linked together through one of the phosphate groups. This is called a **phosphodiester bond**. The carbon atoms in a pentose are numbered in a clockwise direction from the one that carries the base. When many nucleotides become linked together like this, they form a polynucleotide. You can see the diagram of a single **polynucleotide** strand in Figure 4.5a. Notice how the pentoses and phosphates form a sugar–phosphate backbone. As we will be more interested in the organic bases, we can simplify the polynucleotide strand to the structure shown in Figure 4.5b.

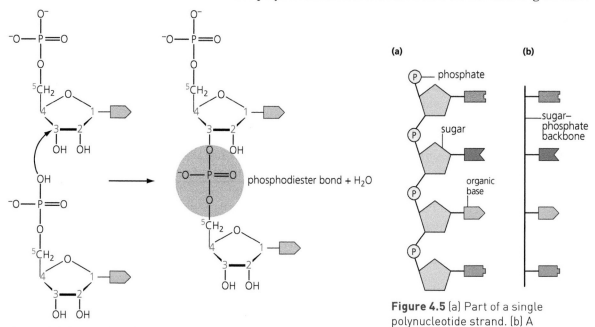

Figure 4.4 Two nucleotides react to form a dinucleotide. Numbers show the positions of the carbon atoms in the pentoses.

Figure 4.5 (a) Part of a single polynucleotide strand. (b) A simpler way to represent the same polynucleotide strand.

Hydrogen bond A chemical bond important in the three-dimensional structure of biological molecules. Hydrogen bonds require relatively little energy to break.

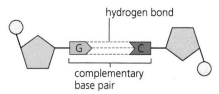

Figure 4.6 Hydrogen bonds form between the complementary bases of two nucleotides, producing a complementary base pair.

> **TIP**
> If you are asked to draw a box around the phosphodiester bond in a diagram, include the phosphate atom and all four oxygen atoms of the phosphate group.

Base pairing

We saw in Figure 4.4 how two nucleotides condense to form a dinucleotide joined by a phosphodiester bond. Figure 4.6 shows a different way that two nucleotides can join together. The two nucleotides with complementary bases (see Table 4.1) are joined by different chemical bonds called hydrogen bonds between the bases. In this way the bases become a complementary base pair. Individually, hydrogen bonds are weaker than the phosphodiester bonds holding the sugar–phosphate backbone together, but collectively they are very strong, which gives stability.

Base pairing occurs only between complementary bases. In DNA, adenine always pairs with thymine, and cytosine always pairs with guanine.

> **TIP**
> You need to remember that A pairs with T, and C pairs with G. A good way to remember this is to think of initials of people you know, or celebrities you admire, such as footballers. Think of people whose initials are AT or TA, and CG or GC.

DNA has two polynucleotide strands

Some types of nucleic acid molecule are made of a single polynucleotide strand, like the one shown in Figure 4.5. DNA is not single-stranded; it is made from two polynucleotide strands. The two strands are held together by hydrogen bonds between the complementary base pairs shown in Table 4.1.

Table 4.1 Complementary base pairs. A and T are complementary bases (A=T pair); C and G are complementary bases (C≡G pair).

Purine base		Pyrimidine base
Adenine (A)	pairs with...	Thymine (T)
Guanine (G)	pairs with...	Cytosine (C)

Look at Figure 4.7, which shows part of a molecule of DNA. You should be able to identify individual nucleotides, the phosphodiester bonds holding together the nucleotides in one polynucleotide strand and hydrogen bonds between complementary base pairs that are holding together the two polynucleotide strands. Look closely at the base sequence of each polynucleotide strand. Notice that one strand has a base sequence that is complementary to the base sequence on the other strand, and that the two strands run in opposite directions. For this reason, we call them **anti-parallel strands**.

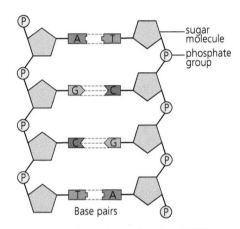

Figure 4.7 Part of a molecule of DNA showing that it has two anti-parallel strands.

Figure 4.8a shows a simpler version of part of a DNA molecule. Here, the sugar–phosphate backbones are shown as single lines. Figure 4.8b shows the final complication of a DNA molecule. The two polynucleotide strands are twisted into a coil called a helix. This diagram shows why a DNA molecule is often referred to as a **double helix**.

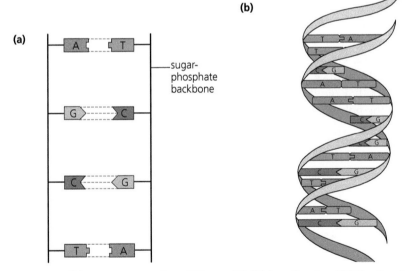

Figure 4.8 (a) A simpler version of Figure 4.7. (b) A molecule of DNA, showing the double helix of polynucleotide strands.

DNA is a stable molecule

Since DNA carries the genetic information that cells use to produce their polypeptides, it is important that a DNA molecule does not change. Two types of chemical bond hold DNA molecules together. The first is the phosphodiester bond that joins the phosphate group of one nucleotide to the sugar of the next. Look again at Figure 4.4 to see this. This is a fairly strong (covalent) bond and is not easily broken. The second is the hydrogen bond between bases in a base pair, as shown in Figure 4.8a. Although hydrogen bonds are relatively weak, a single molecule of DNA might be several thousand nucleotides long. Thousands of hydrogen bonds ensure that the two polynucleotide strands are held firmly together.

○ **TEST YOURSELF**

1 How many nucleotides are shown in Figure 4.7?

2 What makes one DNA nucleotide different from another?

3 A hydrogen bond is a relatively weak bond. Explain why the hydrogen bonds in DNA hold the two strands together relatively strongly.

4 What is a condensation reaction and what type of condensation bond is formed between nucleotides?

5 Because DNA forms a helix shape, this means the molecule is coiled up. Explain the advantage of this.

6 In part of a DNA molecule, 28% of the bases were cytosine. What were the percentages of the other bases?

Scientists were not easily convinced that DNA carries the genetic code. It might be a surprise to hear that, for the first half of the twentieth century, most scientists believed that proteins carried the genetic information. They thought the components of DNA seemed much too simple, as they involved only four bases! In the box below you can see some experiments that changed their minds.

Extension

Experiments showing that DNA is the genetic material

In the first experiment, a bacterium called *Streptococcus pneumoniae* was used. It causes pneumonia in humans and other mammals. The bacterium is rod-shaped and has two strains. On agar plates, colonies of the S strain bacterial cells produce an outer polysaccharide coat and appear smooth. Colonies of the mutant R strain bacterial cells lack the polysaccharide coat and appear rough.

A team of scientists injected mice with different combinations of these two strains of the bacterium.

Figure 4.9 shows that mice died from pneumonia when injected with cells of the S strain.

Heat-killed S strain on its own did not cause mice to die, yet mixed with R strain it did. The first team of scientists concluded that the genetic information of the heat-killed S strain was able to get into the live cells of the R strain and transform them into S-type cells. They did not know the chemical nature of this transforming agent.

4 DNA AND PROTEIN SYNTHESIS

live cells of S strain

mouse contracts pneumonia

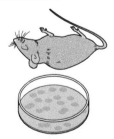

colonies of S strain isolated from tissue of dead mouse

live cells of R strain

mouse remains healthy

colonies of R strain isolated from tissue of healthy mouse

heat-killed cells of S strain

mouse remains healthy

no colonies isolated from tissue of healthy mouse

heat-killed cells of S strain and live cells of R strain

mouse contracts pneumonia

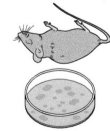

colonies of R strain and S strain from tissue of dead mouse

Figure 4.9 The effect of injecting mice with different combinations of the S strain and R strain of *Streptococcus pneumoniae*.

Some time later, another team of scientists set up an experiment to try to find the nature of the transforming agent. They treated heat-killed samples of the S strain of *S. pneumoniae* with different enzymes. Each enzyme broke down specific molecules within the bacteria. The scientists then mixed each of the extracts from these S strain cells with a different culture of the R strain, and looked at the type of colony that grew on an agar plate. Table 4.2 shows their results.

Table 4.2 The effect of incubation with different enzymes on the ability of the S strain of *S. pneumoniae* to transform the R strain.

Experiment	Enzyme used to treat heat-killed cells of S strain of *S. pneumoniae*	Appearance of R strain colonies growing on agar plates
1	Protease	Smooth
2	Ribonuclease	Smooth
3	Deoxyribonuclease	Rough

This research team suspected that three types of molecule found in cells might be the transforming agent: DNA, RNA or protein. Table 4.2 shows that only hydrolysis of the DNA in the S strain extract prevented the R strain being smooth so only DNA could pass on the information needed for the R strain to produce a polysaccharide coat.

Most scientists remained unconvinced, and these results were largely ignored for many years. The results were criticised for several reasons. These included:

- some contaminating protein could have been left in the protease preparation
- DNA might be only part of a pathway that proteins used for transformation.

It took a Nobel Prize-winning experiment to convince scientists around the world that DNA did carry the genetic code. This experiment used a **bacteriophage**. This is a virus that infects and kills bacteria. The virus was the T2 bacteriophage. This bacteriophage infects *Escherichia coli*, a bacterium that commonly grows in the gut of humans. Figure 4.10 shows a single T2 bacteriophage. Notice its simple structure: it has an outer protein capsule surrounding a molecule of DNA. When a T2 bacteriophage infects an *E. coli* bacterium, it multiplies to produce large numbers of bacteriophages that burst the bacterial cell and are released.

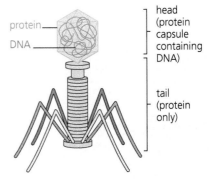

Figure 4.10 T2 bacteriophages are viruses that infect the bacterium *Escherichia coli*.

The proteins contain sulfur but DNA does not. DNA contains phosphorus, but proteins do not. Each of these elements has a radioactive isotope, ^{32}P and ^{35}S. The team of scientists grew some T2 bacteriophage in the presence of ^{32}P, which labels their DNA, and some in the presence of ^{35}S, which labels their proteins. After infecting a culture of *E. coli* with these labelled bacteriophages, the team put samples of the culture in a blender. This removed the bacteriophages from the surface of the bacteria. They then looked to see where the radioactive elements were found. Figure 4.11 shows their results.

Most of the protein was found outside the infected cells and most of the DNA was inside, so it was the DNA that was being used by the infected bacteria to build new bacteriophages. The team went on to show that new bacteriophages released by bursting *E. coli* cells were labelled only with ^{32}P. This finally convinced scientists that DNA did carry the genetic code.

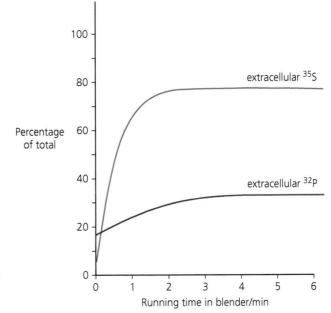

Figure 4.11 The location of radioactivity after removing the T2 bacteriophage from the surface of infected *E. coli* cells.

DNA, chromosomes and genes

The DNA in prokaryotic cells is different from the DNA in eukaryotic cells (see Chapter 3, page 50). Prokaryotes contain a single, circular DNA molecule. In addition, they usually contain one or more much smaller circular DNA molecules called plasmids (see Chapter 3, page 51).

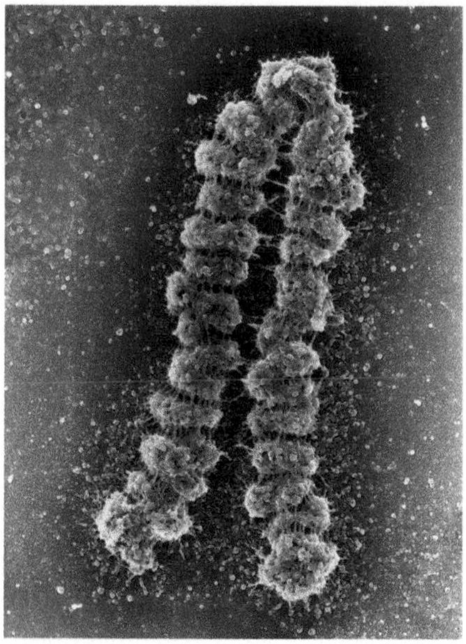

In contrast, a single eukaryotic cell always has many, different, molecules of DNA. In eukaryotic cells each DNA molecule is linear and wrapped around proteins called **histones**, and forms a rod-like structure (with two ends) called a **chromosome**. Figure 4.12 shows a chromosome from a eukaryotic cell. In it, the DNA helix is tightly coiled, so that the chromosome looks much thicker than a DNA molecule. Look closely at Figure 4.12. Can you see some of the polynucleotide strands within the thick chromosome?

Each chromosome carries the genetic information for a large number of polypeptides, as well as the information for building the functional RNA molecules needed for protein synthesis (see page 64). The base sequence of DNA coding for a single polypeptide or a functional RNA is called a gene. Look at Figure 4.13. It shows a chromosome as a long string of genes. Each gene has a specific position on a chromosome, called its **locus**. For example, the human gene that codes for pancreatic amylase (see Chapter 8, page 141) is located at its locus on the short arm of chromosome 1. The sequence of bases in many genes encodes the amino acid sequence of a single polypeptide molecule.

It might surprise you to learn that much of the DNA in eukaryotic cells does not code for polypeptides. In fact, less than 2% of human DNA is thought to code for them. Figure 4.14 shows how some apparently non-coding DNA is found within a gene, and some is found between genes. The non-coding regions of DNA within a gene are called **introns**; the coding regions are called **exons**. In between genes, non-coding DNA often contains the same bases sequences repeated many times called non-coding multiple repeats.

Differences between the DNA found in prokaryotic and eukaryotic cells have been described above. These differences are summarised in Table 4.3. Because of these differences, we should not refer to the genetic material of

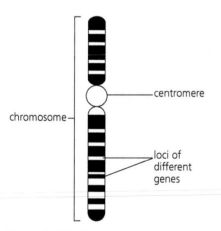

Figure 4.12 An electron micrograph of a chromosome. If you look closely, you can see some bits of the polynucleotide strand that is tightly coiled in the chromosome.

Gene A base sequence of DNA that codes for the amino acid sequence of a polypeptide or a functional RNA molecule.

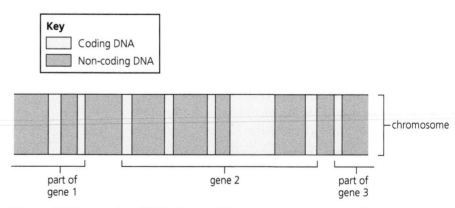

Figure 4.13 A chromosome contains many genes. Each gene occurs on one chromosome only and occupies a fixed position called its locus.

Figure 4.14 Non-coding DNA is found within a gene and between adjacent genes.

a prokaryotic cell as a chromosome. However, as a shorthand description, some people refer to 'bacterial chromosomes'.

Table 4.3 A comparison of the DNA molecules found in prokaryotic and eukaryotic cells.

Feature of DNA	Prokaryotic cells	Eukaryotic cells
Relative length of molecule	Short, i.e. few genes	Long, i.e. many genes
Shape of molecule	Circular, forming a closed loop	Linear, forming part of a chromosome
Number of different molecules per cell	One	More than one
Association with proteins	Not associated with proteins	Associated with proteins, called histones
Non-coding DNA	Absent	Present within genes (introns) and as non-coding multiple repeats between genes

TEST YOURSELF

7 Give three ways in which the DNA of prokaryotic cells is different from DNA in eukaryotic cells.

8 The DNA in a T2 bacteriophage is not associated with histone proteins. Explain why this is important in the investigation shown in Figure 4.11.

9 Explain the difference between:
   a) introns and exons
   b) a gene and a locus.

10 Describe the two types of non-coding DNA found in eukaryotic cells.

DNA and protein synthesis

An organism's genes are made of DNA. Genes contain the information to synthesise proteins. All of an organism's genes, known as its genome, are contained in every one of its cells. The cell uses some, but not all, of these genes to make a set of proteins. Which proteins it actually makes depends on the type of cell it is. The particular range of proteins that a cell produces using its DNA is known as its proteome. However, before you can understand how genes work, you need to learn about another type of nucleic acid, called **ribonucleic acid** (or **RNA**).

Like DNA, a molecule of RNA is a polynucleotide chain. Figure 4.15 shows a DNA nucleotide, which you are already familiar with, and an RNA nucleotide. Can you spot the difference? Look at carbon atom 2 of the five-carbon sugar. In the DNA nucleotide, it lacks an oxygen atom that is present in the RNA nucleotide. There is another difference that is not shown in Figure 4.15. An RNA nucleotide never has thymine as its base; instead it has uracil.

Figure 4.16 (overleaf) shows the structure of DNA and of two different types of RNA: messenger RNA (mRNA) and transfer RNA (tRNA). How many differences in the structures and compositions of DNA, mRNA and tRNA can you spot in Figures 4.15 and 4.16? Table 4.4 summarises these differences for you.

Genome The complete set of genes in a cell.

Proteome The full range of proteins that a cell is able to produce.

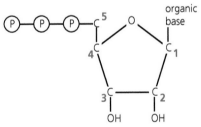

ribonucleotide

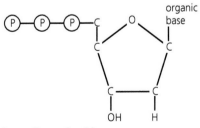

deoxyribonucleotide

Figure 4.15 The molecular structures of an RNA nucleotide and a DNA nucleotide. RNA contains the sugar ribose; DNA contains the sugar deoxyribose (which has one less oxygen than ribose).

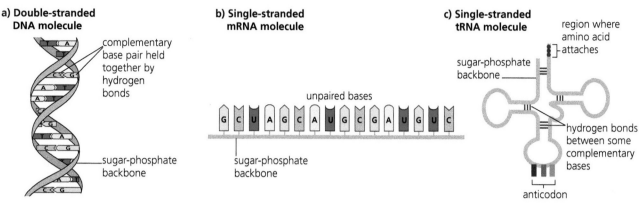

a) Double-stranded DNA molecule
- complementary base pair held together by hydrogen bonds
- sugar-phosphate backbone

b) Single-stranded mRNA molecule
- unpaired bases
- GCUAGCAUGCGAUGUC
- sugar-phosphate backbone

c) Single-stranded tRNA molecule
- region where amino acid attaches
- sugar-phosphate backbone
- hydrogen bonds between some complementary bases
- anticodon

Figure 4.16 (a) Part of a molecule of DNA, (b) part of a molecule of mRNA and (c) a molecule of tRNA. The polynucleotide chains are represented as single lines. The parts are not to the same scale.

There is a third type of RNA called ribosomal RNA (rRNA) that forms part of the structure of ribosomes.

Table 4.4 A comparison of DNA, mRNA and tRNA.

Feature	DNA	mRNA	tRNA
Nucleotide structure			
Pentose sugar	Deoxyribose	Ribose	Ribose
Purine bases	Adenine and guanine	Adenine and guanine	Adenine and guanine
Pyrimidine bases	Cytosine and thymine	Cytosine and uracil	Cytosine and uracil
Polynucleotide chain			
Number of polynucleotide strands	2	1	1
Number of nucleotides in chain	Many millions	Several hundred or thousands	About 75
Hydrogen bonding between complementary base pairs	Present: holds two anti-parallel strands together (A–T, C–G)	Absent	Present in parts of molecule, giving the molecule a clover-leaf shape (A–U, C–G)

Adenosine triphosphate

You will have seen **ATP** (adenosine triphosphate) mentioned several times already. For example, mitochondria make ATP during aerobic respiration in eukaryotic cells (see Chapter 3, page 41). ATP is used by cells whenever a process requires energy. Protein synthesis is an example of an energy-requiring process.

ATP molecules have a structure closely related to the RNA nucleotide containing adenine. Figure 4.17a shows how an ATP molecule has the five-carbon sugar ribose and the base adenine, but has three phosphate groups rather than just one. For this reason, it is called a phosphorylated nucleotide.

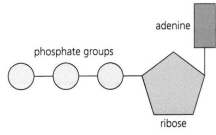

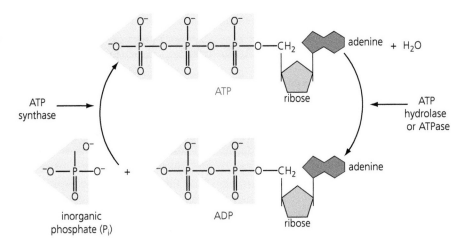

Figure 4.17 (a) Like an RNA nucleotide, an ATP molecule is made from a molecule of ribose and a molecule of adenine but, unlike an RNA nucleotide, it has three phosphate groups. (b) Hydrolysis of ATP to ADP and an inorganic phosphate group (P_i) and resynthesis by condensation of ADP and P_i.

Processes such as protein synthesis require energy from the hydrolysis of ATP. Just like any hydrolysis reaction (see Chapter 1, page 2), water is used in a reaction catalysed by the enzyme **ATP hydrolase (ATPase)** to break the bond between the second and third phosphate groups, releasing energy. The phosphate group is often transferred to other molecules to make them more reactive so that they have the **activation energy** (see Chapter 2, page 20) to take part in another reaction. This transfer is called **phosphorylation**.

Just as quickly as ATP is used up in cells, it is resynthesised by condensing together molecules of ADP and inorganic phosphate groups (P_i) using another enzyme called **ATP synthase**. This requires energy. In all cells, carbohydrate or lipid molecules are hydrolysed during respiration to release the energy required to resynthesise ATP. In some cells, ATP can also be made using light energy during photosynthesis. However it is made, it is important that the cell is able to supply sufficient ATP to meet the demands of all of its energy-requiring processes, including protein synthesis.

How genes are used

The sequence of bases in an organism's DNA carries the genetic information that determines the amino acid sequence of each polypeptide (Chapter 1, page 12) that a cell can produce. One or more very long polypeptides assemble to form proteins. Proteins include haemoglobin found in red blood cells, and enzymes that regulate all the chemical reactions of substances in cells. Because the DNA determines the sequence of amino acids in polypeptides, and hence the nature of proteins formed from them, it thereby indirectly controls how an organism develops and behaves. The fact that the genetic code that determines how information is carried on DNA is common to all living organisms and to viruses is strong evidence for evolution from a common ancestor.

The base sequence on mRNA is used by ribosomes to make polypeptide chains. Before looking at the process in more detail, it is helpful to gain an overview of how genes are used. The sequence of events leading to the production of a polypeptide from the genetic information contained in one gene occurs in two stages: **transcription** and **translation**. Table 4.5 summarises the steps in these two stages. Figure 4.18 (overleaf) is a pictorial summary of the same events in a eukaryotic cell.

Table 4.5 The main steps in transcription and translation in a eukaryotic cell.

Transcription	Translation
1 The sequence of DNA bases in a single gene is used to make a molecule of messenger RNA (mRNA). 2 At the region to be copied, the DNA unwinds. 3 The hydrogen bonds holding the two DNA polynucleotide chains break down, exposing unpaired DNA bases. 4 A molecule of mRNA is made which has a sequence of bases that is complementary to the base sequence of one of the DNA strands.	1 mRNA is used by ribosomes to make a polypeptide chain with a sequence of amino acids encoded by the mRNA base sequence. 2 A molecule of mRNA leaves the DNA and moves to a ribosome in the cytoplasm. One or more ribosomes might attach to each mRNA molecule. 3 Each ribosome travels along the molecule of mRNA, 'reading' its base sequence. As it does so, the ribosome assembles a sequence of amino acids, according to the sequence of bases on the mRNA. 4 Peptide bonds form between the amino acids brought by tRNA, making a polypeptide chain. 5 Once finished, the polypeptide chain detaches from the ribosome. The ribosome now detaches from the mRNA molecule.

Figure 4.18 Transcription and translation in a eukaryotic cell.

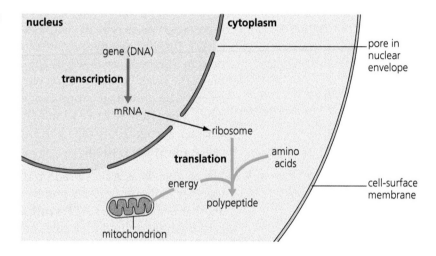

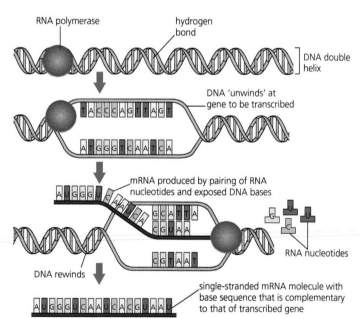

Figure 4.19 During transcription, the base sequence of DNA is copied into the complementary base sequence of messenger RNA (mRNA). A gene is thus copied for use by ribosomes.

Transcription: the production of mRNA from DNA

Figure 4.19 shows in more detail what happens during transcription. At the point at which a gene is to be used, the DNA molecule unwinds and the hydrogen bonds holding the two polynucleotide strands together in a DNA molecule break down. This exposes unpaired bases on the nucleotides of the two DNA strands. RNA nucleotides are already present in the cell. Bases on these RNA nucleotides form new hydrogen bonds between the exposed DNA nucleotides on *one* of the strands of DNA by a process of **complementary base pairing**. This is similar to the complementary base pairing in DNA replication, except that RNA nucleotides have the base uracil in place of the base thymine. You will read more about this on page 75.

Thus, the respective base pairs between DNA and mRNA transcribed from it are:

- adenine–uracil (A–U)
- guanine–cytosine (G–C)
- cytosine–guanine (C–G)
- thymine–adenine (T–A).

In this way, a chain of RNA nucleotides is made which has a complementary base sequence to the DNA making up one gene. An enzyme called **RNA polymerase** joins the ribose-phosphate backbone of these RNA nucleotides to form a molecule of mRNA.

Both tRNA and rRNA are made in the same way as mRNA, by using the sequence of bases in a gene. The difference is that these genes make RNAs that are used in the process of protein synthesis, whereas mRNA is translated to produce a protein. tRNA and rRNA are known as functional RNAs because they have a role in translation. tRNA carries amino acids, whereas rRNA forms part of the structure of ribosomes.

Post-transcriptional processing of mRNA in eukaryotic cells

Earlier in this chapter you saw that not all of the DNA in eukaryotic cells codes for polypeptides or functional RNA and that the non-coding sections of DNA might be:

- between genes: these sections include DNA sequences that are repeated over and over again; they are often referred to as non-coding multiple repeats
- within genes: these non-coding sections of DNA are called **introns** and they separate the coding sequences called **exons**.

During transcription, eukaryotic cells cannot transcribe only the coding sections. Instead the whole gene, including introns and exons, is transcribed into a molecule called **pre-mRNA**. Before it leaves the nucleus, this pre-mRNA is edited. Figure 4.20 shows how this is done by removing the non-coding sections of the pre-mRNA. The coding sections are then 'edited' together to produce mRNA that carries only the coding regions of the gene, in a process called splicing.

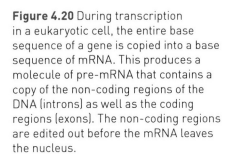

Functional RNA RNA that is not translated into proteins.

TIP
It might help you to remember the difference between exons and introns if you think of **ex**ons as **ex**pressed sections of DNA and **int**rons as **int**ervening sections of DNA.

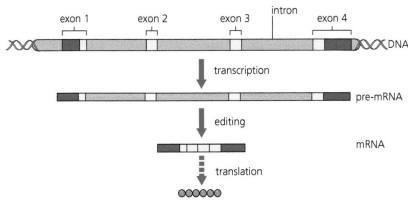

Figure 4.20 During transcription in a eukaryotic cell, the entire base sequence of a gene is copied into a base sequence of mRNA. This produces a molecule of pre-mRNA that contains a copy of the non-coding regions of the DNA (introns) as well as the coding regions (exons). The non-coding regions are edited out before the mRNA leaves the nucleus.

Extension

Exons may be spliced in many alternative ways to form different mature mRNA molecules. For example, early in their development human B cells splice into one of their mRNA molecules an exon that enables a particular protein to be retained in the cell's surface membrane. Later, they stop splicing this exon and, instead, splice into their mRNA an exon that enables the protein to be released from the cell. Different splicing of the same pre-RNA at different times means that a single eukaryotic gene can actually code for more than one polypeptide chain.

ACTIVITY

Types of nucleic acid

1 Copy and complete the table to show the differences between the three types of nucleic acid.

Type of nucleic acid	Hydrogen bonds present or not present?	Number of polynucleotide strands in molecule	Anticodon present or not present?
DNA			
mRNA			
tRNA			

DNA is a double-stranded molecule so there are hydrogen bonds between complementary bases on the two strands. mRNA is single stranded so there is no complementary base pairing and therefore no hydrogen bonds. tRNA folds back on itself so parts of the molecule are double stranded with complementary base pairing and hydrogen bonds. Anticodons are only found on tRNA molecules.

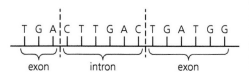

Figure 4.21 Base sequence.

Figure 4.21 shows the bases on one strand of part of a DNA molecule.

2 Give the sequence of bases on the mRNA transcribed from this DNA.

The intron would be removed from the pre-mRNA and the sections complementary to the two exons spliced together to form the mRNA.

Translation

A new molecule of mRNA moves from the DNA to ribosomes in the cytoplasm. In a eukaryotic cell, this involves leaving the nucleus through one of the **nuclear pores** in the envelope surrounding the nucleus. One or more ribosomes attaches to a molecule of mRNA. Each ribosome moves along the molecule of mRNA, 'reads' the base sequence of the mRNA and uses it to assemble a sequence of amino acids. Peptide bonds form between the amino acids, creating a polypeptide. Once this polypeptide is complete, the ribosome releases the polypeptide and detaches from the molecule of mRNA.

A ribosome 'reads' the mRNA base sequence three bases at a time. Each set of three bases codes for a specific amino acid (see Table 4.6 on page 70). For this reason, each set of three mRNA bases is called a codon. Each codon is complementary to three bases, called a triplet, on the DNA from which it was transcribed.

> **Codon** A sequence of three mRNA bases that codes for a specific amino acid.
>
> **Triplet** A sequence of three DNA bases that codes for a specific amino acid.

> **Anticodon** A sequence of three tRNA bases that is complementary to a codon.

Although amino acids can be free in the cytoplasm of a cell, they cannot be used by ribosomes unless they are attached to a molecule of tRNA. At the end of one of its arms, each tRNA molecule has three free bases that can base pair with a complementary codon on mRNA (Figure 4.22). At the end of another of its arms, each tRNA molecule has a site that attaches to an amino acid. The sequence of bases on the tRNA molecule that base pairs with a codon is called an anticodon. A molecule of tRNA with a particular anticodon always attaches to the same specific amino acid.

Figure 4.22 shows a single molecule of mRNA at an early stage of translation. It is attached to a ribosome and three of its bases, a codon, are shown within the ribosome. Ribosomes contain ribosomal RNA (rRNA). A molecule of tRNA is also shown. Notice that the tRNA has an anticodon that is complementary to the codon. The amino acid that is carried by the tRNA molecule is the one encoded by the codon on the mRNA, which in turn was determined by a triplet on the DNA.

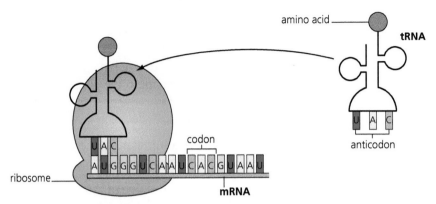

Figure 4.22 A ribosome has attached to one end of a molecule of mRNA. The ribosome is ready to 'read' the first three bases: the mRNA codon. You can also see a molecule of tRNA that has a base sequence that is complementary to the codon; this is the anticodon of the tRNA molecule. The tRNA is attached to an amino acid.

Figure 4.23 shows a molecule of mRNA at a later stage of translation. In this diagram, you can see that the ribosome has moved along the mRNA molecule one codon at a time and has joined more amino acids together, so that a polypeptide chain is being formed by the ribosome.

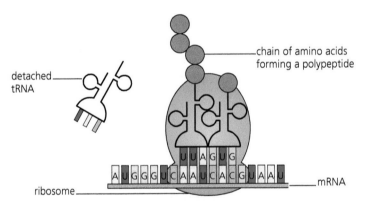

Figure 4.23 The ribosome has moved further along the mRNA molecule and has formed a chain of amino acids that corresponds to the codons along the mRNA molecule.

Once a ribosome has 'read' the entire base sequence of a molecule of mRNA, it releases the polypeptide it has formed. The ribosome then detaches from the mRNA. It might re-attach at the start of the mRNA molecule and produce another molecule of the polypeptide. Several ribosomes can attach to, and move along, a single mRNA molecule at the same time.

TEST YOURSELF

11 Explain the difference between a DNA triplet and an mRNA codon.

12 Why does transcription in prokaryotic cells produce mRNA rather than pre-mRNA?

13 In which direction does a ribosome 'read' an mRNA molecule during translation?

14 Explain how different polypeptides might be made from the same eukaryotic gene.

15 Explain the difference between the genome of a cell and the proteome of a cell.

The genetic code is common, degenerate and non-overlapping

As we have seen, organisms have coded genetic information carried on DNA. The information is carried using the genetic code, in the form of triplets. A triplet is a sequence of bases in a DNA molecule. During the production of a polypeptide, each triplet of a single gene is transcribed from DNA into a molecule of mRNA. The sequence of bases (codon) in mRNA is translated into a sequence of amino acids in a polypeptide. For this reason, the genetic code is often given as mRNA codons rather than DNA triplets.

Table 4.6 summarises the genetic code. We can use the table to show two further important features of the genetic code.

Look at the codons UAA, UAG and UGA in Table 4.6. These are 'reading instructions' for ribosomes. When a ribosome gets to these 'stop' codons it detaches from mRNA and releases its polypeptide chain.

Now look at the codons UCA, UCG, UCC and UCU. They all code for the same amino acid, serine (Ser). This means that it does not matter about the third base: UCX always codes for the amino acid serine. We describe this by saying that the genetic code is **degenerate**: several mRNA codons may encode the same amino acid.

Table 4.6 The genetic code. Three bases (a codon) on a molecule of mRNA encode a specific amino acid. The four bases, adenine (A), cytosine (C), guanine (G) and uracil (U) can form the 64 different codons that are shown in the table. Abbreviations for the names of the amino acids that they encode are shown alongside the codons. **You do not need to remember the contents of this table!**

AAA	Lys	CAA	Gln	GAA	Glu	UAA	STOP
AAG	Lys	CAG	Gln	GAG	Glu	UAG	STOP
AAC	Asn	CAC	His	GAC	Asp	UAC	Tyr
AAU	Asn	CAU	His	GAU	Asp	UAU	Tyr
ACA	Thr	CCA	Pro	GCA	Ala	UCA	Ser
ACG	Thr	CCG	Pro	GCG	Ala	UCG	Ser
ACC	Thr	CCC	Pro	GCC	Ala	UCC	Ser
ACU	Thr	CCU	Pro	GCU	Ala	UCU	Ser
AGA	Arg	CGA	Arg	GGA	Gly	UGA	STOP
AGG	Arg	CGG	Arg	GGG	Gly	UGG	Trp
AGC	Ser	CGC	Arg	GGC	Gly	UGC	Cys
AGU	Ser	CGU	Arg	GGU	Gly	UGU	Cys
AUA	Ile	CUA	Leu	GUA	Val	UUA	Leu
AUG	Met	CUG	Leu	GUG	Val	UUG	Leu
AUC	Ile	CUC	Leu	GUC	Val	UUC	Phe
AUU	Ile	CUU	Leu	GUU	Val	UUU	Phe

How many other amino acids can you find in Table 4.6 that are encoded by more than one codon? Can you find any that are encoded by only one codon?

Having learned that some codons stop translation by a ribosome, you might wonder how a ribosome 'knows' where to start translating the mRNA code. This is slightly more complicated than the STOP codons in Table 4.6. In most organisms, the ribosome must 'recognise' an AUG codon that is followed by a G and has an A preceding it by three nucleotides, for example the sequence AXXAUGG.

TIP

The points you need to understand are that:

- some mRNA base sequences are start and stop messages for translation
- ribosomes only translate the mRNA molecule by taking a whole codon at a time.

You might realise that the sequence of mRNA bases could be 'read' in a variety of ways, but it is important that the ribosome 'reads' the sequence in only one way, so that a polypeptide with the encoded sequence of amino acids is the only one that is produced.

To explain why, let's take the following mRNA base sequence:

UCCCAUGACUCAUUCCCAGGG

If translation starts at the first codon (UCC), the order of encoded amino acids will be

serine – histidine – aspartic acid – serine – phenylalanine – proline – glycine

This results in the correct polypeptide being produced.

If the ribosome missed the first mRNA base, it would now produce a polypeptide with the amino acid sequence

proline – methionine – threonine – histidine – serine – glutamine

In other words, a completely different amino acid sequence would be produced, resulting in a polypeptide that would not have the appropriate function. Having a specific START recognition sequence ensures that the mRNA is always translated from the correct point in the mRNA molecule.

Now suppose that the ribosome were to 'read' the first mRNA base sequence above by jumping one base along each time it had 'read' a codon. The original base sequence would now be 'read' incorrectly as UCC, then CCC, CCA, CAU, AUG, and so on. This is an overlapping sequence and, again, would result in a polypeptide with an inappropriate sequence of amino acids. Translating the first three bases as a single unit, followed by the fourth, fifth and sixth in a second unit, and so on, ensures that the mRNA code is **non-overlapping**. Each base is 'read' only once in the codon of which it is a part. As a result of this non-overlapping nature of translation, an appropriate amino acid sequence is always produced from a molecule of mRNA.

Practice questions

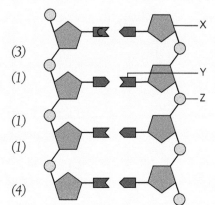

1 The diagram shows part of a DNA molecule.

    a) i) Name structures X, Y and Z.   (3)

       ii) Draw a circle round one nucleotide.   (1)

    b) Name the bonds that form between structure X and the molecule labelled C.   (1)

    c) Name the bonds between the nucleotides in each strand.   (1)

    d) Describe and explain **two** features of DNA that make it useful as an information-storing molecule.   (4)

2 Below is the base sequence of mRNA that codes for part of a polypeptide in a eukaryotic cell.

ACCGUGUCCAUGUAACGU

    a) What is the maximum number of amino acids this sequence could code for?   (1)

    b) Give the anticodon for the second tRNA molecule that would bind to this sequence during translation.   (1)

    c) Write the base sequence for the DNA from which this mRNA was transcribed.   (2)

    d) The length of the section of DNA that codes for the complete polypeptide is longer than the mRNA used to transcribe it. Give two reasons why.   (2)

    e) The table shows the percentage of bases in two different mRNA molecules transcribed from different parts of a chromosome.

Part of chromosome	Percentage of base			
	A	**G**	**C**	**U**
Middle	38		24	18
End	31	22		21

      i) Copy and complete the table with the missing values.   (2)
      ii) Explain why the percentages are different for the two regions of the chromosome.   (2)

Stretch and challenge

3 This chapter began with the story of how the structure of DNA was worked out by Francis Crick and James Watson in 1953. In 1958, Francis Crick published a paper in which he described what he called the Central Dogma of Molecular Biology; that is, once (sequential) information has passed into protein it cannot get out again. This has since been simplified, perhaps oversimplified, to 'DNA makes RNA makes protein'. Since then, further discoveries about DNA and RNA have challenged the Central Dogma. Describe some of these discoveries and evaluate the extent to which you think they undermine either version of Crick's idea.

5 The cell cycle

TEST YOURSELF ON PRIOR KNOWLEDGE

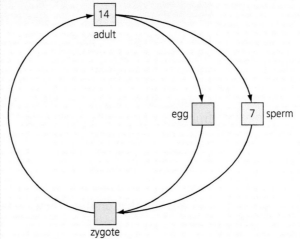

Figure 5.1 A simplified life cycle of one organism. The numbers in the boxes represent the number of chromosomes in one cell of the organism at each stage.

1 In Figure 5.1, what are the missing numbers of chromosomes for
   i) egg and ii) zygote?
2 What are the components of a DNA nucleotide?
3 Name the four bases found in DNA, and identify the bases that form complementary base pairs.

Introduction

If you have visited the Royal Botanic Gardens at Kew, in London, you probably enjoyed the variety of plants on display in the gardens and glasshouses. In addition to the staff who tend the gardens and glasshouses, there are teams of plant scientists working at the Royal Botanic Gardens.

One team works in the micropropagation unit. Members of this team are experts at cloning or propagating plants from very small pieces of plant tissue, hence 'micro'. The containers in Figure 5.2 show plants that are being grown in a special medium containing all the substances they require. They are kept in a sterile environment to make sure that the seedlings do not become infected by moulds, which could kill them. Some plants are grown in very large numbers. Often, they are plants of species from all over the world that are rare and possibly in danger of becoming extinct.

Figure 5.2 A micropropagation unit. Inside each of the containers is a tiny plant that is growing on a sterile growth medium. Some of the plants are grown from seed. Others are grown from small pieces of growing tissue taken from a mature plant.

Cloning plants in this way relies on taking tiny clumps of dividing cells from parts of the plant that are growing. New buds are often used. Each of the cells in the clump has been formed by **mitosis** and contains all the genetic information needed to form roots, stems and flowers in the developing new plant. In this chapter, we will examine how cells pass copies of their genetic information from cell to cell and about the type of cell division that produces new, genetically identical cells.

Replication of DNA

In Chapter 4, we saw that DNA contains two polynucleotide strands that are held together by hydrogen bonds between complementary base pairs. You will see that this base pairing is vital during **DNA replication**; that is, when DNA is copied. We can describe DNA replication using the stages shown in Figure 5.3.

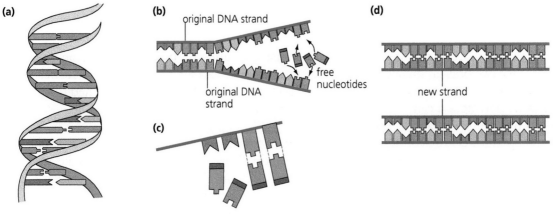

Figure 5.3 The process of DNA replication. (a) The two strands of the DNA molecule begin to unwind when the weak hydrogen bonds between bases break. This happens at the point where the DNA is to be copied, creating a replication fork. (b) Free DNA nucleotides pair with complementary bases that are exposed on each strand of the unwound DNA. (c) Hydrogen bonds form between the new nucleotides and bases of each polynucleotide strand.

Stage 1: the polynucleotide strands of DNA separate

Replication of a DNA molecule begins when its double helix partially unwinds (Figure 5.3a). As it does so, the hydrogen bonds between complementary base pairs break down. If you guessed that an enzyme is involved in breaking the hydrogen bonds, you are correct. The enzyme is called DNA helicase. The breakdown of hydrogen bonds allows the two polynucleotide strands to move apart. The point at which they are separating is called the **replication fork**. Figure 5.3b shows that we now have two separate strands of nucleotides with unpaired bases exposed to free nucleotides.

DNA helicase The enzyme that separates the DNA double helix by breaking its hydrogen bonds.

Stage 2: free DNA nucleotides pair with exposed bases on each polynucleotide strand

The bases on the polynucleotide strands do not remain unpaired for long. Individual DNA nucleotides have already been made in the nucleus. They are attracted to the exposed bases on each polynucleotide strand, and hydrogen bonds form between complementary bases (Figure 5.3c). For example, a free DNA nucleotide that includes the base adenine is attracted to an exposed thymine on one of the polynucleotide strands, and forms hydrogen bonds with it. This happens all along the unwound section of the DNA molecule. As a result, each of the polynucleotide strands of the DNA soon builds up a complementary sequence of nucleotides. Because each original strand is acting as a pattern for the assembly of a new strand, the original strands are known as template strands.

Template strand A DNA strand that is being used as a pattern for the assembly of complementary bases into a new strand.

Stage 3: the new nucleotides bond together

You can see in Figure 5.3c that new nucleotides have formed hydrogen bonds with the bases of each polynucleotide strand. However, these new nucleotides are not joined together themselves. This happens in the final stage of DNA replication, shown in Figure 5.3d. In this stage, phosphodiester bonds are formed between the nucleotides by condensation reactions (see Chapter 4, page 58). This linking of DNA nucleotides is controlled by the enzyme DNA polymerase.

DNA polymerase The enzyme that joins free nucleotides to form a new DNA strand.

75

Extension

Like all enzymes, DNA polymerase is highly specific (see Chapter 2). Because the two original strands are anti-parallel, the two new strands are also anti-parallel (see Chapter 4). The DNA polymerase can form phosphodiester bonds on one new strand as the replication fork opens up and free nucleotides pair up with exposed bases on the template strand. This is called **continuous replication**. But on the other strand, DNA polymerase enzymes must allow the replication fork to open up a short distance before being able to link new nucleotides together because the new strand is pointing the other way. This is called **discontinuous replication** because on this side the process creates the new strand in a series of short sections. These then have to be joined together by other enzymes.

The strand formed by discontinuous replication takes longer to form because the DNA polymerase keeps having to 'catch up' after the replication fork has opened further. For this reason, this strand is called the **lagging strand**, in contrast to the one being built smoothly in the opposite direction, which is called the **leading strand**.

We end up with two new DNA molecules, each formed from one original polynucleotide strand and one new, replicated, polynucleotide strand. We call this **semi-conservative replication**: one original strand remains intact (is conserved) and one new complementary strand is made using the old strand as a template for replication. The DNA molecules rewind as this process is completed. The Watson and Crick model of DNA structure (see page 56) supports the idea that a template strand is used for semi-conservative replication.

Extension

In eukaryotes the DNA molecule in each chromosome is very long, so if replication started at one end and proceeded from one end to the other in a linear way it would be a very slow process. If the DNA molecule unwinds at a number of points along its length, forming more than one replication fork at the same time, replication is speeded up.

TEST YOURSELF

1 One strand of DNA contains the base sequence ATCGACG. What will be the sequence of bases in the complementary DNA strand?
2 Name the type of reaction that is catalysed by DNA polymerase.
3 What is a template strand?
4 Name the enzyme that breaks the hydrogen bonds and unwinds the strands to separate them.
5 What type of bond is formed between the nucleotides in a new strand?

Experimental evidence for semi-conservative replication of DNA

You might wonder how we know that DNA replication is a semi-conservative process. After all, we cannot see DNA actually replicating. The evidence is indirect. Let's look at one experiment that provides evidence about DNA replication.

Look back to Figure 5.3 on page 75. The conserved polynucleotide strands and the new polynucleotide strands are coloured differently to help you to understand the replication process. We can do this in a diagram, but we cannot colour real DNA nucleotides to help identify them.

Scientists in one laboratory came up with a neat way of labelling nucleotides. It depends on the use of two isotopes of nitrogen. The more common isotope has 14 uncharged particles, called neutrons, in the nucleus of each nitrogen atom. The rarer isotope has 15 neutrons in the nucleus of each nitrogen atom. This makes the rarer isotope (15N) heavier than the more common isotope (14N). The difference in mass is tiny. However, there are so many atoms of nitrogen in a strand of DNA that a difference in mass can be detected.

Under laboratory conditions, a bacterium rapidly replicates its DNA and divides into two new cells. Bacteria use nitrogen-containing molecules in their growth medium to make DNA nucleotides. In this experiment, the scientists used two types of growth medium containing nitrogen; in one medium, all the nitrogen atoms were the ^{14}N isotope, and in the other they were all the ^{15}N isotope.

At the start of the experiment, the scientists grew bacteria on a growth medium in which all the nitrogen was the ^{14}N isotope. After many generations of bacteria, they removed DNA from a sample of the bacterial cells, put it into a liquid and spun it in a centrifuge. The DNA formed a band in the liquid, shown in Figure 5.4a.

The scientists then repeated this procedure exactly, but used a growth medium in which all the nitrogen was the ^{15}N isotope. The DNA extracted from bacteria in this experiment formed the band shown in Figure 5.4b.

Finally, the scientists inoculated a sample of bacteria from the medium containing only the ^{15}N isotope into

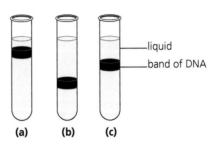

Figure 5.4 The results of an experiment to test the theory that replication of DNA is a semi-conservative process. Each tube shows the position of DNA taken from bacteria after it was centrifuged at the same speed for the same time in a liquid. (a) DNA from bacteria grown for many generations in a medium containing only the ^{14}N isotope of nitrogen. (b) DNA from bacteria grown for many generations in a medium containing only the ^{15}N isotope of nitrogen. (c) DNA from bacteria grown for many generations in a medium containing only the ^{15}N isotope of nitrogen and then a single generation in a medium containing only the ^{14}N isotope of nitrogen.

a medium containing only the ^{14}N isotope. After one generation, they removed DNA from a sample of these bacterial cells and spun it in a centrifuge. This DNA formed the band shown in Figure 5.4c.

DNA replication

Consider the experiment and answer the following questions.

1 In the final part of the experiment, the scientists inoculated bacteria from a medium containing ^{15}N into a medium containing ^{14}N. Explain why.

2 The scientists removed the final sample of bacteria after only one generation in the medium containing ^{14}N. Explain why this timing was important.

3 The scientists went on to conclude that their experiment provided evidence for the theory of semi-conservative replication of DNA. Do you agree with this? In answering this question, consider whether there is another valid conclusion from these results.

DNA replication

Figure 5.5 shows a DNA molecule being replicated. The arrows show the directions in which the enzymes are moving.

1 Name enzymes A and B.
Enzyme A is DNA helicase. This is because it is the enzyme that is separating the two strands of the original DNA molecule by breaking the hydrogen bonds between the bases and unwinding the helix.

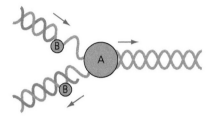

Figure 5.5

Enzyme B is DNA polymerase. This is because it is the enzyme building the new strands by joining on free nucleotides as they are attracted by the exposed bases on the template strands.

2 In eukaryotic DNA replication, a number of replication forks may open up along the DNA molecule. What is the advantage of this?
Having replication happening at a number of points at the same time speeds up the replication of the very long DNA molecules in eukaryotic chromosomes.

Mitosis

According to cell theory, all cells arise from previously existing cells by cell division. Mitosis is a type of cell division that occurs in eukaryotic cells. During mitosis, a parent cell divides to produce two daughter cells. Each of the two daughter cells contains some of the cytoplasm from the parent cell. They also contain a complete copy of the parent cell's DNA, making them genetically identical to the parent cell and to each other.

On pages 74–76 in this chapter we looked at the replication of DNA. The DNA of eukaryotic cells is contained in linear chromosomes. Figure 5.6 shows you that the appearance of a chromosome changes after its DNA has been replicated. Before DNA replication, the chromosome is a single, rod-like structure containing one tightly wound double helix of DNA. DNA replication gives rise to two rod-like structures, each containing identical molecules of DNA. One region of the chromosome, the centromere, holds the two rod-like structures together. While they are held together they are called **chromatids**. We describe the chromatids in a pair as sister chromatids. During mitosis, sister chromatids are separated from each other. At anaphase (see Table 5.1) the chromatids become chromosomes when they separate.

Centromere The structure in a chromosome that holds together chromatids until they are separated by the spindle fibres.

Figure 5.6 (a) Following the replication of DNA, a chromosome appears as a double structure composed of two chromatids. The chromatids are the products of replication of the DNA in the original chromosome. They are temporarily held together by a region called the centromere. (b) Human X (centre) and Y (lower right) sex chromosomes. Each chromosome has replicated and so there are two identical structures (chromatids) joined at the centromere.

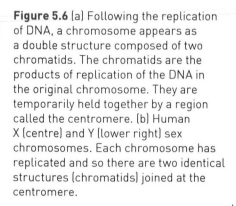

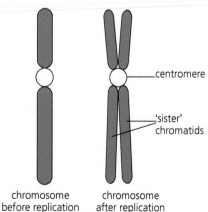

centromere

'sister' chromatids

chromosome before replication

chromosome after replication

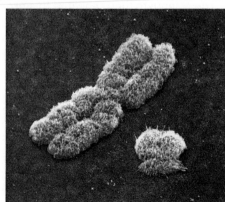

Table 5.1 A summary of the events occurring during each stage of mitosis.

Stage of mitosis	Main events that occur in each stage
Interphase	Cell makes a copy of its chromosomes involving replication of DNA. Cell grows and undergoes its normal physiological functions.
Prophase fibre of spindle	Chromosomes coil, becoming shorter and fatter. We can now see them with an optical microscope. Nuclear envelope disappears. Protein fibres form a spindle in the cell. Each chromosome consists of two chromatids, each containing an identical DNA molecule from DNA replication.
Metaphase	One or more spindle fibres attaches to centromere of each chromosome. Chromosomes line up in the middle of the spindle.
Anaphase	Centromere holding each pair of sister chromatids together divides. Spindle fibres shorten and pull one of each pair of sister chromatids to opposite poles of the spindle. We can now refer to the chromatids as chromosomes.
Telophase	The two sets of separated chromosomes collect at opposite ends of the cell. A new nuclear envelope forms around each set of chromosomes. Chromosomes become long and thin. We can no longer see them clearly with an optical microscope. Cytoplasm divides to form two new cells (cytokinesis).

Although mitosis is a continuous process, it is often described as a series of stages. The names of these stages, and the events that occur within them, are summarised in Table 5.1.

Usually, following telophase, the cell divides into two in a process called cytokinesis. In animal cells, the cell membrane is pulled inwards across the centre of the cell, pinching off the cytoplasm into two equal halves, each containing a new nucleus. In plant cells, vesicles fuse to extend the cell membranes across the cytoplasm and then new cell walls develop between them.

> **TIP**
> The terms 'chromatid' and 'chromosome' can be confusing. Chromatids are replicate chromosomes held together by a centromere. As soon as they separate, they are called chromosomes again.

Cytokinesis Division of the cytoplasm to give two new cells.

> **TEST YOURSELF**
> **6** Mitosis produces clones. What is a clone?
> **7** Explain the difference between a chromosome and a chromatid.
> **8** The chromosomes in Figure 5.6 were taken during prophase. How would they look different if photographed during anaphase?
> **9** What happens to centromeres during anaphase of mitosis?
> **10** Describe what happens during cytokinesis.

REQUIRED PRACTICAL 2

Preparation of stained squashes of cells from plant root tips; set-up and use of an optical microscope to identify the stages of mitosis in these stained squashes and calculation of a mitotic index

This is just one example of how you might tackle this required practical.

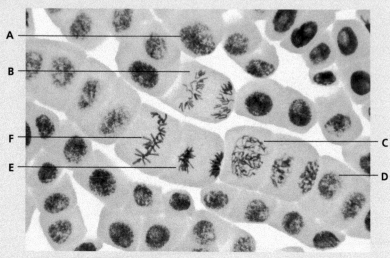

Figure 5.7 Cells from the tip of a plant root. Many of these cells are dividing by mitosis.

A student decided to prepare a stained squash of cells from the root tips of a plant for examining under an optical microscope. She cut off the lower part of growing roots and then placed them in ethanoic alcohol to 'fix' the tissue. This prevented mitosis from continuing in the cells. She then placed the root tips in dilute hydrochloric acid at 60°C. This separated the cells. After this, she placed the tissue on a microscope slide with acetic orcein stain, which stained the chromosomes. Finally, she put a cover slip on top, and pressed down hard on the tissue using a folded paper towel, to obtain a layer of tissue just one cell thick. She then viewed the cells under a microscope.

Figure 5.7 shows a photograph of cells in actively dividing tissue. The chromosomes have been stained so that we can see them. Cells in this tissue were at different stages of mitosis before they were killed and stained. The cells with indistinct nuclei were not dividing at the time the photograph was taken.

1 Look at cell A in Figure 5.7. Its nucleus is clear, but all we can see is a dark-stained nucleolus surrounded by granules that are the tightly coiled regions of DNA. At which stage of mitosis was this cell?
2 Look at cell B in Figure 5.7. It has two groups of thread-like chromosomes. Look closely. Can you see that each chromosome looks V-shaped? This is because a spindle fibre, which you cannot see, is pulling its centromere to the left or right side of the cell. The arms of each chromosome lag behind its centromere, making the 'V'. At which stage of mitosis is this cell?
3 Now look at cell E. It has chromosomes that are visible but are not such clear threads as in cell B. There are two clumps of chromosomes and a clear area is developing between them. At which stage of mitosis is cell E? What is happening in the clear area between the clumps of chromosomes?
4 Cell F has a very distinctive appearance. The chromosomes are in a line across the centre of the cell. In which stage of mitosis is cell F?
5 Cells A, B, C, D, E and F are in different stages of mitosis. Put them into the correct sequence, starting with the earliest stage.
6 Is cell D in prophase or telophase? Explain your answer.
7 The mitotic index of an actively dividing tissue is found by dividing the number of cells seen in mitosis by the total number of cells counted (Chapter 14, page 245). If 320 of these plant root cells were observed and 84 were seen in various stages of mitosis, calculate the mitotic index for this dividing root tissue.

SAFETY

Your teacher will demonstrate to you the correct procedure for focusing the microscope to avoid breaking the slide. Care should be taken when using stains – wear eye protection and make sure you follow all instructions you are given by your teacher.

TIP

Notice how some questions were easy to answer. Anaphase is always very easy to identify. In other cases, you needed to work through a logical pathway before you could answer the question.

The cell cycle

Not all cells in multicellular organisms retain the ability to divide. In actively dividing tissues, the new cells formed by mitosis grow before replicating their DNA and dividing by mitosis again. Thus a cycle is formed, called the cell cycle. Figure 5.8 shows this cell cycle. You can see that the two events we have described in this chapter, namely replication of DNA and mitosis, last only a short time during this cycle.

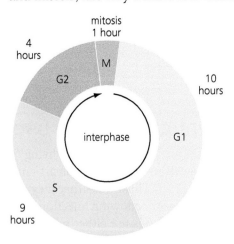

Figure 5.8 The cell cycle in a eukaryotic cell. The times shown represent a cell cycle of 24 hours, which might be found in actively dividing tissues. The actual length of the cycle varies from one type of cell to another.

Extension

Each part of the cycle involves specific cell activities. These are:

- G1 phase: the cell increases in size
- S phase: the cell replicates its DNA
- G2 phase: the cell increases further in size and replicates its cell organelles
- M phase: mitosis.

Cell division in prokaryotes

When prokaryotes such as bacteria divide, the process is simpler than mitosis. Prokaryotes do not have chromosomes, just a single, circular DNA molecule. However, their plasmids are also replicated and passed on. Plasmids are really just much smaller circular DNA molecules. Bacteria can contain more than one copy of a plasmid at any one time and the maximum number of copies they can have of each is tightly regulated.

Binary fission The method of cell division found in prokaryotes.

Cell division in prokaryotes is called binary fission. Figure 5.9 shows how the circular DNA molecule and any plasmids in the cell undergo DNA replication. The cell then divides into two cells, each containing roughly half the cytoplasm, one copy of the circular DNA molecule and some of the plasmid copies. Prokaryotes have no membrane-bound organelles (see Chapter 3, page 51), there is no nuclear envelope to break down and there are no spindle fibres.

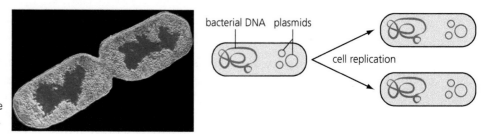

Figure 5.9 (a) Binary fission taking place in bacteria. (b) The single circular DNA molecule in prokaryotes is replicated, together with any plasmids, and the two daughter cells each receive a copy of the single circular DNA molecule and a variable number of plasmids.

Even though there are no spindle fibres involved in binary fission, there are mechanisms to ensure that a copy of the single circular DNA molecule and some copies of each plasmid are in each half of the bacterial cell before division takes place. Obviously each daughter cell must receive a copy of the single circular DNA molecule, but even a daughter cell that fails to receive at least one copy of a plasmid may die.

Replication in viruses

Viruses are acellular and non-living (see Chapter 6, page 99) – they do not have a cell structure, and consist only of DNA or RNA surrounded by a protein coat. This means that they do not carry out cell division. Instead, they replicate following injection of their nucleic acid into a host cell. Different viruses use either prokaryotic cells or eukaryotic animal or plant cells as hosts. Once their nucleic acid is inside a host cell, the host's DNA-replicating and protein-synthesising systems make more virus particles. Eventually, these are released when the whole host cell bursts, or by 'budding' one at a time through the host cell membrane (Figure 5.10). This is why viruses damage host cells and can cause disease.

Figure 5.10 Virus particles leaving a host cell after replication.

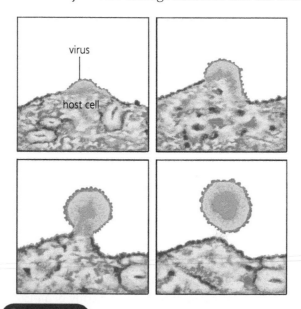

TEST YOURSELF

11 Name two processes involving actively dividing tissue.

12 How many double helices of DNA are present in each chromosome (a) at the start of the G1 phase and (b) at the start of the G2 phase?

13 Apart from the processes mentioned above, name two processes occurring in cells during interphase.

14 List the differences between mitosis in eukaryotes and binary fission in prokaryotes.

Cell differentiation

Not all eukaryotic cells undergo the cell cycle shown in Figure 5.8. Some fungi grow as filaments that contain cells with more than one nucleus in them. As described above, during mitosis the DNA of these fungi replicates and then the new nuclei are formed. However, the cytoplasm does not then divide. Many cells in mature plants, and most cells in mature animals, lose the ability to divide. This is why, in plant micropropagation, described at the start of this chapter, it is easier to use actively dividing cells to obtain explants.

In most multicellular organisms newly formed cells change their properties when they become specialised for specific functions. We call this process **differentiation**, and it does not occur at random. Cells in one part of a mature organism differentiate to form a tissue in which all the cells perform the same function. Tissues are organised into organs and groups of organs form systems, such as the digestive system.

The cell cycle and cancer

During differentiation, most human cells lose the ability to divide by mitosis. Even cells that continue to divide, such as those near the surface of your skin, normally divide only about 20 to 50 times before they die. Losing the ability to divide and programmed cell death are two ways in which cells control their own division and thus numbers. Sometimes these control mechanisms break down. The result is that, if kept supplied with the necessary nutrients, such cells undergo repeated, uncontrolled division. A large mass of these cells is called a tumour. Tumours may become cancerous. Cancer occurs if cells from a tumour are able to break away and form secondary tumours elsewhere.

In about 50% of people with all types of cancer, a gene that helps to control cell growth (called *p53*) has mutated. However, there are many other reasons why the normal control of cell division breaks down. Consequently, there is no single treatment for cancer sufferers. However, many cancer treatments work by controlling the rate of mitosis.

Extension

Cancer treatments that control the rate of mitosis do so in different ways:
- adriamycin and cytoxan are drugs that stop DNA unwinding prior to replication
- methotrexate is a drug that stops cells making DNA nucleotides
- taxol and vincristine are drugs that inhibit formation of the mitotic spindle.

TEST YOURSELF
15 How is a tumour formed?
16 Which process is prevented by methotrexate? Explain your answer.
17 Many drugs that are used to treat cancer have side effects because they also affect healthy cells. Suggest why.

Practice questions

1 a) The following statements describe different stages in mitosis. Put them into the right order. *(1)*

A	Chromatids separate and move to opposite poles of the cell.
B	Chromosomes become shorter and thicker, and the nuclear membrane breaks down.
C	A new nuclear membrane forms around each group of chromosomes.
D	Chromosomes line up along the equator of the spindle.

b) Look at the figure. Put the letters A, B, C, D and E in order to indicate the correct order of the stages of mitosis. The first one has been done for you. *(1)*

C __ __ __ __

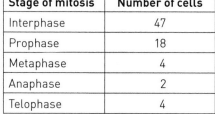

A   B   C

D   E

c) The table shows the number of cells in different stages of mitosis in part of the growing region of an onion root tip.

Stage of mitosis	Number of cells
Interphase	47
Prophase	18
Metaphase	4
Anaphase	2
Telophase	4

i) Calculate the mitotic index for this tissue. *(1)*

ii) If the whole cell cycle for these cells takes 30 hours, calculate how long the cells take to complete mitosis. *(2)*

iii) Calculate how long the cells spend in metaphase. *(2)*

2 The graph shows the mean distance between the centromeres and the poles of the cell during mitosis.

a) i) At what time did metaphase begin? Explain your answer. *(2)*

ii) Calculate the mean rate of chromatid movement during the first 100 minutes of anaphase. *(1)*

iii) Using a tangent on the graph, find the rate of chromatid movement for the second 100 minutes of anaphase. *(2)*

b) Vincristine is a drug used to treat cancer. It inhibits spindle formation.

i) Which stage of mitosis does vincristine inhibit? Explain your answer. *(2)*

ii) Cancer patients who take vincristine can suffer hair loss and anaemia. Suggest why. *(2)*

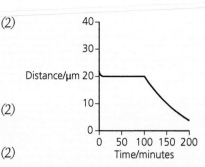

3 Yeast cells are eukaryotic. They reproduce by a form of mitosis called budding. Scientists sampled a yeast culture every hour for 6 hours. They counted how many yeast cells were in each sample and measured the DNA concentration. The results are shown in the graph.

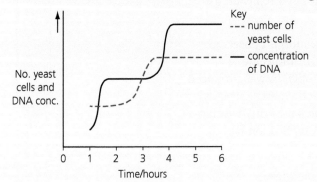

a) What process was happening between the 1 and 2 hour samples? *(1)*

b) In which stage of the cell cycle were many of the yeast cells between 1 and 2 hours? *(1)*

c) Between which samples was cell division happening? *(1)*

d) Yeast cells are single-celled organisms. Cell division by budding results in two genetically identical daughter cells in the same way that binary fission does in prokaryotes. How does budding differ from binary fission? *(1)*

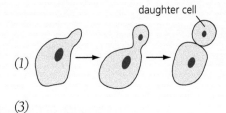

e) Yeast cells are eukaryotic. Using your knowledge, explain how mitosis in yeast differs from binary fission in bacteria. *(3)*

4 The bases in DNA nucleotides contain nitrogen. Scientists grew bacteria on a medium containing ^{15}N ('heavy' nitrogen) for many generations. They then transferred the bacteria to a medium containing ^{14}N ('ordinary' nitrogen). They analysed DNA from the bacteria at three stages:

1 while the bacteria were growing on the ^{15}N medium

2 after one division of the bacteria on the ^{14}N medium

3 after two divisions of the bacteria on the ^{14}N medium.

The DNA was analysed by extracting it from the cells and centrifuging it in a tube. The layer of DNA in the tube formed at different heights because of its density. The diagram shows their results.

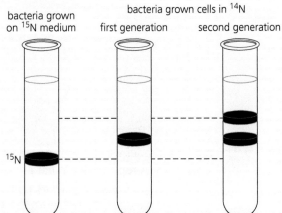

a) **i)** Copy and complete the diagram on the right to show how the tube would appear after one more division of the bacteria on the ^{14}N medium. (2)

ii) Explain how these results confirm that DNA replication is semi-conservative. (3)

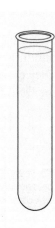

b) An alternative theory was that DNA replication is conservative. This theory suggested that the original DNA molecule remained intact and that a new molecule of DNA was produced without using any of the nucleotides from the parent strands. Complete the diagrams below to show the bands of DNA that would have appeared in the tubes if DNA replication had been conservative. (2)

After one division of the bacteria on the ^{14}N medium.

After two divisions of the bacteria on the ^{14}N medium.

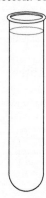

Stretch and challenge

5 The replicon model was first proposed as far back as 1964 to explain prokaryotic DNA replication. What are replicons and how do they differ in prokaryotic and eukaryotic DNA replication?

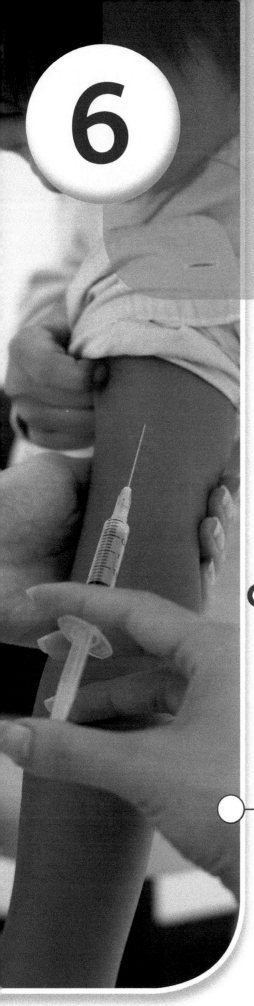

6

The immune system

TEST YOURSELF ON PRIOR KNOWLEDGE

1 Give two ways in which pathogens might affect the body cells they infect.
2 Why are antibodies described as specific?
3 Apart from antibodies, give one way in which white blood cells protect the body against pathogens.
4 If you are vaccinated against measles, you will not suffer from measles. Explain why.
5 Colds are caused by a virus. You might suffer from more than one cold in a single winter. Explain why.

Introduction

Look at the baby in Figure 6.1 (overleaf). He lives in the plastic cage or bubble that you can see around him. The bubble is completely airtight. The air that he breathes has to be filtered before it is pumped into the bubble. The food and water that he consumes must pass through an air lock before he can touch them. Anyone who wishes to touch him must work through the plastic gloves you can see in the photograph. This boy can never leave the plastic bubble. He will not be able to go to school. He will not be able to play out with friends. He will not be able to touch his mother, his father or his siblings. Can you imagine a life like this?

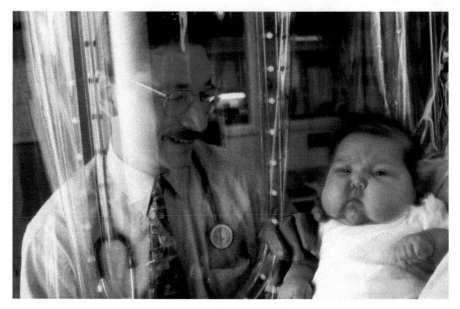

Figure 6.1 This boy suffers from SCID. The plastic bubble in which he lives protects him from infection.

This boy was born with an inherited condition, called severe combined immunodeficiency (SCID). His immune system is unable to recognise foreign cells and does not produce enough white blood cells. White blood cells are part of the immune system, the system that helps to protect us against disease. If he were to suffer any infection he could not fight it off. As a result, even the mildest infection could kill him.

Thankfully, for most of us the body has a number of defences against pathogens. The first is to prevent their entry by various physical and chemical barriers, such as the skin, membranes, tears and saliva. If this fails, the second is for the region invaded by the pathogen to swell and become red, in what is called a non-specific inflammatory response. If this fails, the third defence occurs. Here, the body is able to recognise 'foreign' cells and target particular pathogens in a specific immune response.

Surface proteins are important in enabling the body to recognise its own cells ('self') and the cells of pathogens that are invading the body ('non-self'). Key to this is the way that proteins on the surface membranes of cells are used to identify them (see the fluid mosaic model on page 43). Understanding the role of proteins on the surface membranes of cells is vital to understanding the specific immune response. You will remember that you learned about the proteins in cell membranes in Chapter 3. Some of the surface proteins of white blood cells are receptors that bind to the proteins on the surface of pathogens. One type of white blood cell releases copies of its surface receptors as antibodies.

Phagocytes and lysosomes

If a pathogen gets into your body, an inflammatory response is your second line of defence. This type of response is non-specific, meaning that it is the same for all pathogens. Blood contains two types of blood cell: red cells and white cells. Unlike red cells, there are many different types of white cell. Some of these are **phagocytes**. This means that they can surround and digest microscopic pathogens.

The white blood cell in Figure 6.2 is the most common type of phagocyte in your blood, called a neutrophil. You can see that its cytoplasm is full of lysosomes. You learned about lysosomes in Chapter 3 (see Table 3.1 on page 41). They contain enzymes that can digest proteins, lipids and carbohydrates and are important in destroying pathogens.

The cell-surface membrane of a neutrophil has protein receptors that bind to antigens such as those on the surface of pathogens. Their complementary binding sites enable a neutrophil to bind to a pathogen such as a bacterial cell. You can see in Figure 6.3 how a neutrophil then ingests the pathogen, forming a membrane-bound **vacuole** around it.

Lysosomes move to this vacuole and their membranes fuse with the membrane around the vacuole. This fusion releases enzymes, called lysozymes, onto the ingested pathogen. The lysozymes hydrolyse the pathogen and the neutrophil absorbs the harmless products of digestion. In this way, all phagocytes kill bacteria and other pathogens that have entered the body.

Neutrophils are carried in the blood, but they can also move and leave the blood to attack microscopic pathogens in other tissues. They leave the blood by squeezing through tiny gaps between the cells that form capillary walls (see page 158).

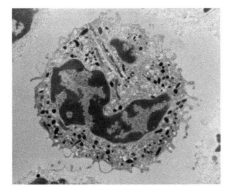

Figure 6.2 This blood cell is a neutrophil, the most common type of phagocyte in your blood. Phagocytes are cells that can ingest and then digest microscopic pathogens. Notice that its cytoplasm is full of lysosomes (coloured red). These contain enzymes, called lysozymes, that are important in destroying ingested pathogens.

Extension

The release of a chemical called histamine helps neutrophils leave the blood by making the walls of capillaries 'leaky'. Histamine is released by another type of white blood cell, called a mast cell, when tissues outside the circulatory system are damaged. It causes capillary walls to become more permeable so that they lose more fluid to their surroundings. This leads to localised swelling.

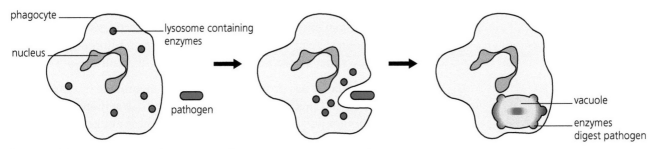

Figure 6.3 How a phagocyte destroys a pathogen.

TEST YOURSELF
1 Explain the meaning of the term 'pathogenic'.
2 The vast numbers of bacteria living in your large intestine help to prevent infection by pathogens. Suggest how.
3 Pathogens are harmful but, once digested by phagocytes, the products of their digestion are not. Explain why.
4 A site of inflammation, such as a cut on your finger, becomes hotter than the surrounding skin. Suggest why.

The specific immune response: lymphocytes

The immune system recognises and destroys any foreign cells, pathogens, abnormal cells or toxins. The immune system does this by recognising molecules on the surface of the body's own cells and identifying them as 'self'. Any cells, toxins or pathogens that have other molecules on their surface are recognised as 'foreign', and these are attacked by the cells of the immune system.

Phagocytes, which you have just learned about, do not respond in a specific way to a microscopic pathogen; they ingest and destroy any. In contrast, **lymphocytes** are specific. Lymphocytes are another type of white blood cell. Each lymphocyte attacks only one type of antigen. Figure 6.4 shows the appearance of a lymphocyte. You can see that these cells do not have the lysosomes in their cytoplasm that we have seen in phagocytes.

Lymphocytes go through a maturing process before they are capable of fighting infection. The maturing process begins before birth, and results in two types of lymphocyte:

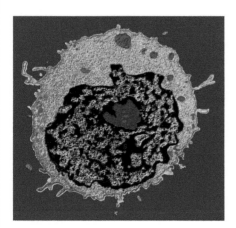

Figure 6.4 Lymphocytes are a type of white blood cell. They are involved in cellular and humoral immunity.

- B lymphocytes, known as **B cells**, mature in bone marrow; they release antibodies into the blood
- T lymphocytes, known as **T cells**, mature in a gland in the chest or base of the neck called the thymus; they cause a **cellular response** to infection and they do not release antibodies into the blood.

Before going further, we need to be clear about three terms used in immunology.

Antigen

An antigen is a molecule that stimulates an immune response, including antibody generation. Small molecules, like amino acids, sugars and triglycerides, do not stimulate antibody generation. Antigens are large, complex molecules, such as proteins, glycoproteins and lipoproteins. Figure 6.5 shows that antigens are located on the outer surface of cells.

Each cell in your body has proteins on its surface membrane (see page 44). Normally, you would not make antibodies against these **self antigens**. In the body of another person or mammal, however, these antigens would stimulate antibody generation.

Antigen A large 'foreign' molecule that stimulates an immune response.

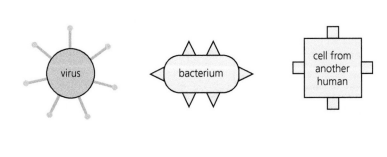

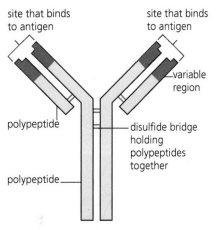

Figure 6.5 Every mammal has self antigens on the surface membranes of its cells. The antigens on the surface of cells from another organism, non-self antigens, trigger antibody generation. (Drawings are not to scale.)

Figure 6.6 This antibody is a Y-shaped molecule made of two long (heavy) and two short (light) polypeptides. The variable region shown in the diagram is the part that binds to an antigen to make an antigen–antibody complex.

Antibody A protein released by a B cell in response to a non-self antigen.

Antigen–antibody complex The complex formed when an antigen binds with a complementary antibody.

TIP
Antigens fit into the receptor site of the specific antibody. Be careful not to call this an active site, because antibodies are not enzymes.

Antibody

An antibody is a protein released by a B cell in response to a non-self antigen. Every antibody has at least one Y-shaped molecule, made of four polypeptide chains (quaternary structure). Figure 6.6 shows one such molecule. Notice that two of the polypeptides are large and two are small. It also shows the two key parts of the antibody molecule: the sites that bind with a specific antigen. Every mammal is able to make millions of different antibodies, each with a different pair of binding sites specific to one type of antigen only.

Antigen–antibody complex

An antigen and an antibody have complementary molecular shapes, meaning that they fit into each other. When an antibody randomly collides with a cell carrying a non-self antigen that has a complementary shape, it binds to the antigen. When this happens, the two molecules form an antigen–antibody complex. This is the first stage in the destruction of a cell carrying a non-self antigen.

Antibodies have at least two sites where they can bind to an antigen, so this means they can bind to more than one bacterium or virus. When this happens, a network of antibodies and particles forms in a clump. This is called **agglutination**. Sometimes the binding of antibodies to the antigen neutralises the pathogen. Sometimes the binding of antibodies to the antigen acts like a 'marker' which attracts phagocytic cells to engulf and destroy the cell bearing it.

B cells and the humoral immune response

A single B cell has a unique type of receptor molecule on its surface membrane. Every day, however, you randomly make millions of B cells, each with a different receptor on its surface membrane. By chance, one of these B cells will have receptors with a shape complementary to the shape of an antigen on a cell that has entered your blood. In that case, the receptors bind to this antigen. Figure 6.7 (overleaf) summarises what then happens.

The B cell divides rapidly to produce a large number of daughter cells. Since the divisions are by mitosis (see Chapter 5), these daughter cells are

Figure 6.7 The diagram shows what happens when, by chance, an antigen collides with a B cell. If this B cell has a complementary receptor, it will bind to the antigen. This stimulates the B cell to divide rapidly, forming a large clone of daughter cells. Most of these are plasma cells (labelled P), which release antibodies into the blood. A few are memory cells (labelled M), which remain dormant in the blood.

genetically identical; that is, they form a **clone**. The majority of cells in the clone become plasma cells, which release antibodies. A smaller number become **memory cells** and remain in circulation long after the antigen is destroyed. They can rapidly divide to produce clones of **plasma cells** if re-exposed to their complementary antigen.

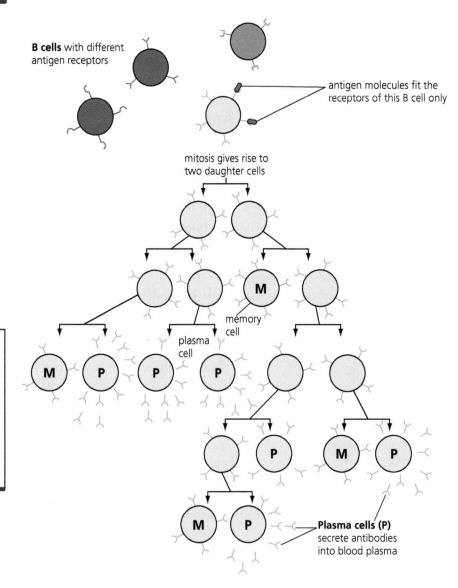

T cells and the cellular immune response

Although T cells are involved in different aspects of the immune response from B cells, they respond in a similar way to exposure to a specific antigen. As Figure 6.8 shows, once it binds to its complementary antigen, a T cell divides rapidly to produce a clone of cells. T cells are, however, different from B cells in three important ways.

The receptor of a T cell has only two polypeptide chains and is never released as an antibody into the blood. T cells respond to an antigen only if this antigen is present on the surface of a macrophage that has become an **antigen-presenting cell** by presenting the antigen from an ingested pathogen, foreign cell or toxin.

Macrophage A type of phagocytic cell (like a neutrophil). Once a macrophage has ingested and partly digested a pathogen, it may transfer some of the pathogen's antigens to its own surface membrane. It then becomes an antigen-presenting cell.

Figure 6.8 T cells will bind to an antigen only if it is on the surface of an antigen-presenting cell. When the surface receptor of a T cell binds to a complementary antigen, the T cell becomes sensitised and starts dividing to form a clone of cells.

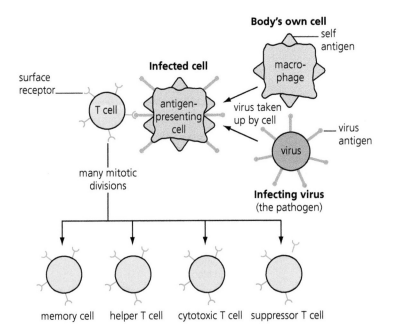

There are several different types of T cell and, following cell division, the cells in each clone differentiate into more of the same type of T cell. Cells of each type have a different function. Memory T cells remain in the blood and cause a rapid increase in the number of T cells when re-exposed to their specific antigen. Helper T cells (T_H cells) assist other white blood cells in the immunological response. For example, by releasing chemical messengers, called cytokines, they stimulate:

- maturation of B cells into plasma cells that secrete antibody
- formation of memory B cells
- activation of cytotoxic T cells (T_C cells), which destroy tumour cells and cells that are infected with viruses
- activation of phagocytes.

T_H cells are extremely important in the immune response, and B cells cannot work without them (see pages 99–100 for what happens during infection with HIV).

TEST YOURSELF

5 Give two ways in which antibodies are different from enzymes.

6 A plasma cell has many mitochondria and extensive rough endoplasmic reticulum. Explain how each of these features is an adaptation for the function of this cell.

7 After recovery from an infection, collision between a B cell and its complementary antigen is less likely than the collision between a memory cell and the same antigen. Explain why.

8 Give two differences in the way that a T_H cell and a B cell react to their respective antigens.

9 In response to infection by a single strain of bacterium, a healthy human produces many, perhaps hundreds of, different antibodies. Explain how this happens.

Why don't we suffer the same infection twice?

If you suffered an illness such as chicken pox when you were a child, you might have wondered why you never caught the disease again. We can explain this using our knowledge of antibodies, plasma cells and memory cells.

Figure 6.9 The effect of repeated injections of antigen on the concentration of antibodies in the blood.

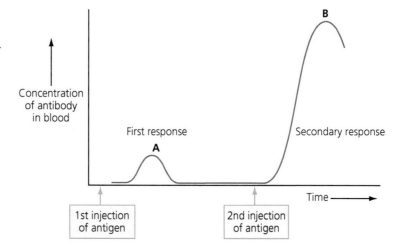

Look at Figure 6.9. It shows the concentration of an antibody in the blood of a person who is injected twice with the same antigen. The part of the curve labelled **A** shows what happens after the first injection. After a delay of several days, there is a small rise in the concentration of antibody in the blood. This is the **primary response**. The antibody concentration quickly falls again. We can use the information from Figure 6.9 to explain this. The short delay represents the time for a complementary, specific B cell to collide randomly with the antigen, bind with the antigen to form an antigen–antibody complex and then divide to form a clone of plasma cells that release their antibodies.

As the antibodies destroy the antigens, so fewer and fewer B cells are made, and the concentration of antibody falls again. The part of the curve labelled B shows what happens after the second injection. This **secondary response**, which may occur weeks, months or even years after the original injection, is more rapid and results in a higher concentration of the antibody in the blood. Again, we can explain this using information from Figure 6.9. After the first response, many memory cells that are specific for the antigen remain in the blood. As a result the memory cells are much more likely to collide with the antigen and bind with it. This means that the memory cells start to divide to produce plasma cells. This explains why the secondary response is more rapid than the first response. In addition, a larger number of plasma cells are produced, so the concentration of antibodies is higher during the secondary response.

The same events occur when we suffer infections. The first time we become infected by a particular pathogen, although we are able to fight the infection and recover, we produce antibodies after a time delay. During this primary response, we exhibit the symptoms of that infection. We talk about catching the disease. However, the next time we become infected by the same pathogen, our response is very rapid and we produce a large concentration of antibodies. We are able to destroy the pathogen before it causes disease.

Why do we get colds and flu every winter?

As we have seen, the surface receptors on lymphocytes have a shape complementary to only one antigen. This is why they are a specific immune response. Memory cells can only be effective against a pathogen that always has the same antigen. If the antigen on a pathogen were to change, the receptors on our memory cells would no longer bind with it.

Some pathogens show **antigen variability**. This means that, as a result of gene mutations, they frequently change the antigens on their surface. The cold virus and the influenza virus show antigen variability in this way. This means that the 'new' cold or flu viruses have antigens that are no longer complementary to the surface receptors of the memory cells remaining from the cold or flu we caught last year. Consequently, we will not have a secondary response and so catch a cold or flu all over again.

Scientists propose explanations for the observations they make. We refer to these explanations as hypotheses. In order to test the validity of a **hypothesis**, scientists use it to make predictions and then devise experiments to test these predictions. If their results are always consistent with their predictions, they become confident in their hypothesis and it becomes a theory.

The **clonal selection hypothesis** proposes an explanation for the way that we produce antibodies against non-self antigens. The box below summarises this hypothesis.

Our immune systems produce millions of different types of B and T cells. Each type has a unique protein receptor on its surface membrane. If, and only if, one of these cells binds with a complementary antigen, it is stimulated to produce large numbers of cells that are identical to itself and, consequently, to each other. We call a group of identical cells a clone. Thus, in response to the presence of a particular antigen, a B or T cell is selected and it forms a clone. Within the clone, each cell has the identical protein receptor on its surface.

Extension

Testing the clonal selection hypothesis

In an experiment to test the clonal selection hypothesis, scientists injected two rats, R and S, with antigens from two different strains of pneumococcal bacteria. They injected each rat twice, with the second injection made 28 days after the first.

1 Why do you think the scientists injected each rat twice, with a gap of 28 days between injections?

This is to allow time for the primary response to occur and memory cells to be made. After the second injection, memory cells will be present that produce very large amounts of antibody against the pathogen.

Figure 6.10 will help you to answer this question. After the first injection, the rats would produce only a small quantity of antibodies against the pathogen (the primary response). After the second injection, the rats would produce much more of the antibody (the secondary response).

2 Which type of immune cell do you think the scientists were attempting to stimulate?

B cells. Hopefully, you spotted that the scientists were going to look for antibodies. This means that the scientists must have been trying to stimulate B cells, since T cells do not release antibodies.

3 The scientists injected rat R with antigens from strain X of the bacterium (type X antigens) and injected rat S with antigens from strain Y of the pneumococcal bacterium (type Y antigens).

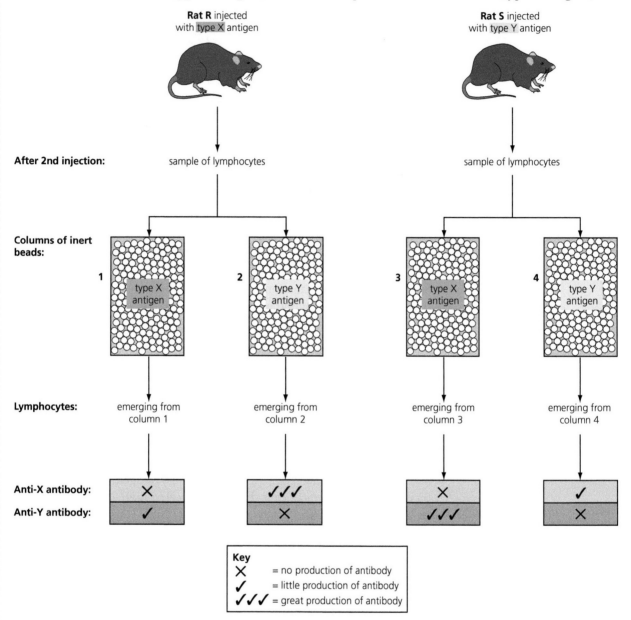

Figure 6.10 This diagram summarises the method and results of an investigation into antibodies from rats injected with antigens.

What do you think the scientists were predicting would happen as a result of these injections?

Using the clonal selection hypothesis, the scientists were predicting that the blood of both rats would already contain one or more B cells with a protein receptor that was complementary to each of the antigens from the two strains of bacterium. If the clonal selection hypothesis is correct, the cells in rat R with a protein receptor complementary to type X antigens should be stimulated to form a clone. However, this should not happen in rat S. Instead, the cells with a protein receptor complementary to type Y antigens should be stimulated to form a clone in rat S.

The scientists now needed a way to find out which cells had been stimulated to produce a clone. The method they chose was rather neat. They coated inert beads with the type X antigens and put the beads into two glass columns (labelled 1 and 3 in Figure 6.10). They then did the same with type Y antigens and put these beads into another two glass columns (labelled 2 and 4 in Figure 6.10). They reasoned that if they washed samples of lymphocytes through these columns, those lymphocytes that had the appropriate complementary protein receptors would bind to the antigens on the inert beads. As a result, they would stay in the glass column and not emerge at the bottom.

4 Why was it important that the beads they used were inert?

You might not have come across the term 'inert' before. Something that is inert will not react with anything. This was important because the scientists did not want any of the lymphocytes to react with the beads and stick to them. They wanted them to bind only with the antigens that coated the beads.

One week after the second injection, the scientists removed a sample of blood from each of the two rats and separated the lymphocytes from the rest of the blood. They put half of each sample into a column of inert beads coated with type X antigen and half into a column of inert beads coated with type Y antigen. They then washed the samples through the columns and collected any lymphocytes that passed through the column.

5 What type of fluid do you think they would use to wash the lymphocytes through the columns?

They would use a solution that has the same water potential as the lymphocytes.

Hopefully, you used your knowledge of water potential and osmosis to answer this question. Since the scientists wanted to recover some cells at the bottom of the column, they would have to use a solution with the same water potential as the lymphocytes. If they had used water, the lymphocytes would have taken up water by osmosis and burst.

Finally, the scientists tested the lymphocytes that had passed through each column to see whether they could make antibodies against either of the two antigens used in the experiment. Figure 6.10 summarises the method they used and their results.

Now let's look at their results and see if we can interpret them. Let's start with the results from column 1. First we need to make sure we have understood what they show by describing them.

6 How would you describe the results from column 1?

The lymphocytes washed through the column could not produce type X antibodies at all, but could produce a small quantity of antibodies against type Y antigens.

Remember that a description translates information from one form to another. At this stage we are not attempting to explain the result. In this case, the answer above is an adequate description.

7 Can you explain the two parts of that description?

The clonal selection hypothesis proposes that, by chance, rat R would have some lymphocytes with protein receptors that were complementary to type X antigen, and other lymphocytes with protein receptors that were complementary to type Y antigen. So, the rat had the potential to make antibodies against both types of antigen. Since rat R was injected with type X antigen, we predict that it would produce a large clone of cells with receptor proteins complementary to that antigen. However, these would bind with the type X antigen on the beads in column 1. That is why none of the lymphocytes emerging from column 1 could produce antibodies against type X antigen (called anti-X antibody in Figure 6.10). The cells with a receptor protein complementary to type Y antigen would not bind with the type X antigen in column 1 and so emerged at the bottom of the column. The cells can produce anti-Y antibody but, since there are so few of them, they only produced a small quantity.

Now let's describe the results from column 2. Here, the emerging cells produced large quantities of anti-X antibody but no anti-Y antibody.

8 Explain why cells emerging from column 2 give large quantities of anti-X antibody but no anti-Y antibody.

We predicted that rat R would produce a large clone of cells with receptor proteins complementary to type X antigen. Since column 2 contained beads coated with type Y antigen, these cells would not bind to this antigen and would emerge at the bottom of the column. Since there were lots of cells, they produced a large quantity of anti-X antibodies. The few cells with protein receptors for type Y antigen bound with the antigen in column 2; none emerged at the bottom of the column.

You should be able to use similar arguments to explain the results from columns 3 and 4. We can then evaluate the experiment. This means we ask whether the experiment was a good test of the clonal selection hypothesis. If the results had not been consistent with this prediction, they would have cast serious doubt on the hypothesis. By using two rats injected with different antigens of the same bacterium, and by using columns with only type X antigen or only type Y antigen, the scientists had built a control into their experiment. Without further details, we must assume that the scientists made sure that the conditions under which the rats were reared were kept constant. You could criticise the scientists for using only two rats, since this was a small sample size. In fact, they used large groups of rats. This account has been simplified to help us understand what was done. Therefore, we can conclude that the experiment was a valid test of the clonal selection hypothesis.

Vaccines

Most of us are lucky not to have suffered potentially lethal diseases in order to become immune to them. You were probably given vaccinations, and as a result, your body made antibodies and memory cells. A **vaccine** is a preparation of antigen from a pathogen. The vaccine can be injected or, in some cases, swallowed. The vaccine contains antigens from the pathogen, but obviously this must be done in a safe way so that you do not become ill.

Vaccines are made harmless in a number of ways. These include:

- killing the pathogen in a way that leaves its antigens unaffected, e.g. vaccines against cholera and whooping cough
- using bacterial toxins (an antigen) to produce less harmful toxoids, e.g. tetanus, or other parts of a pathogen
- weakening the pathogen in a way that leaves its antigens unaffected, e.g. Sabin oral vaccine against polio (these weakened pathogens are said to be **attenuated**)
- using genetically engineered eukaryotic cells, such as yeast, to produce a microbial protein (an antigen), e.g. hepatitis B.

After the first treatment (primary response), you make antibodies against the antigens; you also make memory cells. After the second treatment, you show the secondary response seen in Figure 6.9 (page 94), making large numbers of B cells and memory cells. These memory cells are then able to react rapidly if a pathogen bearing the same antigen as the vaccine enters the body, killing it before it does harm.

It is very important to remember that vaccination can only work before an infection takes place. It cannot be used to treat a person who already has a disease.

Herd immunity

Herd immunity occurs when the vaccination of a significant portion of a population provides some protection for individuals who have not developed immunity. When a high percentage of the population is protected through vaccination against a virus or bacterium, this makes it difficult for a disease to spread because there are so few susceptible people left to infect and fewer people to do the infecting. This means that the spread of disease in a population can be effectively stopped. It is particularly important for protecting people who cannot be vaccinated. These include children who are too young to be vaccinated, people with immune system problems and those who are too ill to receive vaccines (such as some cancer patients).

The proportion of the population which must be vaccinated in order to achieve herd immunity varies for each disease. However, the basic principle is that when enough people are protected, they help to protect vulnerable members in the population by reducing the spread of the disease.

However, when vaccination rates fall the herd immunity can break down, leading to an increase in the number of new cases. There was a measles epidemic in Swansea over the winter of 2012–2013. Most of the people affected were those who had not been vaccinated as babies because of the MMR scare (for more details on MMR, see pages 104–105 for the activity To vaccinate or not? A parent's dilemma). The vaccination rate in the Swansea area had fallen to below 70% of the population. As a result of the epidemic, clinics were set up in hospitals for people to receive the MMR vaccine. During the epidemic more than 1200 people became ill, 88 visited hospital and one person died.

Herd immunity can lead to the eradication of some diseases. The last case of community-acquired smallpox occurred in 1977. Scientists hope to eliminate some other diseases by vaccination, such as polio. However, this is proving more difficult.

Viruses

The structure of a virus

Viruses are about 50 times smaller than bacteria, and they are acellular (they do not have a cellular structure like living organisms do). They do not show any of the features of living things and only replicate when they are inside a living cell. They have a core containing genetic material, which can be either DNA or RNA but not both (see Figure 6.11). Around this there is a protein coat called a capsid. The capsid is made up of protein units called capsomeres. Some viruses have an envelope of lipid and proteins that surrounds the capsid.

HIV: human immunodeficiency virus

The human immunodeficiency virus (HIV) is an example of a virus that is quite complex. You can see its structure in Figure 6.12. It is spherical in shape, and has an envelope made of lipids and glycoproteins. Inside this, there is a cone-shaped capsid containing RNA. It also contains an enzyme called reverse transcriptase.

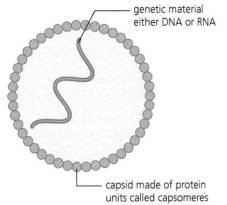

Figure 6.11 The structure of a virus.

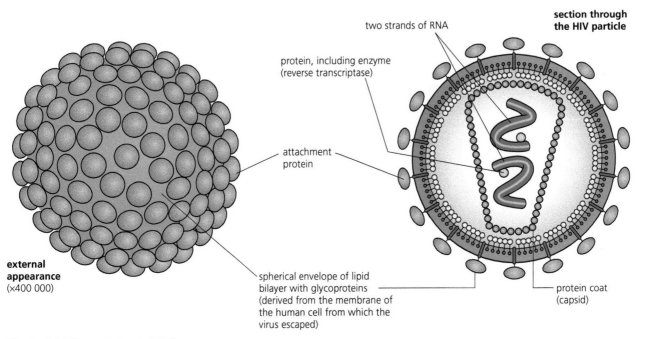

Figure 6.12 The structure of HIV.

How HIV is spread

HIV spreads from an infected person to another person when body fluids mix. The main ways that this can happen is during sexual intercourse, when a blood transfusion from an infected person is given to an uninfected person, when an intravenous drug abuser shares a needle with a person infected with HIV or from an infected mother to her unborn baby across the placenta.

How HIV causes disease

HIV enters the bloodstream. It infects a type of helper T (T_H) cell. The viral RNA enters the cell, and the viral reverse transcriptase enzyme makes a DNA copy of the viral RNA. This DNA copy is inserted into the chromosome of the helper T cell. Every time the helper T cell divides it copies the viral DNA as well, but during this time the cell remains normal. The person is said to be HIV-positive during this stage, because they are infected with the virus and have antibodies against it in their blood.

At some point, which may be many years later, the virus DNA becomes active. It 'takes over' the cell and causes many more HIV particles to be made. This causes the helper T cell to die, releasing thousands of new HIV particles, which infect new helper T cells. Gradually the virus destroys helper T cells. You can see this in Figure 6.13.

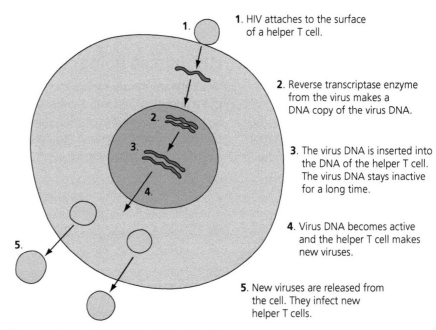

1. HIV attaches to the surface of a helper T cell.

2. Reverse transcriptase enzyme from the virus makes a DNA copy of the virus DNA.

3. The virus DNA is inserted into the DNA of the helper T cell. The virus DNA stays inactive for a long time.

4. Virus DNA becomes active and the helper T cell makes new viruses.

5. New viruses are released from the cell. They infect new helper T cells.

Figure 6.13 How HIV infects helper T cells.

You already know the importance of helper T cells in the way the immune system works (see page 93). With a shortage of helper T cells, B cells are not activated and no antibodies are produced. This stage is called full-blown AIDS, which stands for acquired immunodeficiency syndrome. The person infected with HIV starts to suffer from diseases that would probably not cause problems in a healthy person. These are called opportunistic diseases. Examples of these are tuberculosis and Kaposi's sarcoma (a form of skin cancer that is otherwise very rare). As a result of these diseases, the person dies. There is currently no cure for AIDS, but there are now drugs that can slow down the spread of the virus within the body. With good medical care, a person with HIV can now have a life expectancy that is the same as for a non-infected person.

Antibiotics are not effective against viruses, because viruses are not living. Antibiotics are usually used against bacteria, in which they interfere with metabolism, stop cell wall synthesis, stop proteins being made at the ribosomes, and so on. Because viruses aren't cells, they have neither a cell structure nor their own metabolism, and thus antibiotics cannot affect them.

EXAMPLE

HIV and T cells

The human immunodeficiency virus (HIV) infects and destroys only one type of T cell, the T_H cells.

1 HIV can infect only T_H cells. Explain why.
This is because only T_H cells have receptor proteins in their cell surface membranes that the proteins on the surface of HIV can fit into.

People infected by this virus are HIV-positive. Without treatment, they will develop acquired immunodeficiency syndrome (AIDS). Without their T_H cells, AIDS sufferers lose their ability to overcome infections. As a result, people with AIDS often die from diseases, such as tuberculosis, from which other humans could recover.

2 Without their T_H cells, AIDS sufferers lose their ability to overcome infections. Explain why.
T_H cells are needed to stimulate antibody production by B cells. Without T_H cells, the person's ability to produce antibodies against infections is seriously reduced.

In 2013, the World Health Organisation (WHO) estimated that, worldwide, about 34 million people are either HIV-positive or suffering from AIDS. About 69% of these people live in sub-Saharan Africa.

Some people are at greater risk of HIV infection than others because of their lifestyles. People in 'high-risk' groups include intravenous drug users who share needles and people who have unprotected sex with many partners. Unprotected sex means sexual intercourse without using a condom, which would prevent the virus passing from partner to partner. However, scientists have found to their surprise that in many 'high-risk' groups there are a few people who seem to be resistant to HIV infection.

3 Drug users who share needles are at increased risk of becoming HIV-positive. Explain why.
Used needles will have traces of blood in them. If the person who used the needle is infected with HIV, the virus will be present in this blood. The viruses will then be injected into the blood of the next person who uses the needle.

Medical investigators studied a group of prostitutes at a special clinic in Nairobi, the capital of Kenya. The group included an unusually large number of HIV-resistant women. The investigators tested the blood of all the prostitutes for the presence of HIV.

4 Suggest why the investigators chose to work
a) with prostitutes
b) in sub-Saharan Africa.
The investigators would have chosen to work with prostitutes as they are at high risk of HIV infection because they have sex with many different people. Working in sub-Saharan Africa means that this is an area where many people are infected with HIV.

┌───
TEST YOURSELF

10 Suggest reasons why smallpox could be eliminated by vaccination, but other diseases such as HIV are more difficult.

11 Explain how a person with low numbers of helper T cells is more likely to suffer from infections.

12 Explain how the proteins on the envelope surrounding HIV enables it to infect helper T cells.

13 Suggest ways in which the spread of HIV can be prevented.

14 One kind of anti-HIV drug is similar in shape to a DNA nucleotide. However, it does not contain adenine, thymine, cytosine or guanine. When the viral DNA replicates, the drug molecules are incorporated into the DNA strand. Suggest how this stops the virus replicating.

Using antibodies

Each clone of plasma cells produces only one type of antibody, i.e. **monoclonal antibodies**. Because plasma cells can be cultured in laboratories, scientists can harvest large quantities of monoclonal antibodies. Monoclonal antibodies have a number of uses in the diagnosis and treatment of disease. For example, monoclonal antibodies against:

- a hormone, such as oestrogen, can be used to diagnose hormone deficiency
- an antigen associated with cancer, such as the prostate-specific antigen, can be used to screen for cancer (in this case prostate cancer).

Figure 6.14 shows another use of monoclonal antibodies, known as a 'magic bullet'. A drug has been attached to the molecule of antibody. Since the antibody will attach only to cells with the specific antigen on their surface, the drug will be carried directly and solely to those cells on which it is designed to work.

An example of a 'magic bullet' is a drug called T-DM1, currently undergoing trials for treatment of breast cancer. It contains the monoclonal antibody trastuzumab (Herceptin), which is already approved for treating breast cancer. This binds to receptors on the surface of some kinds of breast cancer cells. Attached to this is a toxic chemical called emtansine, which would harm too many healthy cells if it was not attached to a monoclonal antibody. There is also a third component, a chemical that keeps emtansine inactive until the monoclonal antibody has bound to a cancer cell.

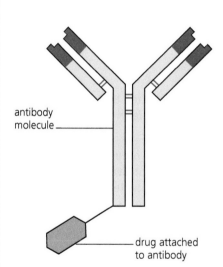

antibody molecule

drug attached to antibody

Figure 6.14 'Magic bullets' are monoclonal antibodies each with a molecule of a drug attached to them. These antibodies attach specifically to cells carrying the complementary antigen on their surface. In this way, the drug gets directly to the cells where it is needed.

ELISA tests

Monoclonal antibodies are also useful in diagnosis. They can be used in test kits to diagnose diseases or conditions, and are very quick and reliable.

Prostate cancer is a cancer of the prostate gland, so it only occurs in men. One way to test for prostate cancer is to test the blood serum for prostate-specific antigen (PSA). If PSA concentration is abnormally high, the patient could have cancer, so further tests are carried out. The test is shown in Figure 6.15.

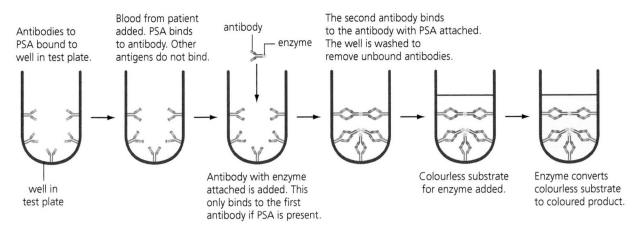

Antibodies to PSA bound to well in test plate.

well in test plate

Blood from patient added. PSA binds to antibody. Other antigens do not bind.

antibody — enzyme

Antibody with enzyme attached is added. This only binds to the first antibody if PSA is present.

The second antibody binds to the antibody with PSA attached. The well is washed to remove unbound antibodies.

Colourless substrate for enzyme added.

Enzyme converts colourless substrate to coloured product.

Figure 6.15 Using monoclonal antibodies to test for prostate cancer.

TEST YOURSELF

15 Why is the well washed after stage 4 in Figure 6.15?

16 Why is the second antibody, with the enzyme attached, needed in the test kit?

17 Explain the result you would get if the patient's blood did not contain PSA.

18 Design a test kit you could use to test a patient's blood for antibodies against HIV.

Monoclonal antibodies can be used in pregnancy testing kits. As soon as the embryo implants in the lining of the uterus, a placenta starts to form. The placenta secretes a hormone called human chorionic gonadotrophin (hCG). This hormone is excreted in the woman's urine. Because the hormone is produced by the placenta, it is only found in the urine of women who are pregnant. There are several different tests available. One kind is a dipstick that is dipped into a sample of the woman's urine. If hCG is present, two blue lines will appear. The required monoclonal antibodies are all present on the dipstick. You can see this in Figure 6.16.

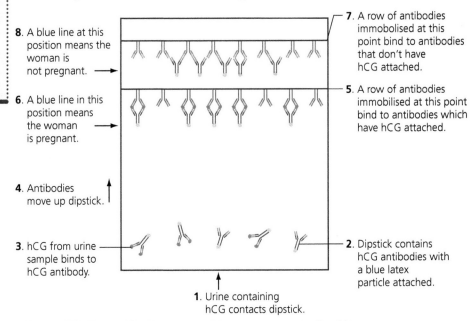

8. A blue line at this position means the woman is not pregnant. →

6. A blue line in this position means the woman is pregnant. →

4. Antibodies move up dipstick.

3. hCG from urine sample binds to hCG antibody.

1. Urine containing hCG contacts dipstick.

7. A row of antibodies immobolised at this point bind to antibodies that don't have hCG attached.

5. A row of antibodies immobilised at this point bind to antibodies which have hCG attached.

2. Dipstick contains hCG antibodies with a blue latex particle attached.

Figure 6.16 How antibodies are used in a pregnancy testing kit.

TEST YOURSELF

19 Explain why a pregnancy testing kit detects hCG and not any other substances present in the urine.

20 Explain why the antibodies have blue latex particles attached to them.

21 Explain why there are two different rows of antibodies immobilised on the dipstick.

22 Explain how the antibodies immobilised in the two different rows are different.

Active and passive immunity

Active immunity is the process that you have learned about in this chapter. It is when the immune system responds to an antigen by producing specific antibodies against the antigen, and memory B cells in a primary immune response to the antigen. This is what happens when a person is exposed to an infectious disease, or when they are given a vaccination.

Passive immunity is when a person is given antibodies. One example of this is when babies are breast-fed. The baby receives antibodies that their mother has made, which protects them against any infection that the mother has been exposed to. However, because the baby has not had an immune response itself, memory cells are not made so the baby has **no lasting immunity**. Another example of passive immunity is when a person is given antibodies to protect them against the effects of the venom (which is an antigen) when they have been bitten by a venomous snake.

ACTIVITY

To vaccinate or not? A parent's dilemma

Parents must make many decisions about the health and safety of their child. Vaccination is one of these decisions. You might think that the decision is obvious: if you have your child vaccinated, you protect it against a potentially lethal disease. However, all vaccinations carry a small risk. Suppose that you believe this risk is too high; would you have your child vaccinated?

Parents recently faced exactly this dilemma. In the 1980s, a new triple vaccine, MMR, was introduced. The MMR vaccine gives protection against three diseases:

- measles, which can lead to severe illness, convulsions, lifelong disability and death
- mumps, which can cause meningitis, permanent deafness, and sterility in males
- rubella, which during pregnancy can affect the fetus by causing deafness, blindness, heart defects and other difficulties.

The triple vaccine is thought to be a better protection than three separate vaccines because it reduces the time over which babies are exposed to rubella, measles and mumps. The MMR vaccination is made in two doses. The first dose is given at 12–15 months. The second dose is given when the child starts school.

1 The first dose of MMR coincides with the time when many breast-fed babies are weaned. What is the advantage of this timing?
2 Rubella can be passed from an infected mother to her unborn fetus. It is considered important to vaccinate boys as well as girls against rubella. Explain why.
3 Look at Table 6.1. What can you conclude from this about the safety of the MMR vaccine?

Table 6.1 The proportion of children affected as a result of getting measles or after their first dose of a vaccine offering protection against measles.

Condition	Proportion of children affected as a result of	
	...getting measles	...their first dose of MMR vaccine
Convulsions	1 in 200	1 in 1000
Brain disease (meningitis or encephalitis)	Between 1 in 200 and 1 in 5000	Less than 1 in a million
Death	Between 1 in 2500 and 1 in 5000, depending on age	0

In February 1988, Dr Wakefield, a British doctor, published a research report suggesting that MMR might cause autism, a behavioural disorder. Dr Wakefield proposed that, in some children, MMR vaccination causes inflammation of the intestine, which causes toxins to leak into the blood. These toxins then pass into the brain, producing the damage that causes autism.

4 Dr Wakefield carried out his initial research on 12 children. How reliable were his findings?

In April 2000, Dr Wakefield and Professor John O'Leary, director of pathology at a Dublin hospital, presented further research findings to the United States Congress. They reported that tests on 25 children with autism revealed that 24 of these children had traces of the measles virus in their guts. Professor O'Leary said this was now 'compelling evidence' of a link between autism and MMR.

5 Does the evidence of Dr Wakefield and Professor O'Leary show that MMR causes autism? Explain your answer.

Many other scientists performed investigations to check these research findings. None could find any evidence to support Dr Wakefield's proposal that MMR caused autism. Despite this, public confidence in the MMR vaccine fell dramatically in the UK. Many parents prevented their children from receiving the MMR vaccine. Some parents of autistic children began to sue the pharmaceutical companies that had produced the MMR vaccine.

6 Dr Wakefield acted as a consultant for some of the parents who were suing the pharmaceutical companies. This led to criticism from his scientific peers. Explain why.

7 Figure 6.17 shows the results of research in California. Do these data support the theory that autism is linked to the use of MMR vaccine? Explain your answer.

8 If everyone was vaccinated against MMR before their second birthday, there would be a correlation between the incidence of autism and being vaccinated against MMR. Explain why.

There is now an overwhelming body of evidence to suggest there is no link between MMR vaccinations and autism. Despite this, the proportion of children in the UK receiving an MMR vaccination by their second birthday fell from 91% in 1997–1998 to 81% in 2004–2005. Would you have had your baby vaccinated with MMR? We must each use the available scientific evidence intelligently to inform our decisions.

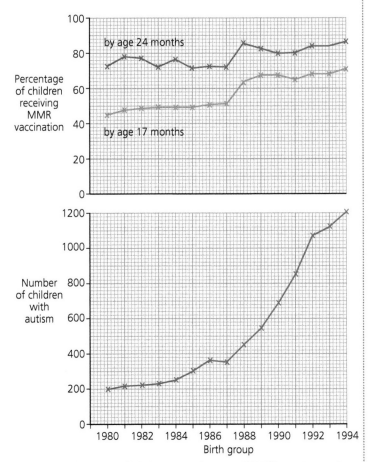

Figure 6.17 Is there a link between the use of MMR vaccine and autism? The upper graph shows the percentage of 2-year-old children who received MMR vaccinations between 1980 and 1994. The lower graph shows the number of reported cases of autism among the children born in these years. The data are from a study in California, USA.

105

Practice questions

1 a) The MMR vaccine contains **antigens**. What is an antigen? (2)

A child was given the MMR vaccine and was given a second dose of the vaccine as a booster later.

b) i) It took more than a week for antibodies to appear in the child's blood after the first vaccination. Explain why. (2)

ii) The concentration of antibodies increased immediately after the second vaccination. Explain why. (2)

2 The diagram shows a human immunodeficiency virus (HIV).

a) i) Name structure P and structure Q. (2)

ii) What is the function of the RNA molecules in this virus? (1)

b) Describe how new viruses are produced after HIV has infected a T cell. (3)

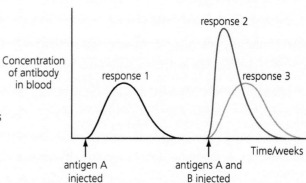

3 A group of doctors carried out an investigation into the immune system. They injected some volunteers were injected with an antigen, A. The concentration of antibodies against antigen A in their blood was monitored for several weeks. Six weeks later, the volunteers were given a second injection containing antigen A and a different antigen, antigen B. The concentration of antibodies against both antigens in the volunteers' blood was monitored for several more weeks. The graph shows their results.

a) i) Complete the table with ticks to indicate whether responses 1, 2 and 3 show a primary immune response or a secondary immune response. (3)

	Primary response	**Secondary response**
Response 1		
Response 2		
Response 3		

ii) Describe and explain the difference between response 1 and response 2. (3)

b) i) Two different responses were shown after the second injection. Explain why. (3)

ii) Suggest a reason for injecting antigen A at the same time as antigen B in the second injection. (1)

4 Ibritumomab is a monoclonal antibody that has a radioactive substance attached. It is used to treat a type of cancer called non-Hodgkin lymphoma.

   a)   **i)** The antibody binds to cancer cells but not other cells. Suggest how. (2)

      ii) Explain the advantage of attaching the radioactive substance to an antibody, rather than injecting it directly. (2)

      iii) The monoclonal antibody is given to the patient through a drip into a vein. Suggest why it is not given as a drug to be swallowed. (2)

   b) A few patients who are given ibritumomab have an immune response to the drug. Explain why. (3)

5 Read the following passage and then answer the questions that relate to it.

When it bites, a black widow spider injects venom, a toxin, into its victim. Although rarely fatal in humans, this venom causes a lot of pain that might last for several days.

There is a cure: people can be injected with antivenin, which destroys the venom from the black widow spider. The antivenin is produced by injecting the venom from black widow spiders into horses, which then produce antibodies against the venom. These antibodies are purified and stored for later use. Only people who seem near to death are normally given the antivenin, though. This is because some patients have a life-threatening reaction to the antivenin.

A new form of the antivenin, called Analatro®, has been produced and is undergoing clinical trials in the USA. Analatro is produced by injecting the venom from black widow spiders into sheep and collecting the antibodies they make. Analatro causes fewer reactions in people than the antivenin.

   a) Suggest the procedure the scientists would have used to ensure each horse produced enough antibodies to be used clinically. (2)

   b) The injected horses produce antibodies against the toxin of the black widow spider. Explain how. (4)

   c) Injecting people who had been bitten by black widow spiders with the horse's antibodies gave them immediate relief from the painful symptoms of the bite. Explain why. (2)

   d) Some people reacted badly to the injection of horse antibodies but not sheep antibodies. Suggest why. (3)

Stretch and challenge

6 Describe an autoimmune disease and explain how the immune system has gone wrong.

7 Explain what happens in the case of an allergic reaction and how people are tested for allergies.

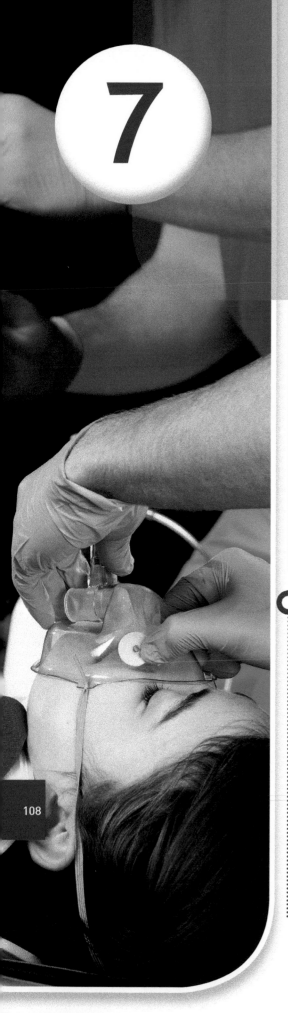

7 Gas exchange

TEST YOURSELF ON PRIOR KNOWLEDGE

1 Where does aerobic respiration take place in living organisms?
2 Name three biochemical processes in living organisms that require ATP from respiration.
3 Complete this word equation for respiration:

   Glucose + ? → ? + water + energy

4 At high altitudes there is much less oxygen in the air than at sea level. When climbing high mountains (such as Mount Everest) climbers get tired and breathless and are unable to move quickly. Explain why.
5 Describe how carbon dioxide produced in the leg muscles is removed from the body.
6 If carbon dioxide is not removed from the body the blood becomes more acidic. This can be fatal. In some enclosed spaces the concentration of carbon dioxide can be quite high. For example, in grain stores the seeds release carbon dioxide, and this can be dangerous for farmers. Use your knowledge of diffusion to explain why.

Introduction

In an emergency the first things to check are that an accident victim is breathing and that their heart is beating. These two factors are closely linked because breathing is how we obtain oxygen for respiration and the circulation of the blood distributes the oxygen around the body. Without a continuous supply of oxygen the biochemical processes that keep us alive will cease. Within a few minutes the brain will be damaged and the victim will die. Fortunately, both breathing and blood circulation are fully automated processes and we do not have to think about either. However, many of us neglect to ensure that our lungs and heart are maintained in tiptop condition and protected from damage.

The need for gas exchange

Respiration A biochemical process by which ATP is produced, using a fuel such as glucose.

The primary function of gas exchange in animals is to supply oxygen for respiration. Respiration is the biochemical process by which ATP is produced, using glucose as a fuel. Respiration can also use other fuels, such as fats. The rate at which cells require energy varies. Active muscle cells need large supplies of energy for contraction. Cells that make digestive enzymes or that are growing rapidly use energy for the synthesis of complex molecules. The electrochemical activities of our brain cells need a constant supply of energy from respiration, and brain cells are damaged if deprived of oxygen for more than a few minutes.

Just as vital as obtaining oxygen is the removal of carbon dioxide, the waste product of respiration. Carbon dioxide produces an acid solution. As carbon dioxide accumulates the pH of the cells and blood is lowered. The pH of the blood plasma and intercellular fluid is normally maintained very close to pH 7.4. Any variation from this upsets the ionic balance, and interferes with enzyme functioning, and can rapidly lead to unconsciousness and death. To avoid this, a build-up of carbon dioxide very quickly stimulates an increase ventilation.

Gas exchange is the process by which organisms take up oxygen for respiration, and excrete the carbon dioxide produced in respiration. In this chapter we look at the structure of the gas exchange systems in different organisms and how they function. We will look at how gas exchange is affected by an organism's size and the environment in which it lives.

Does size matter? Is big better?

A large animal is severely limited by the amount of food it can find to maintain its bulky body. This restricts the number of really large animals that can live in a particular habitat. Other problems include getting enough oxygen for respiration into the body and transporting it to where it is needed, as well as digesting and absorbing food, distributing it to all of the body cells and transporting waste to exchange surfaces, such as the kidney. Larger organisms have a greater variety of specialised cells, tissues, organs and systems, but this does not make them better than smaller organisms.

Tiny organisms are often considered simple and primitive compared with large ones. Yet it is precisely because they are simple and small that they have survived. The fact that there are now far more of them than there are large organisms shows how very successful tiny organisms have been. Each species is adapted to the particular set of conditions in which it exists, and there are many more sets of conditions suitable for small organisms than there are for large ones. Large organisms may seem to us to be dominant, more complex and more advanced, but increasing size brings a variety of problems.

Life as a single cell

Figure 7.1 shows a single-celled organism called *Chlamydomonas*. It lives in fresh-water ponds and ditches. It is roughly spherical and about 20 μm across. It has two flagella, which enable it to swim around. It contains a chloroplast, so it can photosynthesise. Oxygen for respiration and carbon dioxide for photosynthesis are dissolved in the surrounding water and diffuse through the cell wall and the cell-surface membrane. The short diffusion pathway means that *Chlamydomonas* can rely on its cell surface for gas exchange. The maximum distance that oxygen has to diffuse to reach the centre of the cell is about 10 μm, and this takes no more than about a tenth of a second (100 milliseconds).

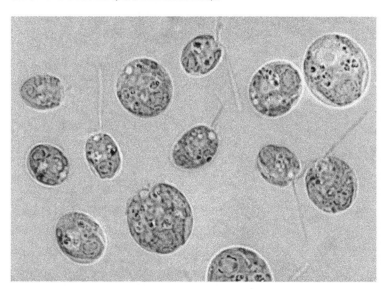

Figure 7.1 *Chlamydomonas* is a single-celled organism that lives in water.

However, if the distance is doubled, the diffusion time is squared, so to diffuse 20 μm would take about 400 milliseconds, and to go just 1 mm would take 100 seconds. These figures are approximate because the actual rate depends on several factors, such as the concentration gradient and the material through which the oxygen is diffusing. However, it illustrates that, while diffusion is sufficiently fast to provide all parts of a small single-celled organism with enough oxygen, it is far too slow to provide enough oxygen to all parts of a larger, multicelled animal.

As the size of an organism increases, the time taken for oxygen to travel by diffusion from the cell-surface membrane to the tissues would become too great. Also, the surface area available for diffusion becomes less and less in proportion to the volume. Imagine a cube-shaped animal in which the length of each side is 1 cm (Figure 7.2). Its volume is 1 cm³, but since it has six faces each with an area of 1 cm², its surface area is 6 cm². The surface area to volume ratio is therefore 6 : 1. If you do the same calculation for 3 cm cube, the ratio is only 2 : 1. If you continue to do the calculations for larger and larger cubes, you will discover that the ratio gets smaller and smaller, so that if you were able to convert a human into a cube shape, the ratio would be less than 1 : 100. Animals are not cubic, of course, but the general rule still applies: the larger the animal the lower the surface area to volume ratio.

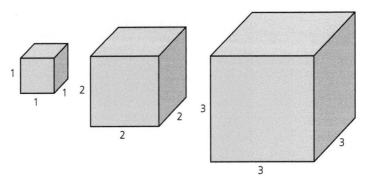

Volume/cm³	1	8	27
Surface area/cm³	6	24	54
Surface area : volume ratio	6:1	3:1	2:1

Figure 7.2 This diagram shows that, as the volume of a cube increases, its surface area to volume ratio decreases.

TIP
See Chapter 14 to find out more about ratios.

Because it is essential to obtain oxygen and expel carbon dioxide, most animals have adaptations that increase the surface area available for gas exchange.

How do insects breathe?

Insects may seem to be small organisms that should have little difficulty in getting enough oxygen for their needs. Many, however, are very active fliers or jumpers and have a high rate of metabolism, as anyone who has chased a buzzing fly round a room will know. A problem for insects is that they have an exoskeleton, which is fairly rigid and coated with a waxy substance. This makes it waterproof, an adaptation of a small creature living in air, to prevent it from drying out. But a waterproof surface is also difficult for gases to diffuse through. One of the reasons insects have been such an evolutionary success is that they evolved a breathing system of tubes that carry oxygen directly to all tissues and organs of their bodies.

Spiracle An opening in the exoskeleton of an insect that connects to the tracheal system.

Air can enter these tubes through a series of openings called spiracles arranged along the side of the body. One downside of these is that valuable water can escape. To help prevent this, the spiracles can be opened and closed using tiny valves. Some insects also have tiny hairs around the spiracles – another adaptation that reduces water loss. In addition, some insects have muscles that control ventilation of the tracheal system. This further reduces water loss.

Tracheae Tubes in the insect respiratory system that carry air.

As you can see from Figures 7.3 and 7.4 (overleaf), air can pass through the spiracles into a system of tracheae and narrower **tracheoles,** so that no cell is more than a short diffusion distance from a tracheole. The tracheae have rigid rings in their walls, similar to the rings of cartilage in the trachea and bronchi of humans, to keep air passages open. The tracheoles penetrate between cells and right into muscle fibres. It is here that gaseous exchange takes place. There are many of these tracheoles, giving a large collective surface area.

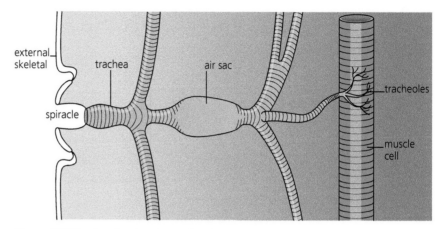

Figure 7.3 The tracheal system of an insect.

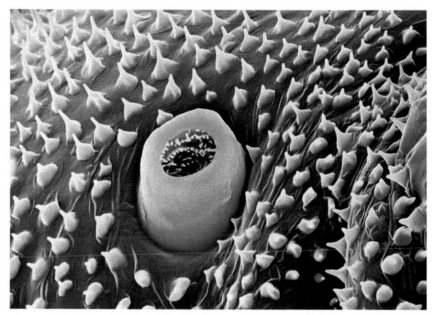

Figure 7.4 The scanning electron micrograph shows a single spiracle in the wall of the caterpillar of the tiger moth. Along the side of the caterpillar's body is a series of spiracles through which air enters the tracheae.

In some of the tiniest insects, this system can provide enough oxygen simply by diffusion. But larger insects, such as houseflies and grasshoppers, take in oxygen more rapidly when active. This is achieved by the spiracles closing and muscles pulling the skeletal plates of the abdominal segments together. This squeezes the tracheal system and pumps the air in the sacs (see Figure 7.3) deeper into the tracheoles. The recoil also lowers the pressure inside the tracheal system so that it is lower than the atmospheric pressure outside. This results in mass flow of air into the insect. If you watch a fly or wasp closely, you can often see these pumping movements that ventilate the tracheal system.

Gas exchange between the tracheoles and tissues is illustrated in Figure 7.5.

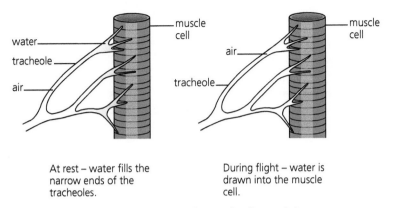

At rest – water fills the narrow ends of the tracheoles.

During flight – water is drawn into the muscle cell.

Figure 7.5 Gas exchange between the tracheoles and tissues.

One extra adaptation helps to get additional oxygen deep into the muscles during flight. When an insect is resting, a little water leaks across the cell membranes of muscle cells. The very narrow ends of the tracheoles fill with water. When the wing muscles are working hard, they respire, partly anaerobically, and produce lactate, a soluble waste product of anaerobic respiration. This lowers the water potential of the muscle cells. As lactate builds up in the muscle cells, water passes by osmosis from the tracheoles into the muscle cells. This draws air in the tracheoles closer to the muscle cells and therefore reduces the diffusion distance for oxygen when it is most needed. This also speeds up diffusion, as diffusion is faster in gases than in liquids.

TEST YOURSELF

1 a) Complete the table to show the surface area and volume of cubes of different sizes.

Length of side of cube/cm	Surface area of cube/cm²	Volume of cube/cm³	Surface area : volume ratio
1			
2			
3			

b) How does surface area to volume ratio change as the size of the cube increases?

2 Explain why an insect would be at risk of drying out if its exoskeleton was not covered with a waterproof substance.

3 Suggest the function of the rigid rings in the walls of the tracheae.

4 Suggest how the breathing system of insects helps to minimise water loss.

5 Explain how the increase in the lactate content of the muscle cells during flight causes the removal of water from the ends of the tracheoles.

How do fish get oxygen out of water?

Oxygen does not dissolve readily in water, and, as water warms up, even less can dissolve. This is bad news for fish living in lakes and rivers that are likely to get warmer as climates change. Fish are adapted to extract oxygen directly from water, unlike marine mammals such as whales and seals that have to come to the surface to take gulps of air. Not surprisingly, fish gills have a large surface area relative to the volume of the fish in contact with water through which to absorb oxygen, and this is provided by gills.

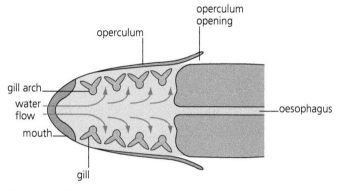

Figure 7.6 The arrangement of the gills of a fish.

Bony fish, such as trout, perch and cod, have a series of gills on each side of the head, as shown in Figure 7.6. Each bony gill arch has two stacks of thin plates called **filaments** that stick out like leaves. On the top of each filament is a row of very thin **lamellae**, which stand out, as you can see in Figure 7.7. The surface of each lamella is a single layer of flattened cells. This covers an extensive network of capillaries so close to the surface that oxygen has only a short distance to diffuse from the water into the blood. Since fish live in water, there is no problem with such thin surfaces drying out.

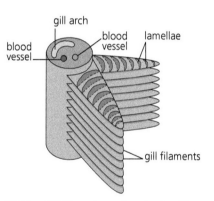

Figure 7.7 Structure of a single gill.

The **operculum** on the side of a fish's head protects the gills from damage. Notice that the operculum opens at the rear edge. The fish takes in water through its mouth and forces it through the gills to maintain a current of water over them. The water then flows out from the back of the operculum, whether the fish is swimming slowly or rapidly.

The blood system in the lamellae is arranged so that the water flows in the opposite direction to the blood flow in the capillaries. This is called a **counter-current** system, as shown in Figure 7.8.

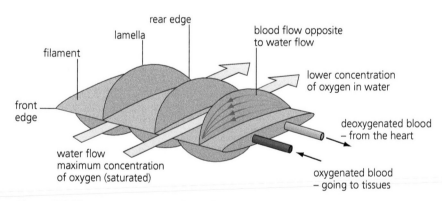

Figure 7.8 The counter-current flow of water over a gill filament.

Generally, the concentration of oxygen in the blood is lower than in the surrounding water, so oxygen diffuses into the blood. The advantage of the counter-current system is that it maintains a diffusion gradient over the full length of the capillary. As Figure 7.8 shows, blood flows from the rear edge of the lamella to the front edge, and surrounding water flows

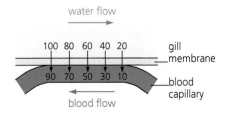

Figure 7.9 The diagram shows that, in the counter-current flow of water over a gill filament, the gradient of oxygen concentration is maintained between water and blood. The figures show the percentage saturation with oxygen.

Figure 7.10 Anglers use holding nets in the water to keep the fish they catch alive.

Figure 7.11 The human breathing system.

in the opposite direction. So the blood at the front will have had the longest time for oxygen diffusion and therefore has the highest oxygen concentration. The front of the lamella is also where the surrounding water is most saturated with oxygen. At the rear edge, the blood has very little oxygen, and though there is now less oxygen in the flowing water, the diffusion gradient is about the same as at the front edge, as Figure 7.9 shows. This system therefore ensures that the concentration gradient is maintained.

> **TEST YOURSELF**
> **6** Make a list of the features of the gas exchange system in fish, under the headings 'large surface area', 'short diffusion distance' and 'large concentration gradient'.
> **7** To keep the fish alive, competition anglers keep the fish they catch in a net in water (Figure 7.10). Their catch is weighed before being safely returned to the river after the competition. Suggest why fish cannot breathe in air, even though there is a much higher percentage of oxygen in air.

Gas exchange in humans

The structure of the lungs

Our lungs almost fill the **chest cavity**, which occupies the upper part of the body enclosed by the ribs. The heart and its major blood vessels tuck in between the lungs and above the diaphragm that separates the chest cavity from the abdomen. The main parts of the human breathing system are shown in Figure 7.11.

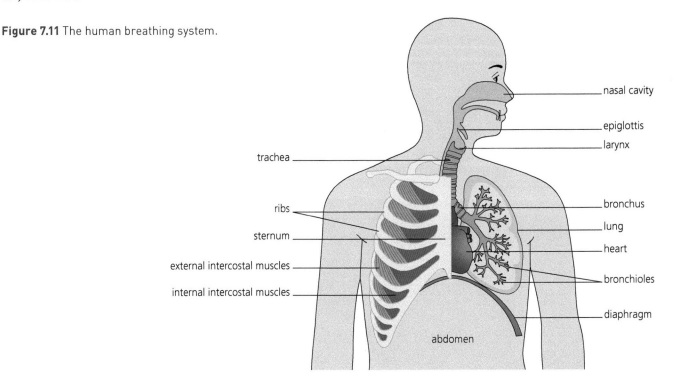

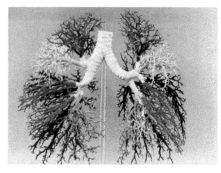

Figure 7.12 The branching network of bronchioles in a lung. The bronchioles have been filled with plastic resin and then the other tissue of the lung has been dissolved away.

(a)

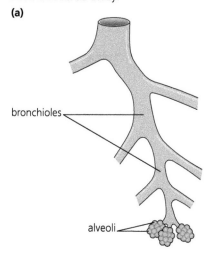

bronchioles

alveoli

Figure 7.13 The diagrams show that the bronchioles have narrow branches, with each branch ending in a cluster of alveoli surrounded by capillaries.

Figure 7.14 An alveolus and blood capillaries. Note that the cells of the alveolus and the capillaries are flattened. This reduces the distance for gas diffusion.

How oxygen gets into the blood

For efficient diffusion, an exchange surface has:

- a large surface area, compared with the volume of the organism
- a short distance for the gas to diffuse
- a large difference in the concentration of gas on opposite sides of the surface.

In human lungs, the large surface area is achieved by having a vast number of very small alveoli. Each lung contains about 350 million alveoli. As shown in Figure 7.12, the airways in the lung, called bronchioles, branch hundreds of times so that the diameter at their ends is tiny. At the end of each branch is a cluster of alveoli. These are rather like bunches of tiny, hollowed-out grapes connected to the network of bronchioles.

(b)

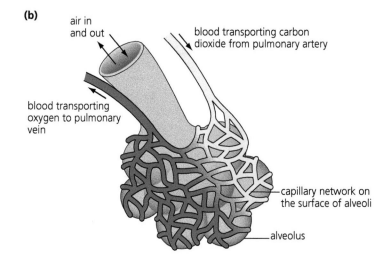

air in and out

blood transporting carbon dioxide from pulmonary artery

blood transporting oxygen to pulmonary vein

capillary network on the surface of alveoli

alveolus

The gas exchange surface

The alveoli are covered with a network of blood capillaries (see Figure 7.13). The walls of the alveoli and of the capillaries form a very thin barrier between the air in the alveoli and the blood (see Figure 7.14). The alveolar wall cells are flattened, with only a thin layer of cytoplasm between their cell-surface membranes. The capillary walls also consist of very thin cells. These cells are curved to form narrow tubes (see cross-section in Figure 7.14). The capillaries are so narrow that the red blood cells, which carry oxygen and carbon dioxide, touch the walls as the blood flows through the capillaries.

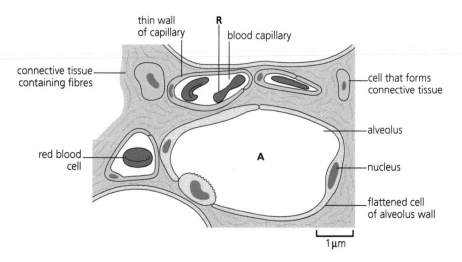

thin wall of capillary

R

blood capillary

connective tissue containing fibres

cell that forms connective tissue

alveolus

red blood cell

nucleus

A

flattened cell of alveolus wall

1 μm

Because of this cell arrangement, the distance between the air in the alveoli and the red blood cells is very short. This minimises the distance that oxygen has to diffuse from air to blood, and carbon dioxide from blood to air.

The inner surface of the alveolus wall is covered in a very thin film of water because the cell-surface membranes of its cells are permeable to water. The rate of diffusion of a gas in water is much lower than its rate of diffusion in air, and the film of water increases the distance that gases have to diffuse. Both these factors slightly lower the rate of diffusion of gases. However, a membrane that is not permeable to water is also not permeable to oxygen. Consequently a moist surface is an unavoidable feature of a gas exchange surface. In infections such as pneumonia, the layer of liquid on the surface of the alveoli gets much thicker. This seriously slows the rate of gas diffusion.

Between the alveoli and the capillaries is a very thin layer of tissue fluid that contains elastic fibres. These fibres help to recoil the lungs, which assists in breathing out. The fibres are made by cells that fit in between the alveoli and form **connective tissue**, as seen in Figure 7.14.

A **concentration gradient** is essential for gas exchange. The gradient is maintained by ventilation in the lungs coupled with the continuous flow of blood. Breathing movements constantly result in a change of air in the alveoli, providing fresh oxygen and removing carbon dioxide. Oxygen diffuses into the red blood cells. As the blood flow rapidly moves the red blood cells on, they are replaced by oxygen-poor cells.

This ensures that the concentration of oxygen in the alveoli is always much higher than the concentration in the blood.

TEST YOURSELF

8 Use the scale line in Figure 7.14 to calculate the minimum distance that oxygen would have to diffuse to pass from the air in the alveolus (A) to the red blood cell (R).

9 Smokers' lungs become lined with tar. Explain how this would affect gas exchange compared with a non-smoker.

10 The capillaries surrounding the alveoli are so narrow that red blood cells pass through one by one, and the cells are pressed against the sides of the capillaries as they pass through. Explain how this helps efficient gas exchange.

The structure of the breathing system

If you could take out your lungs and carefully spread out all the alveoli, they would cover most of the floor area of a typical school laboratory. That is about $70\,m^2$. The total surface area of the skin of an adult is slightly less than $2\,m^2$. The lungs therefore have an area that is roughly 35 times the area of the surface of the body. This shows how large a surface area we need for gas exchange, and it makes up for the low surface area to volume ratio of human bodies. It would obviously be impractical to have such a huge area outside our bodies, billowing out like a massive sail, or as an array of flaps sticking out from the side like external gills. Having the gas exchange surface folded away in the chest has the advantage of protecting the thin surface membrane from damage. It also reduces loss of water, as the moist air in the alveoli does not directly meet the much less moist air outside the body. As the water potential gradient is reduced, evaporation of water is also reduced.

Since the lungs are tucked away inside the chest, it is essential that a good supply of oxygen reaches the gas exchange surface. Breathing movements constantly replace the air in the alveoli by ventilation (mass transport of gases, as opposed to simple diffusion). Without active ventilation, air could not reach our lungs at a rate even close to that needed to supply our oxygen needs. Our energy demands change, for example, when we start to run. So we must also vary the rate at which we ventilate our alveoli.

Air enters our bodies through the nose or mouth and then the:

- trachea (windpipe)
- bronchi
- bronchioles
- alveoli.

Trachea

The **trachea** is a wide tube. Air passes from the throat, through the trachea, down the neck and into the chest. Food must not go down with the air, so when we swallow a flap of cartilage called the **epiglottis** closes over the entrance to the trachea. Normally, this is precisely coordinated by a reflex action. But occasionally some food may go down the wrong way. This stimulates us to cough, which usually expels the food from the trachea and stops us from choking.

While you read, you probably lean forward and bend your neck. You do not choke since the trachea does not kink and close up as a piece of soft rubber tubing would. This is because its structure is more like a shower hose (see Figure 7.15).

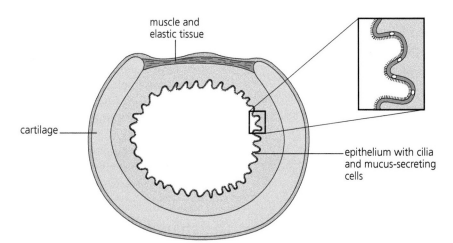

muscle and
elastic tissue

cartilage

epithelium with cilia
and mucus-secreting
cells

Figure 7.15 A section across the trachea. The stiff cartilage keeps the airway open. The muscle and elastic tissue allows flexibility. Cells of the epithelium that line the airway secrete mucus, which traps bacteria and dust particles. Hair-like cilia beat upwards and sweep the mucus to the throat.

The top section of the trachea is adapted to form the **larynx** (the voice box). To produce the sounds that make up our voice, two things happen. We use precise muscular actions to adjust the position of two folds of tissue (the vocal cords) inside the larynx, and at the same time we expel air from the lungs.

Bronchi, bronchioles and alveoli

The trachea branches into two **bronchi**, one to each lung, and these have many branches (look back at Figure 7.11). The smaller branches are called **bronchioles**, themselves repeatedly branched. The smallest branches end in the clusters of alveoli (Figure 7.13). The bronchi and the larger bronchioles also have cartilage in their walls, but here the cartilage is in small sections connected by muscle and elastic fibres. The smaller bronchioles have only muscle and elastic fibres, so that these tubes can both expand and contract easily during ventilation.

Interpreting a photomicrograph of the lung

Figure 7.16 shows lung tissue without magnification. It appears to be quite solid and not obviously full of air. However, you can see some of the bronchioles that transport air in the lungs.

The photomicrograph in Figure 7.17 shows a very thin slice of lung tissue as seen through a microscope. At this magnification, the lung looks rather like a sponge. The sponginess is due to tiny cavities called **alveoli**. When we breathe in, the alveoli fill with air and become roughly spherical. It is in these balloon-shaped cavities that gas exchange occurs.

To obtain a slice for a microscope slide, some lung tissue is first embedded in a waxy substance that makes the tissue rigid. Then very thin slices are cut with the blade of a machine like a small bacon slicer. The alveoli appear to have irregular shapes because they are not inflated as they would be in a living lung after inhalation. The outline of their walls is wavy, not smooth and rounded. As the cells are almost transparent, the section is stained and this shows the nuclei as dark dots.

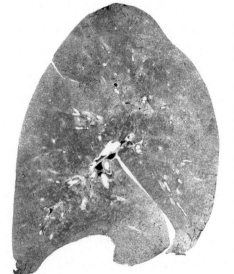

Figure 7.16 A section of part of a lung.

Figure 7.17 A photomicrograph of a small section of lung tissue. Since it is magnified, the spongy texture of the lung is clear.

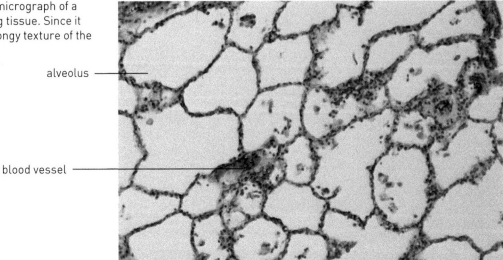

alveolus

blood vessel

100μm

119

Analysing a photomicrograph

Remember when interpreting a micrograph of biological material that what you see through a microscope is only a very thin slice of tissue. Cutting a thin slice may alter the shape of the tissue. The slice may cut across structures in different planes (see also Chapter 3, page 39). For example, a blood vessel may be cut straight across or lengthways. Imagine if you were to cut straight across a bunch of round grapes. Some will be cut through the middle, but others will be cut to one side of the middle and these sections will look much smaller. This is the same as for cells. The tissue may be stained to show structures that would otherwise be transparent. The colours are not natural.

An optical microscope has a maximum magnification of about 1500 times. At this magnification it is impossible to distinguish structures that are less than about 2 µm apart. For example, cell-surface membranes (also called plasma membranes) are too small to be distinguished from adjacent material.

To form an image, an electron microscope uses a beam of electrons instead of light. This allows structures as small as about 1 nanometre (nm) to be distinguished. A nanometre (nm) is one-thousandth of a micrometre; 1 µm = 1000 nm. Specialised equipment is required to make a section thin enough to show structures clearly with an electron microscope. Also, in the electron microscope the specimen has to be placed in a vacuum, and this treatment can distort it.

ACTIVITY

Analysing a photomicrograph

Look closely at the photomicrograph in Figure 7.17 of lung tissue.

1 Why do the alveoli appear to be different sizes?
2 How do we calculate the actual size of an alveolus from the photomicrograph?
3 What is the actual size, in millimetres, of the structure labelled alveolus in Figure 7.17?

Breathing in

The process of breathing in increases the volume in the chest and causes air to enter the lungs. As the volume of the chest begins to increase, the air pressure inside the lungs starts to decrease. It becomes slightly lower than the atmospheric air pressure outside. This small difference in pressure causes air to move down the pressure gradient and rush into the lungs. The air movement is surprisingly rapid and forceful, yet similar small differences in pressure create the strong winds of our weather.

The volume of the chest can be increased in two ways. The chest is separated from the abdomen by a domed sheet rather like a bulging mini-trampoline. This sheet is the **diaphragm**. It is a tough membrane attached by muscles to the inner wall of the chest at the bottom of the rib cage. It seals off the chest and lungs from the organs of the abdomen. When the muscles of the diaphragm contract, the dome flattens. The centre of the diaphragm may be lowered by as much as 10 cm, thus increasing the chest volume considerably. While we are at rest only a small movement of the diaphragm is needed for us to get enough air into the lungs during each breath in.

When we are more active our oxygen requirements increase. Then, as well as the diaphragm becoming flatter, we can also move our ribs to produce a larger increase in volume. As Figure 7.18 shows, the ribs are connected to each other by two layers of **intercostal muscles**. During a deep breath in the muscles in the outer layer, called the external intercostal muscles, contract. They pull the whole rib cage upwards and outwards: each rib swings up from the backbone and its front end moves up and out (see Figure 7.18).

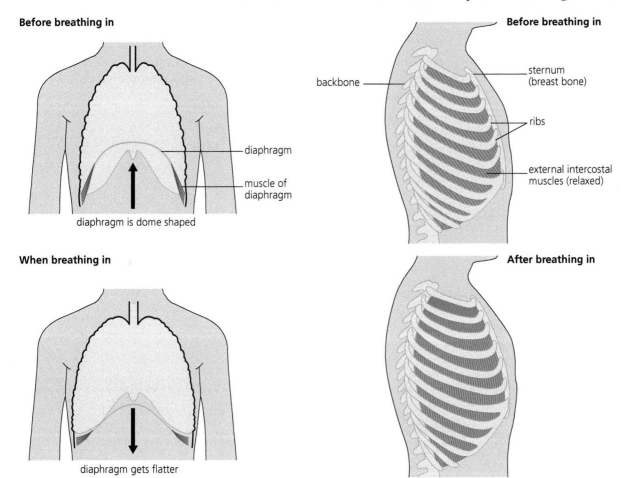

Figure 7.18 The position of the diaphragm (left) and the rib cage (right), before and after taking a deep breath.

Breathing out

At rest, breathing out is mainly due to the lungs recoiling from being stretched. As the external intercostal muscles relax, the elastic fibres round the alveoli recoil and squeeze air out. But when we are exercising it is more important to push air out forcibly. The internal layer of intercostal muscles (Figure 7.18), which slopes in the opposite direction to the external layer, contracts and helps to pull the ribs back down. We may also contract the muscles in the wall of the abdomen. This forces the liver, intestines and stomach upwards against the diaphragm, pushing it back into its domed position, so increasing internal pressure.

Deep breathing

We do not breathe at a constant rate. When we run, our breathing gets deeper and faster. Look at Figure 7.19. It shows the changes in the volume of air in a man's lungs when he changes his activities.

TIP
You do not need to be able to recall the terms 'tidal volume', 'residual volume', 'vital capacity' or 'total lung capacity' in Figure 7.19.

Residual volume The volume of air left in the lungs after as much air as possible is breathed out.

- During the period between A and B, the man is at rest. His breathing is shallow and steady.
- At B he starts to exercise and he takes deeper breaths.
- He stops exercising at C and his breathing starts to return to its resting state.
- At D he breathes out as fully as he can by contracting his abdomen muscles so that the abdominal organs push up against the diaphragm (see page 121). He also uses his internal intercostal muscles to pull down his rib cage as far as possible. This empties his lungs much more than when breathing at rest. That still leaves quite a lot of air in the lungs that cannot be expired. This amount of air is the residual volume of his lungs.
- Then, at E, he breathes in as deeply as he can (his maximum chest expansion).

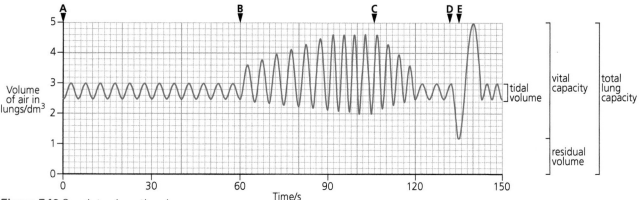

Figure 7.19 Graph to show the changes in rate and depth of breathing of a man who changes his activities.

Extension

The volume of air inspired per breath when at rest is the **tidal volume**. Breathing is at a steady rate and the volume taken in is the same each time. Therefore, to work out the volume of air inspired per minute, the **pulmonary ventilation rate**, simply multiply the tidal volume by the number of breaths per minute, as represented by the equation:

Pulmonary ventilation rate = tidal volume × number of breaths per minute

TEST YOURSELF
11 Use the graph in Figure 7.19 to answer the following questions.
    a) What is the tidal volume for the man when he is at rest?
    b) How many breaths per minute does the resting man take?
    c) Calculate the pulmonary ventilation rate of the resting man.
    d) What was the maximum volume of a single breath while exercising?
    e) Calculate the percentage increase for the maximum volume inspired during exercise compared with the resting tidal volume.
    f) The vital capacity is the maximum volume of air that can be inspired after expiring as fully as possible. What is the vital capacity for this man?
    g) What is the total lung capacity, including the residual volume, that cannot be expired?

ACTIVITY

Interpreting pressure changes during breathing

Look carefully at the two graphs and make sure you understand what they show.

- The upper graph shows the changes in the relative pressure in the alveoli during one breath; that is, during both inspiration and expiration. Pressure is measured in kilopascals (kPa). Zero on the y-axis is when the pressure in the alveoli is the same as the atmospheric pressure outside the body. Atmospheric pressure is normally about 100 kPa, but it varies according to weather conditions.
- The lower graph shows the changes in the volume of the lungs during the same breath. As lung volume increases, pressure falls and air enters the lungs. The volume is measured in cubic decimetres. Remember that 1 dm³ equals 1000 cm³.

1 a) What is the maximum increase in the volume of the lungs?
  b) Do you think these data were measured when the man was at rest or during exercise? Explain the evidence for your answer.

2 Describe the pattern of change in volume of the lungs during inspiration.

3 Describe the pattern of change in pressure in the alveoli during inspiration.

4 Explain what causes the decrease in pressure in the alveoli at the beginning of inspiration.

5 Explain why the pressure in the alveoli returned to zero at the end of inspiration.

6 Describe the pattern of change in volume in the alveoli during expiration.

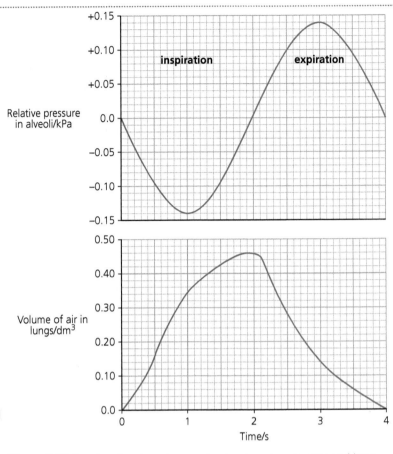

Figure 7.20 Graphs showing changes in alveolar air pressure and lung volume during inspiration and expiration in an adult man.

7 Describe the pattern of change in pressure in the alveoli during expiration.

8 Explain the changes in pressure and volume during expiration.

9 A chest wound (such as a stab or bullet wound) that allows air into the space between the chest wall and the lungs can prevent normal inspiration, even though breathing movements occur. Suggest an explanation for this.

TEST YOURSELF

12 List in order the parts of the breathing system that carbon dioxide goes through as it passes from the blood to the air outside the body.

13 How are the lungs protected from bacterial infection?

14 Give three ways in which the alveoli are adapted for rapid diffusion of oxygen into the blood.

15 Explain how the oxygen diffusion gradient between the air in the alveoli and the blood in the surrounding capillaries is maintained.

16 Suggest why the air we breathe out has more water vapour than the air we breathe in.

17 How is the volume of the chest cavity increased during inspiration?

18 What is the 'tidal volume'?

19 How does the ventilation rate change during exercise?

TIP

See Chapter 14 to find out about interpreting line graphs, taking measurements and describing patterns and trends.

Smoking kills

Environmental pollutants can be difficult to avoid, but many people willingly draw tobacco smoke into their lungs, despite the evidence that this will do long-term damage. Tobacco smoke contains a vast mixture of substances. These include the addictive ingredient nicotine and the highly toxic gas, carbon monoxide. Of over 4000 organic compounds in the tar, at least 60 are known to cause cancer. All of these substances reach the delicate cell lining in a smoker's lungs. Some get swept out again when the smoker exhales, some stay in the lungs and some pass into the blood and reach all parts of the body. After a few years, a smoker's lungs are lined with tar, as the photograph in Figure 7.21 shows.

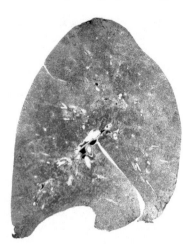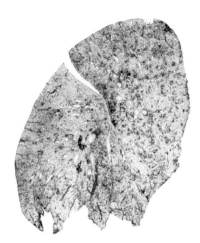

Figure 7.21 Left: section through the clean lung of a non-smoker. Right: section through a tar-filled smoker's lung.

It is estimated that each year in the UK about 115 000 people die early due to their smoking habit. Of these, about 28 000 die from lung cancer, and over 30 000 die from other chronic lung diseases such as emphysema and bronchitis. Smoking accounts for between 80 and 90% of deaths from lung diseases. On average, of those who continue to smoke regularly from their teenage years, about a quarter die prematurely in middle age, 20 years before the normal life expectancy.

Interpreting data relating to the effects of smoking and pollution

Collecting the evidence

Researchers investigating the effects of smoking or other pollution cannot carry out controlled experiments on people. It is not ethically acceptable, for example, to select randomly two groups of people and make one group smoke 20 cigarettes a day for 10 years, while banning the control group from smoking altogether. Evidence must be based on people who were exposed to smoking, or another pollutant, and then comparing them with others who were not exposed.

In **retrospective studies** (collecting data from the past), researchers select groups of people who have already developed a disease such as lung cancer. They then question the people about their past experiences and look for common factors. They may also compare their experiences with the experiences of a control group who do not have the disease being studied. These studies can be unreliable because people may have forgotten details such as how long they have smoked, or they may deliberately deceive or exaggerate.

Extension

In **prospective studies** (collecting data as it accumulates), researchers select groups and then follow what happens to them over a period of years. This makes it easier for the researchers to keep track of changes, such as changes in smoking habits, as long as frequent checks are made. The researchers can also keep records of a wider range of possible variables. However, this adds to the time it takes to get useful results.

Interpreting the results of studies on health risks

You will understand that there are often great difficulties in collecting data on health risks.

- There are often many factors involved.
- Controlled experiments in which one variable only is studied cannot be carried out on people for ethical reasons.
- It is hard to find enough people with similar lifestyles to act as matched control groups.
- It is often several years before the effects of health risks become apparent, and following up groups of people for long periods is difficult and expensive.
- Data obtained by asking people about their past are often unreliable.

When looking at the results of a health risk study, you need to consider the following.

- Find out the number of people who were investigated. You can have more confidence in the evidence if a large number of people were involved than if the number was small.
- Identify the different levels of exposure to the health risk that were investigated; for example, the number of cigarettes smoked per day.
- Assess whether the control group is well matched with the group exposed to the factor being tested. They should, for example, come from similar backgrounds, be of similar ages and so on.
- Assess whether the differences between the results for the two groups are sufficiently large to indicate that the factor thought to be a health risk is indeed a risk.
- Find out whether tests have been done to check that the differences are statistically significant.

If the number of cases (incidence) of the disease or the number of deaths (mortality rate) is given, calculate the **relative risk**; for example, calculate the *difference in the percentages* of the two groups that develop the disease.

Terms used in investigations of the causes of diseases

- **Incidence** The incidence of a disease is the number of cases that occur in a particular group of people in a given time, such as the number of smokers that develop lung cancer in a year. To make it easy to compare how common a disease is in different groups, the incidence is calculated as the number of cases in a standard size of group, for example, number of cases of lung cancer per 1000 smokers per year. When looking for possible effects of pollution on asthma, researchers might compare the incidence of asthma in children living in urban areas with those in rural areas.
- **Mortality rate** The mortality rate is the number of deaths per number of the population per year from a particular disease or other cause, such as

road accidents. A rate of 2.5 lung cancer deaths per 1000 smokers would mean that, out of 10 000 smokers, on average 25 died each year.

- **Correlation** A correlation is an association between two variables. If measurements of both variables increase, there is a positive correlation. For example, there is a well-established correlation between the number of cigarettes smoked per day and the incidence of lung cancer. It is important to note that a positive correlation is *not* proof that one factor is the cause of another. You might find a correlation between baldness and wearing a hat, but that would not mean that wearing a hat causes baldness.

- **Statistically significant** Investigations of the effects of smoking, pollution, drugs or diet on people never give absolutely certain results. You do not find, for example, that everyone who smokes 30 cigarettes a day for 5 years gets lung cancer. People are highly variable and do not always behave consistently. Researchers carry out statistical tests on their results to find out whether it is likely that that they have discovered a genuine effect. These tests check how likely it is that a difference, for example, between smokers and non-smokers, is just due to chance. This will depend on the number of people tested and the size of the differences in results for the two groups. The difference between the results is usually considered to be statistically significant if it shows a more than 95% probability that the difference is real (not due to chance). Even so, this is not absolute proof. There is still a 1 in 20 probability that these particular results were due to chance.

- **Risk factor** A risk factor is something, such as smoking, that correlates with an increased chance of suffering from a particular disease or condition. The relative risk can be calculated by finding the ratio between the incidence of the disease in those exposed to the factor and the incidence in those not exposed. For example, if the incidence of a lung disease in smokers is 30 cases per 1000 per year and in non-smokers it is 10 cases, the relative risk is 3.0, which means that smokers are three times more likely to develop the disease.

What is the evidence for the link between smoking and disease?

Tobacco was introduced into Britain over 400 years ago. It was another 300 years before smoking cigarettes began to be both popular and affordable. By the 1930s, doctors were beginning to suggest that smoking might be damaging to health as they were seeing greatly increased numbers of patients suffering from lung cancer. Their concerns did not affect the smoking behaviour of the general public. The graph in Figure 7.22 shows how the consumption of cigarettes in the UK grew during the twentieth century.

Figure 7.22 Graph of cigarette consumption by men and women in the UK from 1900 to 1980.

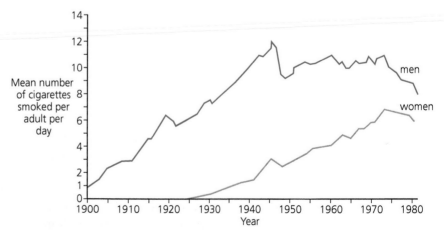

The graph in Figure 7.23 shows the changes in the number of deaths from lung cancer up to 1960 compared with cancers of other organs.

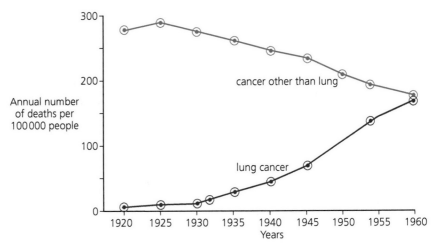

Figure 7.23 Death rates from cigarette lung cancer and other cancers between 1920 and 1960.

Look at Figure 7.22. Notice that the number of cigarettes smoked by men rises steadily from 1900. How does this compare with the number of deaths from lung cancer, shown in Figure 7.23? This also rises, but the rise only becomes steep after about 1930. Then, both the number of cigarettes smoked and the number of lung cancer deaths go up steeply, so there is a **positive correlation** (see page 247) between them.

How could the delay in the rise in lung cancer deaths be explained? Bear in mind that cancer develops only slowly, so it may be 20 years or more before smokers start to show signs of the disease and die.

It is important to note that this positive correlation could not prove that smoking causes lung cancer. Many other factors were also increasing over the same period (such as the number of cars), and it was possible that one of these factors could explain the increase in lung cancer. The correlation did, however, provide one piece of evidence that supported the hypothesis and justified further research.

Much more evidence was needed to prove that smoking really does cause lung cancer. In 1950, two epidemiologists, Richard Doll and Austin Hill, published the results of a study of patients in four hospitals in the UK. Two groups of patients were selected: those who were suffering from lung cancer and those who had been admitted to hospital for other conditions. The patients were questioned about their smoking habits. They were, for example, asked whether they were regular smokers or had given up smoking, and about their daily consumption of cigarettes. Doll and Hill showed that a higher proportion of the lung cancer patients than the control group were regular smokers. Although other studies were showing similar results, they were still not considered to be absolute proof of a link between smoking and lung cancer. Critics suggested that some unknown factor might make people both more likely to take up smoking and to be susceptible to lung cancer without smoking causing the cancer. The research was also criticised on the grounds that the answers to questions about past smoking habits might be unreliable, due to either inaccurate memory or deliberate fibbing.

In 1951, Doll and Hill set about obtaining data that would be more convincing. Instead of checking the past history of people who already had lung cancer, they decided to select a group of healthy people and monitor who developed the disease. They chose British doctors as their subjects. Perhaps surprisingly, many doctors were smokers at the time. Doll and Hill reckoned that doctors would be more honest and reliable in recording their smoking habits. It was also easy to keep track of doctors because they have to be registered with the General Medical Council. In all, the smoking habits and death rate of over 34 000 doctors were recorded for the next 50 years.

EXAMPLE

The evidence that smoking causes disease

Table 7.1 Some results of the Doll and Hill study of doctors.

| Cause of death | Mortality rate of male doctors /deaths per 1000 per year | | | | |
| | Never smoked | Given up smoking | Still smoking: cigarettes per day | | |
			1–14	15–24	25 or more
Lung cancer	0.17	0.68	1.31	2.33	4.17
Cancer of mouth, throat, gullet	0.09	0.26	0.36	0.47	1.06
Other cancers	3.34	3.72	4.21	4.67	5.38
Other chronic lung disease, e.g. emphysema	0.11	0.64	1.04	1.41	2.61
Heart disease	6.19	7.61	9.10	10.07	11.11

Table 7.1 shows some of the results of the Doll and Hill study of doctors. The mortality rate is the average number of deaths per year per 1000 doctors from each cause. A mortality rate of 2.0 means that, on average, two doctors in every thousand died each year. Out of 30 000 doctors, this would mean a total of 60 deaths. However, in this study the rates were calculated separately for each 5 year age group, for example for doctors aged 55–59. The rates were then adjusted because the number of deaths for the younger age groups was much lower than in the older groups. The figures therefore indicate the effect of each cause on both the overall death rate per 1000 doctors and the age of death.

1 What do the data show about the effect of smoking on mortality from lung cancer?
Mortality is increased significantly by smoking. Even smoking fewer than 15 cigarettes per day increases the risk by as much as seven times.

2 What do the data show about the effect of smoking on the other causes of death shown in the table?
Other cancers also increase the mortality rate, especially those associated with the mouth, throat

and gullet. The increased risk is much lower for other cancers. Death from other lung diseases is also significantly increased by smoking. There is also a smaller increased risk of heart disease.

3 For each cause of death, calculate by how much the risk is increased by smoking 25 cigarettes a day compared with not smoking at all.
Lung cancer, ×24.5; mouth/throat/gullet, ×11.8; other cancers, ×1.6; other lung diseases, ×23.7; heart disease, ×1.8.

4 Describe the effects of giving up smoking.
Giving up smoking decreases of lung cancer and associated cancers as well as other lung diseases. It also reduces the chances of dying from other cancers or heart disease.

5 Describe and explain the results for heart disease.
There is higher risk of heart disease because several other lifestyle factors contribute to heart disease, such as diet and lack of exercise. By comparison the risk of lung cancer and other lung diseases is low for non-smokers. However non-smokers are liable to die of other diseases, so the death rate is higher, but still somewhat lower than for smokers.

Analysing the evidence of the study by Doll and Hill

The graph in Figure 7.24 shows one set of results from this investigation. It shows the percentage of smoking and non-smoking doctors who survived to various ages over the 50 years of the study, which ended in 2001. The smoking group was all those doctors who continued to smoke throughout the investigation, whereas the non-smoking group were those who had never smoked.

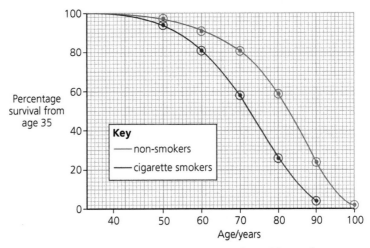

Figure 7.24 Graph showing the percentage of smoking and non-smoking doctors surviving at each decade of age until age 100.

1 Look at the curve in Figure 7.22 for the number of cigarettes smoked by women. Explain how you would expect the data for deaths from lung cancer in women to give evidence supporting the hypothesis that smoking is a cause of lung cancer.

2 As well as deaths from lung cancer, Figure 7.23 shows the number of deaths from other cancers. How do these data support the hypothesis that smoking causes lung cancer?

3 a) Use the graph in Figure 7.24 to determine the percentage of smokers and non-smokers surviving at ages 60, 70 and 80.

   b) For each of these ages, calculate the difference in the percentage surviving.

   c) By approximately how many years was the average lifespan (length of life) reduced in smokers?

Gas exchange in plants

The main function of a leaf is to carry out photosynthesis, and for this it needs a supply of carbon dioxide and water. Also, the chloroplasts require good access to sunlight. Sunlight does not penetrate far, and the leaves are thin. In thin leaves there is a short diffusion pathway from the stomata, and there is also a large surface area because of the air spaces surrounding the cells. However, these properties make them vulnerable to damage and dehydration. The leaves of different species are adapted to meet these requirements in a variety of different ways, depending on the conditions in which they live.

How are leaves adapted for gas exchange?

Figure 7.25 shows the cells and tissues in a section of the leaf of a plant that lives in the relatively moist and cool climate of Britain. Notice how much space there is between the cells in the mesophyll.

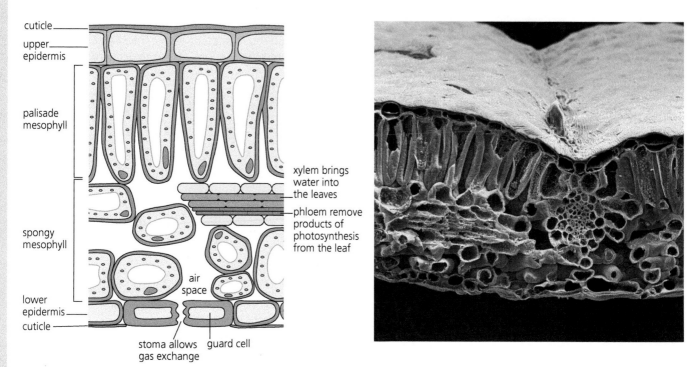

Figure 7.25 A cross-section of a leaf (×210).

The outer cell layer on the upper and lower surface of a leaf is called the **epidermis**. It consists of cells that fit closely together. The outer walls of these cells contain a mixture of a lipid polymer and waxes that make a waterproof **cuticle**. Even in Britain, conditions are usually dry enough for dehydration to be a problem. However, as well as stopping water from escaping, the cuticle prevents most gas exchange. To allow carbon dioxide to get to the photosynthetic cells inside the leaf, there are pores called **stomata** (singular: stoma) in the epidermis. In most leaves, the stomata are mainly on the under-surface of the leaf.

Each stoma is surrounded by a pair of **guard cells**. These are banana-shaped. If the guard cells lose water they become less firm and change shape, closing the stoma, which helps to prevent further water loss. Guard cells are shown in Figure 7.26.

The central tissue of the leaf, called the **mesophyll**, has an extensive network of air spaces, as you can see in Figure 7.25. These spaces allow gases to move to and from the cells by diffusion in the gas phase, which is more rapid than in the liquid phase. There is no active system of ventilation: the thinness of the tissue and distribution of stomata helps to keep the diffusion pathway short. The upper layer, or sometimes two or three layers, is the **palisade mesophyll**. It has elongated cells that contain large numbers of chloroplasts. The **spongy mesophyll**, below it, has more air spaces and the cells have fewer chloroplasts. Water reaches the leaves through the xylem, and we will look in more detail at this process in the Chapter 10.

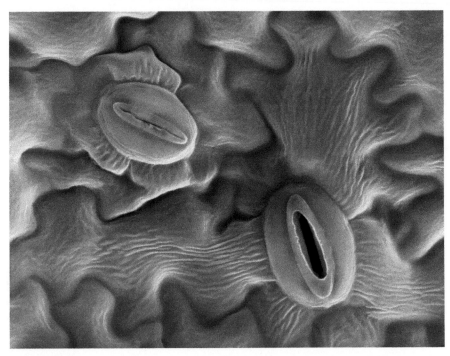

Figure 7.26 The surface view of guard cells of stomata: closed at top left; open at bottom right.

Despite the amount of fossil fuels that we have burned, the proportion of carbon dioxide in the atmosphere is still low, only about 0.036%. In other words, out of every hundred thousand molecules in the air, fewer than 40 are molecules of carbon dioxide. This means that efficient gas exchange is vital for leaves to get enough carbon dioxide for rapid photosynthesis.

Large numbers of stomata dot the lower surface of leaves. The leaves heat up in sunlight and air that has entered the air spaces warms and rises, so becoming trapped there. This humid air has a high water potential, helping to reduce water loss. The air spaces in the mesophyll allow fast carbon dioxide diffusion to the cells. The palisade cells are elongated, so they have a large surface area exposed to the atmosphere inside the leaf. The large number of chloroplasts in the thin layer of cytoplasm next to the cell wall means that carbon dioxide rapidly takes part in photosynthesis and is therefore removed. This maintains a steep diffusion gradient.

The downside is that the adaptations that boost the entry of carbon dioxide also allow water to be lost rapidly. Water evaporates from the mesophyll cells to become water vapour in the air spaces and, although the cuticle and small pores help to prevent water loss, water vapour can diffuse through the stomata down the water potential gradient. In a damp climate, such as Britain's, plants can also overcome this problem of water loss simply by taking in large amounts of water from the soil. On a hot day, an oak tree may absorb over half a tonne of water, and most of this will evaporate from its leaves. During water shortage, the stomata close to reduce evaporation. Although this tends to stop photosynthesis, it avoids the more catastrophic results of dehydration.

ACTIVITY

Counting stomata

A student gathered a sample of leaves from a holly bush. She painted a small area (about the size of a stamp) with clear nail polish on the underside of one leaf. She let this dry. Next she pressed a piece of clear sellotape over the nail-polished area and then carefully peeled it off. She placed this on a clean microscope slide and viewed it under the microscope. She counted the number of stomata visible in the field of view. She repeated this several times. Then she repeated this several times for the top surface of the leaf.

Table 7.2 shows the student's data.

Table 7.2 Student data.

Leaf number	Underside of leaf		Top surface of leaf	
	Number of stomata in field of view	Number of stomata /mm^{-2}	Number of stomata in field of view	Number of stomata/mm^{-2}
1	41	270	0	0
2	55	362	0	0
3	72	474	0	0
4	52	342	0	0
5	40	263	0	0
6	47	309	0	0
7	38	259	0	0
8	59	388	0	0
9	40	263	0	0
10	42	276	0	0
Mean				
SD				

1 Complete the table to find the mean and standard deviation of these data.
2 Explain the advantage of a leaf having no stomata on the upper surface of the leaves.
3 Suggest how the student worked out the number of stomata per square mm of leaf.
4 Explain the advantage of calculating stomatal density per square millimetre of leaf.
5 Suggest how the student could modify this investigation to find out whether stomatal density is influenced by carbon dioxide concentration.

TIP
Find out how to calculate standard deviation in Chapter 14.

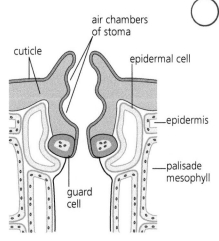

Figure 7.27 Sunken stoma.

How do plants survive in dry environments, such as deserts?

Plants that live in conditions where there is a shortage of fresh water have additional adaptations that enable them to conserve water very effectively. Such plants are called **xerophytes**.

Features that are common in plants growing in environments where there is a shortage of fresh water include:

● thick cuticle
● small or needle-shaped leaves
● few stomata
● stomata sunk into pits in the epidermis (Figure 7.27)
● hairs around the stomata and over the leaf surfaces.

Deserts

Some of the best-known xerophytes are the cacti, which typically populate the deserts of the western United States. These plants are adapted to desert life by having leaves that are reduced to spines. The leaves have also lost the ability to photosynthesise. Instead, the stem has chloroplasts, as you can see from their green colour. The stem has a large diameter, enabling it to store water in its tissues, and it has a thick cuticle. The number of stomata is much reduced. Overall, the cactus has a low surface area to volume ratio, which reduces water loss. The spines may be just as important as deterrents to grazing animals as in reducing the surface area for evaporation. Figure 7.28 shows the features of a cactus that help it to survive in a dry habitat.

> **TIP**
> Xerophytic plants live in a range of dry environments. Deserts and sand dunes are just two examples of such environments but are not required learning.

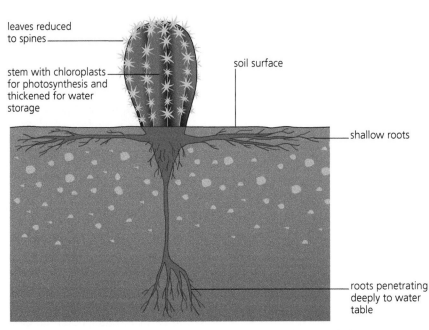

Figure 7.28 Features of a cactus that help it to survive in a dry habitat.

> **TIP**
> You do not need to know specific examples of xerophytes. Make sure you are able to explain the adaptations of any example of xerophyte though.

Sand dunes

Deserts are not the only habitats that require plants to be adapted to dry conditions in order to survive. The water supply in sand dunes is also limited. Marram grass is an example of a plant that is particularly well adapted to life in dry conditions, and as a result it is one of the first species to colonise new sand dunes on a beach.

133

Leaves of marram grass have several adaptations that help to reduce water loss. In Figure 7.29, you can see that:

● the leaf can roll up so that only one surface is exposed to the wind
● the exposed surface has a thick cuticle and no stomata
● when the leaf is rolled up the stomata are protected in deep grooves
● the inner surface has many hairs.

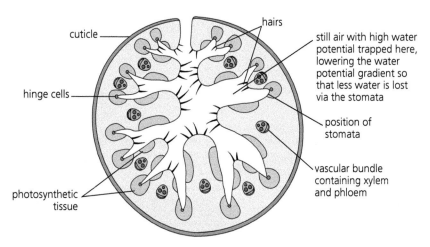

Figure 7.29 Marram grass is very well adapted to restrict water loss, and is an early coloniser of newly formed sand dunes in the UK.

The advantage of the leaf rolling up is that water vapour becomes trapped and kept near the stomata because there is no wind or air current to move the vapour away. The hairs also help to reduce air movement. The air just outside the stomata becomes much more humid, so it has a higher water potential and the water potential gradient and diffusion of water vapour across the stomata is much reduced. The rate of evaporation is therefore much slower. How does the leaf roll up? In dry conditions, the special hinge cells at the base of the grooves lose water rapidly by evaporation. These cells shrink and pull the sides of the grooves together, making the leaf curl up.

Factors affecting water loss

The graph in Figure 7.30 shows the difference between the rate of water loss from a leaf without hairs on it in still and in moving air.

1 What does Figure 7.30 show about the effect of air movement on the rate of water loss from a leaf?
Air movement replaces humid air with dry air, thus increasing the water diffusion gradient out of the plant, and so increasing water loss.

Table 7.3 gives the results of measuring the rate of water loss in leaves from three species of flowering plant with differing amounts of hairiness on the leaves.

Table 7.3 Rate of water loss for flowering plants with leaves of different hairiness.

Name of flowering plant	Hairiness of leaves	Rate of water loss from leaf surface/$g\,cm^{-2}\,h^{-1}$
Sweet violet	Slightly hairy	0.04
Storksbill	Quite hairy	0.09
Woundwort	Densely hairy	0.13

2 What do these results tell you about the effect of hairs on the rate of water loss in these particular plants?
You would expect from the data in the graph of Figure 7.30 that in Table 7.3 the rate of water loss from the hairiest leaves would be lowest because they would trap a layer of still air best. However, the results in the table do not fit with this hypothesis. They clearly do not suggest that the hairs on the woundwort leaves are adaptations to reduce water loss. They may have some other function, such as deterring animals from eating them. Alternatively, any reduction in water loss may be offset by different numbers of stomata, or by some other factor.

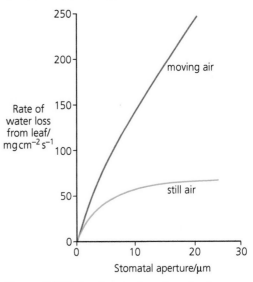

Figure 7.30 Rate of water loss for a leaf without hairs on it in still air and in moving air.

TIP
Look carefully at data before assuming that the answer you are expecting is correct!

TEST YOURSELF

20 Marram grass grows in sand dunes, close to the sea. Explain the advantages of its xerophytic adaptations.
21 Many cacti have stomata that only open at night. This means they take in carbon dioxide at night, and store it for use in photosynthesis by day. Explain the advantage of this.
22 Make a list of the features of gas exchange in a leaf, using the headings 'large surface area', 'short diffusion distance' and 'large concentration gradient'.
23 Deciduous trees that lose their leaves in winter tend to have large, thin and flat leaves. Explain the advantage of this.
24 Coniferous trees, that do not lose their leaves in winter, tend to have needle-like leaves. Explain the advantage of this.

Practice questions

1 Describe and explain three features that reduce transpiration in a xerophytic plant. (3)

2 a) Insect spiracles often have valves that enable them to be closed, and are often surrounded by tiny hairs. Suggest the advantage of these. (2)

   b) Describe how oxygen reaches a respiring muscle cell in an insect. (3)

3 a) The figure shows *Paramecium*, a single-celled organism.

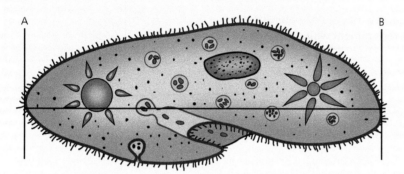

   i) The actual length of this cell from A to B is 240 μm. Use this information to calculate the magnification of the diagram. Show your working. (2)

   ii) *Paramecium* does not have a gas exchange system. Explain how gas exchange occurs in this organism. (2)

   b) Explain how the counter-current principle is important in gas exchange in fish. (2)

4 The table shows some data relating to three different species of fish.

Species	A	B	C
Swimming speed	Fast	Slow	Slow
Water or air breathing	Water	Water	Air
Surface area of gills /cm² g⁻¹	13.5	1.9	0.6

   a) Suggest why the surface area of the gills is given per gram of body mass. (2)

   b) Use the information in the table to explain the advantages of the following features:

   i) the relative surface area of the gills of species A is larger than the relative surface area of the gills of species B. (2)

   ii) the relative surface area of the gills of species B is larger than the relative surface area of the gills of species C. (4)

5 The figure shows part of the tracheal system in an insect that carries out gas exchange.

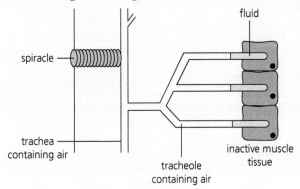

a) i) Describe how oxygen in a tracheole reaches a respiring muscle cell in an insect. (2)

   ii) When insect muscle respires rapidly, products of respiration accumulate in the cells, lowering their water potential. Suggest how this increases the supply of oxygen to active muscle. (3)

b) The figure shows the number of times the spiracles of a flea open when the flea is exposed to various concentrations of oxygen.

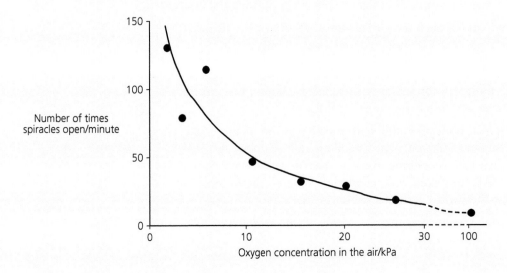

   i) Explain why it is an advantage to an insect, such as a flea, to be able to close its spiracles. (1)

   ii) Describe and explain the results shown in the graph. (3)

6 In the early stages of an asthma attack a man breathed in as deeply as he could. He then breathed out as fast and forcefully as possible through a machine that measured the volume of air as he breathed out. The man then used his inhaler. The inhaler contains a drug that makes muscles relax. Twenty minutes later he did the same test again. The graph overleaf shows the results of the two tests.

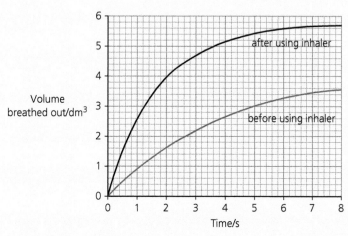

a) Describe the differences between the results before and after using the inhaler. (2)

b) i) The maximum volume of air forcibly breathed out in 1 second is called the FEV1. Use the graph to calculate the FEV1 before and after using the inhaler. (2)

   ii) Describe how you could find the initial rate at which air is expelled from the lungs. (2)

c) Describe what the graph shows about the effect of using the inhaler. (2)

d) Explain why the total volume of air breathed out before using the inhaler was lower than the volume expired afterwards. (1)

e) Suggest how using the inhaler caused the difference. (2)

7 a) Use the following information to calculate the approximate surface area of the alveoli when the lungs are inflated. (2)

Each lung has about 350 million alveoli. The mean diameter of each alveolus is about 0.20 mm. The alveoli are not complete spheres because each has an opening. Assume that the surface area is reduced by about 20% compared with a complete sphere. The formula for calculating the surface area of a sphere is $4\pi r^2$.

b) Suggest why it is unlikely that the whole of this area would be available for diffusion of oxygen. (1)

Stretch and challenge

8 People who have had a spinal cord injury, resulting in paralysis, might need to use a ventilator. Explain the reason for this and how a ventilator works.

9 In the first half of the twentieth century, poliomyelitis was a fairly common disease in the UK. Children who had polio were sometimes so paralysed that they had to be nursed in an iron lung. Explain how an iron lung works. How does an iron lung differ from a ventilator, and what are the advantages of using a ventilator?

10 How does gas exchange occur in molluscs and amphibians? Explain how the gas exchange systems in the two groups of animals are adaptations to their environment.

8 Digestion and absorption

TEST YOURSELF ON PRIOR KNOWLEDGE

1 Explain why muscle is an example of a tissue and the stomach is an example of an organ.

2 Copy and complete this table.

Name of enzyme	Where produced	Reaction catalysed	Part of the gut where reaction occurs
Amylase			
Protease			
Lipase			

3 Explain why it is important that bile contains alkaline substances.

4 Name two factors that affect an enzyme's activity.

Introduction

When your digestive system becomes infected it is not very pleasant. You may have had diarrhoea. After a day or two, however, it is more than likely that you were over the worst of it. By contrast, every year nearly 2 million children die in developing countries when they lose too much fluid and become dehydrated because of diarrhoeal disease. Tragically, we could prevent many of these deaths with a simple mixture of water, glucose and salts. This mixture is called an oral rehydration solution (ORS).

Figure 8.1 In overcrowded refugee camps, the water is often contaminated, and it is difficult to maintain good hygiene. As a consequence, people often die of cholera and other diarrhoeal diseases. During the 1971 war for independence in East Pakistan (now Bangladesh), up to 30% of the refugees in the refugee camps in India died when doctors ran out of medicines. In camps where doctors used oral rehydration solutions, the death rate was only 3%.

What does the digestive system do?

The digestive system consists of the gut, which forms a tube extending from the mouth at one end, through the body, to the anus at the other end. Food is ingested: it is taken in. In the mouth, it is chewed, mixed with saliva, and swallowed. The food is now inside the gut but it is not yet inside the body itself (to truly enter the body, substances have to cross cell-surface membranes). Before it can pass through the wall of the gut and into the blood, it must be digested. The food is mixed with digestive juices secreted by various glands as it is squeezed and pushed along by the muscular walls of the gut.

Figure 8.2 shows the human digestive system. The stomach and the first part of the small intestine are where food is digested. The digestive juices, made by gland cells of the digestive system, contain enzymes. The enzymes act on the large insoluble molecules of protein, starch and fats that are the main components of our food. The enzymes hydrolyse them into smaller soluble molecules that can be transported across cell-surface membranes:

- protein is hydrolysed to amino acids
- starch is hydrolysed to simple sugars
- fats are hydrolysed to a mixture of fatty acids and glycerol.

These small molecules are absorbed through the lining of the small intestine into the blood and transported to the body's cells.

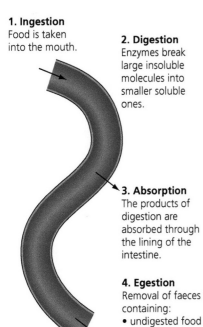

1. Ingestion
Food is taken into the mouth.

2. Digestion
Enzymes break large insoluble molecules into smaller soluble ones.

3. Absorption
The products of digestion are absorbed through the lining of the intestine.

4. Egestion
Removal of faeces containing:
- undigested food
- bacteria
- cells from the intestine lining
- enzymes.

Figure 8.3 Processing food.

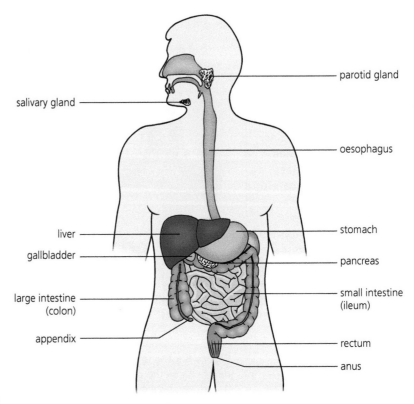

salivary gland

parotid gland

oesophagus

liver

gallbladder

large intestine (colon)

appendix

stomach

pancreas

small intestine (ileum)

rectum

anus

Figure 8.2 The human digestive system.

Some substances in food, such as cellulose, cannot be digested by the human gut. They pass out through the anus, together with cells scraped from the gut lining, enzymes and bacteria, and are egested as faeces. Figure 8.3 summarises these processes.

141

Digesting carbohydrates

Respiration is the biochemical process by which we produce ATP from energy stores such as glucose. The glucose we use in respiration comes mainly from the carbohydrates that we eat and derivatives of lipids and proteins (Figure 8.4).

Any glucose that we eat can be absorbed as soon as it reaches the small intestine because glucose molecules are small and are able to be carried through the cell-surface membranes of the cells that line the intestine and enter the blood. Starch, though, has large insoluble molecules that cannot be absorbed. They must be digested, first to maltose and then to glucose. This involves the enzymes amylase and maltase.

Carbohydrate digestion occurs in the mouth and small intestine. Amylase is secreted in saliva and starts to digest starch until it is denatured by the acidity of the stomach. Amylase is secreted into the small intestine in pancreatic juice from the pancreas. Amylase hydrolyses starch to maltose (a disaccharide). The cells lining the small intestine have the enzyme maltase in their cell-surface membranes. Maltase hydrolyses maltose to glucose (Figure 8.5).

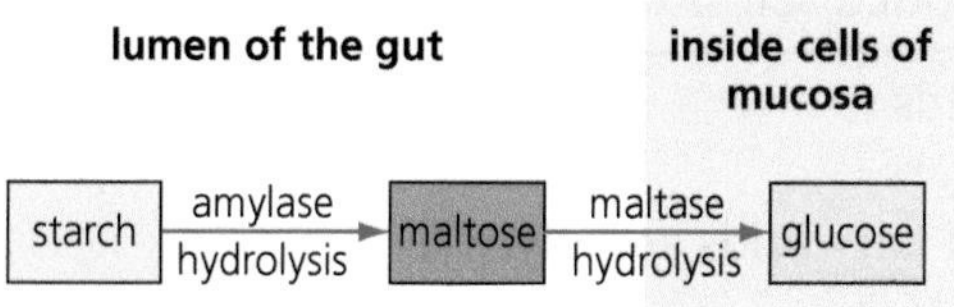

The glucose is released into the epithelial cell. This is shown in Figure 8.6. There are other disaccharidases in the membranes of the epithelial cells in the small intestine. These include sucrase and lactase (in some individuals).

Figure 8.4 We eat different carbohydrates in our diet. Bread, pasta and potatoes all contain starch; fruit contains sucrose, glucose and fructose. Milk contains lactose.

Figure 8.5 Disaccharidases such as maltase are in the cell-surface membrane of epithelial cells in the small intestine.

TIPS

- Notice that most carbohydrates have names that end in -ose, like glucose, maltose, lactose and amylose. But the enzymes that digest them have names that end in -ase, like maltase, amylase and lactase. Be careful to spell them correctly!
- Look back to Chapter 2 to remind yourself about enzymes.
- Look back to Chapter 3 to remind yourself about carrier proteins and facilitated diffusion.

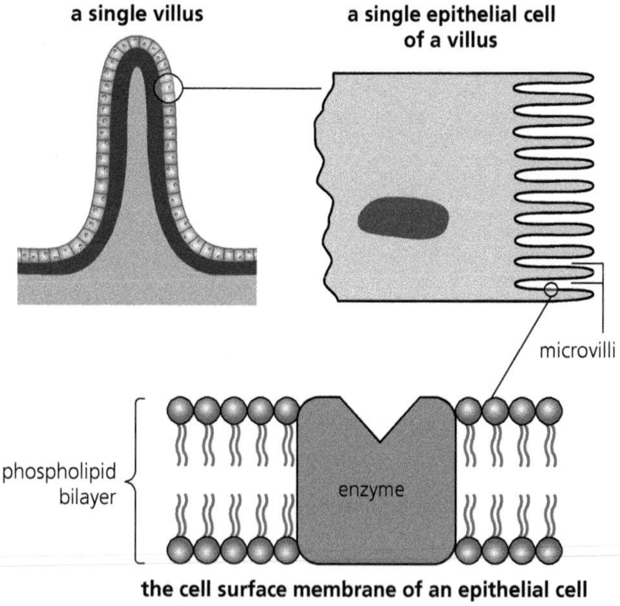

Figure 8.6 The lining of the small intestine, showing a villus, microvilli on the surface of an epithelial cell of a villus and its cell-surface membrane.

The lining of the small intestine is folded into finger-like projections called villi. These increase the surface area for absorption. The epithelium of the villi is made of epithelial cells that have microvilli on their surface (see Chapter 3). There are enzymes in the cell-surface membrane of these epithelial cells. Figure 8.7 summarises carbohydrate digestion.

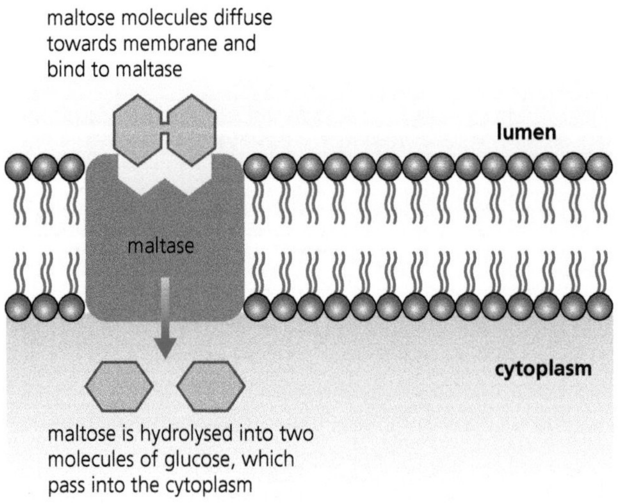

maltose molecules diffuse
towards membrane and
bind to maltase

lumen

maltase

cytoplasm

maltose is hydrolysed into two
molecules of glucose, which
pass into the cytoplasm

Figure 8.7 A summary of carbohydrate digestion in humans.

Co-transport

One of the most important products of carbohydrate digestion is glucose. The intestine absorbs glucose by a combination of facilitated diffusion and active transport. Look at Figure 8.8. There are carrier molecules called **co-transport proteins** in the cell-surface membranes of the epithelial cells. These are carrier molecules that only transport glucose in the presence of sodium ions. Each time a sodium ion is transported into the cell, so is a glucose molecule. This part of the process involves facilitated diffusion. Facilitated diffusion, however, will only work if substances can move down a concentration gradient. This gradient is maintained by actively transporting sodium ions out of the cell into the blood. The glucose molecules pass from the inside of the cells into the blood by facilitated diffusion.

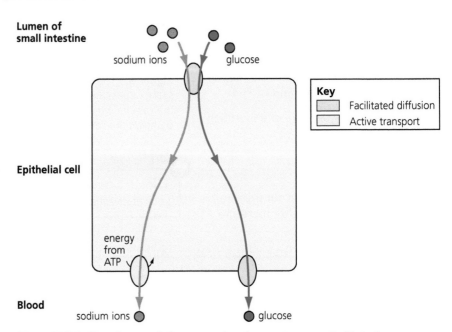

Figure 8.8 Sodium ions and glucose molecules are transported into the epithelial cells lining the small intestine by facilitated diffusion. Sodium and glucose enter the cells through co-transport proteins. The sodium ions are then actively transported from the epithelial cells and into the blood.

TEST YOURSELF

1 The epithelial cells in the small intestine contain many mitochondria. Explain the link between the large number of mitochondria and the transport of sodium ions out of the cells into the blood.
2 Explain why cellulose cannot be digested in the human gut.
3 Draw a molecule of maltose and show how it is hydrolysed into two molecules of glucose.
4 Explain why the carrier protein in Figure 8.8 carries glucose across the membrane but not other molecules such as fructose.

EXAMPLE

Investigating absorption

Scientists investigated how different factors affected the rate of absorption of glucose from a piece of small intestine. The results of their investigation are shown in the graph in Figure 8.9.

1 Look carefully at the graph. What are the independent variables in this investigation?
There are two: the concentration of glucose in the lumen of the intestine, and whether or not the glucose solution was stirred. They are the independent variables because they were the factors that the scientists changed.

2 What is the dependent variable?
The dependent variable is the factor that is measured as a result of changing the independent variable. In this investigation, it is the rate of absorption of glucose from the small intestine.

In this investigation, the scientists measured the rate of glucose absorption.

3 What is meant by the rate of absorption?
The rate is the amount absorbed divided by the time taken. We often use rates in biology because they allow comparisons to be made.

When the scientists carried out the investigation, they kept the temperature constant.

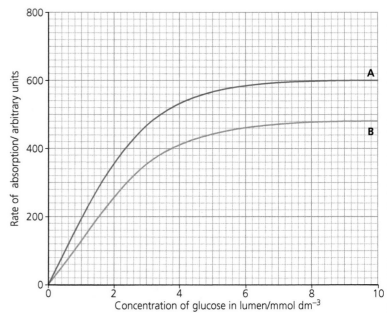

Figure 8.9 The graph shows the effect of glucose concentration on the rate of absorption of glucose from the small intestine. Curve A shows the results when the glucose solution in the intestine was stirred. Curve B shows the results when the glucose solution was not stirred.

TIP
Look at investigating scientific questions in Chapter 15, Practical skills.

---

4 Why did the scientists keep the temperature constant?
Temperature affects the rate of absorption. If the scientists allowed the temperature to vary as well as, say, the concentration of glucose in the lumen of the intestine, they would not know what caused any change in the rate of absorption.

5 Now look at curve B on the graph. Describe how the concentration of the glucose solution in the lumen of the small intestine affects the rate of absorption.
The rate of absorption increases as the concentration of glucose in the small intestine increases, and then gradually levels off. After a concentration of approximately 5 mmol dm⁻³, it remains constant.

6 The rate of absorption is more or less constant above a concentration of 5 mmol dm⁻³. Explain why.
This is another graph where limiting factors are involved (see pages 31–32 where we looked at the effect of substrate concentration on the rate of an enzyme-controlled reaction).

There must be something other than the concentration of glucose in the lumen of the small intestine that is limiting the rate of absorption here. It is probably the number of glucose carrier molecules in the cell-surface membrane of epithelial cells lining the intestine.

7 Describe and explain the effect of stirring on the rate of absorption.
The graph shows that stirring increases the rate of absorption, regardless of the concentration of glucose in the lumen of the small intestine. Think about what happens when the glucose solution has not been stirred. As it is absorbed into the cells, the concentration in the intestine will fall, and the difference in lumen and cell concentrations will become less and less. Obviously, this fall in the concentration gradient will slow the rate of diffusion. Stirring maintains the concentration gradient. This results in a higher rate of absorption.

EXAMPLE

Starch and colon cancer

Scientists investigated the relationship between the food we eat and the **probability** of developing cancer of the colon. The colon is the last part of the digestive system. One of the factors that the scientists looked at was the amount of starch in people's diet. The scattergram in Figure 8.10 shows some of their results.

Probability A mathematical way of expressing the likelihood of a particular event occurring. You could describe the likelihood of a person developing colon cancer as 1 in 1000 (or 0.001) so you should use the term probability. In short, if you could put a number to it, use probability.

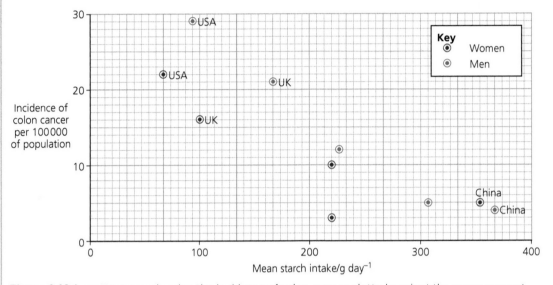

Figure 8.10 A scattergram showing the incidence of colon cancer plotted against the mean amount of starch in the diet for men and women of different nationalities.

TIP

In considering any set of data, one of the first things we do is to look carefully at the data set and make sure that we understand exactly what it shows. We will start here by looking at the axes.

1 The *y*-axis shows the incidence (number of cases) of colon cancer per 100 000 of the population. Why are the figures given per 100 000 of the population?

This is a straightforward question to answer. We want to compare the number of cases of cancer in different groups of people. The only way to do this is to compare like with like. The population of China is around 1 billion. The population of the UK is only about 65 million. In view of this, it is very likely that China will have more cases of colon cancer, simply because more people live there. Looking at the incidence per 100 000 allows us to make a fair comparison.

2 Do you think that giving the starch intake in grams per day lets us make a fair comparison?

It certainly helps because we must make sure that in each case we compare the amount of starch eaten over the same period of time. But people also vary in size. American men, for example, are larger on average than Chinese men. This probably affects the amount they eat. It might have been better to have taken body size into account as well, in which case the figures for starch intake would be grams per day per kilogram of body mass (g day^{-1} kg^{-1}).

3 Why did the scientists plot the figures for men and for women separately?
There are several possible reasons for this, but what they all come down to is that men and women form separate groups. They differ in body size and so probably eat different amounts of starch. There are also other important differences. Women may become pregnant, and they have different concentrations of different hormones circulating in their blood. These are factors that could affect the probability of developing colon cancer. But the scientists did not collect data about these factors. So it is better to treat men and women as separate groups.

4 Is there a correlation between the amount of starch that people eat and the probability of developing colon cancer?
We can find out whether there is such a correlation in several ways. Figure 8.11 shows that we can do this by drawing the line of best fit on the scattergram.

As you can see, the line slopes downwards. It tells us that the more starch people eat, the lower is the probability that they will develop colon cancer. American men eat very little starch. They have the highest incidence of colon cancer. Chinese men, however, eat a lot of starch and they have the lowest incidence of colon cancer.

> **TIP**
> Find out more about correlation and using scattergrams in Chapter 14.

5 Does this mean that eating starch lowers the incidence of colon cancer?
We have to be very careful here. Just because two things are correlated, it doesn't mean that one causes the other. We have seen that there seems to be a clear relationship between the amount of starch in the diet and the incidence of colon cancer, but we cannot say that eating a lot of starch will keep a person free of colon cancer. Other things could be involved. People in the USA probably eat more protein or more fat than those who live in China. Maybe that is the reason for the higher incidence of colon cancer. In other words, there could be other factors involved that we haven't considered.

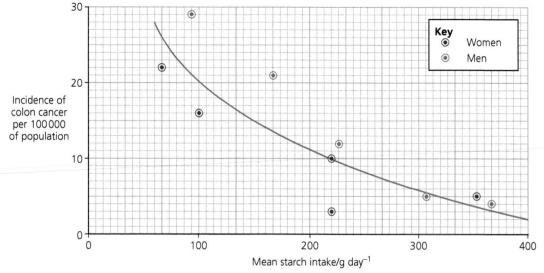

Figure 8.11 A scattergram showing the incidence of colon cancer plotted against the mean amount of starch in the diet for people of different nationalities. The line of best fit has been added.

6 So, does eating starch lower the incidence of colon cancer?
This is where scientists use their biological knowledge to suggest possible explanations for the results they collect. Food such as a banana contains different sorts of starch. Some of the starch in banana is digested only slowly in the human intestines. It is called resistant starch. When resistant starch enters the last part of our digestive system, the colon, it is hydrolysed by the bacteria that live there. They produce substances such as butyric acid when they digest starch.

Resistant starch may help to prevent cancers developing in one of two ways. First, butyric acid is known to kill cancer cells. Second, resistant starch helps to increase the rate of movement of faeces through the colon. This means that any substances in the faeces that could cause cancer spend less time in contact with the cells that line the colon. Before we can say definitely what happens, a lot more work is necessary. On the evidence that we have here, all we can conclude is that it is possible that eating starch lowers the incidence of colon cancer.

Protein digestion

Proteins are digested by enzymes called **proteases**. The process of protein digestion starts in the lumen of the stomach. Here an enzyme called an **endopeptidase** hydrolyses peptide bonds within the protein, hydrolysing it into smaller polypeptide 'chunks'. The endopeptidase in the stomach is secreted with hydrochloric acid, so the pH is very low in the stomach. This was particularly important in early humans as the acidic conditions would have destroyed many of the bacteria and parasites in their food.

After the stomach, the partly digested food passes into the small intestine. Here, pancreatic juice neutralises the acidic mixture that leaves the stomach and contains both endopeptidases and exopeptidases. Exopeptidases hydrolyse near the ends of the polypeptide chains, producing dipeptides. This is shown in Figure 8.12.

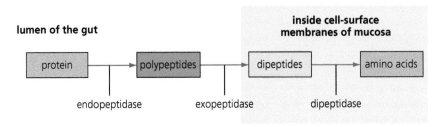

Figure 8.12 A summary of protein digestion in humans.

Finally there are dipeptidase enzymes in the cell-surface membrane of the epithelial cells of the small intestine (Figure 8.13, overleaf). These hydrolyse dipeptides and release amino acids into the cytoplasm of the cell.

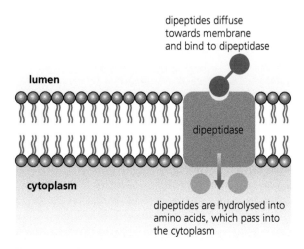

dipeptides diffuse
towards membrane
and bind to dipeptidase

lumen

dipeptidase

cytoplasm

dipeptides are hydrolysed into
amino acids, which pass into
the cytoplasm

Figure 8.13 Dipeptidase enzymes in the cell-surface membrane of the epithelial cells of the small intestine.

In the membrane of the epithelial cells of the small intestine, there are amino acid carrier proteins, similar to the glucose carrier proteins you learned about earlier in the chapter (see page 143). These also rely on sodium ions. The sodium/potassium pump transports sodium ions out of the cell by active transport. Amino acids and sodium ions bind to the carrier protein. As the sodium ions diffuse into the cell, amino acids are carried too. The amino acids then diffuse to the other end of the cell. They are then transferred to the capillaries by facilitated diffusion. This is shown in Figure 8.14.

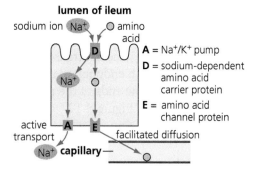

lumen of ileum

sodium ion Na^+ $\bigcirc$ amino acid

Na^+

active transport

Na^+ **capillary**

A = Na^+/K^+ pump
D = sodium-dependent amino acid carrier protein
E = amino acid channel protein

facilitated diffusion

Figure 8.14 Absorption of amino acids from the lumen of the small intestine into the blood capillary.

TEST YOURSELF

5 Explain the advantage of endopeptidase enzymes hydrolysing proteins before exopeptidases.

6 The lining of the stomach is covered in thick mucus. Explain why.

7 Use a diagram to show how a dipeptide is hydrolysed into amino acids.

8 List the similarities and differences between carbohydrate and protein digestion.

ACTIVITY

Chromatography of amino acids

TIP
See Chapter 15 for more information on chromatography.

A student wanted to identify the amino acids present in an unknown food substance, she decided to use a technique called paper chromatography to separate the different amino acids. She used the method below.

She drew an origin line in pencil and placed several crosses on the origin line. She placed a concentrated spot of the mixture on the origin line, and then put concentrated spots of known amino acids on the other origins on the paper. She placed the paper in a tank containing a solvent (butanol and ethanoic acid) for 2 hours. There was only a small amount of solvent in the tank, so that the solvent did not go above the origin line on the paper. When the solvent had almost reached the top of the paper, she quickly marked the solvent front using a pencil. She sprayed the chromatogram with ninhydrin and placed the paper in an oven at 100°C. After this the amino acids showed up as blue/purple spots. A drawing of the student's chromatogram is shown in Figure 8.15.

She calculated the R_f value using this formula:

$$R_f = \frac{\text{Distance moved by the solute}}{\text{Distance moved by the sovent}}$$

The R_f value of each solute is calculated and compared to published values in the same solvent. (R_f values are always less than 1 and have no units.)

1 The student wore plastic gloves while handling the chromatography paper. Suggest why.
2 The origin line was drawn in pencil, not ink. Suggest why.
3 Why was it important that the solvent in the tank did not come above the origin line?
4 Ninhydrin is a locating agent. Suggest why it is needed.
5 Calculate the R_f values of the spots on the chromatogram. Use your calculations to identify the amino acids present in the mixture.

Amino acid	R_f value
Alanine	0.38
Arginine	0.20
Glutamine	0.13
Leucine	0.73
Methionine	0.55
Tyrosine	0.45

TIP
Chromatography is not a required practical until your second year of study.

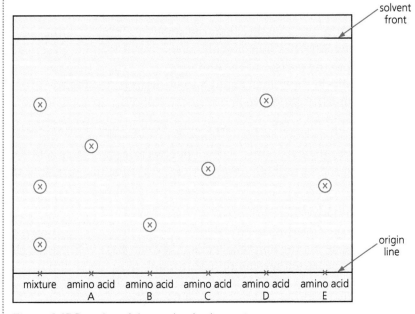

Figure 8.15 Drawing of the student's chromatogram.

Lipid digestion

Lipid digestion only occurs in the lumen of the small intestine. In the stomach, solid lipids are churned into a fatty liquid made of fat droplets, but no digestion takes place in the stomach. Once the fatty liquid enters the first part of the small intestine, bile from the gall bladder, which is connected to the liver, is secreted. This liquid contains bile salts. Bile salts bind to the fat droplets and break them down into smaller fat droplets. This is called **emulsification**. This is not digestion, but a physical process that increases the surface area available for lipase enzymes to digest the lipids. Lipase secreted into the small intestine by the pancreas hydrolyses lipids into fatty acids and glycerol. This is summarised in Figure 8.16.

lumen of the gut

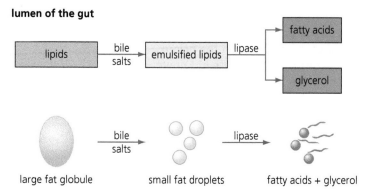

Figure 8.16 A summary of lipid digestion in humans.

Lipid absorption is different from that of proteins and carbohydrates. Lipids are hydrolysed into glycerol, fatty acids and monoglycerides (partly broken down lipids). Monoglycerides and fatty acids associate with bile salts and phopholipids to form **micelles**. Micelles are droplets that are about 200 times smaller than the small fat droplets in the emulsified lipids. Micelles transport the poorly soluble monoglycerides and fatty acids to the surface of the epithelial cells where they can be absorbed. Micelles are small enough to fit between the microvilli. Micelles constantly break down and re-form, building up a small pool of monoglycerides and fatty acids that are in solution. Only freely dissolved monoglycerides and fatty acids can be absorbed, not the micelles. Because they are non-polar, monoglycerides and fatty acids can diffuse through the phospholipid bilayer of the plasma membrane of the epithelial cell.

Fatty acids can have different lengths of hydrocarbon chains in them. Short-chain fatty acids diffuse directly into the blood from the lumen of the small intestine via the epithelial cells. They can pass easily through the membranes because they can diffuse through the phospholipid bilayer. Longer-chain fatty acids, monoglycerides and glycerol diffuse into the epithelial cells where they recombine to form triglycerides again. These triglycerides are packaged with cholesterol and phospholipids to form water-soluble fat droplets called **chylomicrons**. These are transferred to a lymph vessel inside the villus, called a lacteal, by exocytosis. Exocytosis is when a small piece of the cell-surface membrane is wrapped around the lipid droplets and pinched off, so that the fatty droplets are now wrapped in membrane as they enter the lymph vessels. They eventually enter the blood system. This is shown in Figure 8.17.

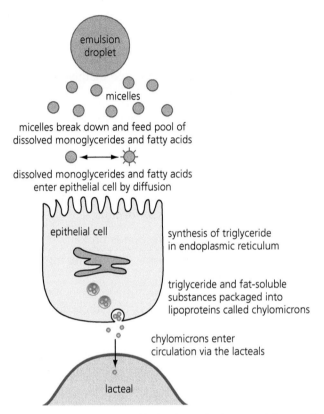

Figure 8.17 Absorption of lipids from the lumen of the small intestine to the blood capillaries and the lymph system.

TEST YOURSELF

9 Explain why the products of lipid digestion can be absorbed by simple diffusion.

10 Explain how bile helps in lipid digestion.

11 How does lipid digestion differ from carbohydrate and protein digestion?

12 Explain why it is important for chylomicrons to be water-soluble.

Practice questions

1 a) Explain the advantage of an epithelial cell from the small intestine

    i) having many microvilli on its surface   (3)

    ii) containing many mitochondria.   (3)

  b) i) Describe how a piece of cheese, composed mainly of fat, is digested in the human gut.   (3)

    ii) Describe how the digestion products of the cheese are absorbed in the gut.   (3)

2 In the food industry, enzymes are used to produce glucose and maltose from starch. Food scientists use the expression dextrose equivalent (D.E.) when they are describing the products formed from the starch. D.E. is calculated from the formula:

$$\text{D.E.} = 100 \times \frac{\text{Number of glycosidic bonds broken}}{\text{Initial number of glycosidic bonds present}}$$

  a) i) The D.E. of glucose is 100. Explain why.   *(1)*

    ii) What would you expect to be the D.E. of maltose formed from starch? Explain how you arrived at your answer.   *(2)*

The flow chart summarises the processes in which enzymes are used to produce glucose syrup and maltose syrup from corn starch.

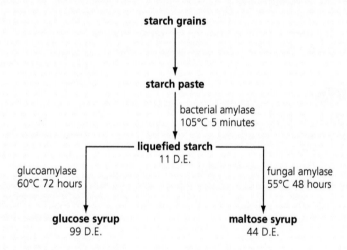

  b) Explain the evidence from the flow chart that:

    i) bacterial amylase is a thermostable enzyme   (2)

    ii) the glucose syrup formed in this process is not pure.   (2)

  c) What does the information in this flow chart suggest about the chemical nature of liquefied starch? Explain your answer.   (3)

  d) Glucose used to be made by hydrolysing starch with acid. The use of enzymes has a number of advantages over using acid.

    i) When the process of converting liquefied starch to glucose is carried out using enzymes, it can be stopped very easily. This is done by heating the mixture to 85°C for 5 minutes. Explain how heating the mixture stops the reaction. *(2)*

ii) Suggest one other advantage of using enzyme hydrolysis rather than acid hydrolysis. *(1)*

3 a) Lactose is a disaccharide found in milk. Describe how it is digested and absorbed in the human gut. *(2)*

b) Many adults are lactose intolerant. This means that they no longer produce the enzyme lactase in their small intestines. Two people were tested for lactose intolerance. Both people had nothing to eat, and only water to drink, for 10 hours. Then both people were given a drink containing 50 g of lactose. The concentration of glucose in their blood was measured over the next 2 hours. The results of this test are shown in the graph.

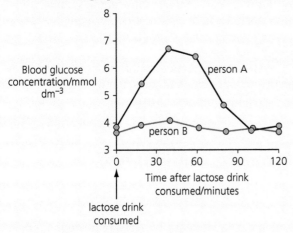

i) The people had nothing to eat, and only water to drink, for 10 hours before the test. Explain why. *(1)*

ii) Which one of these people was lactose intolerant? Explain the reason for your answer. *(2)*

c) How could you calculate the percentage increase in blood glucose concentration for person A in the first 30 minutes after consuming the lactose drink? *(1)*

Stretch and challenge

4 How do animals such as rabbits and cattle survive on a diet consisting mainly of cellulose without producing any cellulase enzymes?

5 Compare the gut of a bird with a human gut. Explain how the differences show adaptations to different diets.

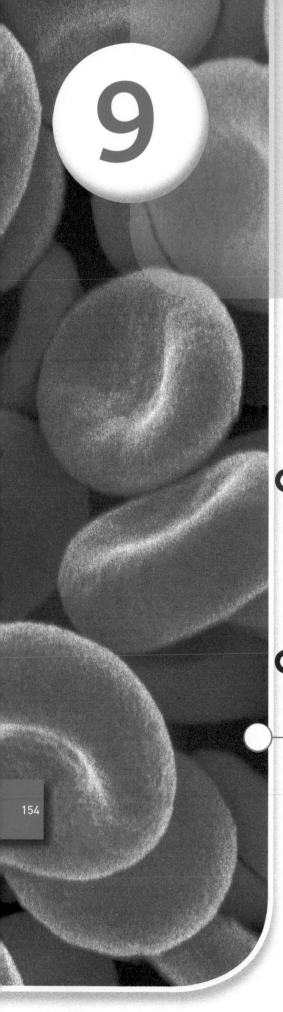

9 Mass transport in animals

TEST YOURSELF ON PRIOR KNOWLEDGE

1 What is meant by cell differentiation?
2 Why do multicellular organisms require transport systems?
3 How many beats has the heart of a man had by the time of his 85th birthday, assuming his heart beats on average throughout his life at a rate of 70 beats per minute? Express your answer in standard form.
4 Name the substances that are transported in blood to the heart muscle in order for it to keep beating.

TIP
See Chapter 14 to see how to express large numbers in standard form.

Introduction

The exchange of substances between the internal and external environments of organisms takes place at **exchange surfaces**. But substances have not really entered or left organisms until they enter or leave their cells by crossing cell-surface membranes. In large multicellular animals, the exchange surfaces you have seen in chapters 7 and 8 are some distance away from most of the rest of their cells. This is also the case for the exchange surfaces of multicellular plants. This means that exchange surfaces such as the lungs and digestive system are linked to mass transport systems that carry substances the relatively large distances to other body cells and from those cells back to the exchange surfaces. Diffusion across such distances

Mass transport The bulk movement of liquids or gases in one direction, usually through a system of tubes or vessels.

would not meet the metabolic demands of cells. But diffusion is still involved. It just takes place at each end of the route that substances travel, at the exchange surfaces and between the cells and their immediate fluid environment. Mass transport carries substances quickly from one to the other but also maintains the diffusion gradients at the exchange surfaces and between the cells and their fluid surroundings. In this way, mass transport helps to keep the immediate fluid environment of cells in multicellular organisms within a suitable metabolic range for effective cell activity.

Blood and circulation

Blood

Blood has two main constituents, the liquid plasma and the cells.

When blood is centrifuged, the red blood cells are forced to the bottom of the tube because they are heaviest. Most of the remainder, pale yellow plasma, is a fluid that makes up just over half the volume. Its role is to transport dissolved glucose, amino acids, urea, mineral ions and hormones. The great majority of cells are the red blood cells (technically called erythrocytes), which transport oxygen and some carbon dioxide. There are many fewer white cells (leucocytes), and these form a barely visible layer on top of the red cells.

The proportions of plasma and cells are important. There has to be enough plasma for the blood to flow easily. On the other hand, an efficient oxygen supply requires a large proportion of red blood cells. A quite small decrease in the proportion of plasma makes the blood more viscous (sticky), and the heart has to pump harder to push it round. This increases blood pressure and places additional strain on the heart. Increased blood viscosity is one way in which dehydration can affect the performance of an athlete.

The human heart is the pump that forces blood round the body. In effect, it is a double pump, because the left half pumps blood to the majority of the body tissues, while the right half pumps blood through the lungs. In this section we consider how the vessels that carry the blood around the body are adapted for their function.

Figure 9.2 is a diagram showing the basic plan of the blood circulation of a mammal. Blood is pumped out of the heart into **arteries**. These branch into narrower **arterioles** within the organs, and the arterioles branch to form the mass of very narrow **capillaries**, which penetrate the tissues. From the capillaries the blood flows into **venules** and **veins** that transport it back to the heart.

Arteries

The arteries have to withstand the full force of the pumping action of the heart's ventricles. They are adapted for this by having thick but flexible walls. You can see from Figure 9.3 (overleaf) that an artery wall has three layers. The thick middle layer consists of a mixture of muscle cells and elastic fibres. Outside this is a layer of tough protein fibres. The innermost layer, the endothelium, consists of flattened cells with an extremely smooth surface. This reduces friction and ensures that blood flows freely and does not stick to the walls. If this surface does get damaged blood clots are liable to form and may block the artery (see page 174).

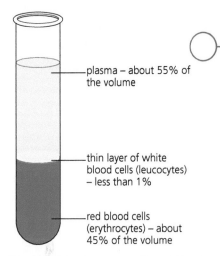

Figure 9.1 A tube of centrifuged blood.

plasma – about 55% of the volume

thin layer of white blood cells (leucocytes) – less than 1%

red blood cells (erythrocytes) – about 45% of the volume

Plasma The liquid component of blood.

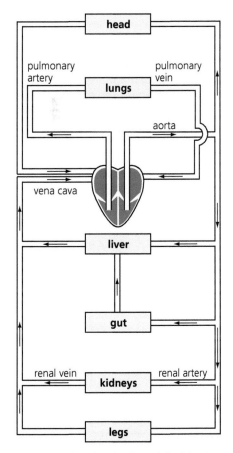

Figure 9.2 The basic plan of the blood circulation of a mammal.

head

pulmonary artery

pulmonary vein

lungs

aorta

vena cava

liver

gut

renal vein

renal artery

kidneys

legs

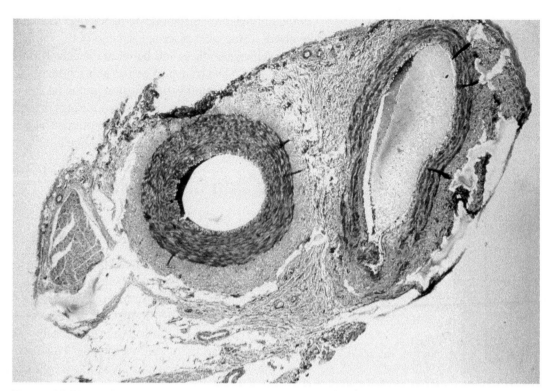

Figure 9.3 Photomicrograph of the cross-section of an artery (left) and a vein (right).

TIP

Arteries and veins are defined by the direction of blood flow. Arteries carry blood away from the heart while veins carry blood back to the heart. You should not define them in terms of carrying oxygenated or oxygen-depleted blood because there are exceptions such as the pulmonary artery and the umbilical artery, which both carry oxygen-depleted blood.

The elastic fibres in the middle layer of the artery wall allow the artery to expand each time that the heart beats. This is better than having a rigid tube because the stretching of the fibres absorbs the shock waves caused by the heart's forceful pumping action. The other advantage of the elasticity is that the fibres recoil to their original length between heartbeats. This smoothes out the changes in pressure and maintains a more constant blood flow. Close to the heart, arteries have a high proportion of elastic fibres and few muscle cells. In the smaller arterioles the balance is the opposite way round. Here, muscle cells can contract and partially shut off blood flow to particular organs. For example, during exercise, blood flow to the stomach and intestines is reduced, allowing greater blood flow to the muscles.

Veins

Veins have much thinner walls than arteries. The blood pressure is much lower and the blood moves more slowly. They have a considerably wider central lumen and so the same volume of blood returns to the heart as leaves it, but the returning flow is slower. It is like a wide river moving slowly compared with a narrow mountain torrent that joins the river. Like the arteries, the veins have walls of three layers, but the middle layer is much less muscular and has only a few elastic fibres. Valves in the veins

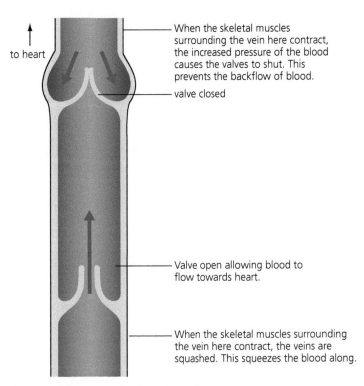

to heart

When the skeletal muscles surrounding the vein here contract, the increased pressure of the blood causes the valves to shut. This prevents the backflow of blood.

valve closed

Valve open allowing blood to flow towards heart.

When the skeletal muscles surrounding the vein here contract, the veins are squashed. This squeezes the blood along.

Figure 9.4 Diagram of valves in a vein.

ensure that the blood can only flow towards the heart (Figure 9.4). Finally, blood is drawn into the heart when the chambers expand and there is a brief period of lower pressure (see Figure 9.32, page 169).

The smallest arterioles and venules are connected by capillaries. A capillary is only about 8 µm in diameter, and it is estimated that an adult human has nearly 100 000 km of capillaries. That is more than twice the distance round the equator! All of the capillaries put together provide an enormous surface area for the exchange of gases, glucose and other substances, and no cell in the body is more than a very short diffusion distance from a capillary. The walls of many capillaries, especially those in muscles and the lungs, consist of a single thin endothelial cell wrapped into a tubular shape, as shown in Figure 9.5. Others have two or three cells linked together, but all have very thin walls only one cell thick, which allows for rapid diffusion but also increases resistance to flow.

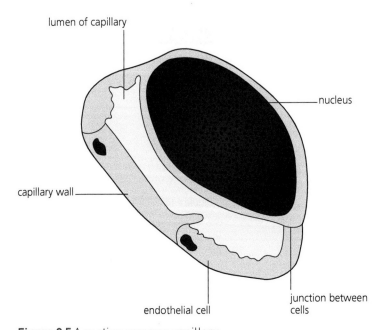

lumen of capillary

nucleus

capillary wall

endothelial cell

junction between cells

Figure 9.5 A section across a capillary.

157

TEST YOURSELF
1 Give the name of the arteries that supply blood to the kidneys.
2 Explain the advantage of having a lot of elastic tissue in the wall of an artery.
3 Explain the role of the muscle tissue in the wall of an arteriole.
4 Explain the function of valves in veins.

Leaky vessels

Capillaries have such thin walls that most substances can easily pass through. Only substances with particularly large molecules, such as most proteins, cannot escape from the blood. This has the advantage that oxygen, carbon dioxide, glucose, amino acids, hormones and many other substances are easily exchanged between the blood and the tissues.

The capillaries also allow some fluid to leak out between the cells. This produces a solution that fills the spaces between cells. This is called tissue fluid. Obviously, water cannot simply drain out of the blood vessels without being replaced. In fact, in the tissues there is constant movement of water into and out of the capillaries. By the time blood enters the capillaries from the arterioles it is at a much lower pressure than when it left the heart in the arteries. Nevertheless, this hydrostatic pressure is still large enough to force water out of capillaries into the tissue fluid. Therefore, as the blood passes along the capillaries, the water content decreases. Small solutes including glucose and ions also pass out, but the larger soluble proteins in the blood plasma cannot go through the walls of the capillaries. As water is lost, the concentration of proteins in the blood plasma increases and the pressure drops. This lowers the **water potential** inside the capillary. The result is that at the venule end of the capillary the water potential of the plasma is lower than the water potential of the tissue fluid. Therefore, water goes back into the capillary by osmosis. This is summarised in Figure 9.6.

Tissue fluid A fluid surrounding cells that is formed from blood plasma without large proteins. It is the immediate environment of each cell.

Hydrostatic pressure Pressure caused by an increased volume of fluid inside a vessel.

> **TIP**
> Refresh your memory of water potential by going back and reading about it in Chapter 3.

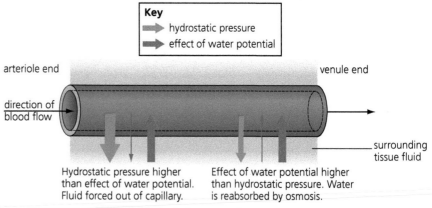

Figure 9.6 The effects of hydrostatic forces and water potential at the arteriole end and the venule end of a capillary.

Not all of the excess water in the tissues is reabsorbed into the blood capillaries by osmosis. The tissues also have a drainage system that consists of tubes slightly wider than capillaries. These tubes are called

Lymph capillary A vessel that helps to drain tissue fluid and return plasma proteins to the blood via the lymphatic system.

lymph capillaries (Figure 9.7). They are not part of a circulatory system like the blood. They have closed ends, which are sufficiently porous for tissue fluid and large molecules to enter. This is important because some **plasma proteins** do escape from the blood capillaries. If they accumulated in the tissue fluid, they would lower the water potential of the tissue fluid until water would no longer be reabsorbed into the blood. Escaped plasma proteins return to the blood via the lymphatic system.

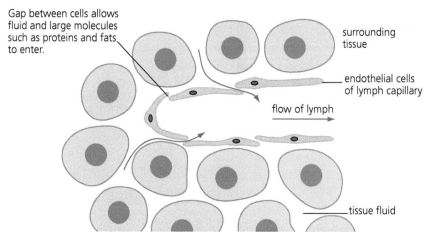

Gap between cells allows fluid and large molecules such as proteins and fats to enter.

surrounding tissue

endothelial cells of lymph capillary

flow of lymph

tissue fluid

Figure 9.7 Section through the end of a lymph capillary.

Figure 9.8 A child with kwashiorkor, a condition caused by malnutrition.

The lymph capillaries drain the excess fluid through a system of larger vessels that connect with large veins beneath the collar bones where the fluid returns to the blood system. The lymph vessels have very thin walls and also have valves similar to those in veins. There are some muscle cells in the walls of the lymph vessels to move the fluid along.

Figure 9.8 shows a child suffering from severe malnutrition. The child's diet contains very little protein. Consequently his blood is very short of plasma proteins. One result is that fluid collects in the tissues, which is also called oedema. This causes the child's belly and limbs to swell, as you can see in the photo.

Figure 9.9 shows a woman suffering from a condition called elephantiasis. The woman has been infected with parasitic worms that have blocked the lymph vessels in one leg. Tissue fluid has accumulated in this leg, making it swell massively so that it looks like an elephant's leg.

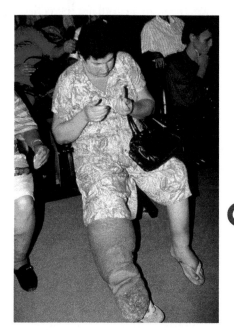

Figure 9.9 A person with elephantiasis, caused by parasitic worms that block lymph vessels.

TEST YOURSELF

5 Explain why the escape of plasma proteins from capillaries could cause fluid to collect in the tissues.

6 Explain how a shortage of protein in the diet can lead to oedema.

7 Explain how a blockage in the lymph vessels, such as a parasitic worm, can cause oedema.

8 People with high blood pressure can have oedema in their hands and feet. Explain why.

How blood is adapted to transport oxygen

Red blood cells have several adaptations for the transport of oxygen. Consider how each of the following features assists in efficient oxygen transport, before reading the explanations below. A red blood cell:

- is small in size: each cell is just over 7 μm in diameter. A blood capillary has an outer diameter of about 8 μm; the diameter of its lumen is rather less
- is shaped like a flattened disc
- has a thin central part of the disc
- has no organelles such as a nucleus or mitochondria
- is filled with haemoglobin.

Small size

The small size allows the red cells to pass through the narrow capillaries. By being about the same diameter as the lumen of the capillary, they touch the sides, which reduces the distance that oxygen has to diffuse once it enters the capillary. The narrowness of the capillaries is an associated adaptation in that they can fit within the spaces between cells.

Flattened disc shape

The flat shape (Figure 9.10) of a red blood cell increases the surface area to volume ratio and greatly increases the area through which oxygen can diffuse. It also results in all the haemoglobin being close to the surface, giving a short diffusion pathway. If the cells were spherical, much of the haemoglobin in the centre would be of little use because it would be too far for oxygen to reach it in the time available.

Thin central part of disc

The thin centre allows the cell to be flexible so that it can bend and squeeze through the narrow capillaries. As blood flows through a capillary, the cell tends to form a dome shape with its edges scraping along the wall of the capillary.

Absence of organelles

The absence of organelles provides maximum space for haemoglobin.

Haemoglobin

The haemoglobin greatly increases the oxygen-carrying capacity of the blood. Oxygen is not very soluble in water, so only small amounts would be transported by plasma alone.

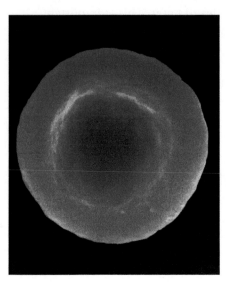

Figure 9.10 The shape of a red blood cell (×15000).

TEST YOURSELF

9 Describe how you could separate the plasma and cells of a sample of blood.

10 Give two ways that red blood cells are adapted to oxygen transport.

11 Red blood cells do contain some enzymes. These are important for carbon dioxide transport. But red cells cannot replace the enzymes. Explain why the enzymes cannot be synthesised in the red cells.

Haemoglobin

The structure of haemoglobin

Haemoglobin is a **quarternary** protein (see Chapter 1, page 13) that consists of four polypeptides called globins, with a haem group tucked in the centre of each (Figure 9.11). It is the haem that is the key to the oxygen-carrying function of haemoglobin.

You will notice from Figure 9.12 that haem contains iron (symbol: Fe) at its centre. The iron exists as an Fe^{2+} ion, which has the remarkable ability to combine reversibly with one oxygen molecule, making it ideal as a means of picking up and delivering oxygen, as we shall see later.

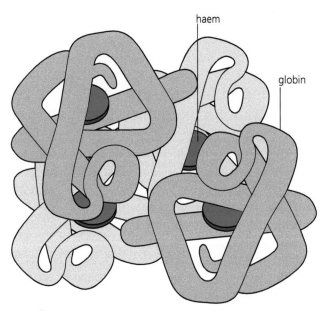

Figure 9.11 Because haemoglobin contains four polypeptide chains, it is said to have a quarternary structure.

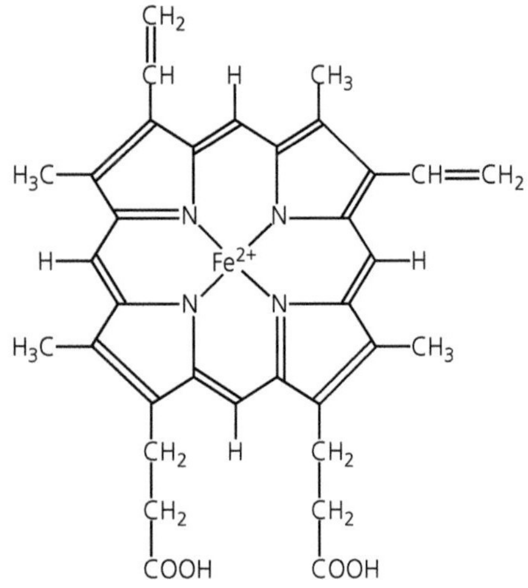

Figure 9.12 The chemical structure of haem.

Haemoglobin molecules combine with oxygen when it is present in high concentrations, but – and this is the important feature – the process is reversed when the concentration is low. This means that adult human haemoglobin takes up oxygen in the lungs, but when it reaches a tissue where there is little oxygen the haemoglobin releases it again.

Different kinds of haemoglobin

The structure of haem is the same in all haemoglobins, but the globin chains vary considerably between species. It is actually the globin component of haemoglobin that determines its precise properties, and many varieties exist.

Different forms of haemoglobin vary both in their oxygen-binding properties and in the conditions in which they take up and release oxygen. For example, llamas have haemoglobin suited to living at high altitude because it combines with oxygen more readily. This is useful because the partial pressure of oxygen at higher altitude is lower than normal. The haemoglobin of a developing baby in the womb differs from the haemoglobin the baby makes after its birth. Foetal haemoglobin is better at absorbing oxygen at low concentrations. This allows it to obtain its oxygen from the mother's blood, which has a much lower concentration than the air in the lungs.

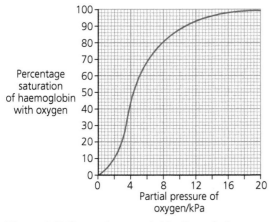

Figure 9.13 The oxyhaemoglobin dissociation curve for adult humans.

Oxyhaemoglobin The complex formed when haemoglobin binds to oxygen.

After birth, the baby starts to make adult haemoglobin, which is better suited for the uptake of oxygen in the lungs.

Haemoglobins can associate and disassociate from oxygen, an exceptionally useful property.

The oxyhaemoglobin dissociation curve

Figure 9.13 shows the oxyhaemoglobin dissociation curve for adult humans. This shows how much oxygen is combined with haemoglobin at different oxygen concentrations. However, if you look at the *x*-axis of the graph, you will see that the scale shows not a concentration but the 'partial pressure' of oxygen. Scientists use **partial pressure** as a measure rather than concentration because it is a more useful indicator of how much oxygen is available to haemoglobin.

Extension

Why is it more useful to measure partial pressure of oxygen in air than its oxygen concentration?

The air is a mixture of gases. Each gas exerts a pressure proportional to its concentration. Atmospheric pressure at sea level is about 100 kPa. (It varies a bit according to weather conditions, so forecasters refer to low-pressure areas and high-pressure areas.) The atmosphere is about 21% oxygen. So, at sea level, oxygen exerts a partial pressure of about 21% of the total atmospheric pressure; that is, about 21 kPa. Up a high mountain, the atmospheric pressure is much lower because the air is much thinner. But the proportion of oxygen is still about 21%. However, there is far less oxygen in each cubic metre of air, so the pressure exerted by the oxygen is much lower. On the summit of Mount Everest it is only about 7 kPa.

● What do we mean if we say that haemoglobin is 100% saturated?

In Figure 9.13, the *y*-axis shows the saturation of haemoglobin with oxygen. Each haemoglobin molecule can combine with a maximum of four molecules of oxygen, making oxyhaemoglobin. If all the haemoglobin molecules in a red blood cell combine with four oxygen molecules, the haemoglobin will be 100% saturated, and obviously it cannot possibly take on any more oxygen. (If 80% of the oxygen is released, the haemoglobin will then be only 20% saturated.)

TIP
Oxyhaemoglobin dissociation curves often make more sense if you read them from right to left. The right-hand side of the curve is when the haemoglobin loads with oxygen at the lungs, then as the curve moves to the left it gradually unloads as the red blood cells reach parts of the body with a lower partial pressure of oxygen.

Let us look at the shape of the curve on the graph. It is like a partly flattened letter S. This is because the haemoglobin molecule changes shape and loads with oxygen more easily once the first oxygen has combined with it. For the first haem group to form a bond with the first oxygen, the four globins have to move a little relative to each other. Once they have moved, the next three oxygens can form bonds with the remaining three haem groups more easily. This is known as **cooperative binding** and explains why the first part of the S shape curves upwards. You will recall the similar idea of induced fit when enzymes bind with their substrate (see Chapter 3).

At the other end of the S shape, the curve flattens off. The graph shows that haemoglobin reaches nearly 100% oxygen saturation when the partial pressure is much lower than atmospheric partial pressure of 21 kPa.

● What is the advantage of this response to partial pressure?

In the alveoli of the lungs, the partial pressure of oxygen is less than the 21 kPa in the air outside the body. This is because the alveolar air contains a lot of water vapour and a relatively high concentration of carbon dioxide. The partial pressure of oxygen is usually about 15 kPa. From the graph you can see that even at this lower partial pressure the haemoglobin becomes almost 100% saturated.

Therefore, as the blood passes through the lung capillaries, the haemoglobin in the red cells becomes loaded with oxygen. This happens extremely quickly: the haemoglobin will become almost fully saturated within a fraction of a second. The blood system carries the red cells through a pulmonary vein into the heart from where they are pumped out through arteries and arterioles to the tissues of the body.

● Why does the oxyhaemoglobin keep its oxygen until it reaches the capillaries in the tissues?

The walls of the veins, arteries and arterioles are too thick to allow oxygen to escape rapidly. The partial pressure around the red cells remains constant, so the haemoglobin stays saturated. The data in Figure 9.14 explain why oxygen is unloaded from the oxyhaemoglobin when the red cells reach the capillaries in the tissues. When the red cells reach a capillary in, for example, a muscle or the brain, the surrounding tissue will have a low partial pressure of oxygen because these tissues will have been using oxygen for respiration.

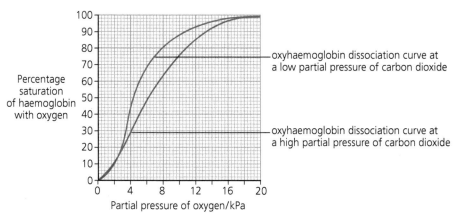

Figure 9.14 The oxyhaemoglobin dissociation curve in low and high partial pressures of carbon dioxide.

163

Suppose that the partial pressure of oxygen in a muscle is 4 kPa. From the graph, at 4 kPa the haemoglobin can be no more than 55% saturated. Therefore, the oxyhaemoglobin rapidly unloads oxygen until the haemoglobin is 55% saturated. This unloading is called **dissociation**. Assuming that the haemoglobin became 100% saturated in the alveoli (in practice it is usually slightly less), it will unload 45% of its oxygen. This will rapidly diffuse into the muscle. Bear in mind that this all happens very fast as the blood circulates. (It is not like a truck that has to stop to load or unload.) So, although it may seem inefficient for oxyhaemoglobin to release only some of its oxygen, the muscle gets a continuous supply.

● Why is more oxygen unloaded in active muscle?

In practice, more than 45% of the oxygen is unloaded in active muscles. An active muscle is respiring and rapidly producing carbon dioxide. The data in the graph in Figure 9.14 explains this. Look at the curve for oxyhaemoglobin dissociation at a high partial pressure of carbon dioxide. You will see that it forms a more forward-sloping S. It is to the right of the curve for oxyhaemoglobin dissociation at a low carbon dioxide partial pressure. This change in position of the dissociation curve due to an increased concentration of carbon dioxide is called the Bohr effect. The increased concentration of carbon dioxide causes a change in the shape of the protein, in the same way that enzymes change shape with a change in pH.

You will recall that muscle has a 4% partial pressure of oxygen, when oxygen in oxyhaemoglobin above 55% saturation is unloaded into the muscle. Now read off the percentage saturation of haemoglobin when the partial pressure of carbon dioxide is high, as it will be in active muscle. It is only 30%. The oxyhaemoglobin will dissociate more completely and release 70% of its oxygen. The advantage of the Bohr effect is that when a muscle is respiring more it receives an increased oxygen supply.

EXAMPLE

Oxyhaemoglobin dissociation in a foetus and its mother

As we have seen, there are different forms of haemoglobin in different animals, adapted for different environmental conditions or different stages in their lives. Figure 9.15 shows the oxyhaemoglobin dissociation curve for the type of haemoglobin that babies have before birth as well as for the mother's adult haemoglobin. While in the womb, a foetus receives its oxygen from the mother's blood by diffusion across the placenta.

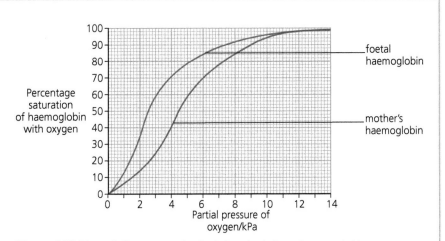

Figure 9.15 Dissociation curves for foetal and adult oxyhaemoglobin.

1 The partial pressure of oxygen in the placenta may be 3 kPa. At that value, what percentage of the oxygen in the mother's oxyhaemoglobin will dissociate?
A line drawn up to the blue curve from 3 kPa will reach 24% on the percentage saturation axis.

2 At the same partial pressure of 3 kPa, what percentage of the foetal haemoglobin will become saturated?
A line drawn up to the red curve from 3 kPa will reach 58% on the percentage saturation axis.

3 Explain the advantage of the foetal haemoglobin having a different oxyhaemoglobin dissociation curve from the mother's haemoglobin.

It means that the foetal haemoglobin can become saturated with oxygen even when the mother's haemoglobin is not fully saturated.

A foetus starts to manufacture more and more adult haemoglobin as it gets closer to the time of birth. After birth, the proportion of foetal haemoglobin in the blood normally declines quite rapidly.

4 Explain why it is important for the baby to have adult haemoglobin after birth.
Once the baby is born, it is breathing in air with a comparatively high oxygen concentration and its haemoglobin will saturate with oxygen in the lungs. It needs adult haemoglobin so that the oxyhaemoglobin will gradually unload its oxygen at the different partial pressures of oxygen in the body.

The function of the heart

The heart is the most obviously active organ in the human body. It is constantly busy, even during sleep. When the heart stops beating we die. Traditionally the heart was thought to be the seat of our emotions: bravery, love, passion, excitement. This is not surprising since we can feel the changes in our heartbeat when we are frightened or get excited.

In reality, the function of the heart is relatively simple: to pump the blood round the body, forcing it to all parts of the body. Strictly, the human heart is two pumps, as it has two halves. One half pumps blood to the lungs, where it is oxygenated. The oxygenated blood returns to the other half of the heart and is then pumped to all the other organs.

The blood transports both oxygen and the glucose required for respiration. When the heart stops beating, the whole of the body is deprived of oxygen and fuel. Cells do not have reserves of oxygen. The brain cells are particularly sensitive to a shortage of oxygen. If the oxygen supply is stopped, within about 5 minutes brain cells will start to die. Rapid first aid to restart the heart is essential if brain damage is to be avoided.

The rate and strength of the heartbeat must be able to change according to the body's requirements. When we are at rest our hearts normally function at a steady 60–70 beats per minute. But, during exercise, when the muscles use oxygen at a much faster rate, the heart rate may more than double. In this chapter we will look at the structure and functions of the different parts of the heart.

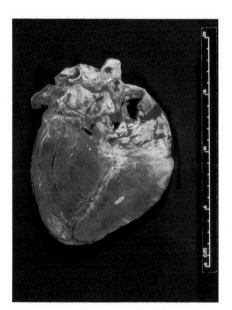

Figure 9.16 The outside of the heart.

The heart as a pump

Look at the diagram of the heart in Figure 9.17 (overleaf).

Notice that the heart has two halves, separated by a wall down the middle. Each half has two sections called chambers. The lower chambers are the ventricles. The right ventricle pumps blood to the lungs, and the left ventricle pumps it to the rest of the body. The two upper chambers are the atria (singular: atrium). The blood that returns from the lungs and the other parts of the body enters the atria. It is the job of the atria to collect and pump this blood into the ventricles.

TIP
The heart maintains a unidirectional flow of blood in the body. This helps to keep a steep concentration gradient at exchange surfaces, for example in the alveoli of the lungs (see Chapter 7).

TIP
Remember that diagrams show you the view from the front of a person, so you see the right side of the heart on the left of the diagram.

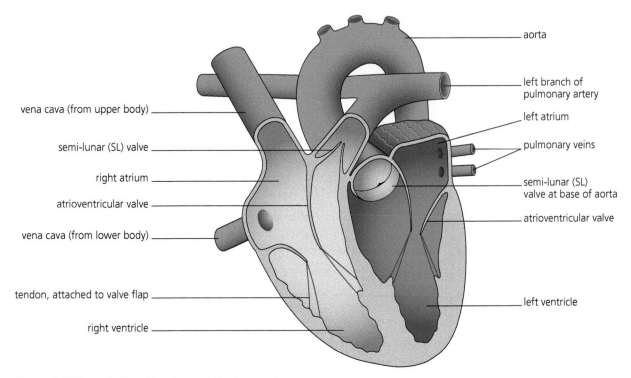

Figure 9.17 A vertical section through the human heart.

Labels on figure:
- aorta
- left branch of pulmonary artery
- left atrium
- pulmonary veins
- semi-lunar (SL) valve at base of aorta
- atrioventricular valve
- left ventricle
- vena cava (from upper body)
- semi-lunar (SL) valve
- right atrium
- atrioventricular valve
- vena cava (from lower body)
- tendon, attached to valve flap
- right ventricle

TIP

Many students wrongly think that the wall of the left ventricle is thick to withstand the pressure in it. This is the wrong way round. It is thick to create the pressure needed to circulate the blood around the body.

TIP

Do not confuse the pulmonary artery and the pulmonary vein. The pulmonary artery, like all **a**rteries, carries blood **a**way from the heart. However, it is the only artery that carries deoxygenated blood. The pulmonary vein, like all ve**in**s, carries blood **in**to the heart. However, it is one of the few veins that carry oxygenated blood.

Remember: **a**rteries carry blood **a**way from the heart, and ve**in**s carry blood **in**to the heart.

Between the atrium and the ventricle on each side is a valve, called an atrioventricular valve (or AV valve). The valve that separates each atrium from its ventricle has flaps made of thin but tough tissue. The atrium muscles relax and the atrium fills with blood from the vena cava. Its muscles then contract. This pushes the flaps down, and blood flows into the emptied ventricle, which has relaxed muscles. Then, when the ventricle muscles contract, the flaps are pushed together. This stops blood from flowing back from the ventricle into the atrium.

In Figure 9.17 you can see that thin tendons join the edges of the valve flaps to the wall of each ventricle. These tendons are like tough pieces of string and do not stretch. The function of the tendons is to ensure that the valve only opens one way and the blood does not flow back into the atrium.

When the ventricles contract, blood is forced from the ventricles into large blood vessels that pass out of the top of the heart. From the left ventricle, blood enters the aorta. Outside the heart, the aorta has branches to the head, arms, intestines, legs, and so on. From the right ventricle, the pulmonary artery has a branch to each lung. Notice that there is a valve at the lower end of both the aorta and the pulmonary artery. These valves stop blood from flowing back into the ventricles when the ventricle muscles relax and the ventricles start to open up again to fill from the atrium.

Blood returns to the heart through large veins. The pulmonary veins return blood from the lungs to the left atrium. Veins called vena cavae bring blood from the upper and lower parts of the body into the right atrium.

Blood is pumped from the left ventricle to organs of the body where it becomes less well oxygenated. The blood returns to the right atrium. It then passes into the right ventricle, which pumps it to the lungs where it is re-oxygenated. It returns to the left atrium and then reaches the left ventricle again (see Figure 9.18).

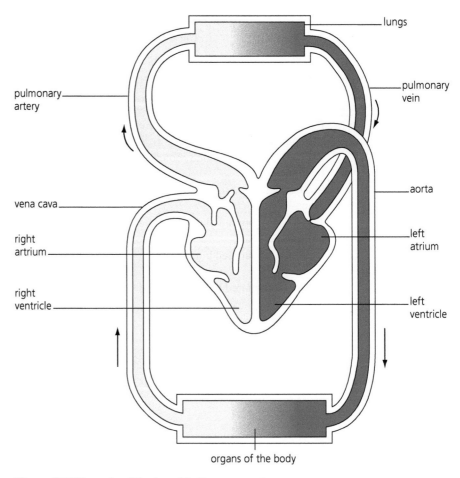

Figure 9.18 The role of the heart in the mammal.

labels on figure: lungs, pulmonary artery, pulmonary vein, vena cava, aorta, right artrium, left atrium, right ventricle, left ventricle, organs of the body

REQUIRED PRACTICAL 5

Dissection of animal or plant gas exchange system or mass transport system or of organ within such a system
This is just one example of how you might tackle this required practical.

Heart dissection

Figure 9.19 shows the heart as seen from the front (an anterior view)

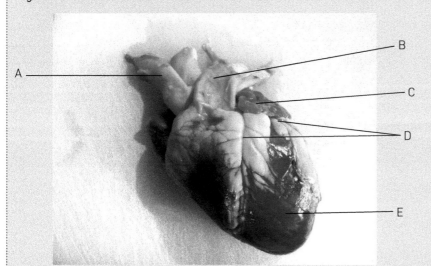

Figure 9.19 The anterior view of the heart.

SAFETY
When performing a dissection, you should only use material that has been obtained from a reputable butcher or supermarket. Wear gloves and follow your teacher's instructions carefully on how to dispose of the material appropriately. Take care when using sharp instruments such as scalpels. Keep your fingers away from the blade, and cut away from you. Dispose of instruments in disinfectant afterwards.

Figure 9.20 shows the heart as seen from the back (a posterior view)

1 Identify structures A–K as shown on Figures 9.19 and 9.20.

2 Copy and complete the table.

Structure	Function
	Receives deoxygenated blood from the body
	Carries oxygenated blood round the body
	Pumps deoxygenated blood to the lungs
	Takes deoxygenated blood from the body to the heart
	Receives oxygenated blood from the lungs
	Pumps oxygenated blood from the heart
	Carries oxygenated blood from the lungs back to the heart
	Carries deoxygenated blood from the heart to the lungs
	Carry oxygenated blood to the heart muscle

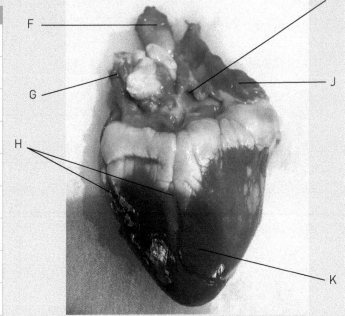

Figure 9.20 The posterior view of the heart.

Figure 9.21 shows the atrioventricular valve (AV) as seen in the right ventricle. Notice the flaps of the valve and the tendons attached to them.

3 Explain how these strong tendons help the valve to work effectively.

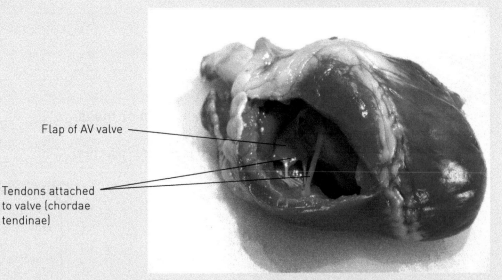

Flap of AV valve

Tendons attached to valve (chordae tendinae)

Figure 9.21 The AV valve in the right ventricle.

Figure 9.22 shows the left ventricle and the left AV valve. Notice that the wall of the left ventricle is much thicker than the wall of the right ventricle.

4 The wall of the left ventricle is much thicker than the wall of the right ventricle. Explain the advantage of this.

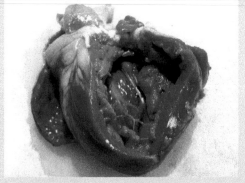

Figure 9.22 The left ventricle and the left AV valve.

Figure 9.23 shows a red pencil inserted into a blood vessel and passing through to the left ventricle.

5 Name the blood vessel and the structures the pencil passes through.

In Figure 9.24 the green pencil has passed from the left ventricle into a blood vessel.

6 Name the structures it passes through.

Figure 9.23 A pencil shown inserted into a blood vessel and passing through to the left ventricle.

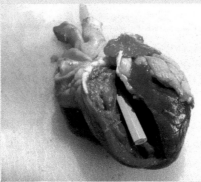

Figure 9.24 A pencil passing from the left ventricle into a blood vessel.

In Figure 9.25 the lilac pencil is inserted into a blood vessel that enters a chamber of the heart.

7 Name the blood vessel and the chamber of the heart it enters.

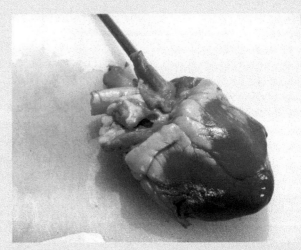

Figure 9.25 A pencil is inserted into a blood vessel that enters a chamber of the heart.

In Figure 9.26 the yellow pencil is passing from one chamber of the heart to another, via a valve.

8 Name the chambers and the valve.

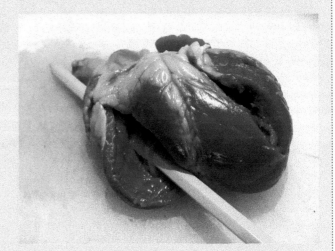

Figure 9.26 A pencil passes from one chamber of the heart to another, via a valve.

The cardiac cycle

Since you are reading this, you must be alive and your heart must be beating. To confirm this, gently press a fingertip on your neck just to one side of your trachea. Here, each time the heart beats, you should feel the pulse caused by the surge of blood through one of the arteries that goes to your head. If you are relaxed, you will feel a pulse roughly every second. The time between each pulse represents the length of the cardiac cycle.

Cardiac cycle The sequence of events that make up one heartbeat.

The cardiac cycle is the sequence of stages that happens during one heartbeat. The term for a stage in the cardiac cycle when the muscles of the heart chambers are contracting is **systole**. The atria contract during atrial systole and the ventricles contract during ventricular systole. The relaxation stage is called **diastole**. Therefore, the three stages shown in Figure 9.27 are atrial systole, ventricular systole and diastole.

TIP
You do not need to be able to recall the terms 'systole' and 'diastole', but if you do use them, make sure you get them the right way round.

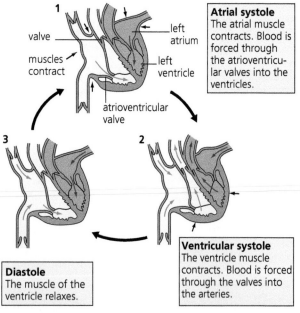

Atrial systole
The atrial muscle contracts. Blood is forced through the atrioventricular valves into the ventricles.

Ventricular systole
The ventricle muscle contracts. Blood is forced through the valves into the arteries.

Diastole
The muscle of the ventricle relaxes.

Figure 9.27 The three stages in the cardiac cycle. The atria contract (atrial systole), the ventricles contract (ventricular systole) and finally the ventricles relax (diastole).

Contraction of the muscles in the walls of the ventricles creates the pressure that circulates the blood. The left and right ventricles contract at the same time.

As the ventricles contract they build up a pressure higher than that in the aorta or the pulmonary artery, which forces open the valves at the base of the aorta and the pulmonary artery and drives blood out through these vessels. When the ventricle muscle relaxes, the ventricle walls recoil. This increases the volume of the ventricle and reduces the pressure so it is lower than the pressure in the aorta and the semi-lunar valve shuts. This means blood doesn't flow back into the ventricle. The pressure in the ventricle then falls below that in the atrium, the atrioventricular valve opens and blood flows into the ventricle.

After passing round the lungs or the rest of the body, the blood flows back into the atria of the heart at a much lower pressure. When the blood flows back in, the muscular walls of the heart are relaxed, so even at low pressure the returning blood expands the relatively thin walls of the atria. As the atria fill, some blood does pass through the valves into the ventricles. However, as the atria fill up, the muscles in their walls contract. This forces the valves to open fully and pushes the remaining blood quickly into the ventricles.

Each time the heart beats, the left ventricle pushes out the same volume of blood as the right ventricle. This may seem surprising, since the lungs are much nearer and smaller than the rest of the body, but otherwise the continuous circulation would not be maintained. If the right ventricle pumped out less blood each time, there would be less returning to the left atrium and therefore less for the left ventricle to pump to the body. The left ventricle would soon run out of blood to send to the body.

Analysing the pressure changes and valve openings and closings in the heart during the cardiac cycle

The graph in Figure 9.28 shows the changes in pressure that occur in the left side of the heart and in the aorta during one cardiac cycle. Let us look first at the curve showing pressure in the left ventricle. Key points on the curve are labelled with the letters A to D.

Figure 9.28 A graph showing changes in pressure during one cardiac cycle for the left atrium and left ventricle and the aorta.

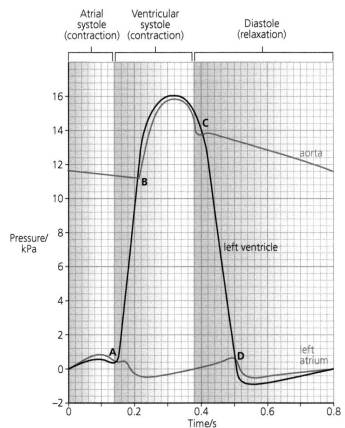

171

- Notice that at the start of atrial contraction the pressure in the left ventricle is more or less 0 kPa. At this time the muscles in the walls of the ventricle are relaxed and therefore not putting any pressure on the blood. During atrial contraction the pressure rises slightly as the left atrium contracts and pushes blood into the ventricle.
- At point A, the start of ventricular contraction, the ventricle walls contract strongly and the pressure shoots up. The pressure increase forces the atrioventricular valve to shut, as the pressure in the ventricle is greater than the pressure in the atrium, preventing blood from being pushed back into the atrium.
- At B, the pressure in the ventricle becomes the same as the pressure in the aorta. As soon as the pressure in the ventricle exceeds the pressure in the aorta, the valve at the base of the aorta is forced open and blood is pushed into the aorta.
- By C, the ventricle has been emptied of blood. The muscles in the ventricle wall relax, pressure in the ventricle falls back to below zero, i.e. lower than the pressure in the aorta, and the aortic valve shuts due to the greater pressure in the aorta.
- At D, for a short time there is a slight period of lower pressure as the ventricle expands and its internal volume increases.

Remember that one cardiac cycle follows another in a continuous process. There is never a gap between cycles, so blood immediately starts to flow into the expanding ventricles from the atria because the pressure in the atria is now greater than the pressure in the empty, expanding ventricles. This in turn causes blood to flow into the atria from the veins. The pressure in the ventricles rises again as they fill.

EXAMPLE

Pressure changes during the cardiac cycle
Look at the curves for the left atrial and aortic pressures shown in Figure 9.28.

1 Describe and explain what happens to the pressure in the atrium between the start of the cycle and A.
The pressure rises as the muscles in the atrium wall contract and force blood through the valve into ventricle. The pressure falls again as the atrium empties.

2 The atrioventricular valve closes at A. When does it open again? Explain your answer.
The valve opens at D. Notice that the curves for pressure in the ventricle and atrium cross at this point, so the pressure on either side of the valve is briefly the same. The pressure in the ventricle then falls slightly below the atrial pressure, so blood again flows through the valve into the ventricle.

3 Soon after A, the pressure falls as the atrium expands after atrial systole. Explain why the pressure then rises again until D.
The pressure of blood in the pulmonary vein, even though it is quite low, is pushing blood into the atrium and raising the pressure in the atrium as it fills up.

4 Explain why the maximum pressure in the atrium is much lower than the maximum pressure in the ventricle.
There is much less muscle in the walls of the atria. They exert much less force when they contract compared with the thick walls of the ventricle.

The pressure in the aorta stays high throughout the cycle. Its walls are elastic and are stretched during ventricular systole. The stretched walls exert pressure on the blood in the aorta, just as an elastic bandage on your leg continues to squeeze the leg.

5 Between which points on the curve is blood entering the aorta from the left ventricle?
Blood enters the aorta between B and C. The valve opens at B and closes again when the pressure in the ventricle falls below the pressure in the aorta.

6 Figure 9.28 shows the pressures on the left side of the heart. How would you expect a similar graph showing pressures on the right side to differ? Explain your answer.

It may help you to refer to the drawing of the heart in Figure 9.17.

The maximum pressure in the right ventricle would be lower because the walls are less muscular and exert less force. In fact, the maximum pressure in the right ventricle is normally less than a quarter of the maximum in the left ventricle. The pressure in the pulmonary artery would also be much lower. However, since the two sides pump together, the timing and pattern of events would be much the same.

EXAMPLE

Calculating cardiac output

Figure 9.29 shows the internal volume of the left ventricle during one cardiac cycle. This graph is on exactly the same time scale as the graph of pressure changes in Figure 9.28. The same positions have been labelled A–D, so you can match up with the changes in pressure in the left ventricle.

You can see that between B and C the volume of the ventricle decreases rapidly. This matches the period when pressure increases as the ventricle contracts. During this stage, blood is being pumped out into the aorta. As the muscles of the ventricle relax after C, its volume increases, the pressure falls and blood flows in from the atrium. At this stage, the muscles of the atrium are still relaxed. Atrial systole, when atrial muscles contract, merely tops up the blood in the ventricle between zero and point A on the curve in this graph.

1 Notice in Figure 9.29 that the volume of the left ventricle stays almost constant between A and B. How can this be explained?
  This is explained because the atrioventricular valve and the aortic valve are both shut. Therefore no blood is entering or leaving the ventricle so its volume stays the same.

The volume of blood pumped out of the left ventricle during one cardiac cycle is called the stroke volume.

2 Use the scale on the y-axis in Figure 9.29 to calculate the stroke volume shown by this graph.
  This is about 75 cm³.

We call the volume of blood that the left ventricle pumps out to the body per minute the cardiac output. To work out the cardiac output, you first need to calculate the heart rate, which is the number of cardiac cycles per minute. To find the number of cycles per minute you have to divide 60 by the time in seconds taken for one cycle.

3 How long is the cardiac cycle shown in Figure 9.29?
  One cardiac cycle lasts 0.8 seconds. You can then use the following equation to find the cardiac output:

  Cardiac output = stroke volume × heart rate

4 Calculate the cardiac output shown in Figure 9.29.
  Heart rate = $\frac{60}{0.8}$ = 75
  Cardiac output = 75 × 75 = 5625 cm³

For a man lying down and doing nothing, the average cardiac output is about 5 dm³ (5000 cm³). During exercise the cardiac output can increase to be about four times as great. This caters for the much higher oxygen and glucose requirements of active muscles. The increase can be produced by changing either the heart rate or the stroke volume, or both.

In an investigation, groups of trained athletes and untrained students were required to ride an exercise bike as fast as they could for a few minutes. Their maximum stroke volume and heart rate were measured. The mean maximum stroke volume for the athletes was

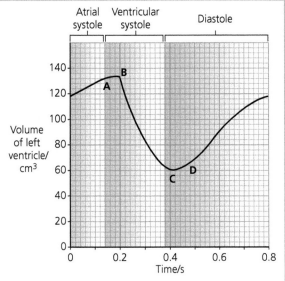

Figure 9.29 Graph of volume changes in the left ventricle in one cardiac cycle.

160 cm³ and for the students 100 cm³. The mean maximum heart rate was 190 beats per minute for the athletes and 200 beats per minute for the students.

5 a) Calculate the mean maximum cardiac output for each group.
  b) Explain which factor accounts for the difference.
  c) Suggest how training may have affected the stroke volume.

  a) For the athletes the mean maximum cardiac output = 160 × 190 = 30 400 cm³. For the students mean maximum cardiac output = 100 × 200 = 20 000 cm³.

  b) The stroke volume accounts for the difference. Both groups have a similar maximum heart rate, but the athletes have a greater stroke volume so they can circulate more oxygenated blood round the body in a minute.

  c) Training has increased the strength of the muscle in the wall of the left ventricle. During exercise, more deoxygenated blood returns to the right side of the heart to be oxygenated at the lungs. In turn, this leads to more oxygenated blood entering the left side of the heart. Therefore the ventricles are fully filled and this stretches their walls. As a result, the ventricles contract more forcefully.

TIPS

- You do not need to learn the formula for cardiac output or memorise the terms 'cardiac output' or 'stroke volume'.
- To find how long one cardiac cycle is on a graph with more than one cycle, find an easily identifiable point on the curve and measure the time taken to reach the same point on the next cycle.
- 1 dm³ is 1000 cm³.

Extension

Coronary heart disease

Coronary heart disease (or CHD) refers to any condition that interferes with the coronary arteries that supply blood to the heart muscle. Being very active, the muscle requires a continuous supply of oxygen and glucose. This supply does not enter the muscle from the blood in the chambers of the heart. It reaches the muscle from arteries that branch from the aorta and spread over the surface of the heart. Small vessels penetrate the heart muscle so that all parts are close enough to the blood supply for the oxygen and glucose to diffuse to the muscle cells.

Problems arise when blood vessels taking blood to the heart muscle become very narrow or blocked so that the supply of oxygen and glucose is reduced or cut off. How serious this is depends on where the blockage occurs and how much of the muscle is affected.

Arteries may become partly blocked by the formation of fatty deposits in their walls when the layer of cells lining the inside of the artery becomes damaged and inflamed. These fatty deposits are called atheroma. The resulting narrowing of the arteries is referred to as atherosclerosis or, in common language, 'hardening of the arteries'. Blood clots often form in a narrowed artery, a condition that is called **thrombosis**. If the blood supply to a major part of the heart muscle is completely blocked, the muscle starts to die due to lack of a blood supply. A heart attack follows. The medical term for heart muscle death is myocardial infarction (an infarction is the death of tissue; myocardial means that it occurs in the heart muscle).

EXAMPLE

Trialling aspirin as a treatment to prevent heart attacks

It was first suggested in the 1960s that aspirin might help to prevent heart attacks when it was discovered that aspirin helps to stop blood clots. This is because aspirin interferes with the action of blood platelets. The platelets are cell fragments that play an important role in blood clotting. Aspirin stops the platelets from sticking together and so reduces the chance that clots will block the vessels taking blood to the heart muscle.

At the time, most doctors were very scornful of the idea that such a cheap and common drug as aspirin could prevent heart attacks. Nevertheless, the Medical Research Council tested the idea, and in 1969 it set up a trial in Cardiff. The researchers decided to test the effect of giving a daily dose of aspirin to men who had recently recovered from a heart attack. One group of the men took a daily pill containing 300 mg of aspirin. The others, the control group, took a dummy pill containing no aspirin. The patients in each group were selected randomly, so neither the doctors nor the patients knew who was being treated with aspirin. Over 1200 men were studied, but only the coded records of the researchers enabled them

to trace which of the men had taken the aspirin. After treatment for a year, 12% fewer of the men taking the aspirin had died of another heart attack compared with the control group. However, the difference between deaths in the two groups was not enough for the results to be statistically significant. This one study had not given convincing evidence for doctors to start using aspirin as a regular part of the treatment for heart attack victims.

The research continued. Evidence was gathered from larger numbers, for longer times after the heart attack, from women as well as men, for different age groups and with different doses of aspirin. The results from these studies were collected and analysed. All showed that treatment with aspirin gave some benefit. In fact, the combined results of about 20 years of studies involved large enough numbers to show that aspirin gave a significant advantage after a first heart attack. During these studies, compared with the control groups, on average 34% fewer of the patients taking a daily dose of aspirin died.

It is now common for people who have had a heart attack to take aspirin. The evidence shows that this simple treatment prevents, or at least delays, many thousands of deaths a year. At the same time, this example illustrates some of the difficulties of developing new treatments in medicine. People vary enormously and respond differently to drugs and other treatments. Trial results rarely show absolutely clear-cut benefits, and some people may have unexpected side effects.

1 The men gave permission to take part in the first trial, but they could not choose to receive the treatment that might help to save their lives.
   a) Do you think they should have been allowed to choose?
   The men had to give permission to receive the aspirin treatment, as they needed to be aware that there might be side effects. It might seem fairer to let the men choose, but if they had been able to choose, this could have damaged the validity of the study.
   b) How might the results of the trial have been affected if they had been able to choose?
   The men needed to be put into groups at random. The kind of person who wants to try the aspirin treatment might be different from those who prefer to have the placebo. For example, they might be more interested in looking after their health. This would mean the two groups would not be random. Also, if a person knows they are receiving a drug treatment, they may actively look for benefits and report these to the doctors carrying out the study.
2 How should dummy pills be made in order to make sure the results are reliable?
   They should contain all the ingredients of the aspirin tablets except the aspirin, so that there is only one variable. They should also look the same as the aspirin tablets so that the people taking them cannot tell whether they are the real tablets or placebo.

3 Why do you think the doctors were not told which patients were getting the aspirin?
   Even though doctors try to be impartial, it is very difficult for them not to look for benefits of a drug if they think it helps patients. They may be more proactive in noticing health benefits in people taking aspirin, if they know which patients these are.

Aspirin can have side effects. For example, if a person also develops a stomach ulcer, it may bleed excessively because the aspirin reduces clotting. Yet the improvements in the survival rate after a heart attack show that the benefits from aspirin treatment are much greater than the risks from this side effect. However, because the benefits of aspirin were publicised, some people started taking aspirin to prevent a heart attack, even though they had never either suffered from one or shown signs of having one.

4 Suggest what advice could be given to the public and to doctors about this use of aspirin.
   This is a 'suggest' question so you need to give your own point of view, backed up with reasons. For example, you may think that people should be informed about the side effects of taking aspirin, so that they understand that it is not beneficial to take aspirin every day unless you have been recommended to do so by your doctor.

TIP
You don't need to learn the risk factors for coronary heart disease. Make sure you can interpret data concerning these risk factors and analyse it though.

Extension
Risk factors for CHD

Many factors increase the risk of suffering from coronary heart disease. Some of these factors are unavoidable, such as increasing age. But others are the result of particular lifestyles, and a person can choose to alter the risk from them. The main risk factors include the following.

- **Age and sex** Deaths from CHD can occur in young adults, but the numbers involved are small. Not surprisingly, the risk increases with age as damage to arteries develops slowly and as the effects of other factors take effect. Men are much more likely than women to get CHD in middle age, but after this the risk becomes fairly similar.
- **Genetic factors** CHD tends to run in families, especially where heart attacks occur in middle age or earlier. This may be partly due to members of families having similar lifestyles, but there is evidence for some genetic causes. For example, the risk of two identical twins both dying from CHD is about four times as great as the risk for non-identical twins.

175

- **Smoking** As we saw in Chapter 7, in addition to its effects on the lungs, smoking can significantly increase the risk of dying from CHD. Table 7.1 on page 128 shows that the risk for doctors who smoked heavily was nearly double that of non-smokers. Working out the precise effect is difficult. For heavy smokers it is likely that smoking will damage both the lungs and the heart. It may be chance as to which is the first to cause death. It is not certain how smoking affects the heart. One possible explanation is that nicotine makes arteries constrict, and this causes an increase in blood pressure, another risk factor.

- **High blood pressure** It is normal for blood pressure to increase during exercise, when the heart beats more forcefully. However, in some people the pressure is high even when they are at rest. The risk factors linked to high blood pressure include genetics, high salt intake, lack of exercise and alcoholism. One effect of high blood pressure is that the arteries develop thicker walls. As the wall of an artery thickens, the lumen gets narrower. As anyone who has attached rubber tubing to a tap and then squeezed will know, the water jet comes out with much greater force. The narrowing of the arteries therefore has the knock-on effect of raising blood pressure even more. This can damage their inner surface, making it more likely that atheroma will develop. It can also result in damage to the heart itself. The ventricles can enlarge, and in the worst cases the beating of the heart can become so irregular that heart failure results.

- **High concentration of low-density lipoproteins in the blood** Low-density lipoproteins (LDL) are involved in the formation of atheroma. Lipoproteins are a complex association of triglycerides, cholesterol and proteins. Although cholesterol has a bad reputation because of its link to heart disease, it is essential for the synthesis of cell membranes. Cholesterol is transported in the blood by LDL from the liver where the lipoproteins are made from fats and cholesterol in the diet. However, research shows that when the diet contains an excess of fats, especially saturated fats, the quantity of LDL in the blood rises and there is a greatly increased risk of developing coronary heart disease. The blood also contains high-density lipoprotein (HDL), which has a higher proportion of protein in their structure. These are the 'goodies' because they absorb excess cholesterol and return it to the liver where it is removed from the blood. People with a high ratio of HDL to LDL have a lower risk of developing heart disease. Blood tests can determine the ratio of HDL to LDL, and the results can be used to advise on protective changes to lifestyle. Drugs called statins may also be used to lower the concentration of LDL.

Measuring blood pressure

Although scientists use kilopascals (kPa) as the units of pressure, doctors still measure blood pressure in old units. These were the units in which a mercury barometer measured atmospheric pressure. Mean atmospheric pressure is 760 millimetres of mercury (mmHg), which is the height of a column of mercury that atmospheric pressure will support. Atmospheric pressure is 100 kPa, so 760 mmHg = 100 kPa.

Blood pressure is measured by finding the maximum and minimum pressures in the artery in the arm. In an adult, a healthy reading is taken to be less than 140 and 90 mmHg for these two measurements.

Risk factors work together

There are some other factors that show a statistical increase in the risk of heart disease, such as obesity, lack of exercise and diabetes. In some cases, the explanation for the increased risk is clearly linked to other factors. For example, obesity is likely to be linked to eating fatty foods, which in turn is likely to lead to high concentrations of cholesterol and LDL in the blood. For many people, their statistical risk of having a heart attack is the result of a combination of different risk factors.

ACTIVITY

Calculating risk of CHD

For each risk factor, statisticians can assess the chance of developing coronary heart disease. Table 9.1 gives points for some risk factors for men in different age groups. The points are based on statistical evidence for the increased risk for healthy men over a 10 year period. For example, the point score of a man aged between 40 and 49 is increased by 5 points if he is a smoker. His total point score in Table 9.1 shows the probability that he will develop CHD during the next 10 years of his life, assuming that he is healthy at the time of assessment.

To calculate the point score, you add together the scores for all of the five categories in Table 9.1. For example, a 42-year-old non-smoking man, with a cholesterol concentration of $180\,mg\,100\,cm^{-3}$ of blood, blood pressure of $18.7\,kPa$ and an HDL concentration of $55\,mg\,100\,cm^{-3}$, has these scores:

Age	3
Cholesterol	3
Smoking	0
Blood pressure	1
HDL	0.
The total score is	7.

From Table 9.2, you can see that this man has a 3% increased risk of suffering from CHD during the following 10 years. This is called his '10 year risk'.

1 Suggest explanations for each of the following.
   a) The point score rises steeply for older age groups.
   b) The point score for blood cholesterol is greatest for the youngest age group.
   c) Blood HDL content above $60\,mg\,100\,cm^{-3}$ has a negative score.
2 Work out the 10 year risk for each of the following people and suggest what advice about their lifestyle should be given in each case.
   a) A male smoker, aged 48, blood cholesterol $265\,mg\,100\,cm^{-3}$, blood pressure $19.5\,kPa$ and HDL $44\,mg\,100\,cm^{-3}$.
   b) A male non-smoker, aged 68, blood cholesterol $188\,mg\,100\,cm^{-3}$, blood pressure $18.7\,kPa$ and HDL $62\,mg\,100\,cm^{-3}$.

Table 9.1 Estimates of the risk of coronary heart disease over the next 10 year period for different age groups of men.

Factor	Range	Points for each age group Age groups/years			
		40 to 49	50 to 59	60 to 69	70 to 79
Age		3	7	10	12
Blood cholesterol/mg 100 cm^{-3}	<160	0	0	0	0
	160–199	3	2	1	0
	200–239	5	3	1	0
	240–279	6	4	2	1
	280+	8	5	3	1
Smoking	Smoker	5	3	1	1
	Non-smoker	0	0	0	0
Systolic blood pressure/kPa	<16	0	0	0	0
	16–17.2	0	0	0	0
	17.3–18.5	1	1	1	1
	18.7–21.2	1	1	1	1
	21.3+	2	2	2	2
Blood HDL content/mg 100 cm^{-3}	<40	2	2	2	2
	40–49	1	1	1	1
	50–59	0	0	0	0
	60+	–1	–1	–1	–1

Table 9.2 Point scores for men and the percentage probability that they will develop CHD in the next 10 years.

Point score	1	2	3	4	5	6	7	8	9	10	11	12	13	14	15	16	17+
Percentage increased risk of CHD over next 10 years	1	1	1	1	2	2	3	4	5	6	8	10	12	16	20	25	30

Practice questions

1 The lugworm is an animal that lives in sandy beaches. It lives in a tube, obtaining oxygen from sea water, which passes through the tube. However, when the tide goes out, very little oxygen is available. Here is the dissociation curve for lugworm oxyhaemoglobin.

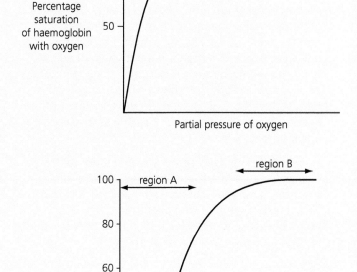

Percentage saturation of haemoglobin with oxygen

Partial pressure of oxygen

a) Explain the advantage to the lugworm of having an oxyhaemoglobin dissociation curve of this kind. (2)

b) Haemoglobin has quaternary structure. Explain what is meant by quaternary structure. (1)

2 The graph on the right shows the dissociation curve for human oxyhaemoglobin at a low carbon dioxide concentration.

a) Explain the advantage of the part of the oxyhaemoglobin dissociation curve in:

   i)   region A (1)

   ii) region B. (1)

b) i) Sketch a line on the figure to show the shape of the oxyhaemoglobin dissociation curve that you would expect if the concentration of carbon dioxide was higher. (2)

   ii) Explain the advantage of the effect of carbon dioxide on the shape of the oxyhaemoglobin dissociation curve. (2)

3 The graph shows the change in volume of the left ventricle during one complete cardiac cycle in one individual.

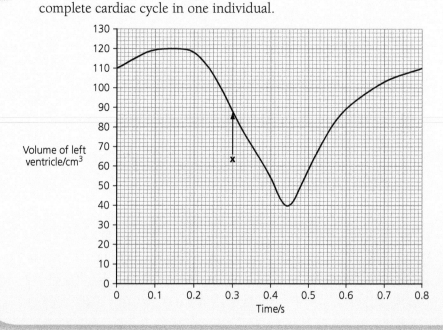

Volume of left ventricle/cm^3

Time/s

a) Copy and complete the table to show whether the following valves are open or closed at time X on the graph. *(1)*

Valve	Open or closed?
Atrioventricular valve	
Semi-lunar valve	

b) **i)** The stroke volume is the volume of blood pumped out of the left ventricle in one cardiac cycle. Use the graph to calculate the stroke volume of this person's heart. *(2)*

ii) The cardiac output is the volume of blood pumped out by the left ventricle in 1 minute. Calculate this person's cardiac output. Show your working. *(2)*

4 Scientists carried out a study to investigate the effects of giving up smoking on the incidence of heart disease. They used large numbers of people who were divided into three groups:

- people who had never smoked
- people who had given up smoking
- people who continued to smoke regularly.

All the people were free of heart disease at the start of the study. They checked the health of each person at regular intervals and tested them for heart disease. The results are shown in the graph.

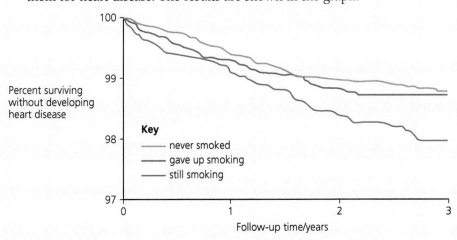

a) Describe the results of this investigation. *(3)*

b) The scientists who carried out the study said 'It's never too late to give up smoking. The risk of heart disease reduces by 50% after 1 year of giving up smoking.' Evaluate this statement. *(4)*

5 The diagram shows some of the risk factors associated with deaths from coronary heart disease (CHD) in people under the age of 75 in the UK during 1998.

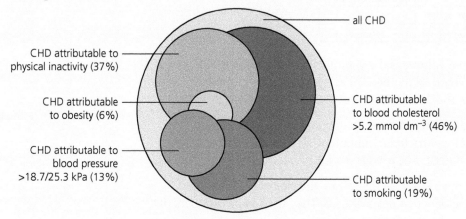

CHD attributable to physical inactivity (37%)

CHD attributable to obesity (6%)

CHD attributable to blood pressure >18.7/25.3 kPa (13%)

all CHD

CHD attributable to blood cholesterol >5.2 mmol dm⁻³ (46%)

CHD attributable to smoking (19%)

a) i) Some of the circles overlap each other. Explain why. *(1)*

ii) Some deaths from CHD are not related to any of the risk factors shown in the diagram. Suggest one possible cause of these deaths. *(1)*

b) Beta-blockers are drugs that reduce blood pressure. Doctors carried out an investigation to find the effect of different doses of a beta-blocker on blood pressure. They used a large number of people who all had high blood pressure in their investigation. The graph shows the results.

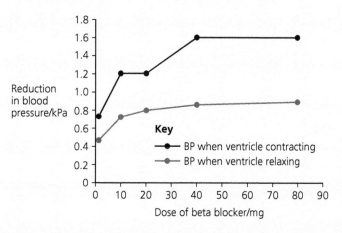

Reduction in blood pressure/kPa

Key
— BP when ventricle contracting
— BP when ventricle relaxing

Dose of beta blocker/mg

c) What dose of the beta-blocker should be used to treat people with high blood pressure? Give reasons for your answer. *(3)*

Stretch and challenge

6 You have seen the pattern of the circulation in a mammal in this chapter. Evaluate the effectiveness of this sort of circulation compared to the pattern of circulation found in fish.

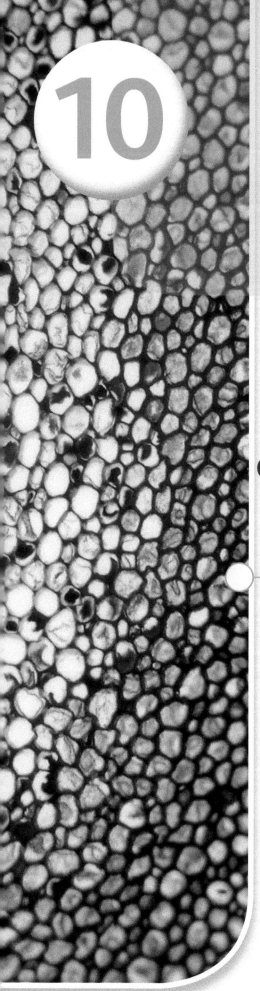

10

Mass transport in plants

TEST YOURSELF ON PRIOR KNOWLEDGE

1 What does xylem transport in plants?
2 What is the advantage of using a plant growth hormone as a weedkiller?

How does water get to the top of a tree?

An oak tree can be at least 30 m high and have a huge array of leaves and branches. Tall trees like this transport water and other materials to the crown from the soil over a vast distance. You may have seen a physics experiment using a pump in which water is 'sucked' up a long tube, perhaps up the side of a school building. No matter how good the pump, the water never goes up more than about 10 m. So, how can a tree get water to the leaves at the top?

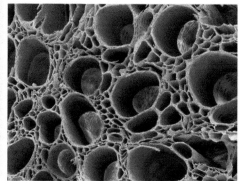 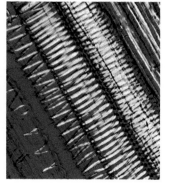

Figure 10.1 (a) Cross-section of a stem showing the xylem tissue. (b) Vertical section of xylem.

In the stem of a plant is the specialised transport system through which water moves rapidly upwards to the leaves. The **xylem vessels** are stained red (Figure 10.1a).

Pulling water up

Transpiration The evaporation of water vapour, mainly through the stomata in the surface of a leaf.

To understand how water reaches the top of the tree, we need to go to the leaves. In Chapter 7 we looked at the structure of leaves. You will remember that leaves have stomata, which open to let in carbon dioxide for photosynthesis. An unavoidable result of this is that water can diffuse out. Normally the amount of water in the air around a leaf is less than that inside the leaf, so water diffuses to the lower water potential of the external air through the stomata whenever they are open. This process is called transpiration (see opposite), and it is a passive process, using energy from the sun, which evaporates water from mesophyll cells. As a result, vast amounts of water can be lost from a large tree. As water vapour diffuses from the air spaces in the mesophyll and through the stomata, it is replaced by water from the mesophyll cells. This in turn is replaced by water from the xylem in the veins of the leaf.

Since the xylem is a continuous system of tubes, water is drawn through to replace the water lost from the uppermost ends of the xylem, a little like the way a drink moves up a straw into your mouth when you suck. However, since water cannot be pulled up more than 10 m, even the most fantastic 'sucker' cannot drink through a straw longer than 10 m. Trees are often much taller than this, so how do they overcome this problem?

The answer is that xylem vessels are very narrow, and water molecules tend to stick together because of the hydrogen bonds between them. This is why water droplets form at the end of a tap: the water does not fall off until the weight is greater than the force holding the molecules together. This property of water is called **cohesion** (see opposite). As water moves out of the xylem in a leaf, it drags other molecules of water behind it. Because the vessels are so narrow, the column of water behind does not break, and water is pulled up all the way from the roots. The pulling force is so great that the column of water is actually being stretched. It is under **tension**, just as an elastic cord being pulled up would be. The tension in the column of water tends to pull the walls of the vessels inwards slightly. However, the lignin in the walls is strong enough to stop the vessels collapsing, just as the cartilages in the trachea and bronchi prevent them from collapsing as we expand the chest to breathe in (see page 118).

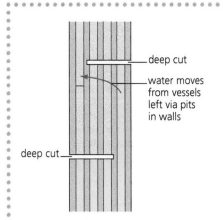

Figure 10.2 Diagram to illustrate the cut stem experiment.

Extension

Vessel damage

Surprisingly, the diameter of a tree trunk is reduced very slightly, but measurably, when the tree is transpiring rapidly on a hot day, because the tension in the xylem is sufficient to pull in the walls of the many vessels just a little. If the column of water in a vessel is broken, for example by an air bubble or a cut, the water will ping apart like elastic, leaving an empty section above and below the bubble or cut. This is why pits are important. They permit water to move from one vessel to a neighbour if a vessel is damaged. A simple experiment showed this. The stem of a young tree was cut over half way across at two positions, one above the other, as seen in Figure 10.2. This ensured that all vessels were cut. Even so, water continued to rise up the stem by moving sideways between the cuts.

The two ideas, of cohesion and tension, used to explain how water is pulled up in a plant (Figure 10.3), are brought together in the **cohesion–tension theory**. The theory explains how water reaches the leaves of a plant from its roots.

Transpiration

Water enters the roots by osmosis and passes up the stem to the leaves, where it evaporates into the air spaces inside the leaf and then passes out to the atmosphere. The air has a low water potential because it normally has a low percentage of water vapour.

Most of the water taken in through the roots of plants living in damp climates simply passes up through the xylem and then out from the leaves. Only a small proportion is used in photosynthesis to manufacture glucose. The loss of so much water by transpiration may seem very wasteful, but it is the unavoidable effect of the need for leaves to take in carbon dioxide for gas exchange. It does, however, have some advantages. The stream of water also transports mineral ions around the plant. The evaporation of water from the leaves has a cooling effect, just as the evaporation of sweat from our skin does. When leaves are exposed to bright sunlight, transpiration can reduce the possibility of the leaves overheating and the enzymes being denatured.

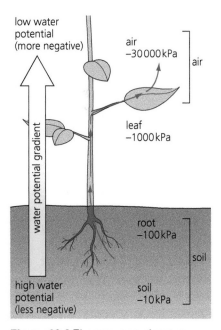

Figure 10.3 The passage of water through a plant.

The properties of water

The movement of water through xylem vessels and the cooling effect of transpiration both depend on the properties of water. Water molecules are polar molecules because the hydrogen atoms have a slight positive charge and the oxygen atom has a slight negative charge. Opposite charges attract one another so water has strong **cohesion** between the molecules (see Figure 10.4).

Polar molecule A molecule with a slight positive charge in one part of the molecule and a slight negative charge in another part.

183

Figure 10.4 Water molecules form hydrogen bonds between molecules, so they 'stick' together.

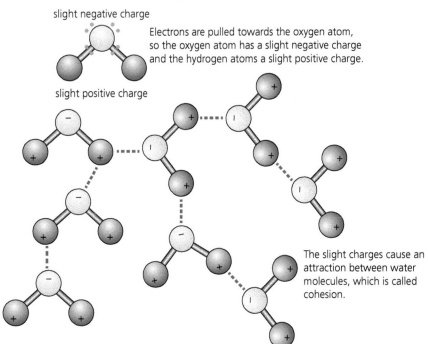

slight negative charge

Electrons are pulled towards the oxygen atom, so the oxygen atom has a slight negative charge and the hydrogen atoms a slight positive charge.

slight positive charge

The slight charges cause an attraction between water molecules, which is called cohesion.

TIP

These properties of water are important in many parts of the course, and you may need to apply these ideas to questions in other topics.

Figure 10.5 Pond skaters are predators that can move around by using surface tension.

Latent heat of vaporisation The heat required to turn liquid at its boiling temperature into gas at the same temperature.

Heat capacity The heat required to raise the temperature of a volume of liquid by 1°C.

The fact that water molecules are drawn towards one another explains why columns of water can be pulled up xylem vessels. The height of the tallest trees is probably limited by the cohesiveness of water. Coastal redwoods (*Sequoia sempervirens*) grow to over 90 m tall. Much taller than this and the long columns of water in their xylem vessels would snap too often to allow water to reach the highest branches.

This property of water also explains why surface tension exists. It is exploited by small specialised animals such as pond skaters, which can move around on the surface of ponds and ditches. Water molecules at the surface are attracted more strongly to one another than those that are completely surrounded by other water molecules. This is because where water and air meet there are no water molecules above them. The stronger cohesion between the layer of surface molecules creates a sort of film on which animals such as pond skaters can stand (see Figure 10.5).

The cooling effect of transpiration that protects leaves from overheating is due to water's relatively large latent heat of vaporisation. The relatively large latent heat of evaporation is due to the fact that water molecules are cohesive and so relatively difficult to separate. Even a small amount of evaporation into the leaf air spaces cools the leaf. This can be critical for a plant's survival in direct sunlight. The same process takes place to cool down an animal's body when sweat evaporates from the skin.

Water has other properties that have biological significance. It has a relatively high heat capacity so large volumes of water such as ponds or lakes do not change temperature very quickly. This resistance to temperature change means that aquatic organisms such as fish live in a very stable habitat. In particular, fish that live in small rock pools on tidal shores depend on this property of water for their survival. Due to the high heat capacity of water, small rock pools do not heat up too much in direct sunlight, protecting the organisms that live there. This also helps larger organisms, which are mainly composed of water, to maintain a stable temperature.

Solvent A liquid able to dissolve other substances.

Water is also the solvent in which biological reactions take place in the cytoplasm of cells and in mass transport fluids. A wide range of substances are soluble in water, which is why so many different solutes can be transported in blood, lymph and plant sap. Water also takes part in many metabolic reactions itself, including hydrolysis and condensation reactions (see Chapter 1, page 2).

TEST YOURSELF

1 List four features of xylem vessels that make them specialised for the transport of water.

2 What property of water molecules does the upward transport of water depend on? Explain your answer.

3 Give two advantages for plants in losing so much water through their leaves.

4 Suggest the combination of environmental factors that would lead to the most rapid rate of transpiration.

Translocation

Water is not the only substance moved by mass transport in plants. A separate transport system moves organic substances around plants. The process is called **translocation** and, unlike xylem vessels, which are tubes made of dead cells linked end to end, it depends on the activity of chains of living cells, called **phloem sieve tubes**.

Like xylem vessels, phloem sieve tubes are elongated cells joined end to end to form a chain, as shown in Figure 10.6a. Where the cells meet, their end walls have holes, which is why they are called sieve plates. The chains of cells run parallel to one another in bundles in the stem, close to similar bundles of xylem vessels. Bundles of phloem sieve tubes in a stem are shown in Figure 10.6b.

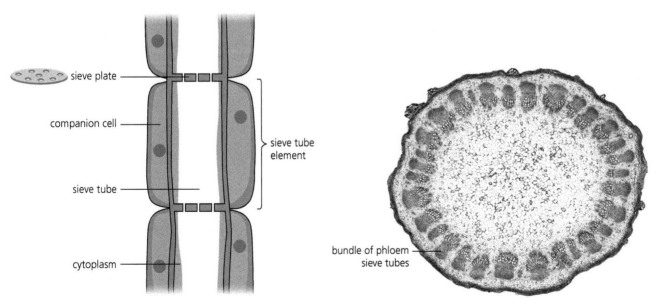

Figure 10.6 (a) Phloem sieve tubes are made of elongated cells joined end to end at sieve plates.

(b) Cross-section of a stem showing phloem tissue.

185

Each separate cell in a phloem sieve tube is called a **sieve element** and has another cell alongside it called a **companion cell**. Companion cells provide metabolic support to the sieve elements. This is important because the cells of sieve tubes lose many of their organelles, including their nucleus, as they become specialised. Although this allows easier flow of phloem sap through the cell, it means that they cannot repair and maintain themselves so well and rely instead on their companion cells for many of these functions.

One of the most important substances translocated in phloem is the sugar sucrose. Sucrose is made in leaves from the products of photosynthesis. It is translocated to parts of the plant that are growing or to places where sucrose can be stored, often used to make starch. Leaves and storage organs such as potato tubers are known as **sources** because they produce sucrose, whereas buds, developing seeds and other sucrose-using parts of the plant are called **sinks**. Translocation occurs from sources to sinks.

Sap The fluid that is transported in phloem.

Mass transport in phloem is explained by the mass flow hypothesis. Evidence suggests that phloem sieve tubes are under pressure rather than under tension. This is why phloem sap leaks out if sieve tubes are punctured. The mass flow hypothesis suggests that phloem sap is pushed from source to sink by a **hydrostatic pressure** gradient caused by osmosis.

Sucrose molecules are too large to simply diffuse across membranes. In the veins of leaves, companion cells known as **transfer cells** actively load sucrose into the phloem against a concentration gradient. Figure 10.7 shows that the loading is similar to that you saw for glucose uptake in the epithelial cells of the intestine (see Chapter 8, page 143). In this case, hydrogen ions rather than sodium ions are actively pumped out of the transfer cell, creating a hydrogen ion gradient. The sucrose is transported in along with hydrogen ions by a co-transport protein and diffuses into the neighbouring sieve element.

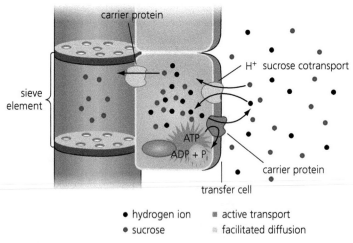

Figure 10.7 Loading of sucrose into phloem sieve elements using co-transport proteins. Hydrogen ions are actively pumped out of the transfer cell.

As sucrose accumulates in the sieve element, the water potential is lowered, so water moves into the sieve tube down a water potential gradient by osmosis from nearby xylem vessels, increasing the hydrostatic pressure. Because the opposite process happens where sucrose is unloaded from phloem, water leaves the sieve tubes by osmosis at sinks, lowering the hydrostatic pressure. The pressure gradient between source and sinks causes the mass flow of phloem sap in one direction along the sieve tube. The mechanism of the mass flow hypothesis is outlined in Figure 10.8.

Figure 10.8 The mechanism of the mass flow hypothesis.

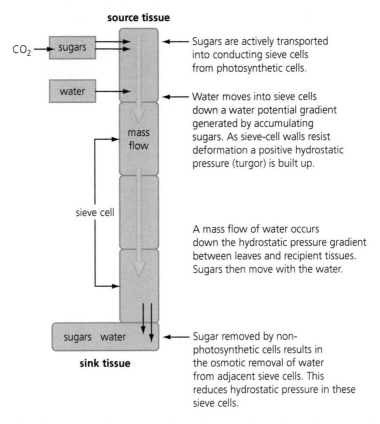

The direction of transport depends on where the sinks are in the plant at any point in time. When growth is taking place, areas undergoing cell division are the main sinks, so sap flows from the leaves both upwards to leaf buds and downwards to growing root tips. When a plant is using materials from a storage organ such as a potato tuber, the sap flows from the storage organ to the buds. A variety of organic substances other than sucrose are translocated by the phloem, including amino acids, plant growth hormones and messenger RNA. Viruses can also be transported in the phloem.

Evaluating the evidence for the mass flow hypothesis

Some simple observations support the mass flow hypothesis. If a phloem sieve tube is punctured then phloem sap oozes out, suggesting it is under pressure. Phloem sap sampled from a source has a higher sucrose concentration than sap sampled from a sink, confirming that different water potentials would cause osmosis into or out of the sieve tubes in the two locations. If a plant virus is applied to well-illuminated leaves, viruses can be detected moving downwards in the phloem towards the roots. But if the virus is applied to leaves in the dark, they are not transported, suggesting that the production of sucrose by photosynthesis is required at the source for translocation to occur.

On the other hand, some evidence contradicts the mass flow hypothesis. Measurements of the rate of translocation of different organic substances suggest that amino acids travel more slowly than sucrose. Some scientists claim to have detected different substances moving in opposite directions in the same sieve tube. If the mass flow hypothesis were correct, bulk flow of phloem sap ought to occur in the same direction at the same rate in any one sieve tube, and probably in most sieve tubes at the same time because they are all connected to the same source: the leaves. These observations suggest a more complex mechanism may be responsible for translocation.

TIP

When you evaluate any data or evidence, you should always consider both sides of the argument. In this case, there are observations that both support and contradict the mass flow hypothesis.

ACTIVITY

Investigating translocation in phloem

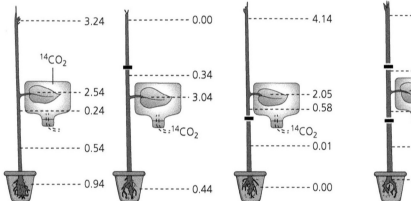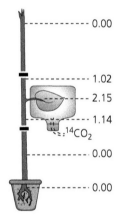

Figure 10.9 Results of a ringing experiment using radioactive carbon dioxide.

Figure 10.9 shows the results of some experiments carried out to investigate translocation in phloem tissue. This type of experiment is called a **ringing** experiment, because it involves the removal of a ring of surface tissues from a plant stem (shown on the diagram by small black bars dividing the stem), leaving the core of the stem intact. You will recall the cross-section of a plant stem from Figure 10.1. The rings are cut deeply enough to remove the phloem, but they leave the xylem intact in the core of the stem. The plant can then be supplied with a radioactive **tracer** to investigate the direction and rate of translocation. Numbers on the diagram indicate the amount of radioactive carbon detected in each part of the plant.

1 Plants in this type of experiment are supplied with radioactive carbon dioxide. Explain how this works as a 'tracer' to detect the translocation of organic substances.

2 When removing a ring of material from a plant stem, why is it important to ensure that the xylem remains intact?

3 What is the purpose of the plant with no rings of tissue removed?

4 How do the results of the ringed plants support the mass flow hypothesis?

Another way to use radioactive **tracers** is to make images of the plant showing where radioactive substances have been transported. After allowing the plant to take up radioactive carbon dioxide and carry out photosynthesis and translocation, the plant is pressed against photographic paper in the dark for several hours. Any radioactive material in the plant tissues creates an image called an **autoradiograph**.

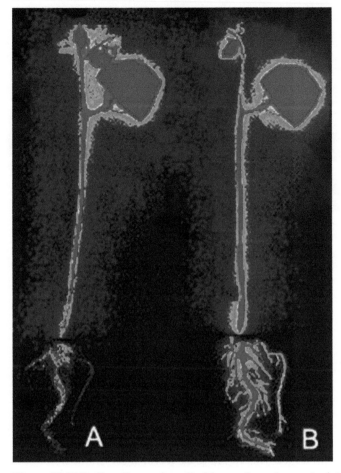

Figure 10.10 Radioactive carbon dioxide used as a tracer to detect the translocation of organic substances from the leaves to the roots.

Autoradiographs taken at a series of times after exposure to radioactive carbon dioxide can be used to calculate how long it takes for sucrose and other organic products of photosynthesis to travel from the leaves to other parts of the plant such as roots, flowers or fruits.

TEST YOURSELF

5 Suggest why transfer cells have numerous mitochondria.

6 Potatoes store sucrose as starch. Describe how the mass flow hypothesis could explain sucrose translocation from leaves to potatoes as they grow.

7 Rabbits and deer often eat the bark of young trees and the phloem tissue beneath it in winter. If they eat all round the tree trunk, this can kill the tree. Suggest why.

8 Plant growth hormone sprayed onto the leaves of a plant is absorbed into the phloem sap. Used as a weedkiller, it can kill the whole plant, including the roots. Suggest how.

Practice questions

1 a) Describe **one** feature of a plant cell that identifies it as a xylem vessel. *(1)*

The figure below shows a potometer.

b) Explain how you could use this apparatus to find the volume of water taken up by this shoot in a minute. *(3)*

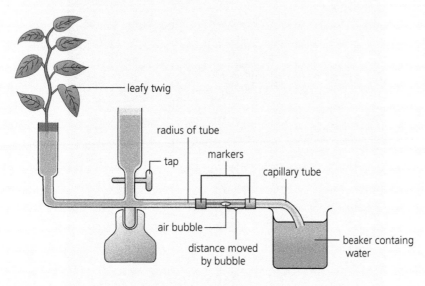

c) You would expect this figure to be lower if the shoot was placed in humid conditions. Explain why. *(2)*

d) If the diameter of the capillary in this potometer were 0.6 mm and the bubble moved 37 mm in 12 minutes, calculate the rate of water loss by the leafy shoot. *(1)*

2 A group of students carried out an investigation. They filled 40 test tubes with tap water until the water was about 1 cm from the top of the tube. They carefully marked the level of the water in the tube by using a pen on the outside of the tube. They covered each tube with a small piece of clingfilm.

Next, they obtained 40 fresh leaves from a plant, checking that all the leaves were similar in size. They inserted one leaf into each test tube, making a small hole in the clingfilm so that the leaves were all in the water. They divided the leaves into four groups of 10. They applied petroleum jelly (a water- and gas-tight substance) to some of the leaves, as described in the table.

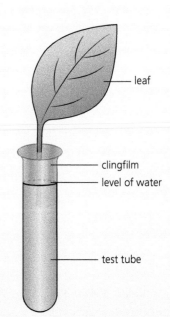

Group	Treatment
A	Petroleum jelly applied to both surfaces of the leaf
B	Petroleum jelly applied to upper surface of leaf only
C	Petroleum jelly applied to underside of leaf only
D	No petroleum jelly applied at all

The test tubes were all placed in test tube racks on a laboratory bench, away from bright sunlight or draughts. They were left for a week. The figure shows one of the tubes after a week.

After a week, the students removed the leaves and clingfilm from the tubes. They carefully measured the volume of water needed to bring the level of water in the tubes back up to the level marked with the pen. They found the mean volume of water needed by the tubes in each group. The results are shown in the table.

Group	Treatment	Mean volume of water added/cm³
A	Petroleum jelly applied to both surfaces of the leaf	3.2
B	Petroleum jelly applied to upper surface of leaf only	4.8
C	Petroleum jelly applied to underside of leaf only	3.7
D	No petroleum jelly applied at all	5.1

a) i) Explain why each test tube had a lid of clingfilm. *(1)*

ii) Explain why it was important to have 10 leaves in each group. *(2)*

b) i) What conclusions could be drawn from these results? Use your knowledge of leaf structure to explain the results. *(4)*

ii) Explain the purpose of group D in this investigation. *(2)*

3 In an investigation, a scientist used a very accurate measuring device to measure the diameter of a tree trunk over a period of several days. Some of the results obtained are shown in the figure.

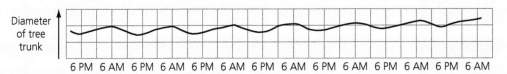

a) i) Describe these results. *(3)*

ii) Use your knowledge of water transport through plants to explain these results. *(4)*

In a different investigation, a scientist set up the apparatus shown in the figure below.

b) i) Parts X and Y of this apparatus represent different parts of a plant. Name the parts of a plant that they represent. *(2)*

ii) As water evaporates from X, the mercury at Y rises up the tube. The scientist said that this is evidence to support the cohesion–tension theory of water transport in plants. Do you agree with this? Give reasons for your answer. *(4)*

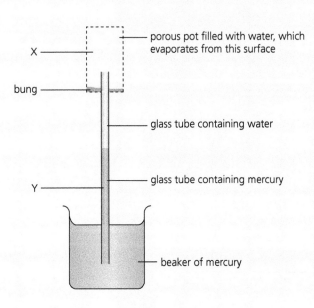

Stretch and challenge question

4 The mechanism of phloem transport remains controversial amongst scientists. In particular, recent criticisms suggest that the mass flow hypothesis does not seem to be adequate to explain translocation in tall trees. Explain why.

Genetic diversity

TEST YOURSELF ON PRIOR KNOWLEDGE

1 Give two reasons why organisms in a population may look different from one another.

2 What is a mutation?

3 Why is the appearance of antibiotic resistance in bacteria such a problem?

Introduction

Scientists analysed DNA samples from the skins of tigers in the collection at the Natural History Museum in London, dating from 1858 to 1947, a period when many tigers were hunted and their skins preserved. They found that the mitochondrial DNA of modern tigers has only 7% of the genetic diversity of historic tigers. Large populations of tigers once existed, but as a result of hunting and habitat loss they have been fragmented into much smaller populations. This means that the number of alleles in the gene pool of each small population is smaller than in the original population. Since they do not exchange alleles by mating with tigers from other populations, their genetic diversity is likely to remain very low. Each time another small population becomes locally extinct, their alleles are lost. A low genetic diversity may prevent the remaining tiger populations from adapting to a changing environment and may threaten their long-term survival just as much as further loss of their habitat.

Allele Alleles are different forms of the same gene.

Genetic diversity A measure of the number of different alleles of genes in a population.

Gene pool All of the different alleles of genes in a population.

Population A group of organisms belonging to the same species found in the same area at the same time and potentially able to interbreed.

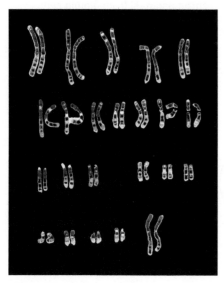

Figure 11.1 The chromosomes in a body cell from a human female.

Future conservation efforts will need to concentrate on trying to connect patches of forest with habitat through which tigers are able to move as well as protecting the patches themselves. Allowing tigers to mix and mate with individuals from other small populations will improve the connectivity of the gene pool and help to maintain what is left of their genetic diversity.

You have seen how the genetic diversity of a species can be reduced if populations become fragmented and the number of organisms in the smaller populations falls markedly. Alleles can be lost from the gene pool and the ability of the population to adapt to change is limited. On the other hand, genetic diversity is increased by random mutation, which results in new alleles being added to the gene pool of the population. Genetic diversity can also be increased by chromosome mutations and by random factors associated with meiosis and fertilisation. Genetic diversity is necessary for natural selection to take place and is crucial in enabling a population to adapt to a changing environment.

Meiosis

Look at Figure 11.1. It shows the chromosomes from one cell of a human female. To make this, a photograph was taken of the cell using a camera and an optical microscope. The image of each chromosome was then cut out of the photograph and the images were arranged as you see them. Notice that they have been arranged in pairs. The members of each pair are the same size and shape. More importantly, they carry genes controlling the same characters in the same order. We call the members of each pair **homologous chromosomes**. You inherited one member of each homologous pair of chromosomes from your mother (**maternal chromosome**) and the other member of the homologous pair from your father (**paternal chromosome**). The fusion of these gametes in sexual reproduction produces genetic diversity in a population because each parent may contribute different alleles of each gene.

A cell with pairs of homologous chromosomes is called diploid. A cell with only one chromosome from each homologous pair is called haploid. We often refer to diploid cells as $2n$ and haploid cells as n. In Figure 11.1 you can see that diploid human cells have 46 chromosomes ($2n = 46$). However, not all our cells are diploid. Our gametes – the egg and sperm cells that we produce – are haploid. Human gametes (the sex cells) have 23 chromosomes ($n = 23$).

The type of cell division that produces haploid cells from diploid cells is called **meiosis**. Most eukaryotic organisms have diploid and haploid cells during their life cycle. This means that meiosis occurs at some stage in the life cycle. Figure 11.2 (overleaf) shows two life cycles. Notice that, although humans produce gametes by meiosis, some organisms such as fungi do not.

Meiosis has one other important effect: the haploid cells produced are genetically different from each other. There are two reasons for this: independent assortment of homologous chromosomes and genetic recombination by crossing over.

Independent assortment of homologous chromosomes

The heading might sound a complicated description of the process, but the sequence is quite easy to follow. Figure 11.3 (overleaf) shows a cell with two pairs of homologous chromosomes. The members of each pair are colour-coded to distinguish the maternal and paternal chromosomes. Before meiosis starts, each chromosome makes a copy of itself by DNA replication. These copies are now called chromatids.

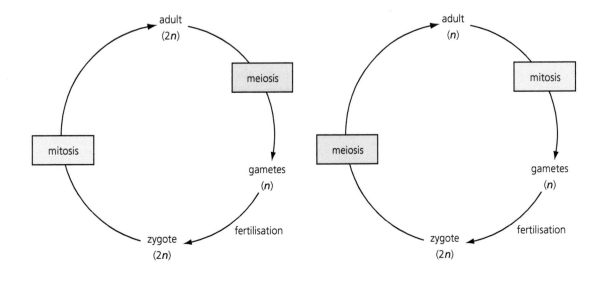

(a) Human life cycle   **(b)** Life cycle of fungus

Figure 11.2 Meiosis produces haploid cells (n chromosomes) from diploid cells ($2n$ chromosomes). (a) In humans, meiosis occurs during gamete production. (b) In fungi, meiosis occurs after fertilisation, not before it.

Figure 11.3 The result of meiosis in a cell with two pairs of homologous chromosomes. Before meiosis occurs, the chromosomes were copied by DNA replication. As a result of meiosis, four cells (haploid gametes) can be formed, and there are two possible combinations. Another cell with the same two pairs of homologous chromosomes can produce another two different combinations (AB and ab), making four possible combinations in total.

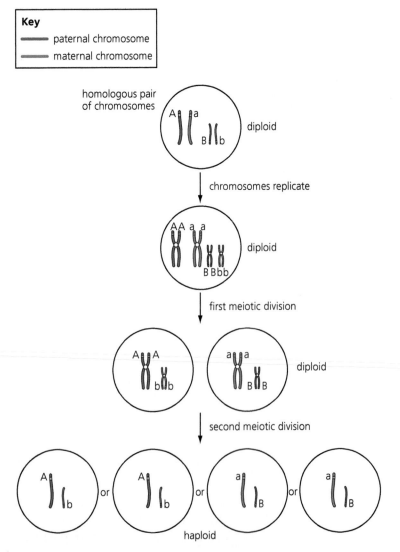

Then two divisions occur:

- the **first meiotic division** separates the chromosomes in each homologous pair
- the **second meiotic division** separates the two chromatids made by DNA replication before meiosis started.

In the first meiotic division, the separation of one pair of homologous chromosomes is not affected by the separation of any other pair. In other words, if the maternal chromosome from one pair moves to one side of the cell, it does not cause the maternal chromosomes of other pairs to go in the same direction. This is independent assortment of chromosomes. The result is that gametes contain a mixture of maternal and paternal chromosomes.

Remember that homologous chromosomes carry genes controlling the same characters in the same order. However, each gene can have different forms, called alleles. To take a simple example, the human gene for haemoglobin has two forms; one leads to sickle cell anaemia and one does not. The maternal and paternal chromosomes in one homologous pair carry different alleles of many of their genes. Therefore, independent assortment produces different combinations of alleles in the cells formed during meiosis. If you look back to Figure 11.3, you will see that a cell with just two pairs of homologous chromosomes could produce haploid cells with any one of four different combinations of maternal and paternal chromosomes. As you know, human cells have 23 pairs of homologous chromosomes. Independent assortment of 23 pairs of homologous chromosomes can result in 2^{23} different chromosome combinations (see page 245) in the haploid egg cells and sperm cells.

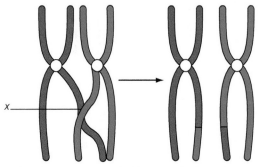

breakage occurs at point X;   result of crossing over
fragments join 'wrong'
chromosome

Figure 11.4 During meiosis, if the chromatids of a homologous pair break at the same point, DNA may be exchanged between the chromatids in opposite members of a pair. This is called crossing over.

Genetic recombination by crossing over

The chromosomes at the end of meiosis in Figure 11.3 are the same colour code as they were at the beginning. Figure 11.4 shows how homologous chromosomes can exchange pieces of their DNA.

During the first meiotic division, the members of each homologous pair lie side by side. If chromatids become tangled with one another, they may break and the broken segments may be rejoined to chromatids in opposite members of the pair. This is called **crossing over** and results in **recombination** or new combinations of alleles that gives rise to genetic diversity.

The effect of crossing over is shown in Figure 11.5 (overleaf). You can see in this diagram that some of the haploid cells have chromosomes with part of the DNA from both the maternal and paternal chromosomes. They have the same genes but a combination of alleles that was not present in either of the parental chromosomes.

195

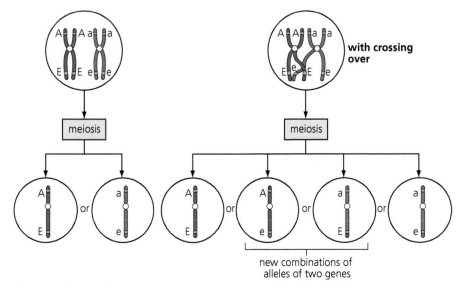

Figure 11.5 Crossing over results in cells with chromosomes that carry new combinations of alleles.

Chromosome mutation A change in the number of chromosomes.

If chromosomes fail to separate properly during meiosis, then a gamete may end up with two copies of a chromosome rather than just one. This is called **non-disjunction.** That gamete would contain one more than the haploid number and, should it take part in fertilisation, it would result in a chromosome mutation, where the organism would have the wrong number of chromosomes in its diploid cells (Figure 11.6).

Figure 11.6 Human chromosomes showing the chromosome mutation that causes Down's syndrome.

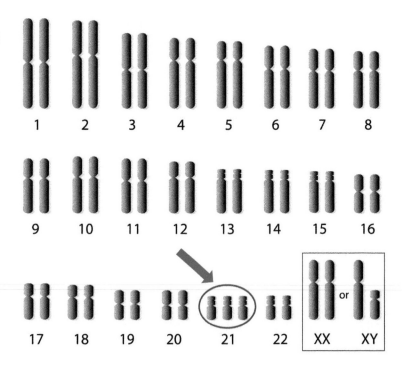

TEST YOURSELF

1 We often use the symbol 2*n* to represent diploid cells. What does *n* represent?

2 Explain one advantage to humans of producing haploid gametes.

3 Use your knowledge of independent assortment of homologous chromosomes to explain why a human couple could have children who are all genetically different from each other.

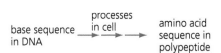

Figure 11.7 A polypeptide is determined by the genetic code held in the DNA base sequence of a gene.

Gene mutation A change in the base sequence of a gene.

Gene mutations

You learned in Chapter 2 that enzymes control reactions in cells and that enzymes are protein molecules. You learned in Chapter 4 that a gene is a base sequence of DNA that codes for the sequence of amino acids in a particular polypeptide, or protein. Figure 11.7 summarises the relationship between DNA and proteins. Although their cell structures are different, prokaryotic cells and eukaryotic cells use their genes to make polypeptides in a similar way.

A change in the DNA base sequence of a gene is called a gene mutation, forming a new version of the gene, or allele. A gene mutation generally happens when DNA is replicated. You learned about DNA replication in Chapter 5. A gene mutation is rare, occurring about once every 1 million times that a gene is copied, and is a completely random and spontaneous event. Sometimes a gene mutation causes the encoded protein to lose its function as an enzyme.

A gene mutation often causes a change in the sequence of amino acids in the encoded protein. Often this has no effect on the function of that protein. These mutations are called **neutral mutations** and are probably the most common type of gene mutation. However, some gene mutations cause a change in the amino acid sequence of the encoded protein, and its function is lost. As a result, these mutations are harmful. It is extremely rare for a gene mutation to produce a beneficial change in the activity of the encoded protein. If it does, it means that the individual in which the mutation has occurred has some sort of advantage over other members of the population. This can lead to the mutation being passed on, or inherited, by its offspring.

Two types of gene mutation that may occur are outlined here.

- **Base deletion**: a base is lost from the base sequence. As a result, the whole base sequence following the deleted base moves back one place (a **frame shift**). This type of mutation often has a significant effect on the encoded protein because it can alter the sequence of all of the codons following the lost base.
- **Base substitution**: the 'wrong' base is included in the base sequence. It does not cause a frame shift but might result in a different amino acid being included in the polypeptide chain. On the other hand, if the substitution results in a triplet that still codes for the same amino acid, it may not change the sequence of amino acids at all because the genetic code is degenerate (see Chapter 4, page 70).

Table 11.1 shows the effects that can arise from base deletion and base substitution.

Table 11.1 The first two rows of this table show the amino acid sequence encoded by part of a molecule of mRNA. The mRNA sequence is shown as individual codons to help you to read the code.

Original base sequence on mRNA	AGA	UAC	GCA	CAC	AUG	CGC
Encoded sequence of amino acids	Arg	Tyr	Ala	His	Met	Arg
mRNA base sequence after base deletion on DNA	AGU	ACG	CAC	ACA	UGC	GCx
Encoded sequence of amino acids	Ser	Thr	His	Thr	Cys	Ala
mRNA base sequence after base substitution on DNA	AGU	UAC	GCA	CAC	AUG	CGC
Encoded sequence of amino acids	Ser	Tyr	Ala	His	Met	Arg

Mutagenic agents

Natural mechanisms occur within cells that identify and repair damage to DNA. These mechanisms become ineffective if the rate of mutation increases above the normal, low rate. Many environmental factors increase the rate of mutation. They are called **mutagenic agents** and include:

- toxic chemicals, for example bromine compounds and peroxides
- ionising radiation, for example gamma rays and X rays
- high-energy radiation, for example ultraviolet light.

Natural selection

A change in the frequency of alleles in a population

When a gene mutates, it has two forms: the original form and the mutated form. We call the different forms of the same gene alleles. If one of the alleles of a gene confers an advantage over the alternative allele of the same gene, it will probably become more common (its frequency will rise) in the population. We call this process **natural selection**, and the allele is known as an **advantageous allele**.

Natural selection can change the frequency of alleles in a population of a species, including in populations of humans.

Figure 11.8 shows two adults of a species of snail called *Cepaea nemoralis*, the banded snail. One of the snails has a yellow shell and the other has a pink shell. The difference in shell colour is controlled by two alleles of the gene for shell colour. The frequency of the alleles for yellow and pink shells is different in populations in different habitats. Table 11.2 summarises how natural selection affects the frequency of these alleles in two different populations. Notice that while natural selection acts on individuals, it leads to change in the gene pool of the whole population.

> **TIP**
> When talking about natural selection, you should refer to alleles rather than genes. The gene is for the character, such as shell colour, whereas alleles are the different forms of the gene that control the pink and yellow colours.

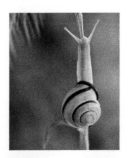

Figure 11.8 The banded snail, *Cepaea nemoralis*, is common throughout Europe. The difference in shell colour is controlled by two alleles of a single gene.

Table 11.2 Natural selection leads to differences in the frequency of two alleles controlling shell colour in populations of the banded snail, *Cepaea nemoralis*.

Habitat in which snail population lives	More frequent allele in population	How natural selection affects frequency of alleles
Beech woodland in England	Pink allele	Snails are eaten by song thrushes, which can distinguish colour. Pink shells are camouflaged amongst the leaf litter but the yellow shells are conspicuous. The song thrushes eat more yellow-shelled snails than pink-shelled snails.
Grassland of the Pyrenees (a mountain range dividing France and Spain)	Yellow allele	No song thrushes live in these mountains. The yellow shells reflect heat from strong sunlight. The pink shells absorb more heat and the snails are more likely to die.

Increased reproductive success

Allele frequencies in a large population generally remain stable from generation to generation. This will not be true if some organisms are:

- more likely to survive until they reproduce
- more likely to grow sufficiently well to reproduce successfully
- more likely to attract a mate.

If the environment changes or a new advantageous allele arises by mutation, some organisms will tend to reproduce more successfully than others and will leave more offspring. We say that they have **increased reproductive success** compared with other individuals in the population.

In beech woodlands, the yellow-shelled snails shown in Figure 11.8 are very conspicuous against the pink leaf litter lying on the ground. The snails with pink shells are better camouflaged and so are more difficult to find. Song thrushes eat banded snails. Like us, song thrushes have colour vision. A large number of investigations have shown that song thrushes find more of the conspicuous yellow-shelled snails than they do pink-shelled snails in beech woodlands. As a result, fewer yellow-shelled snails survive to reproduce. This means that fewer of the alleles for yellow shells are passed on to the next generation. Table 11.3 summarises the process of natural selection.

TIP

The sequence of events on the left-hand column of Table 11.3 can be applied to any example of natural selection. You just need to tailor your explanation by adding details like those in the right-hand column.

Table 11.3 An explanation of natural selection in a beech woodland population of banded snails. The events in the left-hand column can be applied to any example of natural selection.

Sequence of events leading to natural selection	Application of these events to selection of yellow banded snails in beech woodlands
Within a population, a gene has more than one allele. This genetic diversity is due to random mutation.	The two alleles of the shell colour gene result in snails with pink shells and snails with yellow shells.
There is differential reproductive success between the organisms with different alleles of the same gene.	Yellow-shelled snails are more conspicuous than pink-shelled snails among beech litter. Song thrushes find yellow-shelled snails more easily than they find pink-shelled snails. Fewer yellow-shelled snails survive to reproduce.
Organisms with greater reproductive success leave more offspring than those with less reproductive success.	In a beech woodland, pink-shelled snails have more offspring than yellow-shelled snails.
Organisms with greater reproductive success will pass their advantageous allele to their offspring. As a result, the frequency of this allele will increase in the population, i.e. natural selection has occurred.	In a beech woodland, the frequency of the pink allele is higher and the frequency of the yellow allele is lower. Pink-shelled snails are at a selective advantage in beech woodlands.

TEST YOURSELF

4 A gene mutation involves a change in the base sequence of a gene. Where in the DNA would a mutation *not* be likely to cause a change in the encoded protein?

5 Populations of banded snails also live in grassland, where they are preyed on by song thrushes. Which shell colour would you expect to be more common in grasslands? Explain your answer.

6 Explain what an advantageous allele is.

7 Why are advantageous mutations relatively rare?

8 Use the top two rows in Table 11.1 to explain how a base deletion changes the function of an encoded protein.

9 Use the bottom two rows in Table 11.1 to explain why not all base substitutions cause a change in the sequence of encoded amino acids.

Different effects of natural selection

Natural selection operates on most characteristics that organisms possess and results in populations being better adapted to their environment. It can result in the frequency of an advantageous allele increasing in the gene pool (**directional selection**) or it can result in unfavourable alleles becoming rare in the gene pool (**stabilising selection**).

Directional selection

If a gene mutation gives rise to two forms in a population, such antibiotic resistant and non-resistant bacteria, and there is a selection pressure acting on the population, such as the use of antibiotics, **directional selection** acts against one form and favours the other. Organisms with the advantageous allele are more likely to survive. As a result, one form becomes rare and the alternative form becomes common. Figure 11.9 shows how this happens for antibiotic resistance. The frequency of the resistance allele increases over generations.

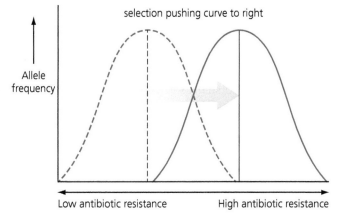

Figure 11.9 Directional selection acts to increase antibiotic resistance.

Stabilising selection

Stabilising selection acts against extreme phenotypes in a population. If some phenotypes are more optimal then the frequency of the alleles for these phenotypes remains the same while the frequencies of other forms that are somehow at a disadvantage are reduced. Stabilising selection occurs on birth mass in humans. Figure 11.10 shows that babies of birth mass within a particular range are at an advantage. Compared with those with alleles that result in very high or very low birth masses, babies with a more optimal birth mass have a lower infant mortality and so the frequencies of alleles for optimal birth masses tend to remain stable over generations.

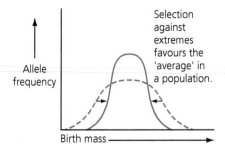

Figure 11.10 Stabilising selection acts to maintain optimal human birth mass.

> **TIP**
> Directional selection usually happens when the environment changes, whereas stabilising selection usually happens when the environment remains the same.

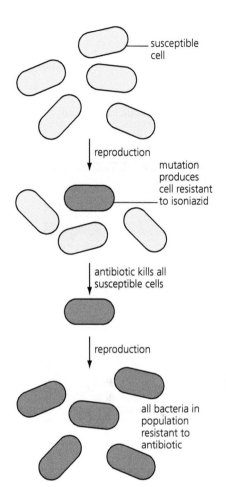

Figure 11.11 Evolution of antibiotic-resistant populations of bacteria in the body of a TB patient. As a result of a gene mutation, a single bacterial cell becomes resistant to the antibiotic isoniazid, and this one cell survives the patient's treatment with the antibiotic. All the offspring of the cell inherit the allele for resistance, and soon the patient's whole population of TB bacteria is resistant to isoniazid.

Natural selection and antibiotic resistance

The increase in antibiotic resistance is explained by natural selection acting on mutations for resistance that arise by chance in bacteria. This presents us with a major medical problem for the effective treatment of many bacterial diseases in the future. Some bacteria, such as those that cause tuberculosis (TB), have already evolved resistance to a range of antibiotics. Imagine a population of bacteria in the body of a TB patient. Figure 11.11 represents a small sample of these bacteria. Initially, none of the bacterial cells is resistant to the antibiotic isoniazid. By chance, a mutation happens to a gene in one bacterium. As a result, it becomes resistant to isoniazid. The patient is being treated with isoniazid, and so cells without the mutation – the susceptible cells – are killed by the antibiotic. Only the one cell carrying the gene mutation by chance – the resistant cell – survives. The patient now has just one bacterial cell. However, as bacteria can double in number in a matter of minutes, the patient soon becomes infected by a population of millions of isoniazid-resistant bacteria. There has been a fundamental change in the bacterial population; all the TB bacteria now carry an advantageous allele that gives a new characteristic beneficial to the bacterium.

TIP

You should refer to bacteria being *resistant* to antibiotics rather than *immune* to them. Immunity involves an immune response, antibodies rather than antibiotics and memory cells (remember Chapter 6). In this case, bacteria simply resist the effects of the antibiotic.

Table 11.4 shows a general example of how natural selection can change the frequency of alleles in a population of bacteria.

Table 11.4 Populations of bacteria become resistant to antibiotics through natural selection.

How natural selection explains antibiotic resistance in bacteria
A chance mutation results in an allele that gives resistance to an antibiotic to bacteria that possess it, so when the antibiotic is present they are at an advantage.
↓
In the presence of this antibiotic, bacteria with the resistance allele are able to survive and reproduce more successfully than those with the original allele.
↓
Bacteria with the resistance allele reproduce.
↓
As the resistance allele is passed on to more offspring, the frequency of the resistance allele will be greater in the population in the next generation than it was in the parental generation.

TIP

As you can see in Table 11.4 the generic sequence of events for natural selection from Table 11.3 has been tailored to fit a bacterial resistance example.

REQUIRED PRACTICAL 6

Use of aseptic techniques to investigate the effect of antimicrobial substances on microbial growth

Note: This is just one example of how you might tackle this required practical.

Bacteria can be tested for their resistance to antibiotics using a method called disc diffusion. A sterile nutrient agar plate is prepared. The bacterium to be tested is cultured in a broth (see Chapter 15, page 263).

1 What is nutrient agar and why must it be sterile?
2 What is a broth and what does cultured mean?

After a period of incubation, the broth is then diluted to a standard concentration, which is normally 1×10^8 CFU mm^{-3}. CFU stands for colony-forming unit. In the case of bacteria, a colony-forming unit means a live bacterial cell capable of dividing and forming a colony on the agar.

The concentration of the diluted broth is given in standard form (see Chapter 14, page 244).

3 Write the number of live bacterial cells in each cubic millimetre of the diluted broth in ordinary form.

A sample of the diluted broth is spread onto the surface of the sterile agar using aseptic technique. Paper discs pre-soaked in a standard concentration of each antibiotic are then pressed lightly onto the surface of the agar. It is important that the discs are spread out evenly and not too close to the sides of the plate or to each other.

4 Describe what is meant by aseptic technique and give examples of some of the steps that might be taken.
5 Why is it important that the concentrations of antibiotic are standardised?

The agar plate is then incubated overnight. During incubation, the antibiotic will diffuse outwards from each disc. This creates a gradient of antibiotic concentration.

The antibiotic is most concentrated nearest the disc. The further from the disc it diffuses, the more the antibiotic concentration decreases.

6 Suggest why the discs must be spread out carefully on the agar surface.

If the bacteria are susceptible to an antibiotic, a clear area is visible around the disc until a concentration is reached where the bacteria are no longer susceptible. The more effective the antibiotic, the lower the

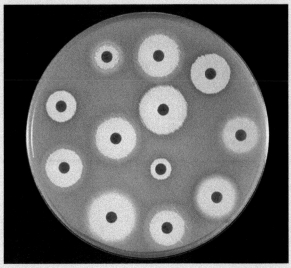

Figure 11.12 Bacteria growing on a plate with antibiotic discs.

concentration at which it will kill the bacteria and the larger the clear zone. Figure 11.12 shows an example of a disc diffusion test. There are no bacteria in the clearer yellow areas.

7 What causes the clear zones?
8 The species of bacteria growing on this plate is not completely resistant to any of the antibiotics. Explain how you know this.
9 How could you tell which antibiotic is least effective against this species of bacteria? Explain your answer.
10 If you wanted to assess the effectiveness of these antibiotics against this species of bacterium, you could compare the radius of each clear zone. However, measuring the diameter would be easier. Suggest why.
11 What problems do you think you might have in making these measurements?
12 Draw a suitable results table for collecting the data from this plate. The Petri dish in Figure 11.12 is life size. Use a ruler to measure the clear zones and record and present your measurements in an appropriate way (see Chapter 14, page 240).
13 Which antibiotic do you think is most effective?

Extension

Variation can be caused by environmental factors as well as by genetic factors

Tall people tend to have tall children. This suggests that height is inherited. However, children born after the introduction of the National Health Service in 1948 grew to be taller, on average, than those born before 1948. It is unlikely that there had been a genetic change in the population. It is more likely that children grew taller because they had a better diet and were free from disease. These are environmental, not genetic, factors.

How easy is it to tell whether variation is caused by genetic or environmental factors? A student measured the height of the students in year 1 of A-level biology. Figure 11.13 summarises her results. What can we conclude from these histograms? You can see that there is no clear-cut 'tall' or 'short' person. The range of heights shows continuous variation (see Chapter 14, page 252). You can see that the distribution of heights is different for males and females. This suggests that these differences result from genetic differences between females and males. You can also see that there is a range of values around the peak values. Is this caused by genetic factors or by environmental factors? Without further information, we cannot easily draw conclusions about which of these two types of factor is the more important.

> **TIP**
> Genetic variation can be inherited or passed on to offspring. Environmental variation is not passed on.

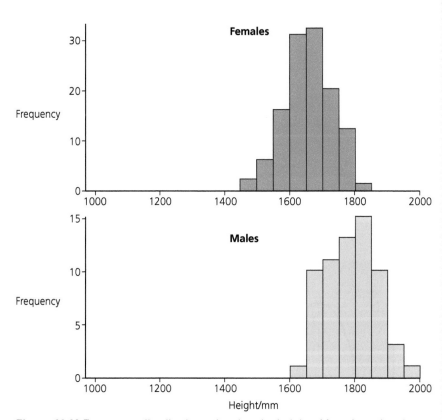

Figure 11.13 Frequency distributions showing the height of female and male students in a group of A-level biology students.

> **TEST YOURSELF**
> 10 Bacteria have been found in samples of ice that was formed in the Antarctic thousands of years ago. Some of these bacteria were found to be resistant to antibiotics. Suggest why.
> 11 Explain why cheetahs have a low genetic diversity and why this concerns scientists.
> 12 Seed banks contain seeds of the ancestors of modern crop plants. Suggest why these seeds are kept.

Investigating variation

The organisms in a population seldom look exactly the same. Although they look broadly similar, they vary slightly in appearance, size and behaviour. Examples of variation include the stripe patterns among zebra, root length in carrots and feeding rate in fiddler crabs. Some kinds of variation can be measured and this is the sort of variation you will probably investigate.

Figure 11.13 may suggest to you the problems that we might have in finding the average height of humans in the UK. We could not measure every human in the country. Instead we would have to measure a small group of the population, called a sample. But would this sample be representative of the country as a whole? For example, if we had more females than males in our sample, this would give us a lower average height.

Any sample might not be representative of the population from which it is taken. This might happen for two reasons.

● Chance: what we commonly call 'luck'. We can reduce the effect of chance by taking several samples and finding their average value, or one very large sample.
● Sampling bias: this happens when the investigator, knowingly or unknowingly, chooses which measurements to include in the sample. We can reduce the effect of sampling bias using a random sampling technique. Random sampling is a technique of selecting the individuals in a sample that removes the investigator's choice, and ensures that the measurements are representative of the whole population and not affected by bias.

Look back to Figure 11.13. Each bar in the histogram shows the number of individuals falling into a particular category. This histogram is called a frequency distribution.

Figure 11.14 shows another frequency distribution. The sample used to make Figure 11.14 was much bigger than that for Figure 11.13, and the distribution is shown as a smooth curve rather than as a histogram. The frequency distribution in Figure 11.14 is of a special type, called a normal distribution. It has several important features, as follows.

● The value at the peak of the curve is the **mode**, or most frequent value. In a normal distribution, the peak is also the **median**, or middle value, and the **mean** value.
● The curve of a normal distribution is symmetrical, with 50% of the values below the peak (to its left) and 50% above the peak (to its right).
● Ninety five per cent of the values are within two standard deviations of the mean.

We have introduced two new terms here: **mean, median, mode** and **standard deviation**. Let's examine them further.

Mean

The mean value is the correct term for what is sometimes referred to as the 'average'. We find it by adding up all the measurements we made and then dividing this total by the number of measurements we made. We can represent this in words as:

$$\text{Mean} = \frac{\text{sum of all measurements}}{\text{number of measurements}}$$

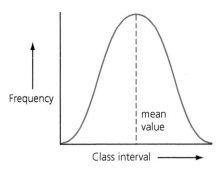

Figure 11.14 A normal distribution curve.

In mathematical notation, this is written as:

$$\bar{x} = \frac{\Sigma x}{n}$$

where $\bar{x}$ (pronounced 'x bar') is the mean value, Σ stands for 'sum of', x represents each measurement made and n is the number of measurements made.

Standard deviation

The standard deviation is a measure of the spread of the data around the mean value. Think back to your practical work in class. You will have carried out an investigation to determine the effect of a variable, say temperature, on the activity of an enzyme-controlled reaction. When you did this, you probably used at least three replicate tubes at each temperature and calculated from them the mean time for the reaction to finish (the end time). Suppose you had found the end times in your three replicate tubes were 44, 45 and 46 seconds. You would probably have felt pleased with these results because they show little variation. You would have been confident in the **precision** of your result. Now suppose your end times were 25, 45 and 65 seconds. Although they give the same mean time as the first example, i.e. 45 seconds, you would probably not have been pleased with these results. The second set of results shows too much variation; they are less precise. You would have doubted that the mean value represented the true value you had tried to measure.

From these two examples, you will realise that the mean itself does not give us enough information about our sample. We want to know how spread out the results were that gave us our mean. If they are not spread out, we are more confident about their reliability than if they are very spread out. This is where the standard deviation is helpful, since it is a measure of the variation around the mean. Figure 11.15 shows two normal distribution curves; they represent the measurements made on two different samples. You can see that the two curves have the same mean but that they have different shapes. Curve a has a narrow spread of measurements, i.e. the measurements in this sample were very similar. Curve b has a much broader spread of measurements, i.e. there was much more variation in the measurements in this sample. Because there is greater variation within the sample in curve b, it will have a larger standard deviation than curve a.

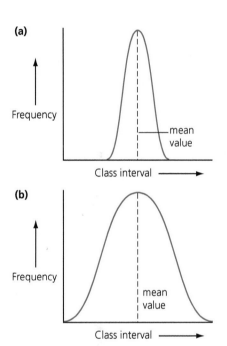

(a)

Frequency — mean value

Class interval

(b)

Frequency — mean value

Class interval

Figure 11.15 These two normal distribution curves have the same mean value. Curve a, with the smaller spread of values, has a smaller standard deviation than curve b.

> ### TEST YOURSELF
>
> **13** Suggest one environmental factor that might account for the variation in the height of females in Figure 11.13.
> **14** Why is it important that samples used to investigate variation are random samples?
> **15** A student measured the heights of two different groups of people. Group A had a mean height of 185 cm and a standard deviation of 3.7 cm. Group B had a mean of 185 cm and a standard deviation of 7.5 cm. What does this information tell us about the two samples?

EXAMPLE

Dog whelks are marine snails that live on rocky shores. Scientists measured the shell length of a random sample of adult dog whelks from a sheltered rocky shore and a rocky shore exposed to large waves. The shell lengths are given in Table 11.5.

1 Calculate the mean length for each sample of dog whelks

The mean length for the sheltered shore sample is:
(22 + 24 + 35 + 30 + 32 + 21 + 26 + 31 + 37 + 24 + 34 + 28 + 29 + 36 + 34 + 28 + 22 + 34 + 23)/19 = 550/19 = 28.947 mm
which you should round to 28.9 mm.
The mean length for the exposed shore sample is
(19 + 22 + 23 + 24 + 21 + 16 + 19 + 21 + 22 + 16 + 23 + 15 + 25 + 21 + 24 + 21 + 23 + 24 + 18)/19 = 397/19 = 20.895 mm
Which you should round to 20.9 mm.

2 Is there a difference in mean length for the dog whelks on the sheltered and exposed shores?

Yes, the mean shell length on the sheltered shore is 8 mm greater than the mean shell length on the exposed shore.

But simply finding the mean values does not necessarily tell you the whole story about the data in each sample. If the standard deviation is calculated for each set of measurements, a summary of the results can be given, as in Table 11.6.

3 What does the standard deviation tell you about the spread of the shell lengths in each sample?

The standard deviations indicate that there is more variation on the measurements in the sheltered shore sample. If you imagine the data plotted as frequency histograms like Figure 11.13, the bars for the sheltered shore would have a larger spread, whereas the bars for the exposed shore would be more clustered together. The overall shapes of the bars would be similar to the two curves in Figure 11.14. This gives you more information than just looking at the mean values. The dog whelks from the exposed shore are smaller, but they also show much less variation in size.

4 Suggest why there are these differences in the variation of shell length on the two shores

One suggestion might be that greater wave action on the exposed shore means that there is a more critical optimum size for dog whelks. Dog whelks that are too small may lack sufficient ability to grip the rocks and if they are too large they may be more prone to being knocked off by the waves. Extreme sizes are less likely to survive in these conditions. On a sheltered shore shell size is less important and a wider range of sizes are able to survive. In addition, in sheltered conditions they may be able to spend more time feeding and so are able to grow larger.

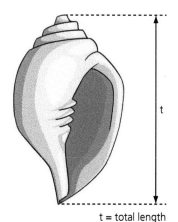

t = total length

Figure 11.16 Dog whelk shell showing length measurements.

Table 11.5 Shell lengths for dog whelk shells from a sheltered and an exposed rocky shore.

Shell length on sheltered shore /mm	Shell length on exposed shore /mm
22	19
24	22
35	23
30	24
32	21
21	16
26	19
31	21
37	22
24	16
34	23
28	15
29	25
36	21
34	24
28	21
22	23
34	24
23	18

Table 11.6

	Mean shell length/mm	Standard deviation
Sheltered shore	28.9	±5.25
Exposed shore	20.9	±2.98

Practice questions

1 a) Complete the table to give the name of the term to which each
definition applies. (4)

	Base that pairs with thymine in the DNA molecule
	The fixed position on a strand of DNA where a particular gene is found
	Protein associated with DNA in a eukaryotic chromosome
	A sequence of nucleotides within a gene that does not code for a polypeptide

b) The figure shows a pair of homologous chromosomes from one
individual, as seen during prophase at the beginning of meiosis.

  i) Describe **two** pieces of evidence from the diagram that these
  are homologous chromosomes. (2)

  ii) Explain why these chromosomes were drawn during prophase
  at the start of meiosis. (2)

c) Most of the gametes from this individual contained alleles
A and B, or alleles a and b. However, a small number of gametes
contained alleles A and b, or a and B. Explain why. (2)

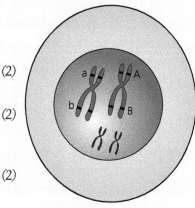

2 A long-term study of antibiotic resistance was carried out between
1978 and 1993 on children suffering from middle ear infections. The
number of children given doses of antibiotic and the percentage of
resistant strains of bacteria identified were recorded each year. The data
are shown in the table.

Year	Annual antibiotic usage/doses per 1000 people day^{-1}	Resistant strains of bacteria/%
1978	0.84	0
1979	0.92	2
1980	1.04	29
1981	0.98	46
1982	1.02	45
1983	1.03	58
1984	0.95	61
1985	1.12	60
1986	1.06	49
1987	1.14	59
1988	1.21	58
1989	1.28	71
1990	1.32	84
1991	1.31	79
1992	1.27	78
1993	1.28	91

a) The scientists concluded that increased antibiotic usage leads
to increased resistance in bacteria. Do these data support this
conclusion? Give reasons for your answer. (5)

b) Explain how natural selection results in antibiotic resistance in bacteria. *(4)*

3 a) i) What is a gamete? *(1)*

ii) How is a gamete different from a normal body cell? *(2)*

The diagram shows the life cycle of a fern. The numbers in the boxes represent the number of chromosomes in one cell at each stage.

b) i) Complete the empty boxes to show the chromosome number at each stage. *(2)*

ii) Mark with the letter M on the diagram a point at which meiosis occurs. *(1)*

c) Explain two ways in which cells produced by meiosis are genetically different from each other. *(4)*

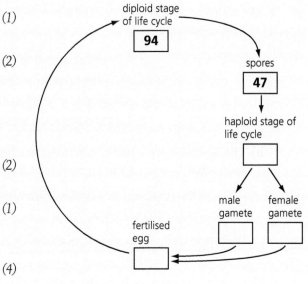

4 In the late nineteenth century, Francis Galton, a half-cousin of Charles Darwin, investigated whether intelligence is inherited or affected by environmental factors. He created a family tree. Part of his family tree is shown in the figure.

Chart showing the inheritance of intelligence

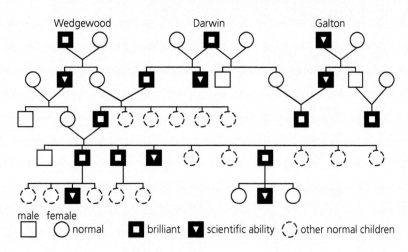

a) Galton concluded from this evidence that intelligence is inherited. Evaluate this conclusion. *(5)*

b) In a modern study, scientists wanted to find out whether height in humans is inherited. They measured the heights of dizygotic same-sex (non-identical) and monozygotic (identical) twins. All the twins grew up in the same home as their twin. The graphs show their results. Do these data show that height in humans is genetic or environmental? Give reasons for your answer.   (4)

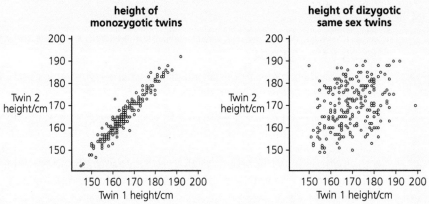

5 a) The Afrikaner population of Dutch settlers in South Africa is descended mainly from a few colonists. Scientists studying the present-day Afrikaner population have found that this population has an unusually high frequency of the gene that causes Huntington's disease. Suggest how this has occurred.   (3)

b) The apple maggot fly originally fed on hawthorn fruits in the United States. About a hundred years ago, it started to become a serious pest of apple orchards and it is now a serious pest throughout the northern United States. Use your knowledge of natural selection to explain how this has happened.   (3)

Stretch and challenge

6 The banded snails you saw in this chapter show a wide range of banding patterns on their shells as well as pink and yellow shells. Some have no bands at all. This means a large variety of different combinations of shell colour and banding pattern can be found in the population as a whole. This is known as a genetic polymorphism. Explain what the polymorphism is in *Cepaea nemoralis* and explain the cause of it.

12 Species and taxonomy

TEST YOURSELF ON PRIOR KNOWLEDGE

1 Name two of the kingdoms that living organisms can be categorised into.
2 What is a DNA nucleotide made up of?
3 Give the complementary base pairs in DNA.
4 What is the maximum number of amino acids that a piece of DNA that is 420 base pairs long could code for?
5 What is functional RNA?

Introduction

Other than both being plants, if you look at a London plane tree and a sacred lotus (see Figure 12.1) you would be forgiven for supposing that they are not very closely related.

One is a large tree with shiny leaves and flaky bark and the other is an aquatic plant with colourful flowers and pepperpot-like seed pods. They could not appear more different. Traditional classifications agreed, placing the lotus with water lilies in the family Nyphaeaceae and the plane tree in the family Platanaceae. But external appearances can be deceptive, especially among living organisms.

Figure 12.1 (a) A London plane tree. (b) A sacred lotus.

More recent work has compared the base sequences of several key plant genes for hundreds of plant species and some of the results have been a real surprise. Based on an analysis of DNA rather than on outward appearance, roses are closely related to figs and nettles, orchids are not related to lilies but to yellow star grasses and, most surprising of all, the lotus is not related to the water lilies, which it most closely resembles, but to the plane trees that line many of London's streets.

Modern approaches to classification now routinely use molecular evidence alongside the appearance and behaviour of organisms, and a great deal of this research goes on in laboratories rather than in the storerooms and herbaria of museums. Reference collections in museums still play an important role in classification, as a tour behind the scenes at the Natural History Museum in London will confirm, but much of the work of a taxonomist now revolves around DNA and protein sequencing.

Why do we classify organisms?

We like to organise and classify things. For example, a cookery book can have recipes grouped into starters and main courses, and meat and vegetarian dishes, and the weekly television guide divides programmes according to the day of the week and the channel. If knowledge wasn't classified in this way, we would never know where to look for information. Imagine, for example, a library full of books that were placed randomly on the shelves. If you wanted information on a particular topic, such as growing roses or the French Revolution, you would waste a lot of time trying to find it.

We really don't know how many different species of organisms there are, but it certainly runs into tens of millions. We need a system that classifies all the organisms we know about to try to establish evolutionary links. To be of real use the system must be universal – it has to be usable by biologists anywhere in the world.

What is a species?

Species A group of organisms that are able to reproduce to give fertile offspring.

The system we use is based on dividing living organisms into species. To use this system we need to have a clear idea of exactly what we mean by a species. There are several things that we consider in defining a species.

First we need to look for similarities and differences. They might involve physical features. For example, in the blackbird and song thrush we see some similarities but differences in the colour of their feathers and their size.

These similarities and differences are reflected in their **binomials** (scientific names). Each species of organism has a unique binomial made up of two words. The binomial of the blackbird is *Turdus merula*. The blackbird is one of a larger grouping of species, the thrushes, which are all related to each other. The thrush group, or **genus** (plural **genera**), is called *Turdus* and all the different species of thrush found in Britain are named *Turdus something*. The song thrush is *Turdus philomelos* and another, the mistle thrush, is *Turdus viscivorus*. The blackbird is closely related to these two species of thrush so it has the same generic name, but its **species** name is *merula*.

TIPS
- When you are writing a binomial, it is either written in italics or underlined. The generic or first part of the name starts with an upper-case letter, and the species or second part of the name is all lower case, e.g. *Homo sapiens*.
- Two or more species that share the same generic name are more closely related to one another than they are to species that have different generic names.

However, it is not just physical features that we should be looking at. The features that distinguish different species are controlled by genes, so we should expect to find differences in the DNA of different species or similarities in the DNA of the same species. Genes code for proteins and functional RNA (see Chapter 4, page 62), and so there will be differences in sequences of proteins such as the haemoglobins that transport oxygen in different species. There will also be differences in their RNA.

Another way that species differ is the diploid number of chromosomes in their cells. A rabbit has 44 chromosomes and a hare has 48, whereas a horse has 64 and a donkey has 62. If haploid gametes from two different species undergo fertilisation, the cell that forms has a different number of chromosomes to either parent. A horse gamete has 32 chromosomes whereas a donkey gamete has 31, so the cells of a mule, which is a **hybrid** of the two, have 63. Cells with an odd number of chromosomes are not viable and usually can't carry out successful meiosis because an odd number of chromosomes cannot form homologous pairs. This means that mules cannot produce gametes, so they are sterile.

This provides us with one way of defining a species. Two organisms belong to the same species if they are able to produce fertile offspring. The offspring of horses and donkeys are sterile, so they are two different species.

When is a species not a species?

We can describe a species, then, as a group of organisms that share certain observable characteristics, and are able to produce fertile offspring. If we use this definition it ought to be easy to decide whether or not two organisms are separate species. Unfortunately, though, it is not always easy to decide whether or not two organisms are the *same* species.

A thorny problem

Hawthorn is a common woody plant. There are two species of hawthorn: *Crataegus monogyna* (common hawthorn) and *Crataegus oxyacanthoides* (midland hawthorn). There are some obvious differences between these two species. Table 12.1 shows some of these.

Table 12.1 Some differences between the two British species of hawthorn.

Feature	*Crataegus monogyna* (common hawthorn)	*Crataegus oxyacanthoides* (midland hawthorn)
General appearance		
Number of seeds in berry	One	Two
Shape of leaf	Many indentations	Few indentations
Hairs on leaf veins	Tufts of hairs present	No tufts of hairs present
Habitat	Along edges of woods and in open areas	In mature woods

The evidence in the table suggests that *C. monogyna* and *C. oxyacanthoides* are different species. But are they? Hawthorns have always been planted as hedges. To grow the earliest hedges, farmers would have taken young hawthorn plants from nearby woodland and planted these. It is likely that they would have collected plants of both types.

Now look at the graph in Figure 12.2. It shows data about the leaves of hawthorn trees growing in a very old hedge, thought to have been planted over 900 years ago. The *x*-axis is an index of indentation of the leaves. This index was calculated by comparing the total depth of indentations with the length of the leaves. The more indentations there are on a leaf, the greater the value of the index of indentation. The *y*-axis shows the number of trees with each value.

We need to make sure that we understand the underlying biology before we look at the graph in detail.

- This hedge was planted over 900 years ago. How old are the hawthorn trees in the hedge now?

The answer to this is that we don't really know. It is very likely that all of the original hawthorn trees have died. Those present now are their descendants and they will have a range of ages. Some will be very old trees; some may be young.

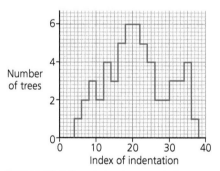

Figure 12.2 Graph to show variation in leaf shape in the hawthorn trees growing in a hedge planted over 900 years ago.

The scientists who collected and analysed the data in the graph calculated the index of indentation by comparing the total depth of indentations with the length of the leaves.

● What was the advantage of presenting the data in this way?

The leaves on a particular tree vary in size. Size is bound to affect the total depth of leaf indentation: the bigger the leaf, the greater the depth of the indentations. Calculating an index like this enabled the scientists to compare leaves of different size.

The scientists calculated the index of indentation for a sample of *C. oxyacanthoides* trees growing in their natural environment. The mean value was 10.

● Explain why some of the trees growing in the hedge had values less than 10.

There will be variation in leaf indentation. When we say that the mean value for *C. oxyacanthoides* was 10, some trees will have a value less than 10 and some will have a value more than 10.

The scientists also calculated the index of indentation for a sample of *C. monogyna* trees growing in their natural environment. The mean value was 34.

● Explain why there are a large number of hawthorn trees in the hedge with an index between 10 and 34.

We can explain some of these intermediate values as being due to variation. A lot of these plants, however, are hybrids between *C. oxyacanthoides* and *C. monogyna*.

When we look at these hybrid plants in more detail, we find that they are intermediate in a wider range of characteristics. What is more, they are fertile and in turn produce fertile offspring.

● Are *C. oxyacanthoides* and *C. monogyna* different species?

This is a difficult question to answer. Table 12.1 shows us that there are some very obvious differences between the two types of hawthorn. But, when they are planted together, they can breed and produce fertile offspring. Using our definition they are the same species but, because they clearly look so different and hybrids are rare, they are currently regarded as separate species. Examples such as the two types of hawthorn, and the two types of duck shown in Figure 12.3, illustrate the problems that biologists have in defining species. Biologists also find that new evidence constantly leads them to reconsider how closely organisms are related. Sometimes this means that they have to change the binomials of organisms. You will read about an example of this, the bug orchid, on page 217.

Figure 12.3a shows a ruddy duck, *Oxyura jamaicensis*. Ruddy ducks were brought to the UK from North America. Then, in 1953, some of them escaped from captivity. Now they are widespread in western Europe. The white-headed duck, *Oxyura leucocephala*, seen in Figure 12.3b, is a native of Europe. It is also now a globally threatened species. Its numbers worldwide have fallen in the last hundred years from perhaps 100 000 birds to just over 5000.

Figure 12.3 (a) The ruddy duck. (b) The white-headed duck.

White-headed ducks face extinction because they produce hybrids with ruddy ducks. Before this was known, the two ducks were thought to belong to different species. There are now about 550 white-headed ducks in the whole of Spain, the largest population in western Europe. To protect them, Spain decided to exterminate all their ruddy ducks. There are no white-headed ducks in the UK, and the policy is to kill all the ruddy ducks here to protect the Spanish white-headed ducks. But should we be doing this? This is the type of question that we have to ask ourselves as biologists.

Sorting out species

In the system we use to classify different organisms, we put them into groups. These groups are based on things that organisms have in common and how they evolved. If two organisms have common features and if there is evidence that they also have the same ancestor, we assume that they are related, and put them into the same group. Often, but not always, the features they have in common reflect their evolutionary history.

Look at the organisms in Figure 12.4. We could divide them into those organisms that can fly and those that are unable to fly. This classification gives us no information about their evolutionary history. However, if we look carefully at the structure of their wings, we can see that, although an albatross wing and a penguin flipper are used for different purposes, they are really very similar. We can recognise the pattern of bones in an albatross wing as being the same as the pattern of bones found in a penguin flipper. This similarity points to the albatross and the penguin having a common ancestor at some stage in the distant past, and so being quite closely related.

Figure 12.4 (a) A dragonfly.

(b) An albatross. It has very long wings and can fly huge distances.

(c) A penguin. It cannot fly, but swims using its flippers as paddles.

....
Phylogenetic classification system
A system of classification based on evolutionary origins and relationships.

Hierarchy The placing of smaller groups within larger groups with no overlap between them.

A dragonfly's wing has a very different structure. Clearly, a dragonfly is not closely related to either an albatross or a penguin. A phylogenetic classification system makes use of features like wing bones that show a common evolutionary history. The classification system groups species together based on their evolutionary history. Similar species are grouped into genera, similar genera into families, and so on. A system like this, where smaller groups are placed into larger groups with no overlap between them, is called a hierarchy. The series of groups are called **taxa**. The highest ranked taxon is the domain, and biologists currently recognise three domains. One possible hierarchy is summarised in Figure 12.5 (overleaf), which shows the classification of the rabbit.

215

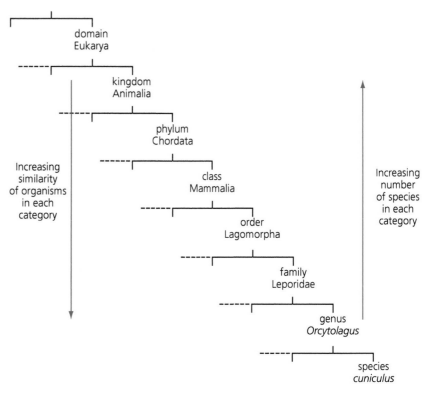

Figure 12.5 How the rabbit, *Oryctolagus cuniculus*, is classified.

Increasing similarity of organisms in each category

Increasing number of species in each category

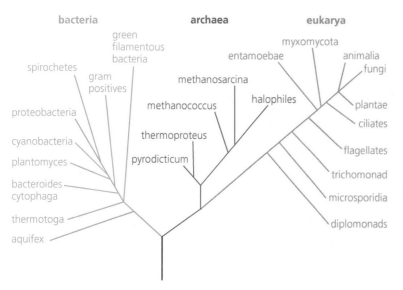

Figure 12.6 Phylogenetic tree showing the three domains based on molecular similarities.

All living organisms are placed in one of three **domains**, the Bacteria, the Archaea and the Eukarya. These reflect some fundamental differences in the RNA sequences found in the organisms within these taxa. A phylogenetic tree represents the evolutionary relationships between taxa. Figure 12.6 shows a phylogenetic tree for the three domains based on molecular similarities suggesting that the Archaea are more closely related to the Eukarya than the Bacteria.

TIP

When you are writing the name of a group, the taxon should have a lower-case letter and the group name should have an upper-case letter; for example, domain Eukarya or kingdom Fungi.

DNA, proteins and classification

The examples you have read about so far in this chapter show that biologists are not always certain whether organisms belong to the same or different species. Relying on physical features to sort organisms into different species can be misleading.

Biologists also have difficulties deciding how closely different species are related. For example, a wild orchid called the bug orchid originally had the scientific name *Orchis coriophora*, and that is the name you will find in many books. In natural conditions it forms hybrids, not just with closely related species, but with orchids of a different genus. Because of this, biologists now think they might have classified it wrongly. They have suggested that it ought to be placed in the same genus as the orchids with which it forms hybrids, so they have changed its scientific name to *Anacamptis coriophora*.

Do two species belong to the same genus? Are two families of organisms closely related? Biological research that finds the answers to questions such as these helps us to understand evolutionary relationships.

TEST YOURSELF

1 The scientific names of three wild cats are shown in the table. What do their scientific names tell you about how these wild cats are related to each other?

Common name	Scientific name
Lion	*Panthera leo*
Leopard	*Panthera pardus*
Clouded leopard	*Neofelis nebulosa*

2 A horse and donkey can interbreed. The offspring is called a mule. Cells in a mule can undergo mitosis but not meiosis. Are a horse and a donkey different species? Give an explanation for your answer.
3 Explain what a hierarchy is.
4 What is a phylogenetic classification based on and what does it show?

The physical features that help us to classify an organism are determined mainly by its genes. A gene is a piece of DNA that codes for a protein, and proteins are the molecules that control the physical features of an organism.

In Chapter 4 you learned about DNA and its structure. Molecular biologists use machines to analyse DNA. They have worked out the complete DNA base sequences of a number of different organisms including humans. They have also found the DNA base sequences of particular genes in many different organisms.

In Chapter 5 you learned how DNA is copied in the process of replication. Errors sometimes arise when base sequences are being copied. When a base sequence in a gene has a copying error, we say there has been a **mutation**. A base may be added to the sequence, replaced by another base, or may be deleted altogether. If this happens in a body cell, it occurs only in that individual. If it happens to sex cells, the next generation inherits the change. Such mutations may either make no difference to the characteristics we see in a species, or they can cause the species to change very slowly over a period of many thousands of years. Either way, the DNA that codes for a particular protein in an organism alive today is slightly different from the DNA that coded for the same protein in its distant ancestor.

Comparing DNA and mRNA base sequences

We can use computer software to compare DNA or mRNA base sequences in different organisms. If the sequences are very similar, it suggests that the organisms concerned are closely related and that they originated from a common ancestor relatively recently. If there are more differences between the sequences, it suggests that the organisms are not so closely related and probably originated from a common ancestor a longer time ago. Because mRNA is derived from the base sequence on DNA, sequencing mRNA gives the base sequence of DNA, and mRNA is sometimes easier than DNA to isolate from the cells of organisms.

Classifying whales

There are two groups of whales. Large whales such as fin whales and humpback whales do not have teeth. Instead, they have baleen plates, which are like huge combs that they use to filter small organisms from the water. They are put into one group. Dolphins and porpoises have teeth and are put into a second group. The sperm whale is a large whale, but it does not have the baleen plates. Instead, it has teeth, so, in the past, biologists classified sperm whales with the dolphins and porpoises.

Figure 12.7 shows this information in a diagram. The fin whale and humpback whale are closely related to each other and they split off from a common ancestor relatively recently. The bottlenose dolphin and the harbour porpoise are related in a similar way. The relationships shown in this diagram have been built up by looking at physical characteristics such as teeth. What do we find when we look at their DNA?

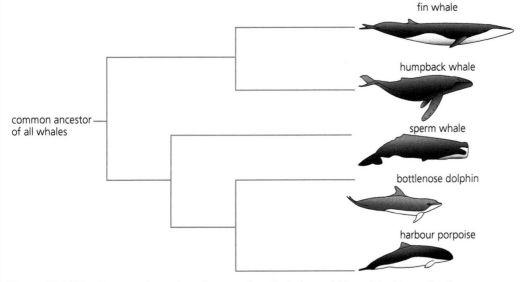

Figure 12.7 This diagram shows how five species of whale could be related to each other.

Now look at Figure 12.8. It shows DNA base sequences in part of a gene that these different species of whales and dolphins have in common.

In this diagram, the bases are lined up with each other so that they match as closely as possible. Biologists use a computer to give the best possible match.

fin whale	TAAACCCCAATAGTCA–CAAAACAAGACTATTCGCCAGAGTACTACTAGCAAC
humpback whale	TAAACCCTAATAGTCA–CAAAACAAGACTATTCGCCAGAGTACTACTAGCAAC
sperm whale	TAAACCCAGGTAGTCA–TAAAACAAGACTATTCGCCAGAGTACTACTAGCAAC
bottlenose dolphin	TAAACTTAAATAATCC–CAAAACAAGATTATTCGCCAGAGTACTATCGGCAAC
harbour porpoise	TAAACCTAAATAGTCC–TAAAACAAGACTATTCGCCAGAGTACTATCGGCAAC

Figure 12.8 Matching DNA base sequences in different whales and dolphins.

We can analyse the differences between these pieces of DNA by using a simple scoring system. We will start by comparing the humpback whale DNA with the DNA from the sperm whale. Some bases match. These have been highlighted in orange. We will score 1 for each match; these two sequences score 48. In other words, there are 48 matches between them. This tells us very little. To get a more detailed picture, we need to count up the matches between all the other pairs of species as well. These scores are shown in Table 12.2.

Table 12.2 Similarities between DNA sequences in some whales and dolphins.

	Fin whale	Humpback whale	Sperm whale	Bottlenose dolphin	Harbour porpoise
Fin whale					
Humpback whale	51				
Sperm whale	48	48			
Bottlenose dolphin	43	43	41		
Harbour porpoise	40	45	45	48	

Let us work through this table and see what it tells us.

1 Using only the evidence from the table, which two species appear to be most closely related?

The fin whale and the humpback whale are most closely related. The score of 51 tells us that they have 51 matching bases. The greater the numbers of matching bases the more closely are two species related.

Here are the scientific names of some whales:

Fin whale	*Balaenoptera physalus*
Sperm whale	*Physeter catodon*
Southern right whale	*Eubalaena australis*
Northern right whale	*Eubalaena glacialis*

2 Suppose you analysed the same piece of DNA in these four species. Between which two would you expect the highest score, and why?

The two right whales belong to the same genus, so they should be the most closely related to each other, and should have the highest score.

The sperm whale is a large whale that has teeth. Figure 12.7 shows the sperm whale classified with the bottlenose dolphin and the harbour porpoise.

3 Does the evidence in the table suggest that this is the best way of classifying the sperm whale?

If you look at the table carefully, you will see that the sperm whale has 48 matches with the fin whale and 48 matches with the humpback whale. It has fewer matches with the dolphin (41) and the porpoise (45). This seems to suggest that Figure 12.7 does not show the best way of classifying the sperm whale. It would be better to classify it in the way shown in Figure 12.9. This diagram suggests that the sperm whale is more closely related to the other large whales than it is to the bottlenose dolphin and the harbour porpoise.

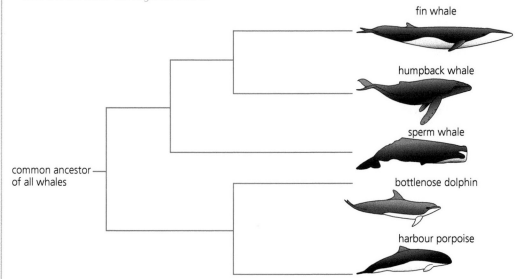

Figure 12.9 An alternative way of showing how the whales and dolphins shown in Figure 12.7 may be classified.

4 We have two different ways of ways of classifying the sperm whale. Which is the correct one?

We have to be very careful about how we interpret evidence about DNA base sequences. Here we have looked only at the similarities and differences in one piece of DNA. Sometimes the evidence we get from another piece of DNA may suggest something else. In coming to their conclusions, biologists have to weigh up data from different sources. In this case, all the available evidence now suggests that Figure 12.9 is a better interpretation than Figure 12.7 of how the five whales should be classified.

ACTIVITY

Another way of looking at differences in DNA: comparing mRNA sequence

Tree shrews (Figure 12.10) are mammals from Southeast Asia. Because of their small size and short reproductive cycle they have been suggested as a possible alternative to non-human primates in biomedical research. For example, they have been used to investigate the way in which the influenza H5N1, or bird flu virus, infects organisms. Scientists hope that this will provide useful information about how the H5N1 virus infects humans. But their suitability depends on how closely they are related to primates. There has been some disagreement among scientists as to whether tree shrews are insectivores or primates because they share features of both groups.

Scientists have investigated this by comparing the base sequences of some of their mRNA molecules to those of other primates. One study investigated tree shrew mRNA for interleukin 7 (IL-7), a protein involved in the immune system. First the scientists found the sequence for tree shrew IL-7 mRNA from an online database. Then they used the database to find the mRNA sequence for the same protein for another 14 mammal species and used computer software to compare all the sequences. The results are shown in Figure 12.11.

1 Does Figure 12.11 suggest that the tree shrew is a good alternative to non-human primates in biomedical research?

2 The mouse and the pig are both currently used as alternatives to non-human primates in biomedical research. Use Figure 12.11 to explain why.

Figure 12.10 Tree shrews share the characteristics of insectivores and primates.

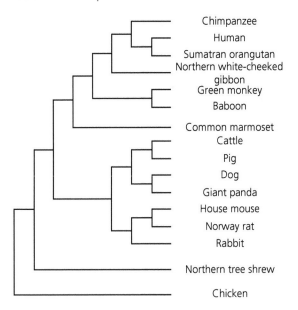

Figure 12.11 The relationship of tree shrews to some other animals in terms of their interleukin 7 mRNA base sequences.

Comparing amino acid sequences

The human body contains many different proteins. You will already have come across some of them, such as enzymes, carrier molecules and antibodies. Each protein is made up from a particular sequence of amino acids. The amino acid sequence reflects the base sequence on the DNA, but proteins are sometimes easier to isolate from the cells of organisms than DNA. If we look at a protein present in several species, we may see that the sequence of amino acids in this protein differs slightly from one species to another. We can find out more about how closely humans are related to different species of apes by comparing the amino acids in their proteins.

We will start by looking at human haemoglobin. Every haemoglobin molecule in an adult human contains four polypeptides: two identical α chains and two identical β chains. Each α chain consists of 141 amino acids joined to each other by peptide bonds. The β chains are slightly longer. They each contain 146 amino acids. Together we have 141 + 146, or 287 amino acid positions that we can compare. Surprisingly, if we take the four species that we looked at in the previous section, 282 of these amino acids are in exactly the same sequences in all four. There are differences in the positions of only five amino acids, and these are shown in Table 12.3.

Table 12.3 Differences in amino acids in the haemoglobin of humans and three species of ape.

Species	α chain			β chain	
	Position 11	Position 23	Position 87	Position 104	Position 115
Human	Alanine	Glutamic acid	Threonine	Arginine	Proline
Chimpanzee	Alanine	Glutamic acid	Threonine	Arginine	Proline
Gorilla	Alanine	Asparagine	Threonine	Lysine	Proline
Orang utan	Threonine	Asparagine	Lysine	Arginine	Glutamine

In the table, the shading shows the places where amino acids are the same as on the human α and β chains. You can see that both chains in chimpanzee haemoglobin have exactly the same sequences of amino acids as both chains in human haemoglobin.

We need to be careful how we interpret this. It doesn't mean that humans and chimpanzees are the same species! Remember, the data are for only one molecule, haemoglobin. A much more likely explanation is that it takes a very long time for differences in the amino acid sequence of haemoglobin to evolve. It is possible that only a few million years have passed since humans and chimpanzees split apart from a common ancestor. Perhaps this is too short a time for differences in their haemoglobin to have evolved.

Finding relationships with RNA sequence data

Figure 12.12 shows the phylogenetic relationship between five species of fruit fly from the genus *Drosophila*.

1 How many millions of years ago did *Drosophila yakuba* and *Drosophila melanogaster* last share a common ancestor?

   They last had a common ancestor 8 million years ago, because this is the time of the last point at which they were connected on the tree.

2 Which two fruit fly species are most closely related?

   Drosophila mauritania and Drosophila sechellia, because they share a common branch of the tree.

3 This diagram is based on mRNA sequence data rather than DNA sequence data. Explain why it is often easier to obtain mRNA sequence data from tissue samples.

   mRNA molecules can be found in the cytoplasm of cells, and there are often many molecules of the same mRNA present because of repeated transcription. Isolating these is often easier than finding a gene in the DNA in the nucleus.

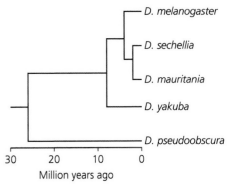

Figure 12.12 The phylogenetic relationship between five species of fruit fly.

TEST YOURSELF

5 Proteins are polymers of amino acids. Give two ways in which the primary structure of proteins may be different.

6 Looking at differences in the amino acid sequence of haemoglobin is not useful for studying classification of all organisms. Explain why.

7 Which three types of sequence data can be used to investigate the relationships between species?

Practice questions

1 a) Complete the table to show the classification of the dandelion, *Taraxacum officinale.*　　　　　　(2)

	Eukarya
kingdom	Plants
	Anthophyta
	Magnoliopsida
	Asterales
	Compositae
genus	
species	

b) Some scientists found a group of dandelions growing on the side of a hill. These dandelions had much smaller leaves than the dandelions they found in other areas. They wondered if this group of dandelions was a new species. Describe how they could find out whether these dandelions were a different species from those with larger leaves.　(2)

2 The table shows the number of amino acid differences in the protein cytochrome *c*, taken from different organisms. Cytochrome *c* is a protein used in aerobic respiration.

Organism	Number of amino acid differences
Human	0
Monkey	1
Pig	10
Dog	11
Duck	11
Turtle	15
Tuna	21
Moth	31
Yeast	51

a) i) Give one reason why cytochrome *c* might be a better protein to use for examining evolutionary relationships than haemoglobin.　　　　　　(1)

ii) These data suggest that monkeys are more closely related to humans than pigs. Explain how.　　　　　　(2)

b) A student looked at these data and suggested that they show that dogs are closely related to ducks. Is this a valid conclusion? Explain your answer.　　　　　　(2)

3 Scientists wanted to investigate the evolutionary relationships between several different species of viper, *Bitis* spp. This was done as follows.

A They extracted pure samples of the protein albumen from the blood of each species.

B They injected a sample of the albumen into rabbits and extracted antibodies against each type of albumen.

C They mixed samples of each of the antibodies produced with each of the different kinds of albumen.

D They used these results to produce the diagram shown in the figure.

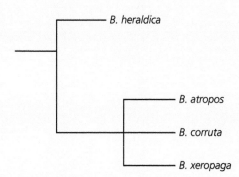

a) Describe how the rabbits produced antibodies against the viper albumin when it was injected into their blood in stage B. *(4)*

b) i) Describe what the results show about the relationships between *Bitis corruta, Bitis atropos* and *Bitis heraldica.* *(2)*

ii) Describe how the results from stage C could be used to produce the results shown in the diagram. *(3)*

c) i) At least two rabbits were used to produce each kind of antibody in stage B. Suggest why. *(1)*

ii) If the scientists compared the base sequences of the gene that codes for albumen from these different species of snake they might have obtained slightly different results. Suggest why. *(1)*

Stretch and challenge

4 The original work that underpins the idea of the three domains was carried out by a scientist called Carl Woese. What made him come to the conclusion that the kingdom Prokaryotae should be separated into two domains?

13

Biodiversity within a community

TEST YOURSELF ON PRIOR KNOWLEDGE

1 What is a community?
2 Give an example of an animal or plant species that is adapted to a particular environment.
3 Describe how two plant species might compete with one another.
4 Explain why animals are dependent on plants.

Introduction

The photograph on the left shows a wood at the edge of Hauxley Nature Reserve in Northumberland. The area the wood occupies was once part of an open-cast coal mine. When the mine closed, the whole coal-mining area was landscaped to include a lake with islands. Then in 1983 the Northumberland Wildlife Trust bought the land to develop into a wildlife reserve. The Trust decided to plant trees and create the small wood that you see in the photograph.

Questions the Trust asked were:

● What trees should we plant?
● Where should we plant them?

These decisions would influence the number and kinds of other organisms that came to live in the wood, including birds, insects and mammals, and the Trust wanted to make the mix of species in their wood – its biodiversity – as wide as possible.

Habitat The environment in which an organism or population of organisms usually lives.

Wherever habitats such as at Hauxley Nature Reserve's wood are developed, compromises need to be made. One compromise is that making a wood will inevitably destroy open ground, which is the habitat of species that do not live in woods. Also, particular species of tree have specific ecological needs and this limits the choice of suitable trees. Alder and willow grow in moist soil; birch will not grow in heavy shade; and beech requires alkaline soils. Clearly, not all species of trees will grow in a particular area. In the Trust's project and others like it, the second compromise is to plant different tree species to encourage a wide range of animal species, yet to plant only tree species that will grow well in the area.

Investigating biodiversity

Scientists carry out investigations that can help to answer questions such as those posed for Hauxley Wood. We will look at the results of some of the research that has been done on woodland birds. These results are shown in Figure 13.1. We have a much better picture of the factors determining the different species of birds that live in a wood than we have for most other groups of animals and plants. This is because many scientists have studied bird populations. Also, it is generally true that a wood that supports many different species of birds will also support many different species of other organisms. This makes studies on birds very useful when selecting types of trees and deciding where to plant them to form a new wood.

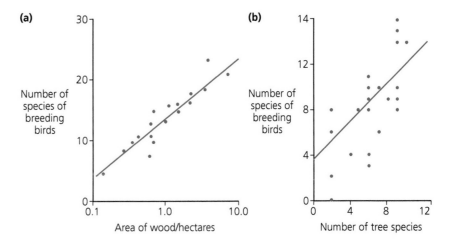

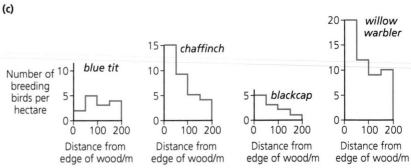

Figure 13.1 Graphs showing the results of some research on woodland birds. (a) Number of species found in woods of different sizes. (b) The relationship between the number of species of birds and the number of species of tree growing in a wood. (c) The relationship between distance from the middle to the edge of the wood on the breeding populations of blue tit, chaffinch, blackcap and willow warbler.

We will start by looking at the data in Figure 13.1a. The graph shows the number of species of birds found in woods of different sizes. On the *x*-axis in this graph, the numbers for the size of the wood have been plotted on a logarithmic scale. This means that the values on the *x*-axis increase by a factor of 10 each time. Instead of going up in equal increments (0, 0.2, 0.4, 0.6 hectares, etc.) the scale goes from 0.1 to 1.0 to 10.0 hectares. The advantage in plotting data this way is that we can show a far greater range of sizes and see more easily if there is a correlation.

A lot of studies have been carried out on the effect that wood size has on the number of bird species. They have all produced similar results to those in Figure 13.1a. The line drawn through the points is called the line of best fit. It shows the overall trend of the data. Since the slope is upwards, the trend is upwards, so there is a **positive correlation** (see page 249) between number of species and wood area. This means that we can conclude that as the size of the wood increases so does the number of bird species.

Figure 13.1b shows data collected from a second investigation. In this graph, the number of breeding species of birds has been plotted against the number of different species of tree present. Again, there is a clear correlation between the two. The more species of trees that are present, the more species of birds that breed.

Finally, we look at Figure 13.1c. The scientists who collected the data for these graphs investigated the effect of distance from the edge of the wood on the number of times that particular species of birds were recorded. The graphs show the results for four different species. Notice that records for blue tits are spread roughly evenly. For the other three species, however, more birds are found near the edge of the wood, and there are fewer in the interior.

The results from scientific investigations such as those shown in the graphs help us to make decisions about conserving biodiversity, such as those made by the Northumberland Wildlife Trust at Hauxley Nature Reserve. The results suggest three things, as follows.

- As large an area as possible should be planted with trees because there is a clear correlation between the area of a wood and the number of species of birds.
- Different tree species should be planted. Again, there is a clear correlation between the number of species of tree and the number of breeding bird species.
- The wood should have an irregular shape or should have open areas within it. This will allow for more woodland edge, and this is the habitat favoured by most species of bird.

Measuring diversity

Biodiversity can be considered at a range of scales, from a small habitat such as a ditch or patch of meadow to entire countries and even the Earth. Look up the word **diversity** in a dictionary and you will see that it simply means 'being different'. So species diversity means 'the mix of different species'. The simplest way of measuring species diversity in a community is to count the number of species present. This is called species richness. A community with more species has a richer mixture of species. The

Community A group of interacting populations of different species living in the same place at the same time.

Species richness The number of different species in a community.

data in both Figures 13.1a and 13.1b shows species richness for the bird community in a wood. Producing a list of species and then simply counting them up can, however, be misleading if you are comparing communities.

If you were to look walk through a typical deciduous wood in the UK and look carefully at the trees you would find quite a lot of species present. Some will be more common than others but they would all count to the same extent in a species richness score. Look at Table 13.1 (opposite). The species richness is 8 but it would still be 8 even if there were 100 oak trees rather than just 6. The numbers of each species give an idea of the evenness of the mix of the community. A wood dominated by oak trees would not have such an even mix as the one in Table 13.1.

A more useful measurement of species diversity for the trees in the wood would take into account both the number of species present (richness) and the number of individuals of each species (evenness). We call such a measurement an index of diversity. It describes the relationship between the number of species present and how each species contributes to the total number of organisms in that community. Because of the fairly even mix of quite a few tree species, this index would be high for the wood in question. Let us look at how a biologist who gathered information about the diversity of trees in the wood used the formula for an index of diversity.

> **Index of diversity** An index giving the relationship between the number of species in a community and the number of individuals in each species.

Calculating an index of diversity

Here is the formula for calculating an index of diversity, d:

$$d = \frac{N(N-1)}{\Sigma n(n-1)}$$

> **TIP**
>
> As you will see in other formulae in Chapter 14, the symbol Σ means 'the sum of'. So in this formula it means the sum of all the different values of $n \times (n-1)$.

In this formula:

N = the total number of organisms in the community

n = the number of organisms of each species in the community.

> **TIP**
>
> If you look up the term index of diversity elsewhere you will find that there are many kinds. The formula above is a type of Simpson's index of diversity. There are even two versions of this, each of which can be expressed in three different ways. You need to stick to the formula shown here. With this formula, the larger the number you obtain the greater the diversity.

The biologist walked along a path through the wood and recorded all trees within 5 m of the path along a 50 m length. His results are shown in Table 13.1.

Table 13.1 The numbers of trees of different species along a path through a wood.

Species	Number of trees
Beech	2
Cherry	1
Hawthorn	1
Hazel	4
Holly	7
Lime	14
Oak	6
Rowan	1
Total	36

Then he calculated the index of diversity (d). We can substitute the figures for N and n from the data in the table:

$$d = \frac{36(36-1)}{2(2-1) + 1(1-1) + 1(1-1) + 4(4-1) + 7(7-1) + 14(14-1) + 6(6-1) + 1(1-1)}$$

$$d = \frac{36 \times 35}{(2 \times 1) + (1 \times 0) + (1 \times 0) + (4 \times 3) + (7 \times 6) + (14 \times 13) + (6 \times 5) + (1 \times 0)}$$

$$d = \frac{1260}{268} = 4.7$$

How would the biologist use this information?

On its own, the index of diversity of the wood does not mean a lot. Its value is that it allows us to compare the diversity of the trees in this wood with diversity of trees in different woods and other habitats. Biologists concerned with species diversity find it very valuable to make comparisons like this.

A higher index of diversity indicates a richer community where ideal conditions allow more species to be more equally successful. A lower index of diversity indicates fewer successful species and may indicate more challenging or stressful conditions such as a restricted range of food sources, fewer habitats or pollution. In such conditions, one or two well-adapted species may thrive, but they will dominate the community. If an index of diversity decreases over time, that is especially significant.

EXAMPLE

Investigating heathland diversity

Lowland heath is a very diverse community of plants and animals and a unique UK habitat. The dominant plant species are heather, gorse and ling and the plant community supports a variety of insects such as the heath grasshopper and the potter wasp. Bracken is found on many heaths and is a potentially invasive species that is shade tolerant.

A scientist investigated the diversity of plants in an area of heath. The table shows the data the scientist collected.

Species	Number of plants/m²
Ling	3
Bell heather	8
Heath bedstraw	6
Bird's foot trefoil	2
Bracken	5
Tormentil	5
Mat-grass	10

1 Calculate an index of diversity using the formula:

$$d = \frac{N(N-1)}{\Sigma n(n-1)}$$

where:

N = total number of organisms of all species in the community

n = total number of organisms of a particular species in the community.

To find N, total up the number of each species in the table. This comes to 39. To find $\Sigma n(n-1)$, take each species and multiply the number found by the number found minus 1. Remember that Σ means add each of these together. So $\Sigma n(n-1) = 3(3-1) + 8(8-1) + 6(6-1) + 2(2-1) + 5(5-1) + 5(5-1) + 10(10-1)$. This comes to 222. So:

d = 39(39 - 1)/222

d = 1482/222

d = 6.68

This is a relatively high index of diversity.

2 If heathland is not grazed by animals, trees such as birch can become established. Suggest why this would reduce the diversity of insects.

Trees are taller than the low-growing plant species in the table. They will create shade, which would reduce the ability of many of the plant species to compete for light so they would not grow effectively. Bracken can tolerate shade more than the other species so it would spread and become dominant, further out-competing the other plant species for light. The variety of different food sources and habitats for insects found in the diverse plant community would both be reduced. In turn, this would reduce the number of species of insect able to survive and reproduce on the heath so the diversity of insects would also fall.

TEST YOURSELF

1 Suggest two reasons why a large wood often contains many more bird species than a small wood.
2 A student was asked to compare the diversity of birds in the middle of a wood and at the edge of the same wood. How should she do this?
3 Willow trees are planted in some areas to provide biofuel. These trees are planted close together. Would you expect the diversity of trees in such a willow plantation to be higher or lower than that for the wood discussed earlier? Explain how you arrived at your answer.
4 The index of diversity for the invertebrate community in a stream declines at the point of a treated sewage outfall and then slowly recovers further downstream. Explain why.

Farming and biodiversity

Since the Second World War there have been huge changes in the way that land has been farmed. Agricultural machines have become larger and more powerful, and to work efficiently they need very big fields. Farmers rear more productive varieties of livestock, and grow more productive varieties of plants. They control insect pests and weeds with chemical pesticides. As a result of these changes, farming has become more intensive, with more food produced per hectare. Also, farms have become more specialised, concentrating for example on just crops or just livestock. All this adds up to larger quantities of food being produced more cheaply.

Farmers are under pressure to use all available land for food production. In the drier east of England, for example, most farmers now concentrate on growing crops such as cereals. They have removed hedges to create larger fields, drained wetter grazing land and have filled in most farm ponds because they no longer have farm animals that need water. Changes in farming practices like these have reduced the diversity of wild plants and animals found on farmland. We will look at some specific examples of how changes in farming have affected diversity.

Improving grazing land

Intensive production of milk requires good quality grass for grazing. Ideal grazing pasture to support large numbers milk cows has to be fast growing and nutritious. A dairy cow can eat 19 kg of dry plant biomass per day.

For high milk yield, the quality of the food is just as important as the quantity. If there is too much **cellulose** in the food, it takes longer to digest (see Chapter 1, page 7) and so the cow eats less. It therefore has less energy available to put into milk production. Figure 13.2 shows that high-quality food (feed quality index of 1.0 or less) can generate as much as 25 kg of milk per cow per day.

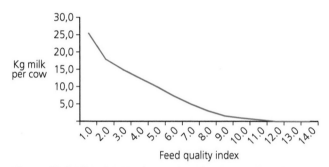

Figure 13.2 Milk yield is determined by the quality of the food eaten by a cow.

The best-quality grazing for milk can be obtained by growing a mixture of ryegrass and white clover. These species will only grow on well-drained and fertilised land. Where possible, dairy farmers improve traditional pasture by draining and re-seeding. Whereas a few other plant species might colonise this grassland over time, the community has a low species richness. Regular use of fertiliser encourages new, vigorous growth of these two species at the expense of any others so although the food quality is high, the diversity of plants remains low (Figure 13.3).

TIP

The feed index is an example of a ratio (see Chapter 14). Feed indices show the ratio of digestible to non-digestible biomass in the feed.

Figure 13.3 Improved grassland, showing predominantly ryegrass and white clover.

Unimproved grazing pasture, especially where the land is poorly drained, looks very different (Figure 13.4). It contains a mixture of more than one grass species and a number of other flowering plant species, many adapted specifically to tolerate damp conditions. Such wet grassland may support 10–15 plant species in a square metre.

Figure 13.4 Unimproved wet grassland; notice how many different types of plants there are.

These different species and their different heights, flowers, seeds and leaves offer a far wider range of food sources and habitats to insects such as grasshoppers, butterflies and beetles. In turn, these insects provide food for insectivorous birds and mammals. The diversity of insects, mammals and birds on and around unimproved wet grassland is higher. But dairy cows grazed on such land would be eating older, tougher grass and a variety of other stalky plants with a higher cellulose content which would not allow them to produce as much milk.

Because of grassland improvement for dairy farming, large areas of grazing land have been improved to support high milk yields. There is very little wet grassland left in some parts of the country and the specialised plant species adapted to their damp conditions have become uncommon.

Estimates suggest that 97% of flower-rich grassland has been lost since 1930. It is thought that this could be one factor behind the current decline in animal diversity of intensively farmed countryside. Any remaining pockets of wet grassland now form **biodiversity hotspots** in the agricultural landscape (Figure 13.5).

Figure 13.5 A pocket of unimproved grassland in an agricultural landscape. Notice where the biodiversity hotspot is.

Spring sowing of cereal crops

For bird species that live on arable land, the two most important events in the farming year are ploughing and harvesting. When a field is ploughed, the soil is turned over and many invertebrates are brought to the surface. Large numbers of birds are often attracted to newly ploughed fields to feed on these invertebrates.

This supply of food, however, only lasts for a day or two after ploughing. Harvesting is also a time when food is plentiful. Seed-eating species of birds can feed on spilt grain and seeds from weeds growing in the crop. Insect-feeding species also benefit because soil-living invertebrates become more accessible once the cover provided by the crop plants has been removed.

One of the main changes in arable farming is a switch from spring planting to autumn planting of cereal crops. Look at Table 13.2 in the Activity box below. You will see that spring sowing involves ploughing the soil in early March. The grain is sown in March or early April. It germinates in April and is ready for harvest in September. Fields are left with stubble over winter before being ploughed again the following March. Autumn sowing involves ploughing immediately after a harvest in July. The grain is then sown in September and starts to germinate. The young plants grow rapidly once spring comes, and the grain is ready for harvesting in late June or early July. Some bird species have declined as a result of this change. This may lead to a reduction in the diversity of farmland birds in the future.

Arable land Land used to grow crops.

> **TIP**
>
> Improving grassland and sowing crops in the autumn instead of spring are simply examples to illustrate the impacts of changing farming techniques on biodiversity, which you might be asked to interpret in an exam. **You do not need to memorise these examples.**

ACTIVITY

The effect of autumn sowing of cereal on rooks

Rooks are omnivorous birds; their food includes plant and animal material: they often feed in fields, eating soil invertebrates; they also eat seeds, including newly planted and spilt grain. Surveys of rook numbers were carried out by biologists and showed a decrease, particularly in eastern England. The biologists thought that this decrease was due to the change from spring to autumn sowing of grain crops.

1 Look at the first column in Table 13.2. This shows information about spring sowing. In which months would you be unlikely to find rooks feeding in arable fields?
2 When they were feeding in spring-sown fields, on what do you think the rooks would be feeding in (a) October and (b) March?
3 With autumn-sown crops, for what proportion of the year is food available in arable fields?
4 Between 1975 and 1980, rook numbers decreased in south-east England but remained more or less the same elsewhere. Use the data in the table to suggest an explanation for this.

Table 13.2 This table summarises the differences in the timing of the spring and autumn sowing of cereal. The shading shows when rooks rely heavily on arable fields for their food.

Spring sowing	Month	Autumn sowing
Stubble	January	
	February	
Ploughing	March	
Sowing and germination	April	
	May	
	June	
	July	Harvesting
	August	Ploughing
Harvesting	September	Sowing and germination
	October	
Stubble	November	
	December	

233

TEST YOURSELF

5 Why do dairy farmers prefer to improve grassland for grazing?
6 Explain why improving grassland for grazing reduces the diversity of plants
7 Why does a reduction in the diversity of plants also reduce the diversity of insects?
8 What are the advantages and disadvantages of autumn sowing of cereals?

Organic farming

Organic farms, where pesticide and inorganic fertiliser use is restricted, are thought to benefit biodiversity. Apart from the fact that the food they produce is free from harmful chemicals, one of the reasons that customers may pay a premium for organic produce is the idea that they are supporting farming practices that support higher plant and animal diversity on and around organic farms. Although this is a widely held view, there is surprisingly little rigorous scientific evidence to support it.

ACTIVITY

The effect of organic farming on butterflies

One study in Sweden compared the butterflies found on organic farms with those found on conventional farms. Twelve pairs of farms were selected for the study. Six pairs were in a mixed landscape where the mean percentage of arable land was only 15%. The other six pairs were located in a more uniform arable landscape where the mean percentage of land used for growing crops was 70%. In each case, an organic farm was paired with a conventional farm of similar area. The scientists tried to match pairs of farms as closely as possible in terms of landscape feature such as hedges, stone walls and patches of woodland.

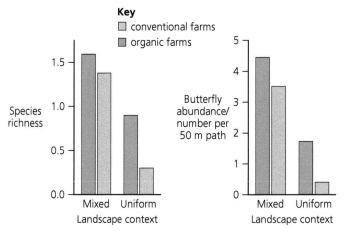

Figure 13.6 Butterfly abundance and species richness in relation to farming practice.

1 Why were the pairs of farms matched as closely as possible for landscape features?
2 What was the reason for selecting six pairs of farms in a mixed landscape and six pairs in a more uniform landscape?

The results for the butterfly surveys are shown in Figure 13.6.

3 What effect do organic farming practices have on (a) butterfly species richness and (b) butterfly abundance in mixed compared to uniform landscapes?

4 Why might organic practices have had more impact on butterfly species richness in uniform landscapes?
5 Suggest how the restrictions on pesticide and inorganic fertiliser use on organic farms might explain the differences in butterfly species richness and abundance.
6 The scientists did not calculate an index of diversity in their study. Could you use information from the graphs to calculate an index of diversity for each group of farms? Explain your answer.

Disappearing bumblebees

Recent concern has been expressed over the possible link between intensive crop production and the extinction or near extinction of some bumblebee species. Some 24% of the 68 species of bumblebee found in Europe are threatened with extinction. This includes two species found in the UK, the great yellow bumblebee (*Bombus distinguendus*) and the shrill carder bee (*Bombus sylvarum*). A third UK species, the short-haired bumblebee, was declared extinct in 2000, although a recent attempt at re-introduction has been made in Kent.

This rapid reduction in the diversity of bumblebees is causing concern. It is not just because scientists are interested in bumblebees. Bumblebees are vital pollinators of arable crops such as oilseed rape, peas, strawberries and apples. Insect pollinators are estimated to contribute over £400 million to the UK economy every year. Bumblebees play a large part in this free ecosystem service. Without it the cost of some fruit and vegetables would rise considerably.

Ideal bumblebee habitats have a high diversity of plants. This ensures a supply of pollen and nectar at different and overlapping times throughout the bumblebees' feeding season. In agricultural areas, ideal habitats include species-rich grassland, field margins, hedgerows and ditches. Farmers have to manage these habitats sensitively to help bumblebee species survive (Figure 13.7).

Figure 13.7 This species-rich field margin is suitable for bumblebees.

Bumblebees depend on a high diversity of flowering plants for a reliable source of food, whereas wild plant species depend on bumblebees for their pollination. A variety of plant species is threatened by the potential loss of bumblebee species, including arable crops. The biodiversity of bumblebees and plants are interlinked.

A laboratory study highlighted another possible factor in the decline of bumblebees. This investigated the impact of some pesticides on bumblebee foraging effectiveness. It found that chronic (long-term) exposure of bumblebees to two neonicotinoid pesticides impairs their ability to navigate and fly to find food.

Neonicotinoids are a relatively new kind of insecticide that affects the nervous system of insects, resulting in paralysis and death. They protect the young crop plants from insect pests from the very start of their life and are now widely used to treat seeds before planting. It is estimated that neonicotinoids have improved winter wheat yields in the UK by as much as 20%. Oilseed rape production now depends on them to combat pests such as the rape flea beetle. However, when used as seed treatments, neonicotinoids move to the pollen and nectar of the adult plants through **translocation** (see Chapter 10, pages 185–189), which is how they may affect bees and other pollinators as well as the target pest species.

The laboratory study gave rise to calls for these sorts of insecticides to be banned. This caused widespread concern among farmers. But there is no conclusive evidence from preliminary field trials to support the laboratory study and it is clear that the pesticides are extremely useful to farmers. This example clearly illustrates the difficulty of balancing the interests of productive arable farming with biodiversity conservation.

TEST YOURSELF

9 Give two reasons why organic farming is thought to increase biodiversity.

10 Explain why maintaining the diversity of pollinators on arable farmland is so important.

11 Why are neonicotinoid pesticides so useful to arable farmers?

12 How might neonicotinoid pesticides used on seeds be causing problems for bumblebees?

TIP

When you write about environmental issues such as those described in this chapter, remember that you are a scientist and you should write as a scientist. Use scientific terms and use them accurately. It is not a good idea to write, for example, about animals 'losing their homes' and 'having nowhere to live' because their habitat has been lost.

Practice questions

1 Scientists investigated the number of different bird species present in hedges of different lengths. The graph shows their results. Black circles represent hedges surrounding conventional fields, and white circles represent hedges surrounding organic fields where no pesticides had been used.

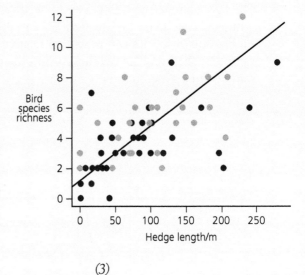

a) Describe the relationship between the length of hedges and the number of bird species present. (2)

b) Suggest why the length of a hedge might have an effect on the number of bird species present. (2)

c) The scientists counted the number of different species present rather than calculating the index of diversity. Was this a good decision? Give reasons for your answer. (3)

d) Do these data show that organic farming is better for encouraging the diversity of birds? Give reasons for your answer. (4)

2 Two students carried out a survey of bird species in two different woodlands in the same area of the country on the same day. They made a list of all the birds they saw or heard singing over a period of 1 hour in each woodland. Their results are shown in the table.

Species	Number at site A	Number at site B
Blackbird	7	3
Blue tit	2	5
Bullfinch	0	1
Carrion crow	1	0
Chaffinch	3	0
Chiffchaff	0	2
Goldfinch	2	0
Greenfinch	0	2
Great tit	4	2
Magpie	0	3
Robin	4	6
Starling	2	7
Tree sparrow	12	8
Wood pigeon	6	4

a) i) Give one additional variable that the students should have kept the same, explaining the reason for this. (2)

ii) Explain one possible limitation of this method and the effect this might have on the results obtained. (2)

b) i) Calculate an index of diversity for site A. (2)

ii) Suggest an advantage of calculating an index of diversity rather than just counting the number of birds seen or heard. (1)

3 a) Explain what is meant by biodiversity. (2)

The figure shows the changes in the relative abundance of different groups of species in the UK from 1970 to 2010. The abundance of each group of species was measured relative to the first year they were sampled. The numbers in brackets indicate the number of species sampled in each group of organisms.

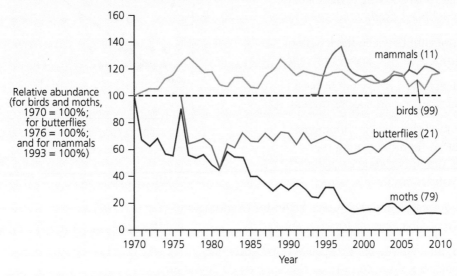

b) A student concluded that moth populations were much lower in 2010 than butterfly populations. Suggest two reasons why this conclusion might be wrong. (4)

c) Explain how the relative abundance of mammals in 1998 would have been calculated. (3)

Stretch and challenge

4 Conservationists set up nature reserves such as Hauxley Wood to maintain biodiversity as high as possible. In order to achieve this, what can seem to be destructive activities are carried out, such as cutting down trees and bushes, putting grazing animals onto the reserve and cutting reeds in ponds. Explain how these activities can help to maintain biodiversity rather than reduce it.

14 Energy transfer

- Radiation from the Sun is the source of energy for most communities of living organisms.
- Green plants and algae absorb a small amount of the light that reaches them. The transfer from light energy to chemical energy occurs during photosynthesis. This energy is stored in the substances that make up the cells of the plants and algae.
- Respiration supplies the energy needs for living processes, including movement. Much of this energy is eventually transferred to the surroundings.
- In all cells, carbohydrate or lipid molecules are broken down during respiration to release the energy required to resynthesise ATP. In some cells, ATP can also be made using light energy during photosynthesis.
- The amounts of material and energy contained in the biomass of organisms are reduced at each successive stage in a food chain because some materials and energy are always lost in organisms' waste materials.

TEST YOURSELF ON PRIOR KNOWLEDGE

1 Name two substances that make up the cells of a plant.
2 A mouse is completely at rest in a comfortable environmental temperature. Name **two** processes during which it uses ATP.
3 Name the enzyme involved in resynthesising ATP.
4 In what form is energy released by respiration eventually transferred to the surroundings?

Introduction

Sugar cane is grown as a crop in more than 70 countries and provides around 80% of the world's sugar. It is one of the most efficient crop plants in cultivation. Originally from south Asia, sugar cane is a tropical grass. Along with some other tropical grasses such as elephant grass, it has evolved a way of photosynthesising that allows it to make the most of the tropical sun.

Most plants use a form of photosynthesis called C_3 photosynthesis. In these plants, the first stable product of photosynthesis is a three-carbon molecule. The reaction is catalysed by an enzyme called ribulose bisphosphate carboxylase (rubisco for short). But rubisco has a surprising flaw. Although it catalyses the reaction with carbon dioxide, it will also catalyse a reaction with oxygen instead, resulting in a different outcome and inhibiting the enzyme's contribution to photosynthesis.

When plants close their stomata in hot, dry conditions, the carbon dioxide concentration in their leaves falls and oxygen competes more successfully with carbon dioxide for rubisco. The efficiency of photosynthesis is then reduced. Plants in tropical conditions, especially grasses, face hot, dry conditions more often than those in other areas. Some, like sugar cane, evolved a form of photosynthesis, called C_4 photosynthesis, that reduces this problem.

Sugar cane does not grow as effectively in the UK. But you may have seen fields of elephant grass being grown in the UK as a biofuel for electricity generation. Elephant grass is also a C_4 plant and its rapid growth and high annual biomass yield makes it useful as a renewable energy source. Unlike sugar cane and elephant grass, most plant species in the UK carry out the form of photosynthesis you will learn about in this chapter.

Life on Earth depends on the continuous transfer of energy through photosynthesis, feeding and respiration. Plants and other chlorophyll-containing organisms photosynthesise, absorbing light. In this process, some of the energy of light is conserved in the production of ATP and ultimately in carbohydrates and other biological molecules.

These biological molecules can be used directly by the plants, mostly in respiration and to make the other biological molecules they require, or they can be transferred to other organisms by the animals that feed on them and by saprobionts that decompose them.

During respiration, various respiratory substrates are oxidised and some of their chemical energy is conserved in the production of ATP. ATP can then be used in various forms of biological work, such as movement, synthesis of large biological molecules and active transport.

Photoautotrophic organisms Organisms that synthesise their own biological molecules using light energy.

These energy transfers are fundamental to life. Respiration is common to all organisms and photosynthesis is common to all photoautotrophic organisms. The fact that these processes are so widespread suggests that these organisms all evolved from common ancestors. This is indirect evidence for evolution.

Respiratory substrates Biological molecules used as fuel in respiration.

Biomass The mass of carbon in biological molecules or dry mass of tissue per given area.

Energy and efficiency

Neither photosynthesis nor respiration is totally efficient. During respiration, for example, not all the chemical energy from a molecule of respiratory substrate is transferred into molecules of ATP. Some of the energy will inevitably be lost as heat. This is obviously important when we come to look at the transfer of biomass and its chemical energy from one organism to another along a food chain or through a food web.

Figure 14.1 summarises the three ways in which energy is transferred within and between different organisms.

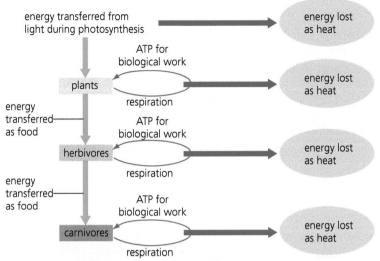

Figure 14.1 A summary of the ways in which energy is transferred within and between organisms. Note that, each time a transfer occurs, some energy is lost as heat.

Photosynthesis

Organisms must have a continuous supply of biological molecules for respiration and with which to build new cells and tissues. For animals, these biological molecules come from food: this is sometimes from other animals but ultimately from plants. It is only by photosynthesis that light energy can be transferred to chemical energy.

Photosynthesis is a complex process involving a number of separate reactions. It is useful to get an idea of the overall process before we look at the detail. Figure 14.2 shows the two basic steps.

- Light-dependent reaction; light energy is absorbed by chlorophyll and some of this energy is transferred to chemical energy in ATP. A second substance is also produced. This is reduced NADP (see page 253 for a definition of reduction and oxidation). In order to produce these substances, a molecule of water is split and oxygen is given off as a waste product.
- Light-independent reaction: ATP and reduced NADP are involved in the use of carbon dioxide to make carbohydrate.

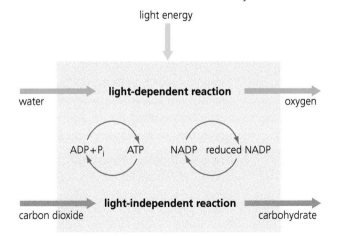

$$\text{carbon dioxide} + \text{water} \xrightarrow{\text{light energy}} \text{carbohydrate} + \text{oxygen}$$

Figure 14.2 The main steps in photosynthesis. The substances entering and leaving the leaf can be arranged to give the basic equation for photosynthesis.

Structure of chloroplasts

You should remember from the first year of the course that chloroplasts are the site of photosynthesis (see Chapter 3 page 42). Each chloroplast (Figure 14.3 below) is surrounded by two membranes. Both the outer and inner membranes of a chloroplast are smooth. Inside the chloroplast there are a series of disc-shaped, membrane-bound structures called thylakoids. In some places the thylakoids are arranged in stacks called grana. The membranes that form the grana provide a very large surface area for chlorophyll molecules and other light-absorbing pigments.

Thylakoids Disc-shaped, membrane-bound structures found inside chloroplasts.

Grana Groups of thylakoids arranged in stacks.

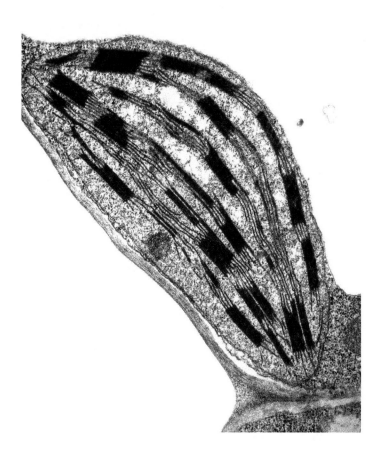

Figure 14.3 A transmission electron micrograph of a chloroplast. This chloroplast is approximately 5 μm in diameter.

The light-dependent reaction

If you crush up some nettle leaves in an organic **solvent** such as ethanol you can make a chlorophyll solution. If you shine a bright light on this solution it fluoresces (emits light), but, instead of appearing green, it looks red. Light falling on the solution causes electrons to leave some of the chlorophyll molecules. This is because of photoionisation.

Photoionisation The process by which a chlorophyll molecule becomes positively charged as a result of losing two electrons when it absorbs light.

In a solution of chlorophyll the electrons have nowhere to go. This is why, when we shine light on the solution, it fluoresces red. The electrons lose most of their energy as light of a different wavelength as they fall back into their places in the chlorophyll molecules.

In a chloroplast, however, these electrons do not return to the chlorophyll molecule from which they came. They pass down a series of electron carriers, losing energy as they go. In chloroplasts, this energy is conserved in the production of ATP and reduced NADP.

REQUIRED PRACTICAL 7

Use of chromatography to investigate the pigments isolated from leaves of different plants

This is just one example of how you might tackle this required practical.

A chlorophyll solution contains a mixture of chlorophyll and other pigments. The different pigments can be separated using chromatography.

Figure 14.4 shows the result of using thin layer chromatography (TLC) to separate the pigments extracted from nettle leaves.

1 What is a solvent?
2 Suggest why the origin line was placed a little way above the base of the plate.
3 How could the origin spot have been made sufficiently concentrated?

The different pigments in the mixture can be identified by finding their R_f values. The R_f is an example of a ratio and is calculated as:

$$R_f = \frac{\text{distance moved by the component spot}}{\text{distance moved by the solvent from the origin}}$$

The standard R_f values for these pigments using this solvent are:

chlorophyll *a*   0.31   chlorophyll *b*   0.15
carotene   0.96   xanthophyll   0.63

(a)

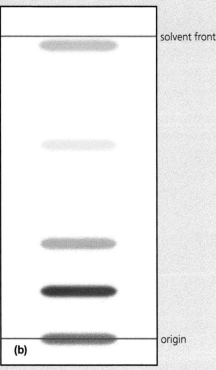
(b)

Figure 14.4 (a) Nettles. (b) Thin layer chromatography plate following separation of nettle leaf extract.

> **TIP**
> You do not need to know about R_f values, but using them is a good way to practise your maths skills.

> **TIP**
> Look at Chapter 27 for information on how to carry out thin layer/paper chromatography.

4 Identify the different pigments on the thin layer chromatography plate in Figure 14.4b by measuring the distances with a ruler and calculating their R_f values.
5 Give one possible source of error in finding the R_f values.

The light-dependent reaction is described on the next page and is summarised in Figure 14.5 below.

Figure 14.5 A summary of the light-dependent reaction of photosynthesis.

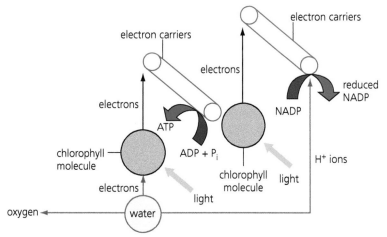

- Light strikes a chlorophyll molecule, causing photoionisation. Two electrons leave the chlorophyll molecule and pass to an electron carrier.
- The electrons are transferred along a series of electron carriers, which forms an electron transfer chain within the thylakoid membrane. The electrons lose energy as they are passed from one electron carrier to the next. This energy is used to produce ATP.
- **Photolysis** takes place. This is the breakdown of a water molecule to release protons, electrons and oxygen.

$$\underset{\text{water}}{H_2O} \quad \rightarrow \quad \underset{\text{protons}}{2H^+} \quad + \quad \underset{\text{electrons}}{2e^-} \quad + \quad \underset{\text{oxygen}}{\tfrac{1}{2}O_2}$$

- The two electrons from photolysis replace those lost from the chlorophyll molecule during photoionisation. Oxygen is released as a waste product.
- Light strikes a second chlorophyll molecule, causing photoionisation. Two electrons leave the second chlorophyll molecule and pass to an electron carrier. They pass along another electron transfer chain. They are used, together with the protons from the photolysis of water, to produce reduced NADP (see page 253 for a definition of reduction and oxidation).

$$NADP + 2H^+ + 2e^- \rightarrow \text{reduced NADP}$$

TIP
Notice that the oxygen produced in photosynthesis comes from photolysis of water, rather than from the carbon dioxide that plants take up from their environment.

Electron transfer chains and ATP production

As each pair of electrons passes along an electron transfer chain, a small amount of energy is released. This energy enables carrier proteins (see Chapter 3 page 46) within the thylakoid membranes to actively transport protons from the stroma across the thylakoid membrane and into the spaces between the thylakoids. This develops a higher concentration of protons inside the thylakoid spaces than in the stroma. As a result, protons diffuse down their concentration gradient from the thylakoid spaces to the stroma. They diffuse through molecules of ATP synthase embedded in the thylakoid membranes, and the resulting change in environment results in a change in the protein, causing the ATP synthase molecules to spin. This spinning provides energy for the synthesis of ATP from ADP and inorganic phosphate.

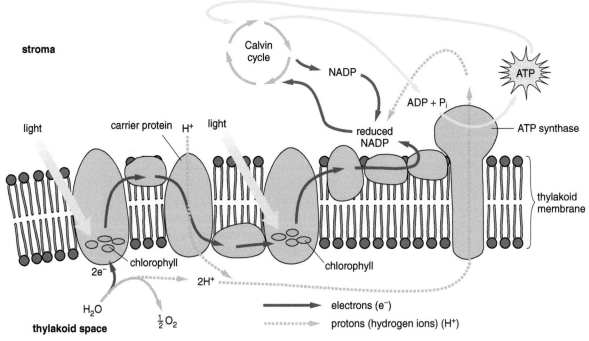

Figure 14.6 How the electron transfer system and ATP synthase molecules are embedded in the thylakoid membrane.

Chloroplasts isolated from plant leaves can be used to investigate the light-dependent reaction. If isolated chloroplasts are placed into a dilute buffer, they absorb water by osmosis and burst, releasing their thylakoids into the solution. If a blue dye called dichlorophenolindophenol (DCPIP) is added, electrons from active electron transfer chains are transferred to DCPIP molecules, reducing them rather than NADP molecules. When DCPIP is reduced, it changes colour from blue to colourless.

REQUIRED PRACTICAL 8

Investigation into the effect of a named factor on the rate of dehydrogenase activity in extracts of chloroplasts

This is just one example of how you might tackle this required practical.

Figure 14.7 shows the results for isolated chloroplasts mixed with dilute buffer and DCPIP and placed into different conditions.

The tubes were set up as follows:

Tube 1: chloroplast extract, buffer, DCPIP
Tube 2: chloroplast extract, buffer, DCPIP, completely wrapped in foil
Tube 3: boiled chloroplast extract, buffer, DCPIP
Tube 4: boiled chloroplast extract, buffer, DCPIP, completely wrapped in foil
Tube 5: chloroplast extract, buffer
Tube 6: chloroplast extract, buffer, completely wrapped in foil

Figure 14.7 Results of an investigation using spinach chloroplast extract with DCPIP.

All were then placed under a bright light for several hours.

Using what you now know about the light-dependent reaction, explain the results.

Rubisco, or ribulose bisphosphate carboxylase. The enzyme that catalyses the reaction between carbon dioxide and ribulose bisphosphate in the Calvin cycle.

The light-independent reaction

The light-independent reaction of photosynthesis comprises a cycle of reactions called the **Calvin cycle** (see Figure 14.8). The main steps in the cycle are as follows:

- Carbon dioxide reacts with **ribulose bisphosphate (RuBP)** to form two molecules of **glycerate 3-phosphate (GP)**. This reaction is catalysed by an enzyme called ribulose bisphosphate carboxylase (rubisco).
- GP is then reduced to **triose phosphate**. This is a reduction reaction and requires the two substances formed during the light-dependent reaction: reduced NADP and ATP. The reduced NADP is used to reduce GP. ATP provides additional energy for the reaction.
- Triose phosphate is a simple sugar. Some triose phosphate is converted to useful organic substances, such as sucrose for transport or cellulose for storage, or into amino acids and triglycerides.
- Most of the triose phosphate is converted into glucose and used by the plant as a respiratory substrate.
- Some triose phosphate is used to regenerate more RuBP.

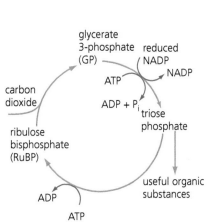

Figure 14.8 The main steps in the Calvin cycle.

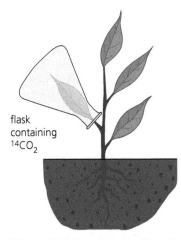

Figure 14.9 Supplying radioactive carbon dioxide to a plant leaf.

The Calvin cycle can be investigated by supplying a plant with radioactively labelled carbon dioxide. A plant leaf can be enclosed in a flask containing radioactive carbon dioxide (Figure 14.9). A series of leaves can be left in radioactive carbon dioxide for different amounts of time. The leaves can then be removed from the plant and analysed for radioactive substances.

The results of such an investigation are shown as a graph in Figure 14.10. The first radioactive substance detected in the leaf is GP. RuBP is detected more slowly. This indicates that the radioactive carbon dioxide is used to form GP and that the GP is then used to form RuBP.

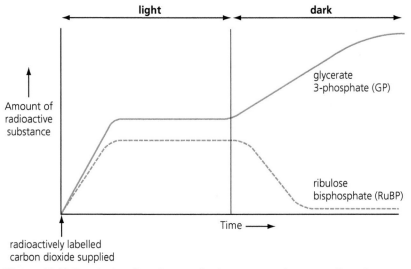

Figure 14.10 Graph showing changes in the amount of some radioactive substances in a leaf in light and dark conditions, after being given radioactively labelled carbon dioxide.

After a certain time in light, the amount of radioactive GP and RuBP both become constant. This is because GP and RuBP are being formed in a cycle. As fast as GP is being formed from carbon dioxide and RuBP, it is being used to regenerate RuBP.

If the light is switched off, the amount of GP increases while the amount of RuBP decreases. This is because the light-dependent reaction can no longer produce ATP and reduced NADP. If you look at Figure 14.8, you will see that ATP and reduced NADP are required to convert GP into triose phosphate but not to convert RuBP into GP. In the dark, RuBP is converted to GP, so the amount of RuBP decreases while the amount of GP increases.

TEST YOURSELF

1 What is photoionisation?
2 In the light-dependent reaction of photosynthesis, what happens to the electrons that come from:
   a) a water molecule
   b) the first chlorophyll molecule struck by light
   c) the second chlorophyll molecule struck by light?
3 Name the three-carbon sugar produced in the Calvin cycle.
4 Look at Figure 14.10. Use the information to explain the changes in the amounts of radioactive substances in the dark.

Limiting factors and photosynthesis

Yield The biomass of the part of the crop that is harvested.

Plants rely on photosynthesis to produce their respiratory substrates and other biological molecules they need to form their biomass. The greater a plant's rate of photosynthesis, the greater its rate of growth and, if we are considering crop plants, the higher the yield. Among the environmental factors that may affect the rate of photosynthesis are:

● light intensity
● carbon dioxide concentration
● temperature
● availability of water in the soil.

247

The graph in Figure 14.11 shows the effects of light intensity, carbon dioxide concentration and temperature on the rate of photosynthesis.

TIP
The shape of curve A in Figure 14.11 is very similar to the one for the rate of an enzyme-catalysed reaction against substrate concentration (see Chapter 2 page 32) because both involve limiting factors.

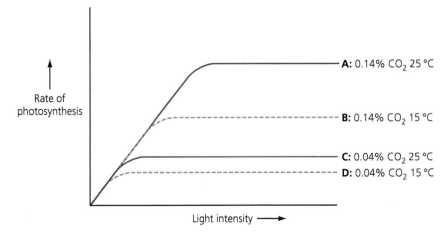

Figure 14.11 Limiting factors: the effects of light intensity, carbon dioxide concentration and temperature on the rate of photosynthesis.

Over the first part of the curve A, the rate of photosynthesis is directly proportional to light intensity. Because of this, we can describe light as limiting the rate of photosynthesis over this part of the curve (it is a **limiting factor**). As light intensity continues to increase, the curve starts to flatten out: light intensity is no longer limiting the rate of photosynthesis.

However, the graph also shows two other environmental factors that may interact with light intensity. For example, increasing carbon dioxide concentration or temperature will not increase the rate of photosynthesis when light intensity is low, such as on a winter's day in the UK.

Carbon dioxide limits the rate of photosynthesis in bright conditions. Increasing the concentration of carbon dioxide in the atmosphere from its normal level of approximately 0.04% to 0.14% and keeping the temperature constant has a much greater effect on the rate of photosynthesis than increasing the temperature from 15 °C to 25 °C and keeping the carbon dioxide concentration constant.

You can also see from the graph that increasing the temperature from 15 °C to 25 °C increases the rate of photosynthesis. The effect is greatest when neither light intensity nor carbon dioxide concentration are limiting.

By understanding how environmental factors can limit photosynthesis, farmers are able to take steps to overcome their effects and improve the yield of their crops. This is easier in closed environments such as glasshouses, but some simple practices, such as limiting the shade from high hedges or irrigating crops, can be worthwhile for field crops too.

TIP
The amount of water available to a plant can limit the rate of photosynthesis if lack of water causes the stomata to close. The rate of photosynthesis is actually then limited by reduced carbon dioxide availability, not by a lack of water for photosynthesis itself.

EXAMPLE

Carbon dioxide concentration and crop production

In bright conditions, the concentration of carbon dioxide usually limits the rate of photosynthesis. An increase in carbon dioxide concentration should therefore increase the rate of photosynthesis. Look at Figure 14.12. It shows the results of a laboratory experiment in which scientists investigated the effect of an increase in carbon dioxide concentration on the rate of photosynthesis of wheat. The scientists made sure that no other factors were limiting the rate of photosynthesis.

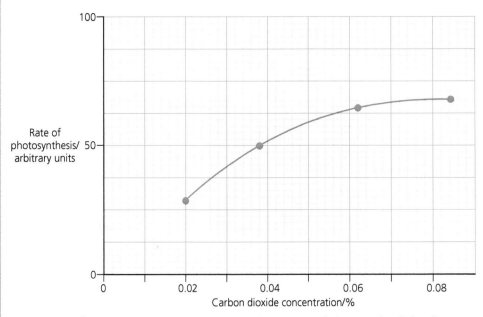

Figure 14.12 Effect of carbon dioxide concentration on rate of photosynthesis in wheat.

1 Suppose we doubled the concentration of carbon dioxide from 0.033% to 0.066%. What would happen to the rate of photosynthesis in wheat?
  The rate of photosynthesis would increase from 45 to 66 arbitrary units.

2 Can you conclude from this graph that an increase in the concentration of carbon dioxide in the atmosphere will result in an increase in the growth of wheat plants? Explain your answer.
  Yes, a higher rate of photosynthesis will produce more triose phosphate, which can then be converted to other useful substances, resulting in more growth.

3 Will an increase in carbon dioxide affect the rate of photosynthesis of all crop plants in the same way?
  No, it depends how the crops are affected by other limiting factors.

Look at the data in Table 14.1. The data predict the effect of doubling current atmospheric concentrations of carbon dioxide.

4 What general conclusions can you draw from the data in this table?
  All the plants will grow faster, including the weeds that compete with crops. But the increased growth in the harvested part of the crop will not be as great as the increased growth of the whole plant.

5 Use the data in Table 14.1 to suggest how doubling the current atmospheric concentration of carbon dioxide would affect the marketable yield of cabbages growing in a field.
  About half the increased growth of cabbages would be in the harvested part of the plants, the rest would be roots and outside leaves.

Table 14.1 The effect of doubling current atmospheric concentrations of carbon dioxide on some different crop plants.

Plant	Percentage increase in biomass	Percentage increase in marketable yield
Cotton	124	104
Tomato	40	21
Cabbage	37	19
Weeds	34	not applicable

How efficient is photosynthesis?

A lot of light falls on the surface of the Earth. Only a small part of this is used during photosynthesis. Look at the wheat crop shown in Figure 14.13.

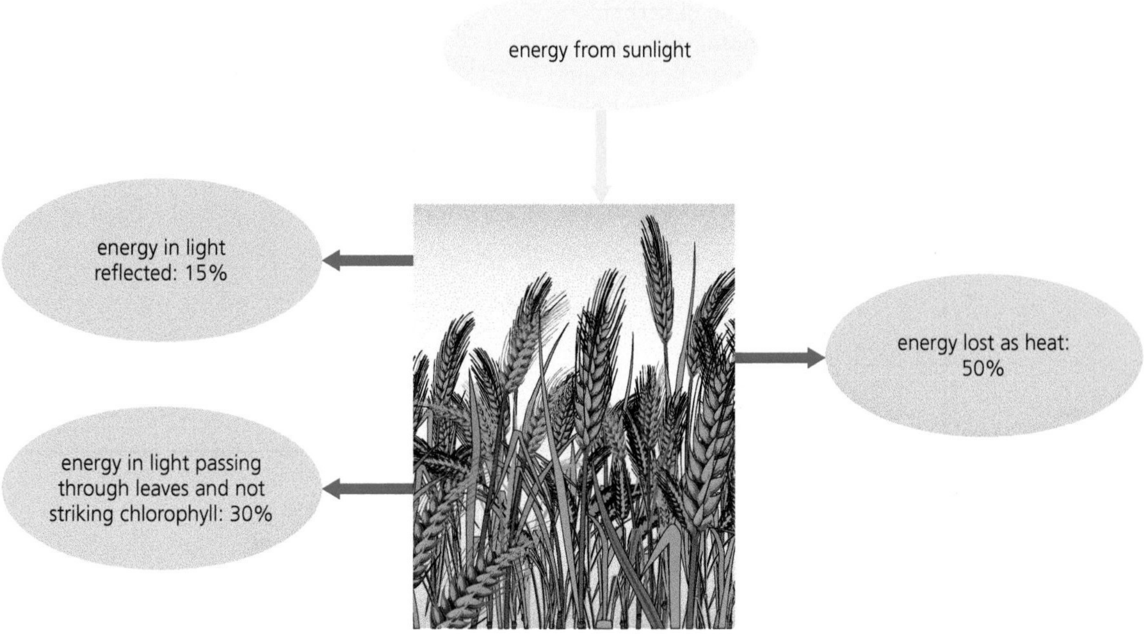

Figure 14.13 This wheat crop conserves only a small percentage of the incident light energy as chemical energy in organic substances.

- About 50% of the light energy absorbed by the plants in this crop is lost as heat. Much of this heat evaporates water from leaves during transpiration.
- Approximately 15% of the light is **reflected** from the leaf surface.
- Almost a third of the light, approximately 30%, is **transmitted**. It passes directly through the plants without striking any chlorophyll molecules.

You can see from this that only a very small percentage of the light energy can be conserved as chemical energy in the biological molecules produced by photosynthesis.

The chemical energy stored in the plant biomass in a given area is the gross primary production (GPP). The rate at which plants are able to store this chemical energy as a result of photosynthesis is called **gross primary productivity**. Since this is a rate, the units must include time. Like gross primary production, the units also include the area or volume of plants being measured. For land plants, area is used (usually a square metre), whereas for aquatic algae, volume is used. Standardising the quantity of plant material being measured by area allows gross primary production and gross primary productivity values for different plant populations or communities to be compared.

To calculate GPP, the biomass must first be found as either

- **dry mass** of tissue, measured in units such as $g\,m^{-2}$, or
- mass of carbon, measured in units such as $\mu g\,m^{-2}$.

The dry mass of plant material is found by warming it in an oven to evaporate all the water in the tissues. At intervals, the plant material is weighed. When three successive mass measurements are the same there is no further water left to evaporate. This must be done slowly. It is important not to overheat the plant material otherwise it may burn and lose dry mass by combustion rather than just evaporation.

TIP
See Chapter 10 page 184 to revise evaporative cooling.

Gross primary production (GPP) The chemical energy store in plant biomass, in a given area or volume.

TIP
Dry mass is more valid than simply weighing fresh plant material because fresh plant material can contain very different quantities of water depending on the environmental conditions when it was collected.

The chemical energy stored in the biomass can then be estimated by calorimetry. The GPP would then be expressed in $kJ\,m^{-2}$ (or $kJ\,m^{-3}$ if it were aquatic algae). Finding the value again for the same area after a period of time allows the increase to be found and allows the gross primary productivity to be calculated, expressed in, for example, $kJ\,m^{-2}\,year^{-1}$.

Some of the substances formed during photosynthesis are not used to form new cells and tissues. They are used as respiratory substrates and consequently, some energy is lost as heat to the environment. The difference between GPP and respiratory losses to the environment (R) is called net primary production (NPP). We can calculate net primary production from the equation

$$NPP = GPP - R$$

Net primary production is important because it represents the amount of energy available to the primary consumers (herbivores) and decomposers at other trophic levels in a food web (see page 261). Finding the value again for the same area after a period of time allows the rate of increase, or **net primary productivity** to be calculated.

> **Net primary production (NPP)** The chemical energy store in plant biomass after respiratory losses to the environment have been taken into account.

Efficiency of energy transfer by heather

Figure 14.14 shows heather plants. Scientists have found heather particularly useful in studying the efficiency of photosynthesis for the following reasons.

- In moorland areas, heather grows in large clumps called stands. Each stand is almost pure heather, so we do not have to consider what fraction of the total visible-light energy is used by other species.
- In many moorland areas, heather is managed. This means that it is burned at regular intervals. Old woody plants are replaced by young plants, which provide a better food supply for game birds, such as grouse. Estate managers keep records so we usually know the age of a particular stand of heather.
- Many investigations have been carried out using heather. By sharing the findings of their research, scientists have been able to replicate and further test their work. This increases the reliability of the conclusions that they draw.

Figure 14.14 Heather.

We can calculate the efficiency of energy transfer in heather from the formula:

$$\text{Efficiency of energy transfer} = \frac{\text{chemical energy increase in an area of heather plants in a year}}{\text{light energy falling on this area of heather plants in a year}}$$

To calculate the figure on the top line of the equation we need to multiply the chemical energy in 1 g of heather by the biomass of heather produced in $g\,m^{-2}\,year^{-1}$.

The chemical energy in 1 g of heather can be found by using a calorimeter (see overleaf).

> **TIP**
> This is just an example of calculating the efficiency of energy transfer. You may be asked to calculate the efficiency of transfer in other situations, such as in farm crops or animals.

Finding the chemical energy in 1 g of heather

We can find this by burning a sample of heather in a **calorimeter** (Figure 14.15) and measuring the energy released as heat.

Look at Figure 14.15 and answer the questions.

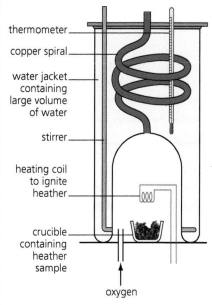

Figure 14.15 A calorimeter used to measure the energy released when a sample of heather is burned.

1 An oxygen supply is connected to the apparatus. What is the advantage of burning the heather in oxygen and not in air?
Oxygen ensures that the sample of heather is burned completely.

2 A copper spiral is attached to the top of the combustion chamber. What is the function of this copper spiral?
The copper spiral provides a large surface area for heat exchange with the water.

3 The water jacket contains a large volume of water. The total rise in temperature of the water in this jacket when the heather is burned is only small. Explain the advantage of a small rise in temperature.
A small temperature rise means less heat will be lost to the surroundings.

Table 14.2 Shows some typical results obtained from the apparatus in Figure 14.15.

Table 14.2 Results obtained from burning a sample of heather in a calorimeter.

Volume of water in water jacket/cm³	650
Temperature of water before the heather sample was burned/°C	18
Temperature of water after the heather sample was burned/°C	23
Mass of heather/g	0.5

4 The amount of energy needed to raise the temperature of $1\,cm^3$ of water by $1\,°C$ is 4.2 joules. Calculate the amount of energy released by burning 1 g of heather.
In this investigation, the energy released has raised the temperature of $650\,cm^3$ of water by $5\,°C$. So, the total amount of energy released by the heather sample is $4.2 \times 650 \times 5\,J$.
This is the amount of energy released by 0.5 g of heather. We need to divide the total amount of energy released by 0.5 to get the amount of energy released per gram:

$$4.2 \times 650 \times 5/0.5\,J\,g^{-1}$$

This gives us a figure of $27\,300\,J\,g^{-1}$ or $27.3\,kJ\,g^{-1}$.

5 Do you think that the figure that we have calculated is an overestimate or an underestimate? Explain your answer.
It is probably an underestimate. There are several reasons for this. Not all the heather will have been burned. The ash that is left in the crucible may contain some chemical energy that has not been released. In addition, it is unlikely that all the heat from the combustion will have been transferred to the water in the water jacket.

TEST YOURSELF

5 Give three reasons for the relatively low efficiency of photosynthesis.
6 Describe how the dry mass of a sample of plant material would be found.
7 What is the equation for finding net primary production?
8 Suggest suitable units for the net primary production of algae in a lake.
9 Explain how the net primary production of trees is made available to decomposers in a wood.

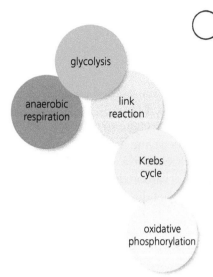

Figure 14.16 The breakdown of glucose in respiration. Aerobic respiration takes place in the presence of oxygen. Respiration can continue when there is no oxygen. This is anaerobic respiration.

Phosphorylation The addition of a phosphate group to a molecule.

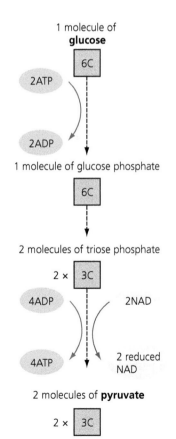

Figure 14.17 A summary of glycolysis, the first step in respiration. The boxes show the number of carbon atoms in the different molecules. There is a net gain of two molecules of ATP from each molecule of glucose.

Respiration

Respiration takes place in all living cells, providing indirect evidence for evolution. It is a biochemical process in which biological molecules called respiratory substrates are used as fuel. They are broken down in a series of stages and the chemical energy they contain is transferred to ATP. We can summarise the process with the equation:

$$C_6H_{12}O_6 + 6O_2 \rightarrow 6CO_2 + 6H_2O + energy$$

Unfortunately this equation is misleading in a number of ways. It shows the fuel as glucose. In many cells the main fuel is glucose, but fatty acids, glycerol and amino acids are also respiratory substrates and can be used for respiration. The equation also shows that oxygen is required: this is correct when **aerobic respiration** is occurring but respiration can also take place anaerobically. **Anaerobic respiration** means respiration without oxygen. It is also important to understand that the oxygen is not directly used to make the carbon dioxide shown in the equation. Finally, the equation shows respiration as a single reaction. It isn't a single reaction. It involves a number of reactions in which the respiratory substrate is broken down in a series of steps, releasing a small amount of energy each time. The steps involved in the respiration of glucose are summarised in Figure 14.16.

Glycolysis

Glycolysis (Figure 14.17) is the first step in the biochemical pathway of respiration.

- Glucose contains a lot of chemical energy. In order to release this energy, some additional energy from ATP is required to achieve the activation energy for the reaction (see Chapter 2 page 20). In the first stage of glycolysis, a molecule of glucose is converted into glucose phosphate. This requires two molecules of ATP and is called phosphorylation.
- Each molecule of glucose phosphate is then oxidised to two molecules of triose phosphate.
- Each molecule of triose phosphate is then converted to pyruvate. This reaction produces ATP. A total of four molecules of ATP are produced, two for each triose phosphate molecule. During glycolysis, then, there is a net gain of two molecules of ATP for each molecule of glucose.
- The conversion of triose phosphate to pyruvate is an oxidation reaction and involves the removal of hydrogen to reduce a coenzyme called NAD. NAD is converted to reduced NAD as a result.

NAD, oxidation and reduction

Oxidation is sometimes represented as the addition of oxygen to a substance but, more accurately, it is any reaction in which electrons are *removed*. **Reduction**, on the other hand, involves the gain of electrons. Whenever one substance is oxidised another must be reduced. In simple terms, if one substance loses electrons, another must gain them. We often use the term **oxidation–reduction reaction** for a reaction in which one substance is oxidised and another is reduced. In glycolysis, the conversion of triose phosphate to pyruvate is an oxidation reaction in which pyruvate loses electrons and NAD gains them, becoming reduced NAD.

The link reaction

Pyruvate still contains a lot of chemical energy. When oxygen is available, this energy can be made available in a series of reactions

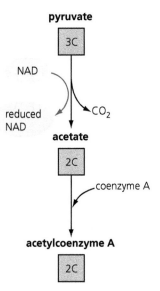

Figure 14.18 The link reaction. The boxes show the number of carbon atoms in the molecules involved in the respiratory pathway.

Figure 14.19 The Krebs cycle plays a very important part in respiration. It is the main source of reduced coenzymes, which are used to produce ATP in the electron transfer chain.

known as the Krebs cycle. The **link reaction** is a term used to describe the reaction linking glycolysis and the Krebs cycle. It is summarised in Figure 14.18.

- Pyruvate is oxidised to acetate. This is an oxidation reaction and, like glycolysis, also involves the production of reduced NAD.
- Acetate combines with coenzyme A to produce acetylcoenzyme A.
- Pyruvate contains three carbon atoms. Acetylcoenzyme A contains several carbon atoms but only two of these enter the Krebs cycle. One of the carbon atoms from pyruvate goes to form a molecule of carbon dioxide. (Remember that the production of this carbon dioxide does not directly use oxygen.)

The Krebs cycle

Figure 14.19 is a simple diagram that summarises the essential features of the Krebs cycle.

- Acetylcoenzyme A, produced in the link reaction, is fed into the cycle. It combines with a 4-carbon compound to produce a 6-carbon compound.
- In a series of oxidation-reduction reactions, the 6-carbon compound is converted back to the 4-carbon compound. In this process two molecules of carbon dioxide are given off. (Again, remember that the production of this carbon dioxide does not directly use oxygen.)

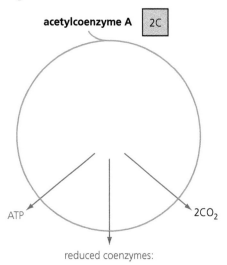

- In a series of oxidation reactions the Krebs cycle generates reduced coenzymes and ATP. For each complete cycle:
  - one molecule of ATP is produced. This is called substrate-level phosphorylation, because the ATP is formed as a result of a one of the Krebs cycle reactions. It is linked to the reaction of one of the substrates.
  - three molecules of reduced NAD and one molecule of another reduced coenzyme, reduced FAD, are produced.
- The most important function of the Krebs cycle in respiration is the production of reduced coenzymes. They are passed to the electron transfer chain, where the chemical energy that these molecules contain is used to produce ATP.

As well as the pyruvate formed from glucose by glycolysis, the breakdown products of other respiratory substrates such as lipids and amino acids can also enter the Krebs cycle and form ATP and reduced coenzymes. This means that some cells can respire lipid for some of the time instead of glucose.

Cells do not usually respire the breakdown products of amino acids unless there is no glucose or lipid available.

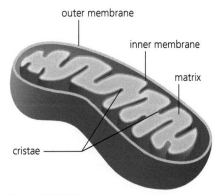

Figure 14.20 The structure of a mitochondrion.

> **TIP**
> The reason why it is more correct to describe mitochondria as the site of aerobic respiration is because only the electron transfer chains use oxygen and these are in the mitochondria. Glycolysis is an anaerobic process, but it takes place in the cytoplasm, not in mitochondria.

Mitochondria and electron transfer chains

You should remember that **mitochondria** are the site of aerobic respiration (see Chapter 3, page 41). Each mitochondrion (Figure 14.20) is surrounded by two membranes, a folded inner membrane and a smooth outer one. The folds on the inner membrane form many projections, called cristae.

The different steps of respiration take place in different locations. Glycolysis takes place outside of mitochondria, in the cytoplasm. The Krebs cycle occurs in the matrix of the mitochondria. Like the thylakoids in chloroplasts, the cristae provide a large surface area in which the electron carriers of electron transfer chains are embedded.

Reduced coenzymes have so far been produced in glycolysis and the link reaction. More reduced coenzyme is produced by the Krebs cycle.

Reduced coenzymes transfer electrons to chains of protein molecules embedded in the inner membranes of mitochondria. These proteins act as **electron carriers** and form electron transfer chains similar to those in chloroplasts.

Figure 14.21 shows how electrons from reduced coenzyme pass from one protein to the next along electron transfer chains.

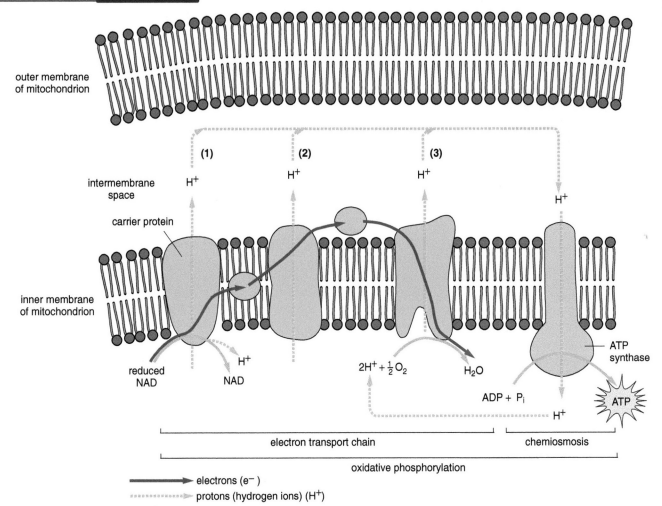

Figure 14.21 Energy is released as electrons pass from one carrier to the next in an electron transfer chain. Some of this energy is lost as heat but a lot of it goes to produce ATP.

255

Oxidative phosphorylation The production of ATP by an electron transfer chain using oxygen as the final electron acceptor.

Figure 14.22 When a cuckoo pint flowers, the temperature of the spadix increases until it is as much as 15°C above the temperature of the environment. This increase in temperature causes molecules of a substance similar to substances found in faeces and decaying bodies to be released. This attracts the flies that pollinate cuckoo pint flowers.

When these electrons are transferred, energy is released. Some of this energy is lost as heat but some is used by carrier proteins in the active transport of protons across the inner mitochondrial membrane into the space between the inner membrane and the outer membrane. This develops a higher concentration of protons in the space between the membranes than there is in the matrix of the mitochondrion.

As a result, protons diffuse down their concentration gradient from the space between the membranes to the matrix (this is an example of facilitated diffusion; see Chapter 3 page 46). They diffuse through molecules of ATP synthase embedded in the inner mitochondrial membrane, causing the ATP synthase molecules to spin. This spinning provides energy for the synthesis of ATP from ADP and inorganic phosphate. The last molecule in the electron transfer chain is oxygen. Oxygen combines with protons and electrons to produce water. ATP production by the electron transfer chain is called oxidative phosphorylation.

The heat released during respiration can be used to good effect by organisms. Endothermic animals use it to raise their body temperature above that of their environment. Although plants have much slower rates of respiration and produce less heat, they sometimes do this too. An interesting example is shown in Figure 14.22.

Anaerobic respiration

Sometimes there is not enough oxygen for an organism to respire only aerobically using the pathways just described, and also some organisms are adapted to live without oxygen. Under these conditions ATP is produced by anaerobic respiration. The only stage in the anaerobic pathway that produces ATP is glycolysis, so although it is fast the process is not as efficient as aerobic respiration, because there is an incomplete breakdown of glucose.

Look back at Figure 14.17. You will see that during glycolysis the coenzyme NAD is reduced. Reduced NAD is normally converted back to oxidised NAD when its electrons are passed to the electron carriers in the electron transfer chain. This can only happen when oxygen is present. Obviously, if all the oxidised NAD in a cell was converted to reduced NAD, the process of respiration, would stop.

In anaerobic respiration in animals pyruvate is converted to **lactate**. In plants and microorganisms, such as yeast, it is converted to **ethanol** and **carbon dioxide**. In both of these pathways (Figure 14.23), reduced NAD is converted back to oxidised NAD. This allows glycolysis to continue.

Figure 14.23 Anaerobic respiration allows organisms to produce ATP in the absence of oxygen.

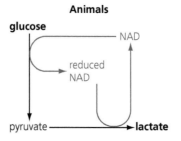

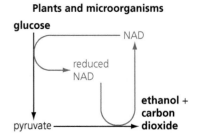

REQUIRED PRACTICAL 9

Investigation into the effect of a named variable on the rate of respiration of cultures of singled-celled organisms

This is just one example of how you might tackle this required practical.

Some data for the production of ethanol by yeast is shown in Table 14.3. The rate at which ethanol is produced can be used as a measure of the rate of anaerobic respiration.
Yeast cultures were grown in solutions containing different concentrations of glucose under anaerobic conditions. At certain time intervals, the concentration of ethanol was measured.

Table 14.3 Ethanol production by yeast cultures with different concentrations of respiratory substrate

Time/hours	Ethanol concentration in culture/g dm^{-3}			
	10% glucose	20% glucose	30% glucose	40% glucose
6	5	4	2	1
12	22	17	7	5
24	31	32	34	22
48	42	52	67	37
72	41	54	83	42
96	41	53	84	42

TIP
Look at Chapter 26 page 490 to find out about tangents.

1 Plot a suitable graph to show the data in Table 14.3.
2 Use tangents to find the initial rate of ethanol production at each glucose concentration.
3 What is the relationship between the initial rate of ethanol production and glucose concentration?
4 What is the relationship between total ethanol production and glucose concentration?
5 Suggest why the yeast cultures respond differently to different glucose concentrations.

TEST YOURSELF
10 Where does glycolysis take place?
11 What are the products of glycolysis and how is each used?
12 Explain the difference between substrate-level phosphorylation and oxidative phosphorylation.
13 What happens to pyruvate if conditions in an animal cell are anaerobic?
14 Rice is grown in swampy conditions. The cells in rice roots are very tolerant to high concentrations of ethanol. Explain how this is an advantage to a rice plant.

Energy transfer

Once energy has entered an ecosystem, it is transferred through food chains and food webs. Ecologists study feeding habits of organisms so that they can investigate food webs. Some species can be watched directly.

Figure 14.24 Badgers are large conspicuous animals but they are nocturnal and difficult to watch feeding. Badger faeces are easily recognisable and contain undigested remains, such as these plum stones.

The larvae of butterflies and moths – caterpillars – can be observed feeding on the leaves of particular species of plants. The food of other species, such as the badger, can be identified by studying the remains in faeces (Figure 14.24).

With other animals, particularly small invertebrates, it is very difficult to make direct observations of feeding behaviour. You would be very lucky to see, for example, a ground beetle catch and eat its prey. Even if you did see such an event, you would not know whether it was normal feeding behaviour or whether it was unusual. Ecologists sometimes make use of antibodies in an ELISA test to provide information about feeding behaviour. You will remember the how an ELISA tests works from the first year of your course (see Chapter 6 page 102).

Extension

Another source of energy

On page 241 we saw that photosynthesis was the main route by which energy enters an ecosystem. There are other ways, however, by which it can enter. Consider the bottom of the Pacific Ocean. It is pitch dark. No light penetrates its depths. It is also cold. The water remains just above freezing all year round. In a few areas volcanic vents bubble out a mixture of sulfur-rich gases. You might think that life could not possibly exist in these conditions, but it can!

Around the volcanic vents, bacteria are found. These bacteria use the sulfur-containing substances bubbling from the vents. They obtain energy from chemical reactions involving these substances and use it to synthesise the biological molecules that make up their cells. The bacteria support large worms and other invertebrate animals. A community of living organisms exists because chemical energy in biological molecules is transferred from one organism to another through a food web, even though there are no photoautotrophic organisms present.

EXAMPLE

Using antibodies to investigate feeding

Ecologists can dissect some animals and identify the food remains in their guts. But suppose they wanted to know if an animal, such as a ground beetle, fed on slugs. There wouldn't be any hard parts to identify in the beetle's gut. Some of the proteins that made up the slug's body, however, would be present.

Ecologists can use an **enzyme-linked immunosorbent assay (ELISA)** to confirm that these proteins come

TIP
You do not need to be able to recall the details of this example for your exam.

from a specific organism. Figure 14.25 shows how an ELISA test is used to find out whether ground beetles eat slugs.

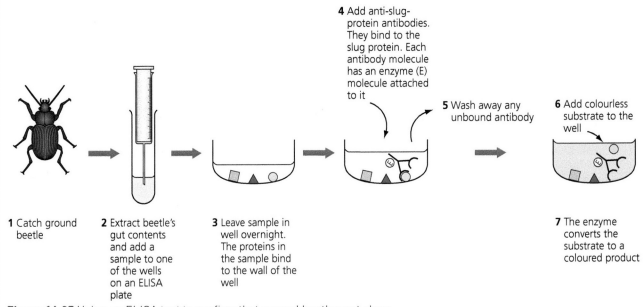

4 Add anti-slug-protein antibodies. They bind to the slug protein. Each antibody molecule has an enzyme (E) molecule attached to it

5 Wash away any unbound antibody

6 Add colourless substrate to the well

1 Catch ground beetle

2 Extract beetle's gut contents and add a sample to one of the wells on an ELISA plate

3 Leave sample in well overnight. The proteins in the sample bind to the wall of the well

7 The enzyme converts the substrate to a coloured product

Figure 14.25 Using an ELISA test to confirm that ground beetles eat slugs.

Before we look at the diagram in detail, there are two important principles about which we must remind ourselves. We first met these principles in Chapter 6 and Chapter 12.

● Different species contain different proteins (see Chapter 12 page 221). Suppose our ground beetle had been feeding on slugs and, say, earthworms. Some of the proteins found in the slugs would probably be very similar to proteins found in the earthworms so we probably couldn't tell which animal the protein concerned came from. Some, however, would be different. These specific slug proteins would have specific sequences of amino acids and their molecules would have specific tertiary structures.

● Antibodies are also proteins. They have specific binding sites. These binding sites mean that they will only bind to molecules that have a complementary shape (see Chapter 6 page 91). An anti-slug-protein antibody will only bind to one particular slug protein. It won't bind to proteins from any other species unless they are identical to the slug protein.

Steps 1 and 2 in Figure 14.25 should be easy enough to understand but we may need to explain some of the other steps. We will start with step 3 where the sample of gut contents has been left overnight in one of the wells on the ELISA plate.

1 Three protein molecules are shown attached to the wall of the well. Why do these protein molecules have different shapes?

Each shape represents a different protein, with a different amino acid sequence, so there are three different proteins shown here. All three could be slug proteins, but it is possible that one or two of them might have come from other animals that the ground beetle had eaten.

2 Step 4 shows anti-slug-protein antibody binding only to a slug protein. Why does this antibody bind only to slug protein?
This is the point made earlier. Antibodies are specific and an anti-slug-protein antibody will only bind to the protein shown as a blue circle. This is a slug protein.

3 Why is the unbound antibody washed away (step 5)?
If we don't wash the unbound antibody away, it will remain in the well. The enzyme on the unbound antibody will result in the coloured product being formed even if no slug protein is present.

4 Not all ground beetles eat slugs. Explain how you would be able to tell if the ground beetle from which you had obtained the gut sample had not been eating slugs.
There would be no slug proteins attached to the wall of the well to bind to the anti-slug-protein antibodies. Therefore there would be no enzyme to catalyse the reaction in which the colourless substrate was converted to a coloured product.

5 How could you use an ELISA plate to find out whether slugs were important items of food for ground beetles?
You could add samples from different ground beetle guts to different wells on the ELISA plate. By counting the number of wells where there was a colour change, you could find the percentage of ground beetles that ate slugs.

Trophic levels

Trophic levels The feeding positions
organisms occupy in a food web.

In any ecosystem different organisms gain their food in different ways
(Figure 14.26). They feed at different trophic levels. Green plants are
primary producers. They produce biological molecules from carbon
dioxide, water and mineral ions. They rely on photosynthesis to transfer
light energy to chemical energy in biological molecules. The other
organisms that make up the community rely either directly or indirectly on
the biological molecules produced by the producers. Primary consumers
(**herbivores**) feed on producers. Secondary consumers feed on primary
consumers and tertiary consumers feed on secondary consumers.
Organisms that are not eaten eventually die. Another group of organisms,
the **saprobiotic decomposers**, digest dead tissues and use the biological
molecules that make up these tissues as a source of chemical energy.

> **TIP**
>
> Some organisms, such as the
> dark green bush cricket, feed
> at different trophic levels and
> some organisms feed on different
> sources of food when they are
> larvae and when they are adults.
> The caterpillar of the peacock
> butterfly, for example, eats nettle
> leaves. The adult butterfly feeds
> on nectar produced by flowers.

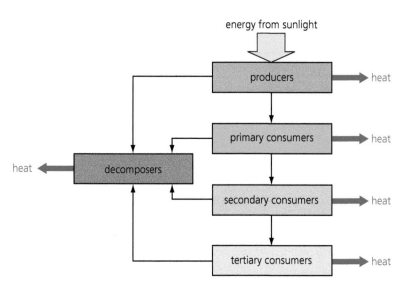

Figure 14.26 The transfer of energy in an ecosystem. The boxes represent trophic
levels. The arrows show the direction in which energy is transferred.

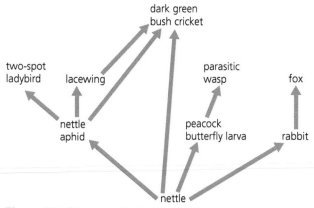

Figure 14.27 A simplified food web showing some of the
organisms that feed on nettles.

We often talk about **food chains**, suggesting perhaps that
we frequently encounter situations where animal B feeds
only on plant A. In turn, animal C only eats animal B, and
animal D only eats animal C. This hardly ever happens
under natural conditions. Food chains are linked to each
other to form complex **food webs**. Figure 14.27 shows a
possible food web associated with a nettle patch.

Many farming practices are based on an understanding
of the energy losses between trophic levels and
attempt to reduce them. For example, pests that eat
crops divert energy away from the human food chain.
Reducing pest populations on crops by the use of chemical
pesticides (see Chapter 13 pages 235–236) minimises the
energy losses. This increases the yield.

> **TIP**
>
> Refresh your memory about the impacts of pesticides in Chapter 13.

Conversion efficiencies among consumers

Intensive rearing of animals for food also involves keeping energy losses to a minimum. On page 249 we considered the efficiency with which energy is transferred to plants in photosynthesis. In this section we will look at the efficiency with which energy is transferred to consumers.

Figure 14.28 shows the percentage of energy transferred between different trophic levels. We can look at this in a different way. If we take a figure of 2% as representing the percentage of light energy conserved as chemical energy in plants, then for every 10 000 kJ of light energy absorbed by the producers, 200 kJ will be incorporated into their tissues.

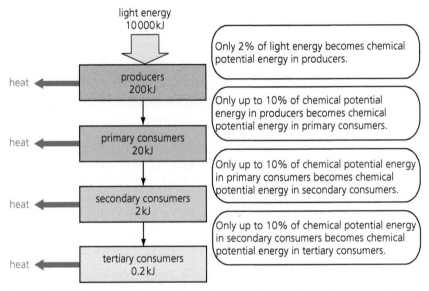

Figure 14.28 Only a small percentage of energy is transferred from one trophic level to the next. The rest is used to make ATP in the process of respiration and is eventually lost as heat.

Similarly, if we assume that about 10% of the chemical energy in producers becomes chemical energy in primary consumers, then 20 kJ from our original 10 000 kJ will be transferred to the tissues of primary consumers. At each step less chemical energy will be transferred.

Note, however, that the values on Figure 14.28 are only generalisations. There are many factors that influence exactly how much energy is transferred at each stage, as the following examples show.

- Mammals are **endothermic**. This means that they are able to keep their body temperature more or less constant at a value between approximately 35 and 40 °C, depending on the species. This high temperature is a result of heat produced during metabolism.

- Crocodiles are found in many parts of the tropics. Unlike mammals, they rely on their environment to maintain a high body temperature. More of the food they eat can therefore be converted into new cells and tissues, and less chemical energy will be lost in maintaining body temperature.

- The surface area to volume ratio of a small mammal or bird, such as the humming bird in Figure 14.29, is much bigger than that of a large mammal or bird. Small mammals and birds consequently lose a lot more heat relative to their size and cannot convert as much of the food that they eat into new cells and tissues.

Figure 14.29 Humming birds have a larger surface area to volume ratio than other larger mammals or birds and consequently lose a lot more heat relative to their size.

- In general, carnivores convert the food they eat into new tissue more efficiently than do herbivores. Herbivores feed on plant material and plants contain a lot of substances, such as cellulose and lignin, that are difficult to digest. A much higher proportion of the food that a herbivore eats passes through the gut and is lost as faeces.

Net production in consumers

We can calculate net production (N) of consumers from the equation

$$N = I - (F + R)$$

where I represents the chemical energy in ingested food, F represents the chemical energy lost in faeces and urine and R represents the energy loss through respiration to the environment.

Net production of consumers, or secondary production, is the chemical energy stored in animal biomass after respiratory losses to the environment have been taken into account, measured in $kJ\,m^{-2}$. **Secondary productivity** is the rate of secondary production, measured in, for example, $kJ\,m^{-2}\,year^{-1}$.

Rearing livestock is a commercial business that needs to be profitable. Clearly, if a farmer is to run a successful business, he or she needs the maximum yields of milk from the milking herd, eggs from the hens or meat from the livestock. In biological terms, the farmer wants maximum net production.

Look at the equation for net production given above. At its simplest, achieving maximum net production involves manipulating conditions so that the animal's food is as digestible as possible (so reducing the amount of faeces) and the loss through respiration is as low as possible. In this next section, we will look at some of the factors associated with ensuring maximum net production of chickens.

Net production and poultry farming

Commercially, chickens are reared to either produce meat or lay eggs. Chickens reared to produce meat are known as broilers. Figure 14.30 shows growth curves for male and female broilers.

TIP
Endothermic animals that maintain a higher body temperature than their environment will usually have comparatively lower net production because maintaining a high body temperature involves a high rate of respiration.

Figure 14.30 Growth curves for male and female broilers.

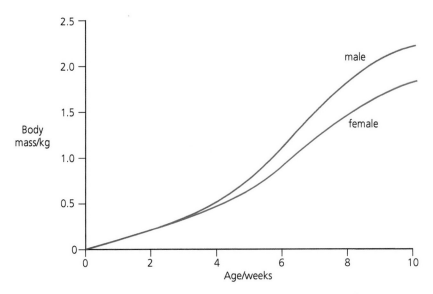

You can see from this graph that intensively reared broilers, fed with high-energy, high-protein food, grow very rapidly. A modern broiler may be ready for marketing 7–8 weeks after hatching. At this age it is still growing and may not have even reached its maximum growth rate. The reason for slaughtering birds at this age can be seen if you look at the data in Table 14.4. Look at the figures for male birds. You can see from the last column that the mean mass of food eaten per kilogram gain in mass rises steadily. In other words, efficiency of food conversion falls. This is mainly because the bird produces less protein-rich muscle and more body fat as it gets older.

Table 14.4 also shows that, by the time they have reached 10 weeks, female broilers have a smaller mean body mass and their efficiency of food conversion is lower than that of males. The difference in efficiency of food conversion may not seem very much – 2.5 kg of food per kilogram gain in body mass compared with 2.4 kg in males – but differences like this have a considerable influence on overall profit.

Table 14.4 Growth and food consumption in broiler chickens.

Sex	Age/weeks	Mean body mass/kg	Mean cumulative mass of food eaten/kg	Mean mass of food eaten per kilogram gain in mass/kg
Male	2	0.2	0.3	1.5
	6	1.0	2.0	2.0
	10	2.2	5.4	2.4
Female	2	0.2	0.3	1.5
	6	0.9	1.8	2.0
	10	1.8	4.5	2.5

> **TIP**
> Growth and food consumption in broiler chickens is simply an example to illustrate the concept of net production in consumers. You do not need to memorise this example.

TEST YOURSELF

15 State the equation for the net production of consumers.

16 Explain how energy is lost in faeces.

17 It is rare for there to be more than five trophic levels in a food web. Explain why.

18 Trout grown in fish farms are fed special pelleted food. Give two reasons why pelleted food enables farmed fish to convert food into new tissue more efficiently than free-range chickens, which find their own food.

19 Intensively reared broilers are kept under temperature-controlled conditions. Suggest how controlling temperature may increase net production.

Practice questions

1 The diagram shows some of the steps in respiration.

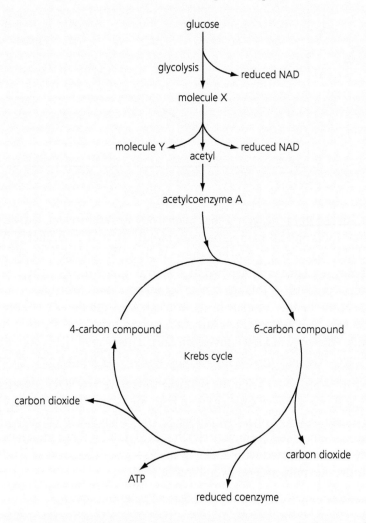

a) Where in a cell does glycolysis occur? *(1)*

b) i) Name molecule X. *(1)*

   ii) Name molecule Y. *(1)*

c) Name the step in respiration that produces acetylcoenzyme A. *(1)*

d) What type of phosphorylation produces the ATP in the Krebs cycle? *(1)*

e) Describe how the reduced coenzyme produced by the Krebs cycle is used. *(2)*

2 The diagram summarises the steps involved in photosynthesis.

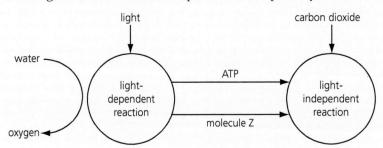

a) Name molecule Z. *(1)*

b) Name the three-carbon sugar produced by the light-independent reaction. *(1)*

c) i) Which enzyme catalyses the reaction between carbon dioxide and ribulose bisphosphate in the Calvin cycle? *(1)*

    ii) What is the product of this reaction? *(1)*

d) Describe and explain what happens to the ribulose bisphosphate concentration if a plant is placed in the dark. *(2)*

3 The trophic levels in a food web can be numbered, starting with 1.0 for primary producers, 2.0 for primary consumers and so on. The table shows the mean trophic levels of marine fish caught for human food in the period 1950 to 2000.

Year	Mean trophic level
1950	3.37
1960	3.36
1970	3.39
1980	3.29
1990	3.26
2000	3.21

a) i) Phytoplankton consists of single-celled photosynthetic organisms that float in the surface water. A species of fish feeds only on phytoplankton. Identify the trophic level of this species of fish. *(1)*

    ii) Another species of fish has a trophic level of 3.0. State whether this fish is a primary consumer, a secondary consumer or a tertiary consumer. *(1)*

b) i) Describe how the mean trophic level of the marine fish catch has changed over the period 1950 to 2000. *(2)*

    ii) Suggest an explanation for the change you described in your answer to question 3b(i). *(2)*

c) Over the same period of time, more cattle and sheep have been fed on protein concentrates as well as grass. These protein concentrates are often made from animal material. Suggest how the mean trophic levels of farm animals have changed over the period shown in the table. *(2)*

4 The graph shows the mean biomass of heather from plants of different ages growing on an area of moorland in Yorkshire.

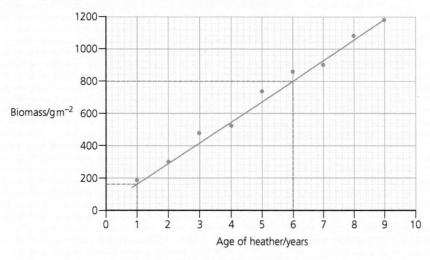

Biomass/g m⁻²

Age of heather/years

a) Give two ways that biomass can be measured. (1)

b) The biomass of the heather is given as dry mass.

   i) Describe how you would measure the dry mass of a heather sample. (2)

   ii) How would you make sure that your value for dry mass was valid? (1)

   iii) What is the advantage of measuring biomass as dry mass? (1)

c) Describe how you would use the graph to calculate the mean annual increase in dry biomass of the heather plants. (2)

d) What is the evidence from the graph that the heather did not increase in biomass by the same amount each year? (1)

e) Suggest why the heather did not increase in biomass by the same amount each year. (1)

Scientists estimated the total amount of light energy falling on 1 m² of the heather moorland in Yorkshire where the samples were collected to be 1 415 000 kJ. They based this estimate on the amount of light falling on the heather in the growing season, so it does not include the winter months when the temperatures are too low for growth.

f) Use all the figures in this section to calculate the efficiency of photosynthesis in these heather plants. (3)

Stretch and challenge

5 Examine the structure of marine ecosystems around deep-sea volcanic vents. Discuss how chemical potential energy in large molecules is transferred from one organism to another through a food web, even though there are no photoautotrophs present.

6 Some plants have alternative photosynthesis pathways, e.g. crassulacean acid metabolism (CAM) and C4 photosynthesis. Contrast these with the C3 photosynthesis that most plants use. To what extent are these alternative pathways an adaptation to the environment?

15 Nutrient cycles

TEST YOURSELF ON PRIOR KNOWLEDGE

1 What conditions accelerate decay?
2 Give two types of microorganism that are involved in decay.
3 Other than proteins, name one group of biological molecules that contain nitrogen.
4 Describe the way in which phosphate groups form part of the structure of ATP.

Introduction

One of the problems that almost all human communities face is how to get rid of the enormous quantities of waste they produce. Some of this waste can be composted. Composting uses microorganisms to digest organic material. The compost that is produced can be added to the soil as a natural fertiliser, providing useful nutrients to plants. Composting has been carried out for hundreds of years – the ancient Romans left written records of composting – but scientific research has led to discoveries that mean that we can now make compost more efficiently and on a much larger scale (Figure 15.1).

Figure 15.1 Composting on a commercial scale.

Most commercial composting involves a similar process. Figure 15.2 shows how compost is made.

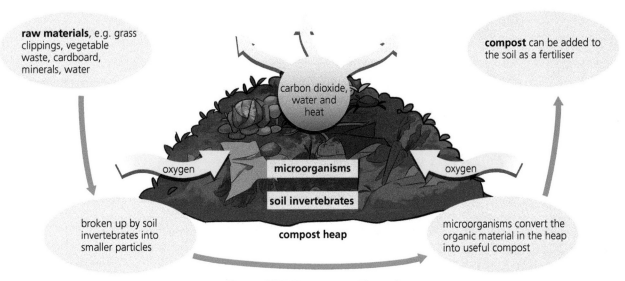

raw materials, e.g. grass clippings, vegetable waste, cardboard, minerals, water

carbon dioxide, water and heat

compost can be added to the soil as a fertiliser

oxygen

microorganisms

soil invertebrates

oxygen

broken up by soil invertebrates into smaller particles

compost heap

microorganisms convert the organic material in the heap into useful compost

Figure 15.2 How compost is made.

- Suitable organic material, for example grass clippings, vegetable waste and even coffee grounds and cardboard, is collected into a heap. In small-scale garden compost heaps, soil invertebrates such as worms, slugs and millipedes break this material up into smaller particles. In large-scale composting, it is broken up by mechanical grinders and choppers.
- Soil microorganisms colonise the heap. They are called mesophils because they live at temperatures of between 10 °C and 45 °C. The compost heap heats up as they multiply and respire.
- The mesophils are gradually replaced by thermophils. Thermophils are microorganisms that live in conditions where the temperature is high; in this case, between approximately 50 °C and 65 °C. This is the active phase of composting and these thermophilic microorganisms are mainly responsible for converting organic material in the heap into useful compost.
- The last stage is the curing phase. The temperature falls and mesophils again colonise the heap. During the curing phase the compost matures until it finally becomes suitable for adding to the soil.

TIP
You do not need to be able to recall the steps involved in making compost.

Huge quantities of compost are made from green waste in the UK every year and sold as a plant fertiliser. In the process, nutrients in the waste material are recycled and made available to plants.

In this chapter, we shall look at the natural recycling of nutrients in ecosystems. We shall follow the passage of phosphorus and nitrogen through different trophic levels and consider the role of microorganisms in converting organic substances into the inorganic substances and ions that are taken up by plants.

Decomposition

Although the invertebrates in a compost heap help to break up the waste material into finer pieces by feeding on it, they are not true decomposers. Worms, millipedes and slugs feed on the waste material, or detritus. They digest some of it in their digestive system and produce more waste material in the form of faeces.

True decomposers are saprobionts. These are mainly fungi and micro-organisms such as bacteria. Unlike detritivores they do not ingest their food. Instead, saprobionts secrete enzymes onto the waste material. These enzymes are therefore outside the saprobionts when they hydrolyse biological molecules. Some of the products of hydrolysis are then absorbed by the saprobionts across their cell membranes (see Chapter 3), but many remain in the surroundings and may then be absorbed by other organisms.

Saprobionts secrete a very wide range of enzymes that allows them to hydrolyse a large variety of biological molecules, in some cases resulting in mineral ions such as ammonium and phosphate ions as products. Some saprobionts also excrete these mineral ions as waste products of metabolism. This is why microorganisms are vital in recycling nutrients. Instead of using all the breakdown products of decomposition themselves, some are made available for other organisms.

If nutrients were not recycled by decomposition they could remain trapped in dead material for very long periods of time and plant growth would be limited by a lack of available nutrients in the soil.

Dead leaves and decomposition

We saw in the last section that in small-scale garden compost heaps soil invertebrates break up waste plant material into smaller particles. This also happens in natural habitats, such as woodlands. These invertebrates include arthropods, such as mites, insects and woodlice, nematode worms and earthworms. The graph in Figure 15.3 compares the biomass of these organisms and of fungi and bacteria in a British woodland at different times of the year.

Saprobionts Organisms that secrete digestive enzymes onto the dead remains of other organisms, digest the biological molecules in these dead remains and then absorb some of the products of this digestion.

Decomposition The process in which the biological molecules in dead material and waste products are digested, producing carbon dioxide, water and inorganic ions.

TIP
There are a number of words that could be used in place of decomposition, such as decay and rotting, but these words also cover the idea of physical fragmentation of the material. Decomposition refers to the chemical process of hydrolysis of complex biological molecules to smaller inorganic molecules or ions.

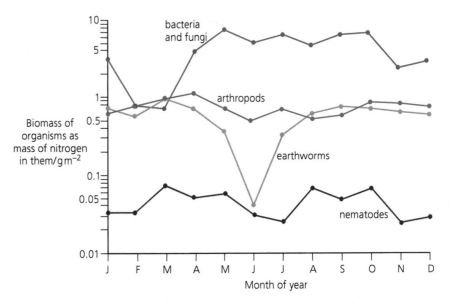

Figure 15.3 Variation in the nitrogen content of different groups of organisms involved in breaking down dead leaves at different times of the year in a woodland.

This graph is quite complex. Before we look at the conclusions that we can draw from it, we need to make sure that we understand how the data have been presented (see the Example box below).

EXAMPLE

Investigating the decomposition of leaves

> **TIP**
> This specific example is not required learning for the exam.

The graph in Figure 15.3 compares the biomass of these organisms and of fungi and bacteria in a British woodland at different times of the year.

1   The biomass has been given as the mass of nitrogen present in the organisms. Explain why nitrogen content can be used as a measure of biomass.
   Nitrogen is present in many of the substances that make up the structure of soil organisms. These substances include protein, DNA and chitin, which forms the hard outer layer of insects. The proportion of nitrogen-containing substances is more or less constant in these organisms, so it is an accurate measure of body mass.

2   The data in the graph have been plotted on a logarithmic scale. Suggest the advantage of using a log scale.
   A log scale lets a greater range of data be plotted than an arithmetic scale. If you look carefully at the y-axis,

you will see that it runs from 0.01 to 10 g m^{-2}. It also shows that the mass of nitrogen present in nematodes fluctuates from approximately 0.02 to 0.08 g m^{-2}. If the data on nematodes had been plotted together with all the other information on a scale with an arithmetic axis, they would have appeared as a straight line. You wouldn't have been able to see the fluctuations.

3   Suggest an explanation for the change in biomass of bacteria and fungi between February and June.
   Leaves fall and accumulate in the autumn so there will be plenty of dead leaves available. The most likely explanation for the increase in biomass is that the soil is getting warmer and the enzyme-controlled processes associated with leaf breakdown and the growth of microorganisms are faster in these conditions.

Ecologists investigated the importance of arthropods and microorganisms in the decomposition of leaves. Different numbers of woodlice were added to containers, each with the same mass of dead oak leaves. The ecologists measured the rates of respiration of the microorganisms in these containers and in a control container. The results are shown in Figure 15.4.

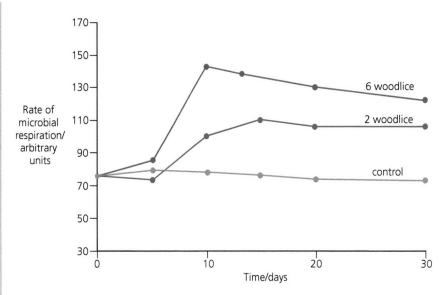

Figure 15.4 The effect of woodlouse activity on the microbial breakdown of leaves.

4 Describe how the control container should have been set up in this investigation.
It should be set up exactly the same as the other containers and exposed to the same environmental conditions. The only difference should have been that it did not contain any woodlice.

5 What does Figure 15.4 show about the effect of woodlouse activity on the microbial breakdown of leaves?
To answer this question you need to compare the containers in which woodlice were added with the control container. You can see that microbial respiration in the control chamber remains more or less constant. After 5 days, the rate of microbial

respiration in both of the containers with woodlice increases and is higher than in the control. In addition, there is a greater rate of microbial respiration when six woodlice have been added than when there are only two woodlice in the container.

6 Suggest how the presence of woodlice could have caused the increase in the microbial respiration rate that occurred between 5 and 10 days.
The woodlice break up the leaves into smaller fragments by chewing them so there is more surface area to be colonised by microorganisms. This means that they can feed faster and gain respiratory substrates more quickly.

TEST YOURSELF

1 Why is nutrient recycling in natural ecosystems vital for plant growth?
2 Explain what is meant by decomposition.
3 What is a saprobiont?
4 Describe how the method of digestion used by saprobionts results in some nutrients being made available to plant roots.

Nutrient cycles

Living organisms, such as animals and plants, require many different chemical elements. Plants take up many of these elements as ions from the soil. Inside the plant they are involved in various chemical reactions and are eventually incorporated into biological molecules that form plant cells and tissues. Consumers obtain most of their supplies of these elements from plants or from other animals that feed on plants. A snail, for example, digests the biological molecules in its plant food and absorbs the products through its gut wall (see Chapter 8). In this way, elements are passed from organism to organism along the food chains that make up a food web (Figure 15.5).

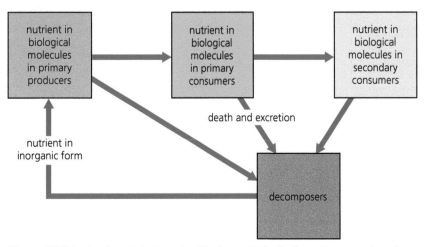

Figure 15.5 A simple nutrient cycle. All elements including nitrogen, phosphorus and iron are cycled in this basic way. The detail of the process is different in each nutrient cycle.

Those organisms, or parts of organisms, that are not eaten as food eventually die and are decomposed. Saprobionts, such as fungi and bacteria, break down the biological molecules that form the dead material (Figure 15.6). They absorb some of the products but the rest are released, some as mineral ions that can be taken up by plants.

This is the basic nutrient cycle. Elements are taken up by plants as ions and some are incorporated into biological molecules and pass from organism to organism in the various trophic levels. Death and decomposition result in microorganisms making these elements available to plants again.

Nutrient cycle How a chemical element moves from the abiotic environment into living organisms and then back into the abiotic environment.

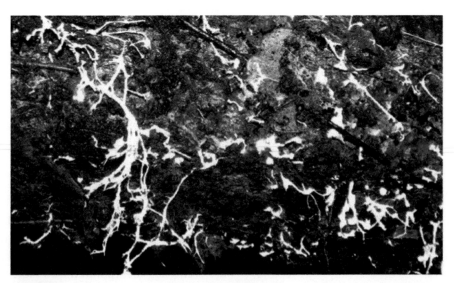

Figure 15.6 These white thread-like structures are fungal hyphae covering the surface of a piece of rotting wood. The fungi are saprobionts. Their hyphae secrete enzymes that hydrolyse biological molecules such as lignin in wood.

The phosphorus cycle

The general features of the **phosphorus cycle** differ very little from the basic nutrient cycle shown in Figure 15.5. Phosphate ions are released from rocks on land by chemical weathering and washed into soils by rain. They are absorbed by plant roots by **active transport** and are used to produce ATP and nucleic acids in plant cells.

Primary consumers feed on plants and the biological molecules in their food are digested to smaller phosphate-containing molecules such as nucleotides. Phosphate ions may also be present in their food. Nucleotides and phosphate ions are absorbed in the small intestine and are used to produce nucleic acids and ATP in animal cells. The same thing happens when a secondary consumer eats a primary consumer.

When organisms produce faeces or die, the phosphorus-containing substances in their tissues or faeces are digested by **saprobiotic** bacteria releasing phosphate ions, which can then be taken up again by plants. Animal urine also contains excreted phosphate ions. Phosphorus can cycle like this within terrestrial communities for centuries. You can trace this terrestrial phosphate loop in Figure 15.7.

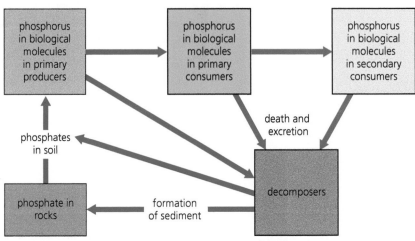

Figure 15.7 The phosphorus cycle.

Some phosphate ions are washed from soils into streams, and eventually reach lakes or the sea. Here they may be taken up by algae near the surface and incorporated into their biological molecules. The algae may be eaten by fish, passing the phosphorus from one trophic level to another. But when aquatic organisms die they tend to sink to the bottom of lakes or the sea, where they decompose and phosphate ions become trapped in aquatic sediments.

Over very long periods of geological time, aquatic sediments form sedimentary rocks and may eventually become exposed again on land, where chemical weathering releases phosphate back into the soil. This part of the cycle is extremely slow and takes millions of years.

The nitrogen cycle

The nitrogen cycle is more complex than the phosphorus cycle. Plants take up nitrate ions from the soil. The nitrates are absorbed into the roots by **active transport** and are used to produce amino acids and then

Ammonification The decomposition of amino acids in proteins, releasing ammonia as a product.

Nitrification The two-step oxidation of ammonium ions firstly to nitrite ions and then to nitrate ions.

> **TIP**
> Although the product of ammonification is ammonia (NH_3), ammonia ionises in soil water and in aquatic environments to give ammonium ions (NH_4^+).

proteins and other nitrogen-containing substances in plant cells, including the nitrogenous bases in nucleotides. Primary consumers feed on plants and the proteins in their food are digested to release amino acids. These amino acids are absorbed from the gut and built up into the proteins that form the tissues of the primary consumers. The same thing happens when a secondary consumer eats a primary consumer. In this way, nitrogen is passed from one trophic level to the next through the food web.

When organisms die, the nitrogen-containing substances they contain are digested by saprobiotic bacteria. Nitrogen from consumers is also made available to saprobiotic bacteria through nitrogen-containing excretory products, such as urea in urine. These saprobiotic bacteria release ammonia, so the process is called ammonification. Another group of bacteria, the **nitrifying** bacteria, then convert ammonia to nitrites and nitrates in a process called nitrification. The complete nitrogen cycle is summarised in Figure 15.8.

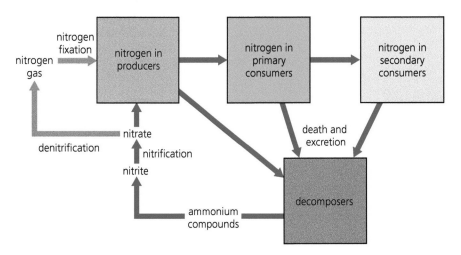

Figure 15.8 The nitrogen cycle.

Denitrification The reduction of nitrate ions to nitrogen gas.

Under anaerobic conditions, such as those that are found when soil becomes water-logged, **denitrifying** bacteria are found in large numbers. These bacteria are able to use nitrate in place of oxygen as an electron acceptor in the respiratory pathway. This reaction, called denitrification, involves reduction of nitrate to nitrogen gas. This nitrogen escapes into the atmosphere, so it is no longer available to plants.

Nitrogen gas may be made available to plants again by nitrogen fixation.

Nitrogen fixation The reduction of nitrogen gas to ammonia.

Some species of microorganism are able to fix nitrogen to form ammonia. Some live free in the soil. Others are associated with the roots of leguminous plants, such as peas and beans, clover and lupins. The biochemical reactions associated with nitrogen fixation are complex, but they can be summarised by the following equation.

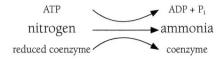

Nitrogen is reduced to ammonia. The reaction is catalysed by the enzyme nitrogenase. Nitrogenase, however, does not function in the presence of oxygen and many nitrogen-fixing organisms have adaptations that ensure that anaerobic conditions exist in the parts of cells involved in nitrogen fixation.

Mycorrhizae and ion uptake

So far in this chapter we have seen how microorganisms have key roles in decomposition and in the different processes in nutrient cycles. There is a third way in which microorganisms are crucial in recycling chemical elements, by assisting plants with their uptake of inorganic ions and water.

Mycorrhizae Associations between the plant roots and beneficial fungi found in nearly all plants on Earth.

The root systems of most plant species have fungi, called mycorrhizae (see Figure 15.9), growing in and around them. The fungi are often highly specific to the plant species and have evolved with them. This is an example of mutualism, a relationship between two species where both gain a nutritional advantage.

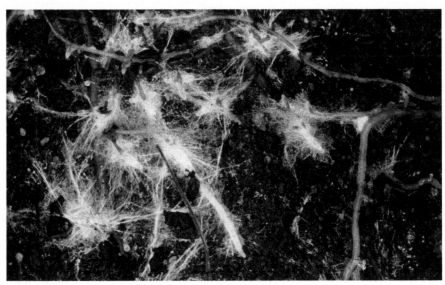

Figure 15.9 Mycorrhizae around the roots of a young pine tree.

Each fungus consists of microscopic threads called hyphae. The fungus colonises the roots from the soil by hyphae growing on, and often into, the root tissues. Once established, many hyphae extend out from the root surface into the surrounding soil. Because there are so many hyphae, they vastly increase the surface area of the plant for uptake of water and ions, including phosphates and nitrates. This is shown in Figure 15.10.

The hyphae absorb ions and water from the soil and transport them into the plant roots. In return, the part of the fungus growing inside the root is able to obtain carbohydrates translocated from the plant leaves. This supplements their food supply.

Because fungi are saprobiotic, their hyphae still secrete enzymes and hydrolyse biological molecules in leaf litter and other organic detritus in the soil. This releases ions, which can then be absorbed and transported back along the hyphae to the plant roots. Scientists have shown that in some circumstances plant roots would be incapable of accessing phosphate ions at all without the help of mycorrhizae.

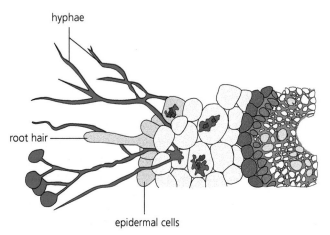

hyphae

root hair

epidermal cells

Figure 15.10 Fungal hyphae extend into the soil and increase the root surface area.

> **TEST YOURSELF**
> **5** Describe the ways that the nitrogen cycle shown in Figure 15.8 is similar to the carbon cycle that you already know about.
> **6** Describe the ways that the phosphorus cycle shown in Figure 15.7 is different from the nitrogen cycle.
> **7** What is nitrification and why is it important for plants?
> **8** What are mycorrhizae and how do they benefit many plant species?

Fertilisers and net primary production

When organisms living in natural ecosystems die, they decompose. Soil microorganisms convert, for example, the nitrogen in organic substances such as proteins and nucleic acids to nitrates, and these are taken up by the producers. There is a continual recycling of nutrients. In agricultural ecosystems, however, a large part of the biomass that is produced is harvested and removed. This is true whether we are considering crop plants, such as wheat or potatoes, or animals, such as dairy cattle and sheep. Unless the mineral ions in this biomass are replaced, their concentration in the soil decreases and crop yield or milk yield falls.

Fertilisers can be used to add mineral ions such as those of nitrogen, phosphorus and potassium to the soil. We can either use **artificial fertilisers** or **natural fertilisers**, such as farmyard manure.

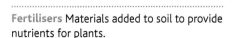

Fertilisers Materials added to soil to provide nutrients for plants.

Artificial fertilisers:

- have a guaranteed composition, making it easier to determine rates of application and predict the effect on crop yield
- are concentrated sources of nutrients and can therefore be applied in smaller amounts; this saves on transport costs and on the damage done by heavy machinery compacting the soil and crushing the crop
- are clean and convenient to handle and apply evenly.

Natural fertilisers, on the other hand:

● are mixtures of substances and may contain trace elements, substances that are important to plants in small amounts
● add organic matter to the crop; this may improve soil structure, reducing erosion and improving water-holding properties
● release the nutrients they contain over a longer period of time.

It is clear from various investigations that regular dressings with farmyard manure can benefit a crop and produce increases in yield similar to those obtained with artificial fertiliser. Table 15.1 shows the results obtained from a number of different investigations.

The first column in the table gives the name of the scientist or scientists responsible for carrying out the investigation and publishing the results. Comparing the work of different scientists allows us to see that similar findings have been reported by others. This helps to make sure that the conclusions that we draw are robust, or reliable.

Table 15.1 Yields of crops with long-term applications of farmyard manure or artificial fertiliser.

| Investigation | Crop | Yield in tonnes per hectare | | |
		Control	Farmyard manure	Artificial fertiliser
Dyke (1964)	Wheat	2.08	3.50	3.11
Trist and Boyd (1966)	Wheat	1.28	2.38	2.43
	Barley	1.03	2.03	2.26
Johnston and Poulton (1977)	Barley	1.59	3.03	2.87
Warren and Johnston (1962)	Sugar beet	3.80	15.60	15.60
	Mangolds	3.80	22.30	30.90

Look at the data on yield in the other columns. You can see that in all cases the addition of either farmyard manure or artificial fertiliser led to an increase when compared to the control.

How much fertiliser should be applied?

Farmers can improve the net primary production (see page 251) of a crop by adding fertiliser, but they must apply the right amount at the right time. There are recommendations available for how much fertiliser to apply to a particular crop, but these are only general estimates. Factors such as the previous crop that was grown and the type of soil mean that these recommendations have to be modified if they are going to be applied to a particular crop growing in a particular field.

Figure 15.11 shows the effect of adding different amounts of two types of nitrogen-containing fertiliser on the yield of maize in a number of trial plots.

Curve B shows the effect of adding a fertiliser containing nitrogen but no potassium. The shape of this curve is typical of a yield response to added fertiliser. The yield increases with increasing application of fertiliser. It reaches a peak and then falls. This pattern is sometimes referred to as the law of diminishing returns, because beyond a certain point the addition of more fertiliser results in very little extra gain, or even reduction.

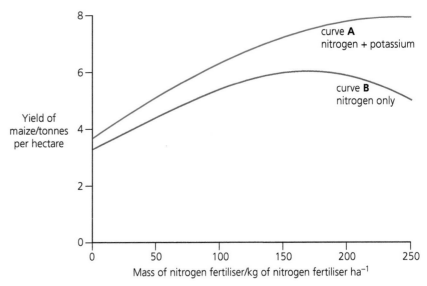

Figure 15.11 The effect of adding nitrogen-containing fertiliser on the yield of maize.

Now look at the curve between applications of 0 and 180 kg of nitrogen per hectare. You have seen curves with this shape frequently during your A-level course. What it is showing you here is that, as the curve rises, the amount of nitrogen added in the fertiliser is limiting the yield. At approximately 160 kg of nitrogen per hectare, the curve flattens out. Something else is limiting the yield. Curve A suggests that this may be potassium.

When should fertiliser be applied to a crop?

In the UK, winter wheat is an important crop. Figure 15.12 shows when a winter wheat crop is sown, grown and harvested.

We will look at some of the issues relating to the application of nitrogen-containing fertiliser to this crop. The timing of any application must take the following points into consideration.

- Maximum uptake of soil nitrates occurs early in the growth of the plants.
- Time is required for nutrients in fertilisers to dissolve and reach the roots.
- Wastage through loss from the soil must be kept to a minimum.
- Weather conditions need to be taken into account, and fertilisers should not be applied while it is raining.

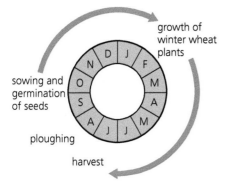

Figure 15.12 Growth, development and harvest of a winter wheat crop. The letters in the circle correspond to the months of the year.

EXAMPLE

Timing the application of artificial fertiliser

Most of the nitrogen in artificial fertiliser is supplied either as ammonium ions or as nitrate ions.

1 Use your knowledge of the nitrogen cycle to explain why there is a rapid decrease in the concentration of the ammonium ions added in the fertiliser.
   Nitrifying bacteria oxidise ammonium ions to nitrates and then to nitrates.

2 Nitrate added to the soil at the time of ploughing may be lost rapidly by denitrification. At this time of the year, heavy rain is likely. Explain why the rate of loss through denitrification is rapid:
   a) after a period of heavy rain
      Denitrifying bacteria are more active in anaerobic conditions and, following heavy rain, water-logged soil has less air, and therefore less oxygen, in the soil.

b) when straw from the previous crop is ploughed into the soil and is decomposing

Saprobiotic bacteria decomposing the straw use up soil oxygen in respiration, creating the anaerobic conditions in which denitrifying bacteria are more active.

c) when the soil is still warm

Denitrifying bacteria are more active in warmer conditions and diffusion of nitrogen gas from soil air spaces into the atmosphere is more rapid at higher temperatures.

In the UK, the risk of nitrogen loss in autumn is high, so most nitrogen-containing fertiliser is added in spring.

3 Use Figure 15.12 to explain the advantage of applying fertiliser to a winter wheat crop in early spring.

The plants are starting to grow more rapidly again as air temperature and day length increase following the winter, so use phosphate and nitrate ions more rapidly for protein and nucleic acid synthesis, resulting in less leaching of ions from the soil (see page 280)

4 Suggest disadvantages of applying fertiliser in early spring.

Initial growth of the seedlings in autumn does not have the benefit of the fertiliser. In early spring, the soil may still be fairly cold and so root respiration will be slow, so there may not be much ATP being produced for active transport of ions into roots. Earlier growing weeds may also take up some of the ions instead of the crop.

TEST YOURSELF

9 Why is it necessary to use fertilisers when (a) growing crops and (b) on grass when rearing dairy cattle and sheep?

10 Explain why natural fertilisers release the nutrients they contain over a longer period of time than do artificial fertilisers.

11 Explain why the curves in Figure 15.11 do not start at the origin of the graph.

12 From Figure 15.11, what mass of nitrogen-only fertiliser would you recommend a farmer use per hectare for growing maize? Explain your answer.

Fertilisers and the environment

The stream shown in Figure 15.13 runs through an area of farmland. It contains a high concentration of nitrate and phosphate ions. The stream is in a rural area and environmental scientists suspected that these ions came from fertiliser applied to the surrounding farmland.

Figure 15.13 The growth of algae and water plants in this stream is the result of pollution by nitrate and phosphate ions.

279

Eutrophication The addition of extra nutrients such as nitrate or phosphate ions to aquatic ecosystems.

Leaching The process in which soluble ions dissolved in soil water drain through the soil into aquatic ecosystems.

Table 15.2 Total mass of nitrogen in fertiliser added to surrounding fields and the mean concentration of nitrate in a stream.

Total mass of nitrogen in fertiliser added to surrounding fields/kg ha^{-1}yr^{-1}	Mean concentration of nitrate in stream/mg dm^{-3}
41	1.2
41	1.3
51	1.5
56	1.8
63	1.6
69	1.9
72	2.0

Table 15.2 shows data the scientists collected for the same stream over a number of years. It suggests that as the total mass of nitrogen in fertiliser added to fields increases, the mean concentration of nitrate in nearby streams also increases. Because these data consist of two sets of measured variables, the best way to present it would be on a scattergram. The line of best fit would slope upwards, indicating a positive correlation. You would use a Spearman's rank correlation test to find out whether the correlation between these two variables is significant or not.

Eutrophication

The addition of extra nutrients to aquatic ecosystems is is called eutrophication. The word is generally used when freshwater streams or lakes, such as the stream in Figure 15.13, are enriched with nitrate and phosphate ions because of leaching. Because nitrate and phosphate ions are so soluble they can be carried in soil water as it drains from fields into ditches.

If fertilisers are used on agricultural land there is a risk of eutrophication in nearby aquatic ecosystems. The risk is higher if the fertilisers are artificial and especially high if the fertiliser is used on bare fields or before heavy rain. This is because artificial fertilisers contain soluble salts such as ammonium phosphate, which readily dissolve in soil water. If they are not quickly absorbed by plant roots, they remain in the soil water and from there they can be leached into ditches and streams by rainwater.

Natural fertiliser such as compost or manure decomposes and releases nutrients slowly. Plants usually absorb them as fast as they are released, so the risk of nutrients leaching into aquatic ecosystems is lower when using natural fertilisers.

If nutrients such as nitrate or phosphate ions reach aquatic ecosystems such as stream or lakes they cause an increase in the growth of aquatic plants and algae. This is because the growth of many aquatic plants and algae is usually limited by nitrate and phosphate ion concentration.

Increased biomass of aquatic plants, and especially algae, reduces the amount of light entering the ecosystem. Dense aquatic plant foliage at the surface can shade the plants beneath them. Large numbers of algae in the water turn it murky green, or turbid. This is sometimes called an algal bloom.

The reduced light available further underwater means that some aquatic plants and algae die. They are decomposed by saprobionts in the water, mostly bacteria. The bacteria grow and reproduce rapidly due to the large amount of dead plant material available. Since the bacteria respire aerobically, they use most or all of the oxygen dissolved in the water.

Aquatic organisms such as stonefly and mayfly larvae, which require a relatively high concentration of oxygen, die. In extreme cases fish may also die. The few animal species that might survive are those tolerant of low oxygen concentrations. However, the diversity of animal species will be much reduced (see Chapter 13).

Figure 15.14 Stonefly larvae are especially sensitive to depleted oxygen concentration and die if eutrophication occurs in a stream.

Figure 15.15 summarises how eutrophication can affect an ecosystem.

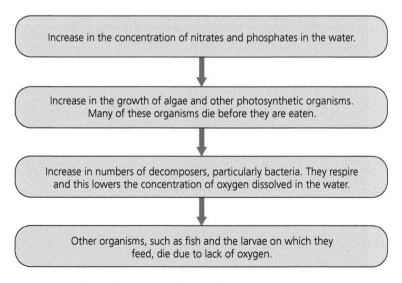

Increase in the concentration of nitrates and phosphates in the water.

Increase in the growth of algae and other photosynthetic organisms. Many of these organisms die before they are eaten.

Increase in numbers of decomposers, particularly bacteria. They respire and this lowers the concentration of oxygen dissolved in the water.

Other organisms, such as fish and the larvae on which they feed, die due to lack of oxygen.

Figure 15.15 The effect of eutrophication on freshwater ecosystems.

Scientists can use two different approaches to study the effects of eutrophication. The first is a fieldwork-based approach. This involves either comparing polluted and unpolluted water, or looking at data collected at a single place over a long period of time. Problems arise with trying to explain the results of investigations such as these, however. It is often very difficult to link an effect to a particular cause because situations are often very complex and many factors may change at the same time.

The second approach is to use an experimental approach, in either the laboratory or the field, to test the effect of changing a specific variable. In this case, scientists have to take care in applying results obtained under carefully managed conditions to natural situations.

Data from fieldwork

The Norfolk Broads are a series of shallow lakes along the lower reaches of three of the main East Anglian rivers. The Broads are surrounded by agricultural land. Some of the land is used for grazing and some for growing crops. Table 15.3 shows figures for the maximum concentration of phosphates, nitrates and algae in the water of seven different lakes in a particular year.

Table 15.3 Maximum concentrations of phosphates, nitrates and algae in seven lakes in the Norfolk Broads.

Lake	Maximum phosphate concentration/$g\,dm^{-3}$	Maximum nitrate concentration/$mg\,dm^{-3}$	Maximum concentration of algae given as chlorophyll concentration/$\mu g\,dm^{-3}$
A	340	2	460
B	200	4	230
C	320	3	410
D	240	8	370
E	160	6	220
F	190	8	370
G	190	3	200

ACTIVITY

The effect of phosphate concentration on the growth of algae using data from fieldwork

Start by looking at the data on phosphate concentration and the concentration of algae. Is there a significant correlation between these two variables? If you carry out a Spearman's rank correlation test on these data you should get a value for R_S of 0.85. Look this up in the table of probability values (see Table 26.7 on page 495) and you will see that this value is greater than the critical value for seven pairs of measurements. You can therefore reject your null hypothesis and can conclude that there is a significant correlation between the maximum phosphate concentration and the maximum concentration of algae in the water.

Now look at the relationship between nitrate concentration and the concentration of algae. In this case Spearman's rank correlation test gives an R_S value of –0.28. The minus sign shows us that, in this case, we are looking at a negative correlation. In other words, the greater the concentration of nitrate in the water, the lower the maximum concentration of algae. If you ignore the minus sign and look up this

value in the table you will see that it is lower than the critical value. There is a greater than 0.05 probability that this correlation arose by chance. You should therefore accept the null hypothesis and conclude that the correlation between nitrate concentration and concentration of algae is not significant.

The data shown in Table 15.3 confirm what many other scientists have found. In most freshwater ecosystems, the factor limiting growth is very likely to be phosphate concentration. Increasing the concentration of phosphate is often associated with an increase in algal growth. As pointed out earlier, however, scientists must take care in interpreting data like these. There is obviously a strong correlation between phosphate concentration and the concentration of algae. This does not mean that it is the increase in phosphate concentration that causes the rise in the algal population. There may be other ecological factors that vary between these lakes, and any one of them could have affected the algal population.

ACTIVITY

The effect of phosphate concentration on the growth of algae using data from an experiment

A Lund tube is a large rubber tube used to investigate freshwater ecology. It is designed to isolate large volumes of water from the lake outside. The top of the tube is surrounded by a large inflatable ring that floats on the surface, while the bottom of the tube sinks into the mud on the lake floor.

In one experiment, phosphate was added to the water inside a Lund tube in October. The populations of algae in the water in the tube and in the surrounding lake water were measured at regular intervals over the next few months. The results of this investigation are shown in the graph in Figure 15.16.

Look at the data in Figure 15.16 and answer these questions.

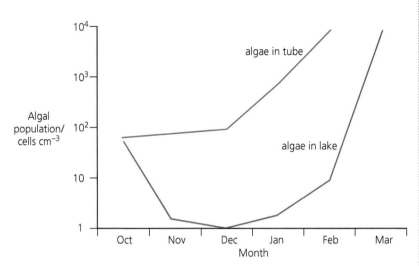

Figure 15.16 The effect of adding phosphate to the population of algae inside a Lund tube.

1 By how many times is the population of algae in the tube greater than the population in the rest of the lake, in the middle of February?
2 a) Describe the difference in the shapes of the two curves shown on the graph.
 b) Explain this difference, in terms of phosphate concentration.

3 The Lund tube used in this investigation was approximately 45 m in diameter. Other than the presence of added phosphate, suggest how the water inside the tube might have differed from the surrounding lake water.
4 Use your answers to the questions to explain why scientists have to take care in applying the results of this investigation to natural situations.

TEST YOURSELF

13 What is leaching and why is it a risk to the environment?
14 Why is the risk of leaching often higher if artificial fertilisers are used on farmland rather than natural fertilisers?
15 Describe the effect of eutrophication on a typical freshwater ecosystem.
16 Explain why it was important that all the lakes in Table 15.3 were sampled in the same year.

Practice questions

1 The sequence below illustrates the events after the addition of nitrate and phosphate ions to a stream flowing through farmland.

Increased nitrate and phosphate ion concentration → Increased growth of algae → Death and decay of aquatic plants → Reduced oxygen concentration in the water

a) **i)** Suggest why the increased growth of algae results in the death of other aquatic plants. (2)

ii) Describe what causes the reduced oxygen concentration of the water. (2)

b) Explain the impact of a reduced oxygen concentration on the animal community in the stream. (2)

2 The diagram shows the phosphorus cycle.

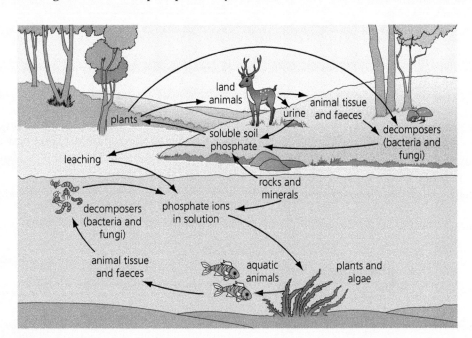

a) Explain how the herbivore obtains phosphorus from the biological molecules in its plant food. (2)

b) Describe two ways that the herbivore would use phosphorus. (2)

c) Explain why the cycle shown in the diagram depends on saprobionts. (2)

3 Sometimes farmers grow mustard plants on land in autumn when there would otherwise be no crop. The mustard takes up nitrate ions, which might be leached from the soil at this time of the year. When it is time to plant seeds of a new crop in the spring, the mustard is ploughed into the soil as a natural fertiliser.

a) Explain why the mustard crop is used to remove nitrate ions from the soil. (1)

b) When the mustard plants are ploughed into the soil, nitrogen contained in biological molecules in their tissues become available for the new crop. Describe the role of saprobionts in this process. (3)

c) Explain why using mustard plants as a natural fertiliser would reduce the risk of leaching in the spring. (2)

d) Plants also absorb phosphate ions from the soil. Describe why phosphate ions would be needed by a growing crop. (2)

4 Scientists compared the decomposition of beech and fir leaves in leaf litter for one year. The table shows some of their results.

	Beech	Fir
Biomass lost (%)	13.2	23.6
Organic C (%)	45.7	47.9
Total N (%)	0.63	1.57
Lignin (%)	36.1	28.1

a) Explain why it was appropriate that leaf biomass lost was determined as the percentage dry mass lost. (2)

b) Along with cellulose, lignin is a component of plant cell walls. Saprobionts find lignin far less digestible than cellulose. Describe and explain the relationship between the lignin content of the leaves and their decomposition rate. (3)

Organisms that decompose organic material use carbon-containing biological molecules as respiratory substrates and nitrogen-containing biological molecules for protein and nucleic acid synthesis. They need more carbon than nitrogen. Microorganisms require a carbon:nitrogen ratio of about 30 : 1.

c) Explain how the above information relates to the data in the table. (3)

Stretch and challenge

5 Discuss the role of mycorrhizae on the growth of certain plants, especially orchids. Evaluate the recent work by scientists at the University of Maryland in the USA which claims that one of the key factors that determine if certain species of orchid can grow in a particular place is the presence or absence of their mycorrhizal fungi in the soil. Explain possible reasons for this and examine its importance for orchid conservation.

16 Response

The idea that some plants respond to music or being talked to is not new but until now has been regarded with scepticism. But recent experiments with a small cabbage-like plant called *Arabidopsis* have shown that they do indeed respond to sound vibrations. More than that, they can detect the difference between different sorts of sounds.

If you think about the possible ecological significance of sound in the life of plants, this is not so unusual as it might seem. Plants are constantly exposed to the feeding of herbivores, from large mammals to a much wider range of smaller animals, especially insects. The chewing, munching and tunnelling of herbivorous animals creates a constant barrage of sound. Since being damaged by feeding animals is disadvantageous to plants, it is actually not surprising that they can detect and respond to the sound of this happening.

Extension

Scientists at the University of Missouri recorded the sounds of feeding caterpillars by shining a laser at a small reflective patch on an *Arabidopsis* leaf and allowing a caterpillar to chew it. The tiny vibrations detected by the laser beam were recorded and then played back to other *Arabidopsis* plants. A second group of plants were kept in silence.

When caterpillars were then allowed to feed on both sets of plants, those that had been exposed to the sound of feeding had more defensive chemicals, substances that caterpillars find distasteful, in their leaves than those that had been kept in silence. Caterpillars respond to the defensive chemicals by moving to another plant.

The advantage of this response to *Arabidopsis* plants is clear. Although plants have other chemical-mediated responses to feeding animals, sound vibrations are a faster way for distant parts of the plant to detect herbivores feeding and produce defensive chemicals. Further experiments showed that other vibrations, such as those caused by wind moving the leaves, did not stimulate this response.

So some plants can detect and respond to sound. Future developments based on this work might include genetic modifications to crop plants that boost the production of defensive chemicals, making them more resistant to damage by pests. These experiments also show that plants respond to more of the same stimuli as animals than you might think, even though their responses may be rather different.

Survival and response

How does a blackbird find an earthworm in a lawn? How do grass leaves grow towards the light? How does a cat catch a blackbird? How does a blackbird recognise a cat as a dangerous predator? How does a worm escape from a blackbird? Getting food, avoiding being eaten and finding favourable conditions to live in are all essential requirements for survival. Any species that does not have the ability to respond to these requirements will die out.

Figure 16.1 Blackbird finding earthworms in a lawn.

287

Stimulus A change in the internal or external environment.

As we shall see, responses vary in complexity according to circumstances and from plants to animals. Detection requires a stimulus that can be detected by **receptors**. Some receptors are cells that secrete substances in response to stimuli such as β cells in the pancreas (see Chapter 19, page 342). Some receptors are specialised cells that produce electrical activity in nerve cells. For example, a sudden movement by a cat may be the stimulus that is detected by receptors in a bird's eyes. Processing involves nerve impulses being conducted to a coordinator, either the brain (or in a worm to a sort of mini-brain) or to the spinal cord, and from there to the parts of the body that will produce the appropriate **response**. The response is carried out by **effectors**. In the bird the effectors will be the muscles that operate the wings. So, the full sequence is:

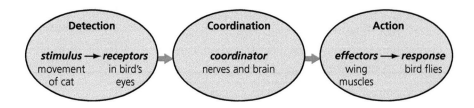

This response obviously has vital survival value to the bird. It is important that the response should be rapid, and a reflex action achieves this. At the same time it cannot be an absolutely fixed and automatic action. A successful escape flight should be controlled so that it carries the bird away from the cat to reach a safe place. After the initial reflex, the bird is able to take control and undertake much more complex behaviour than just flapping its wings.

Simple reflexes

In humans some automatic responses to an external stimuli are called simple reflexes. If, for example, you accidentally touch a hotplate on a cooker you will very rapidly pull your hand away. You won't have to think about your action and you won't be able to stop this response if you are not prepared for the heat. This has the obvious advantage that it minimises the damage that might be caused.

Simple reflex An unlearned, fixed response to a stimulus.

Figure 16.2 shows the **neurones** involved in a reflex arc that produces a rapid response. The high temperature of the hotplate stimulates pain receptors close to the surface of the skin. These are actually thin branches at the end of a **sensory neurone** that has a long extension all the way through the arm to the spinal cord. The receptors trigger nerve impulses that are conducted to the opposite end of the sensory neurone. This end of the axon of the sensory neurone has tiny branches that almost touch similar branches on a **relay neurone**. A junction between neurones is called a **synapse** (see page 315). The relay neurone has another synapse linking to a **motor neurone**. The axon of the motor neurone conducts nerve impulses to a muscle in the arm and stimulates it to contract. This produces the response that causes the arm to be quickly pulled away.

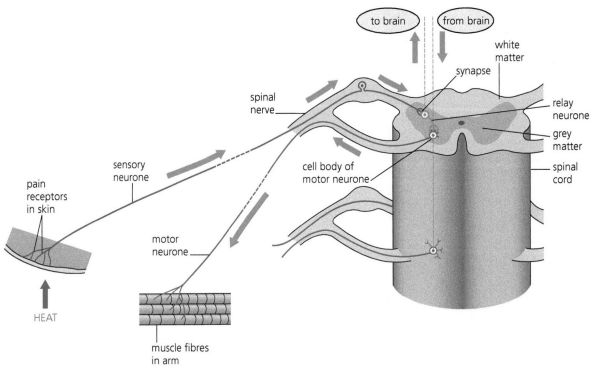

Figure 16.2 A three-neurone reflex arc.

A **three-neurone** reflex arc is an example of a mechanism that results in a totally automatic reflex response. In practice, however, the response of pulling the arm off a hotplate includes much more than a single reflex arc. Touching the hot surface stimulates not just one but a large number of receptors, so many sensory neurones send nerve impulses to the spinal cord. The response involves the coordinated contraction of several muscles in the arm. Therefore the relay neurones despatch nerve impulses not just to one muscle but to several, which is why they have synapses connecting with many other neurones. Nerve impulses also pass to the brain, making us conscious of pain and enabling us to take further action, as well as to shout 'ouch'.

Finding the right environment

Kinesis A non-directional response to a stimulus.

Taxis A directional response to a stimulus.

The ability to respond to environmental stimuli occurs in all living organisms. Even seemingly simple organisms that do not have a complex nervous system can find favourable conditions or get out of trouble. Motile organisms move more or faster in response to a stimulus, or turn less frequently when they experience less favourable conditions. This is called kinesis, which simply means movement. A kinesis is a non-directional response (relative to the direction of the stimulus) in which the rate of movement is affected by the intensity of the stimulus. A second way is to move directly towards or away from a stimulus. This is called a taxis.

Example of kinesis

Planarians are carnivorous flatworms that live in shallow streams and ponds. They have a network of neurones and simple 'eyes' that have light-sensitive cells but no lens.

Planarians are often found clustered on the underside of stones, where they normally remain hidden during daylight. If a stone is turned over, the flatworms immediately start moving around in random directions. When their movements bring them back into the darkness they stop moving. This behaviour helps to protect them from predators. Laboratory experiments show that the brighter the light the less frequently they change direction. The flatworms move around randomly until they happen to get to a darker environment, so this is directionless movement. But in the dark they change direction more frequently. This tends to keep them in a darker environment.

Figure 16.3 The planarian *Dendrocoelum lacteum* is about 15 mm long.

TIP
Bacteria, viruses, plants and small organisms such as flatworms do not think like humans, so avoid using language that suggests that they 'want' or 'prefer' things or 'need to' do something.

Example of taxis

Euglena viridis is a single-celled organism that that lives in small ponds (Figure 16.4). It has chloroplasts and so is able to photosynthesise. It also has a long flagellum that it uses for swimming. There is a receptor near the base of this flagellum that is sensitive to light. You can also see a red spot close to the light-sensitive area.

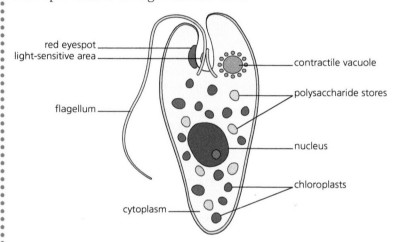

Figure 16.4 Structure of *Euglena*.

Euglena responds to light by swimming towards it. Waves pass along the flagellum from the base to the tip and pull the *Euglena* forward. Because there is only one flagellum the waves also make the cell rotate as it moves forward. If the light is shining from one side the red spot will shade the light-sensitive area (the receptor) each time the cell rotates. When it is moving straight towards the light the area will stay illuminated all the time. The molecules of the light-sensitive pigments are aligned so that when they are illuminated they stimulate the flagellum to beat. This results in the *Euglena* moving directly towards the light. This response is therefore a taxis and because it is movement towards light it is called a phototaxis.

It is amazing that such seemingly complex behaviour can be coordinated by tiny organelles within a single cell. Nevertheless, the ability to move to and away from light is common in single-celled organisms. In shallow seas there is a mass migration of vast numbers of plant photosynthetic plankton towards the surface in the daytime and then at night-time back to deeper levels where there are higher concentrations of mineral nutrients.

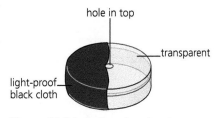

Figure 16.5 A choice chamber is a small container with sub-sections in which different abiotic conditions can be set up, into which small motile animals can be introduced and observed.

TIPS
- In an exam you might be asked to recognise the type of data you have been given and to select an appropriate statistical test.
- Look at Chapter 26 on maths skills to find out more about statistical tests and to see a worked example of this test.
- You won't need to do statistical calculations in a written paper.

TIPS
To be able to decide conclusively if a response is a kinesis or a taxis, the animals must be observed during the experiment to see if their rate of movement or turning frequency is different in the different conditions.

Investigating taxes and kineses

It is quite likely that you have already carried out a behaviour experiment using apparatus similar to the choice chamber shown in Figure 16.5. These are suitable for studying taxes and kineses in small animals, such as woodlice and maggots.

A student was investigating the hypothesis that woodlice exhibit negative phototaxis; that is, they move away from the bright light. The student covered one half of a choice chamber with light-proof black cloth (Figure 16.5). The other half of the chamber was illuminated by a light bulb fixed about 20 cm above it. Ten woodlice were put into the chamber through the hole in the centre of the top. Five minutes later the student counted how many woodlice could be seen on the illuminated side (and by simple subtraction worked out how many were on the shaded side). This procedure was repeated 10 times using different woodlice each time. Table 16.1 gives the student's results.

Table 16.1 Results of an experiment to investigate negative phototaxis in woodlice using a choice chamber.

Trial number	Number of woodlice after 5 minutes	
	Bright light	Shade
1	3	7
2	5	5
3	4	6
4	4	6
5	3	7
6	5	5
7	6	4
8	3	7
9	2	8
10	4	6

Looking at these results the student concluded that woodlice do move away from bright light and therefore may show negative phototaxis, since there are clearly more woodlice found in the shade than in the bright light. It fitted with the student's expectations, since woodlice are normally found in damp, dark places, such as under logs or stones. However, it is possible that the response was actually a kinesis. The student only counted the woodlice at the end of 5 minutes. There is no evidence that they were observed during this period and without being able to tell if they showed non-directional responses to end up where they were the experiment is not conclusive.

Because the student was dealing with two discrete categories, bright light and shade, the best way to present the totals would be on a bar graph. The bars would be different heights indicating that more woodlice moved to the shaded side. You would use a chi-squared (χ^2) test to find out whether the difference in the number moving to each side is significant (see page 491).

Investigation into the effect of an environmental variable on the movement of an animal using either a choice chamber or a maze

This is just one example of how you might tackle this required practical.

Some of the first experiments on woodlouse behaviour were carried out by J. Cloudsley-Thompson. In one series of experiments he studied the response of woodlice to humidity using large choice chambers similar to the one shown in Figure 16.6. The humidity on each side was controlled. Under the gauze on one side was a dish of a drying agent that absorbed moisture from the air. On the other side was distilled water that maintained a high humidity above it.

Cloudsley-Thompson compared three sets of conditions:

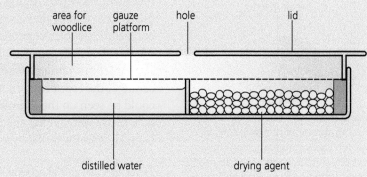

Figure 16.6 Apparatus used by Cloudsley-Thompson to investigate woodlouse behaviour.

A choice chamber in the light using woodlice that had been in the light for several hours before the experiment

B choice chamber in the dark using woodlice that had been in the light for several hours before the experiment

C choice chamber in the dark using woodlice that had been in the dark for several days before the experiment.

In each experiment, he put five woodlice into the choice chamber and then recorded the positions of the woodlice after 15 minutes (Table 16.2).

Table 16.2 The results of Cloudsley-Thompson's three experiments (A, B and C) in which he investigated three different sets of conditions.

Woodlouse behaviour in choice chamber	Number of woodlice		
	A	B	C
Moving around	113	40	17
Stationary on the dry side	21	35	53
Stationary in the central area between the dry and moist sides	49	47	86
Stationary on the moist side	317	378	344

1 What do the results show about the response to humidity?
2 Suggest how this response might be advantageous to the woodlice.
3 Cloudsley-Thompson noticed that after putting the woodlice into the chamber they usually moved around for a short time before becoming stationary. What does this suggest about the type of behaviour shown by the woodlice?
4 Calculate the percentage change in woodlice still moving around between experiments A and C and show the results on a suitable graph. Suggest a possible explanation for the difference.
5 What does your analysis of the results suggest about the response of woodlice to humidity in the light compared to the dark? Suggest how this might this increase the chances of survival of woodlice.

Plant responses

Plants may just seem to sit around and do nothing, but in fact their survival is as dependent as that of animals on being able to respond to environmental conditions. Whichever way a seed lands in the soil, the shoot will grow upwards and the roots will grow down. Common observations show examples of responses. A pot plant growing on a window sill will grow towards the light unless it is turned frequently. Many flowers close at night and then open again in the morning. You may have come across so-called sensitive plants that respond to a touch. Lightly brushing against the end of a *Mimosa* leaf will stimulate the leaflets to close up like the ripple of a Mexican wave.

Some trees can defend themselves if herbivores damage their leaves. Like the *Arabidopsis* plants in the introduction to this chapter, they respond by producing nasty-tasting or poisonous substances. This ability can pass from affected to unaffected parts of the tree by the use of chemical messengers. Some trees can even pass this chemical message to neighbouring trees, which then produce the same noxious substances before the attackers move in. You might wonder why the trees don't just produce the noxious substances before any herbivores cause some damage. The answer is probably that the production of the poison requires energy and resources, so it is more economical to wait until the threat of attack is real.

Tropism A growth response to a stimulus.

Plants don't have a nervous system, so how are they able to respond to stimuli? Responses such as bending towards or away from light or gravity result from uneven growth. The seedlings in Figure 16.7 are bending towards the light because they have been stimulated to grow slightly faster on the more shady side of the stem. Roots placed horizontally will begin to grow vertically downwards. This sort of growth response to a stimulus from a particular direction is called a tropism. Response to light is referred to as **phototropism** and that to the force of gravity as **gravitropism.** Both may be either positive – towards the stimulus – or negative – away from the stimulus.

A growth response depends on chemical substances released in response to a stimulus. These are known as **specific growth factors**, and they act a little like hormones in animals. Although this is much slower than the electrical activity of nerves or the response of some plants to sound vibrations, it can be surprisingly rapid. For example, the phototropic response of plant shoots can be detected within minutes of exposure to light.

The first specific growth factor to be discovered was **indoleacetic acid (IAA)**. Several other substances that affect growth in plants have been discovered since. But there is still uncertainty and disagreement about exactly how these substances work. One of the problems is that the actual concentrations of the substances present in the plant tissues are extremely low.

Figure 16.7 Increased growth on the shady side of these seedlings results in a positive phototropic response to a light source.

293

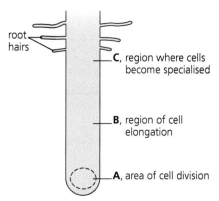
Growth of a root or shoot has two distinct stages (Figure 16.8).

● First, cell division (mitosis) takes place at or near the tip.
● Second, the vacuoles of the new cells expand as they take up water by osmosis, causing the cells to elongate. This starts in the cells a short distance behind the tip in an area called the region of elongation. This is where the most obvious increase in length of a root or shoot occurs.

IAA has its main effect in the elongating region. It is synthesised in the young root or shoot cells at the tip. As it moves into the elongating region it attaches to protein receptors on the membranes of cells. Exactly how it works is still unclear, but one of its effects is to lower the pH by the release of hydrogen ions. These break some of the bonds between the microfibrils in the cellulose walls (see Chapter 1, page 6), making the microfibrils more easily stretched by the increasing turgor of the cells.

IAA and phototropism

Possible explanations for the phototropic response in plant shoot tips include:

● IAA is destroyed by the light on the illuminated side
● extra IAA is produced on the shaded side
● IAA moves away from the illuminated side.

Figure 16.9 shows results from experiments to investigate these explanations in shoot tips. In each case, the shoot tips were cut off and then placed on thin blocks of agar. In some experiments, very thin slices of mica, which are impermeable to water, were used to separate the two sides of the tips and/or the agar blocks. After 3 hours, the concentration of IAA in each agar block was measured.

root hairs

C, region where cells become specialised

B, region of cell elongation

A, area of cell division

Figure 16.8 Growing regions in a root.

Figure 16.9 Results of four experiments on shoot tips to investigate possible explanations for their phototropic response. The figures show the concentration of IAA in arbitrary units.

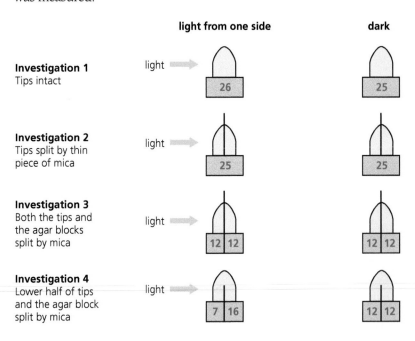

In investigations 1 and 2, the concentration of IAA in the agar blocks is about the same whether in the light or dark. If it was destroyed on the illuminated side you might expect an overall decrease in the light. If more was produced on the shaded side you might expect an increase in the dark.

In investigation 3, the total concentration of IAA collected in the agar was about the same as in investigations 1 and 2. Also, equal concentrations of IAA were found in the agar either side of the mica sheets. Again, these observations support the idea that light does not destroy IAA and darkness does not cause more IAA to be produced.

In investigation 4, the total concentrations of IAA collected in the agar blocks were about the same as in the previous investigations. In darkness, equal concentrations of IAA were found in the agar either side of the mica sheets. Under unilateral illumination, however, less IAA was found in the block nearest the light and more was found in the agar block on the shaded side. This suggests that when the tip is illuminated the IAA produced is actively moved towards the shaded side, unless prevented by the mica, as in investigation 3.

The difference in the concentration of IAA moving down each side of the block in the light in investigation 4 explains why the intact shoot would bend towards the light. More IAA would reach the region of elongation on the shaded side. As we have already seen, IAA causes the cell wall to be more easily stretched by the expansion of the cell vacuole due to the uptake of water by osmosis in the cells on this side of the shoot. It elongates faster than the lit side, and because of this uneven growth the shoot would bend like those in Figure 16.7.

IAA and gravitropism

Unequal IAA concentrations either side of the root tip are also why roots bend towards the force of gravity. Cells called columellar cells near the root tip contain dense organelles called **amyloplasts**. These are packed with starch, which makes them heavy, and so sink to the bottom of the cells they are in. When a root is moved from the vertical to the horizontal the amyloplasts fall to what is now the bottom of the columellar cells. This enables these cells to detect the direction of the force of gravity. IAA seems to be actively transported to the side of the root to which the amyloplasts sink (Figure 16.10).

As IAA moves from the root tip towards the region of elongation there is more on the lower side. **Unlike shoots, in roots a higher IAA concentration inhibits elongation.** This means that the lower side elongates slower than the upper side, causing the root to bend downwards.

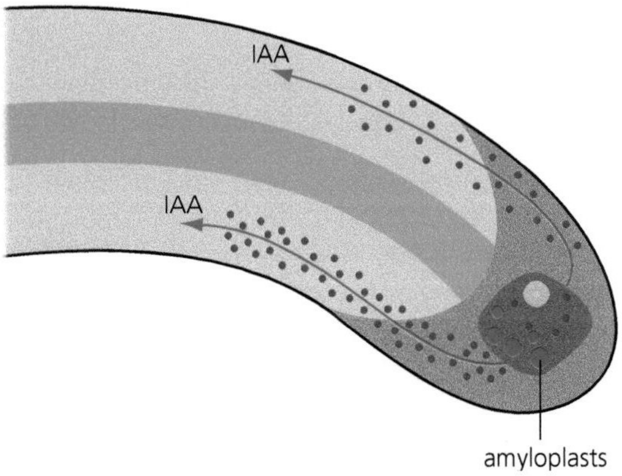

> **TIP**
> Remember that the response of roots to higher IAA concentration is the opposite to that of shoots.

amyloplasts

Figure 16.10 The mechanism of the gravitropic response in a root tip.

TEST YOURSELF

1 Describe the components of a three-neurone reflex arc.
2 Maggots that feed on dead animals tend to move away from the light. What type of response is this and why?
3 Would the experiment shown in Table 16.1 distinguish whether the response was a taxis or a kinesis? Explain your answer.
4 If plants have no muscles, how does a plant shoot or root bend towards or away from a stimulus?
5 Describe how amyloplasts allow some root cells to detect a change in the direction of the force of gravity.

Responses to internal stimuli

So far you have seen examples of responses of organisms to external stimuli. Some activities over which we have no conscious control result from responses to internal stimuli. A good example is the control of heart rate.

The muscle of the heart is amazing. For a start, it doesn't get tired. Even in someone who is not very active it can go on steadily contracting and relaxing 70 times a minute, 24 hours a day for 80 years or more. (That works out at about 3 billion contractions in a lifetime, assuming a life with very little exercise or excitement.)

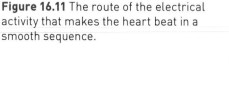

Myogenic Muscle cells that are able to contract without nervous stimulation.

Secondly, it can carry on contracting and relaxing rhythmically without any nerve impulses from the brain. This ability to work on its own is called myogenic. A heart can be removed from the body and, as long as it is given an oxygen and nutrient supply, it will carry on beating.

The sequence of muscle contraction in the heart is initiated by a group of modified heart-muscle cells called the sinoatrial node (**SAN**) near the top of the right atrium. These cells produce regular waves of electrical activity, similar to nerve impulses. The rate at which the SAN produces these waves determines the heart rate because they start off contraction. For this reason the SAN is often called the heart's pacemaker.

During one cardiac cycle, a wave of electrical activity spreads over the walls of both atria, as shown in Figure 16.11. This makes the muscles in the atrial walls contract. Notice that contraction spreads outwards from the top of the atria, squeezing blood towards the ventricles.

Figure 16.11 The route of the electrical activity that makes the heart beat in a smooth sequence.

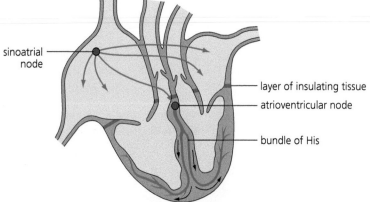

sinoatrial node

layer of insulating tissue

atrioventricular node

bundle of His

The electrical activity cannot pass directly from the walls of the atria to the walls of the ventricles, because it is stopped by a layer of insulating fibrous tissue (Figure 16.11). At the lower end of the wall that separates the atria is another group of specialised cells, the atrioventricular node (**AVN**). These cells can detect the electrical activity passing across the atria.

After a short delay, the AVN triggers electrical activity in specialised muscle cells called **Purkyne tissue**, which in turn conduct the electrical activity rapidly down the wall between the ventricles to the bottom of the heart. The delay allows time for the ventricles to fill completely with blood. Initially, these specialised muscle cells are bunched in a single group, called the **bundle of His**. This then divides into two branches that extend back up the walls of the two ventricles. Electrical activity conducted along these specialised muscle cells stimulates the muscle of the ventricles to contract rapidly from the base of the heart upwards.

However, there are times when the heart rate increases, such as during exercise. For this to happen, the brain is involved, although of course we do not have to *think* about changing our heart rate. In the brain is a special area that controls the heart rate called the cardioregulatory centre and it is situated in the **medulla**. The medulla is tucked away in the base of the brain at the top of the spinal cord (Figure 16.12).

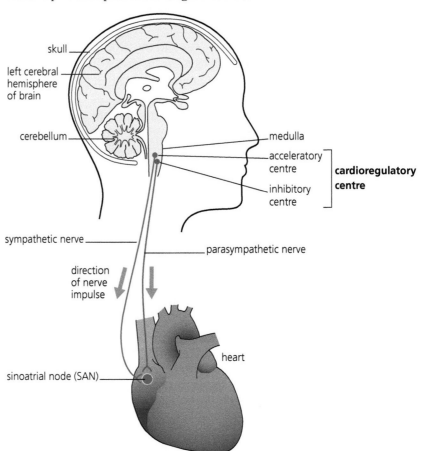

Figure 16.12 The position of the medulla in the brain and its role as a regulatory centre in controlling the heart rate.

The cardioregulatory centre in the medulla actually consists of two discrete parts:

- the acceleratory centre – responsible for speeding up the heartbeat
- the inhibitory centre – responsible for slowing down the heartbeat.

TIP

You do not need to be able to recall the names of the neurotransmitters involved in the control of heart rate.

A variety of factors can stimulate these centres and we will come back to these shortly. Each of these centres is connected to the SAN via nerves. These nerves are quite separate from the nerves that control our conscious activities. They form part of what is called the **autonomic nervous system**, which means the 'self-controlling' system. When the acceleratory centre is activated, impulses pass along **sympathetic** neurones to the SAN. At the synapse with the SAN, noradrenaline is secreted. It is this substance that causes the SAN to increase the frequency with which it produces waves of electrical activity. Noradrenaline is chemically very similar to adrenaline, which is a hormone produced by the adrenal glands. **Adrenaline** is well known as the hormone that is produced when we are frightened or stressed and which prepares us for a 'fight-or-flight' response. Adrenaline also increases the heart rate and is responsible for the thumping heart when we experience fear. We will find out more about adrenaline in Chapter 19.

Activating the inhibitory centre sends impulses along **parasympathetic** neurones that cause the SAN to decrease the frequency with which it produces waves of electrical activity. This returns the heartbeat to its normal or resting state. Like the sympathetic neurones, the parasympathetic neurones secrete a substance at synapses with the SAN. In this case it is acetylcholine, which has the effect of inhibiting the myogenic activity of the SAN.

So, how do the internal stimuli from exercise lead to activation of the acceleratory centre in the medulla? This question does not have a simple answer and the processes involved are still not fully understood. The onset of exercise increases the concentration of carbon dioxide in the blood and initially causes a fall in blood pressure as muscle arterioles dilate. These internal stimuli are detected by **chemoreceptors** and **pressure receptors**. Both are situated in the aorta close to the heart and in the carotid arteries that pass through the neck to the brain. An increase or decrease in the frequency of nerve impulses from these receptors activates the acceleratory centre and contributes to the increase or decrease in heart rate (Figure 16.13).

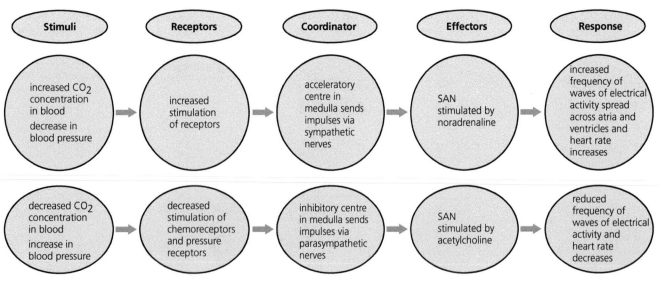

Figure 16.13 Summary of the factors involved in the control of heart rate.

Imagine a hare that suddenly notices a fox closing in on it (Figure 16.14). A rapid response will allow the hare to avoid becoming the fox's prey. The hare immediately leaps into action and escapes at high speed. As long as it can start off quickly it can outrun the fox, which can only reach its maximum speed over short distances. The hare's leg muscles must go to maximum activity as soon as it sees the fox. It is therefore an advantage for the heart rate to increase almost instantaneously, thus speeding up the blood supply and hence the supply of oxygen and glucose for respiration. If the heart responded only *after* the effect of exercise on carbon dioxide production had been detected, vital seconds might be lost. This is the role of the hormone adrenaline, which is released when the hare sees the fox and acts almost immediately on the SAN to increase heart rate.

Figure 16.14 The hare's flight response on seeing the fox must be immediate if it is to get away.

EXAMPLE

Responding to danger

Although we may take part in exercise for sport or pleasure, most other mammals only engage in vigorous exercise to escape danger or to chase food. As well as the increase in the rate of heartbeat there are several other changes that take place in the heart and circulatory system during exercise. Some of the changes are described below and are followed by questions about them. In answering some of the questions you may want to look back to Chapter 6 about the heart.

1 The cardiac output increases by 100–200% during exercise, or even more in fully trained athletes. Explain how the cardiac output is increased and the advantage of the increase.
Cardiac output is heart rate multiplied by stroke volume (the volume of blood ejected each beat) so it is not just heart rate that increases during exercise. If the volume of blood being pumped each beat increases too, then the increase in cardiac output can be even greater. The increase delivers blood faster to the tissues, especially muscle tissue.

2 The blood pressure increases in the arteries during heart contraction. Explain how this increase is produced.
More forceful contraction of the ventricles results in a stronger pulse.

3 The arterioles carrying blood into skeletal muscles dilate (get wider). Explain the advantage of this.
More blood is supplied to the active muscle tissue,

delivering oxygen and glucose faster, and removing carbon dioxide faster, enabling a higher rate of aerobic respiration to continue.

4 The arterioles carrying blood to the abdominal organs and skin constrict (get narrower). What is the advantage of this?
The blood temporarily diverted away from these less critical tissues during exercise can be directed to muscle tissue instead.

5 In the muscles, the higher blood pressure forces open many more of the capillaries. What is the advantage of this?
A larger total surface area becomes available for capillary exchange to occur between the blood and muscle cells.

6 The blood pressure in the major veins returning blood from the body to the heart increases. Explain what causes this increase in blood pressure.
Greater compression of the veins by more muscle contraction taking place in the arms and legs squeezes the blood and increases the pressure.

7 The saturation of haemoglobin with oxygen in the arteries to the muscles is higher than in the veins. During exercise this difference in saturation of haemoglobin with oxygen is much increased. Explain what causes this.
Actively respiring muscle cells use more oxygen so the haemoglobin unloads more oxygen as it passes through muscles, lowering saturation of haemoglobin with oxygen by more than at rest.

Receptors

We have **receptors** that can respond to a wide variety of stimuli, such as light, temperature, chemicals and mechanical effects including pressure, stretching and vibration. Each type of receptor normally only responds to one particular sort of stimulus. This is obviously important, as it enables us to distinguish between a large number of different environmental conditions, both outside the body (external) and inside the body (internal).

Although we often speak of having five senses, we are in fact able to recognise many more different stimuli than this. In the olfactory area of the nose, for example, there are several hundred slightly different receptors that enable us to distinguish as many as 10 000 different smells. Some of these receptors are incredibly sensitive, being able to respond to just a few dissolved molecules of a particular substance.

Consider the sense of touch, which is what we usually think of as the sense to which our skin responds. You will easily be able to distinguish between a very light touch and pressure that pushes your skin in a little. Different degrees of pressure feel quite different. This is not just due to different areas of your skin being touched. You can also tell whether your skin is surrounded by warm or cold air. You will be able to detect a slight movement of a hair, and you will certainly get a very different sensation from a jab with a sharp pin – the sensation of pain. All these different sensations are detected by different types of receptors in the skin (Figure 16.15).

Figure 16.15 Pacinian corpuscles are one of a range of types of receptors in the skin. In an actual area of skin, there are many of each type mixed together. The numbers of receptors vary in different areas; for example there are many more touch receptors in the fingertips than in the middle of the back. The precise structure and functions of some types of receptors is still not understood.

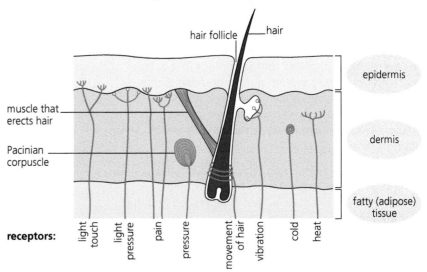

You will see from Figure 16.15 that receptors in the skin have different shapes and positions. Some of the receptors, such as pain receptors, are not separate cells but just the fine branches at the ends of the fibres of sensory neurones. Other receptors, such as the **Pacinian corpuscles** that respond to changes in pressure on the skin, are more complex structures (Figure 16.16) but are still not separate cells.

Pacinian corpuscles

Pacinian corpuscles are quite large compared with most receptors, so it has been possible to study the way in which these work. You can see from Figure 16.16 that a Pacinian corpuscle consists of many layers of membrane, rather like a tiny onion.

Figure 16.16 Coloured photomicrograph of a Pacinian corpuscle.

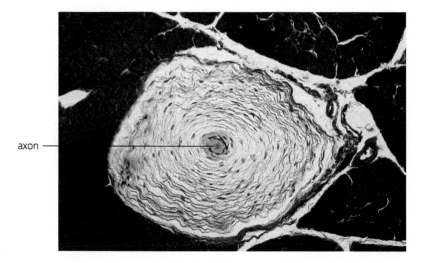

axon

These layers surround the end of the axon of a sensory neurone. Normally there is an excess of positive sodium ions (Na⁺) outside the axon. But when pressure on the Pacinian corpuscle is increased the layers are distorted and proteins called sodium channels in the membrane of the axon are opened. It is why these particular sodium channels are called **stretch-mediated sodium channels**. Previously you learned about the ways that ions can pass through **carrier proteins** in the membrane. It may be helpful to refer to Chapter 3, page 46.

This allows sodium ions to move into the axon by facilitated diffusion. This changes the electrical potential difference across the membrane, as shown in Figure 16.17. It is this that triggers impulses that can pass along the axon of the neurone to the central nervous system. For this reason it is called a generator potential. We shall see in Chapter 17 how nerve impulses pass along the axons of neurones.

Generator potential The change in electrical potential in a receptor when it is stimulated.

> **TIP**
> Remember that you learned about facilitated diffusion and different types of carrier protein during the first year of your course.

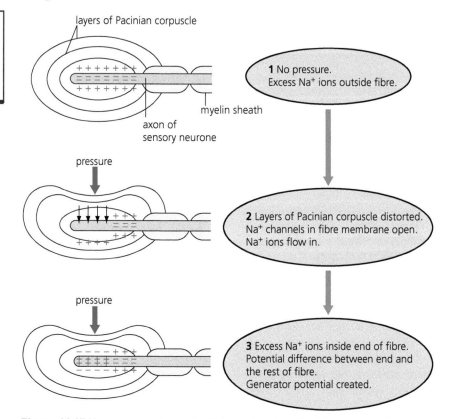

Figure 16.17 How a generator potential is set up in a Pacinian corpuscle.

TEST YOURSELF

6 What is a generator potential?

7 What happens to stretch-mediated sodium channels in response to pressure?

8 What does myogenic mean in relation to the sinoatrial node?

9 What is the role of the AVN in coordinating heart activity?

10 Where are the chemoreceptors that are involved in the control of heart rate?

Vision

You have probably studied the eye in an earlier stage of your science education. You may remember that images are focused on to the retina at the back of the eye. Figure 16.18 will remind you of the structure of the eye.

TIP

You do not need to be able to recall the detailed structure of the eye, but this is a useful diagram to remind you of the location of the retina.

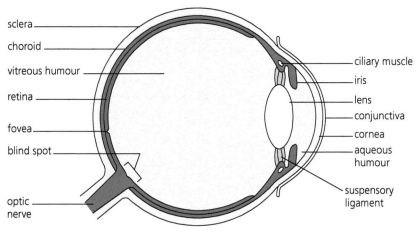

Figure 16.18 Structure of the eye.

The receptors that respond to light are situated in the **retina**. We are going to concentrate on how the different types of receptor cell in the retina helps us to see in different conditions. We have two different types of light-sensitive receptor cell in the retina, the rods and the cones.

Sensitivity

The rods and cones (Figure 16.19) both contain optical pigments, which absorb light (the stimulus) and are broken down. This results in the production of a generator potential. The pigment within rods is broken down in dim light whereas the pigments within cones are only broken down in bright light.

Sensitivity (to light) Describes how much light is needed to stimulate the receptor. Rods are more sensitive so are stimulated in much dimmer light than cones.

- **Rods** are sensitive to a very low intensity of light and therefore enable us to distinguish light from dark in very dim light. They are so sensitive that they can produce a generator potential in response to just one photon. (A photon is a 'particle' of light that is emitted when electrons are excited and jump from one orbit to another in an atom.) Rods do not permit us to distinguish different colours.

- **Cones** are not as sensitive to light as rods. They are, however, sensitive to light of different wavelengths. Human eyes have three types of cone each containing one of three different optical pigments. One pigment is sensitive to the wavelengths of light corresponding to red light, one to green and one to blue. When combinations of the three types of cone are stimulated we perceive the range of other colours in the visible spectrum.

TIP

The three types of cone are not actually coloured so do not call them blue, red or green cones; they are blue-sensitive, red-sensitive or green-sensitive cones.

Visual acuity

Unlike the receptors in the skin, rods and cones are not directly connected to the central nervous system. As you can see from Figure 16.19, they have synapses connecting them to bipolar neurones that are also in the retina. These neurones have synapses connecting to ganglion cells that have axons extending via the **optic nerve** all the way to the brain.

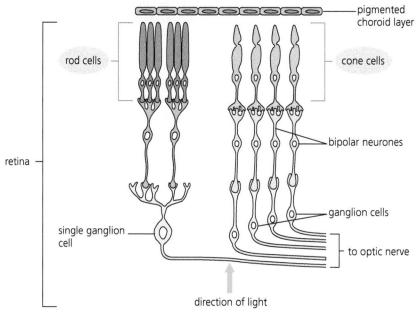

Figure 16.19 How the rods and cones are connected to other cells in the retina.

Each of our eyes has about 125 million rods and 7 million cones. However, they are not evenly distributed in the retina. Most of the cones are concentrated in the centre of the retina, at the position called the **fovea** (look at Figure 16.18). This is where the main part of an image is formed when we look straight at an object. The rods are spread much more widely and they cover most of the back part of the eye. One exception is the area where the optic nerve leaves the eye, the so-called blind spot.

The retina contains about 132 million receptors, each of which can be stimulated by a ray of light falling on it. You can imagine, therefore, that when an image falls on the retina it will have an effect rather like a computer screen, which consist of a mass of tiny dots, called pixels.

Where light falls, the receptors will be stimulated, and where there is a dark patch there will be no stimulation. Each receptor that is stimulated can pass impulses to the brain and the brain can interpret the pattern. At the fovea all of the receptors are cones. Each cone in the fovea connects through a single bipolar cell to one ganglion cell in the optic nerve.

The optic nerve has about 1.2 million ganglion cells. Since there are 132 million receptors there obviously cannot be individual connections to the brain for all the receptors. You can see from Figure 16.19 that several rod cells synapse with a single bipolar cell. In turn, more than one of these bipolar cells synapses with a single ganglion cell. Although the simplified diagram shows only a few synapses, in practice there must be an average of about 100 rod cells with synaptic connections to each ganglion cell. In

Visual acuity How far apart two spots of light must be to be seen separately.

contrast, each cone shown in Figure 16.19 synapses with a single bipolar cell which, in turn, synapses with a single ganglion cell. This affects the **visual acuity**, the amount of detail, or resolution, that can be perceived in an image on the retina.

A ray of light that falls on just one cone in the fovea will show up as a spot of light, as long as it is bright enough to stimulate the cone. This is because the cone is connected to a single ganglion cell in the optic nerve via one bipolar cell (Figure 16.20a). If another ray of light falls on another cone, as in Figure 16.20b, the brain will interpret this as two separate spots.

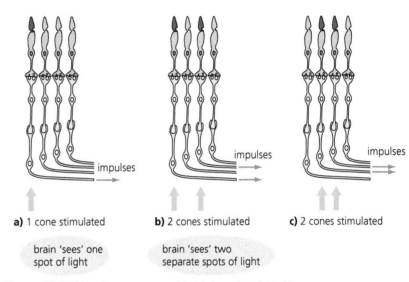

a) 1 cone stimulated

brain 'sees' one spot of light

b) 2 cones stimulated

brain 'sees' two separate spots of light

c) 2 cones stimulated

Figure 16.20 How the cones provide higher visual acuity.

Now consider what will be seen when a ray of light falls on a single rod. Figure 16.19 shows three rods with synapses to each bipolar cell and six rods that connect via the bipolar cells to a single ganglion cell. Assuming that the stimulation of one rod cell is sufficient to send an impulse to the brain, there is no way in which the brain can interpret which of the six rods had been stimulated. In reality there may be many more than six rods with synapses to a single ganglion cell. The rods therefore provide lower visual acuity than the cones.

So, what is the advantage of having many rods connecting to a single ganglion cell in the optic nerve? As we shall see in Chapter 17, synapses can act as barriers to the transmission of impulses. Although the rod is very sensitive to light, a single rod is unlikely to produce a sufficiently large generator potential to be able to stimulate the bipolar cell to conduct nerve impulses. However, if a group of rods is stimulated by dim light at the same time, the combined generator potentials will reach the threshold required to cause nerve impulses to be conducted in the bipolar cell and the ganglion cell in the optic nerve. This process is called **summation**, because the effect of several cells is added together. Although the image is less sharp, we are able to see in much dimmer light than would be possible with cones alone. The eyes of most nocturnal mammals have only, or mostly, rods. For an animal that is hunting or hunted at night it is more useful to see a slightly fuzzy image than to be unable to see at all.

TIP

No animal can see using their eyes in complete darkness. Rods enable vision in very low light intensities, not in the dark.

TEST YOURSELF

11 Give one difference in the structure and one difference in the distribution of rod and cone cells.

12 Which are more sensitive to light, rods or cones? Explain your answer.

13 Explain what is meant by visual acuity.

14 Describe how the way that rods and cones are connected to ganglion cells affects visual acuity.

15 How do humans detect colour?

Practice questions

1 Brook lampreys are aquatic animals. Their larvae live buried in mud at the bottom of rivers. Brook lamprey larvae were kept in tanks with different light intensities. They were each observed for 20-minute periods and the time they spent moving around was recorded. The results are shown on the graph as the mean percentage of the observation period when the lampreys were moving.

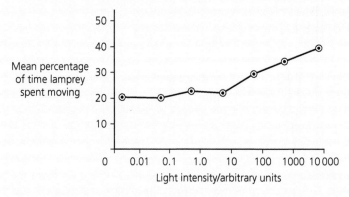

a) **i)** How many minutes did the lamprey larvae spend moving in the brightest light intensity? *(1)*

ii) Identify the scale used on the *x*-axis. *(1)*

iii) Give one advantage of using this kind of scale. *(1)*

b) **i)** Identify the type of response lamprey larvae show to light. *(1)*

ii) Give a reason for your answer to b(i). *(1)*

c) Suggest how this response might be an advantage to the lamprey larvae. *(2)*

2 The diagrams show the results of four different treatments of shoot tips. The tips were placed on blocks of agar and the mean concentration of indoleacetic acid (IAA) that diffused into the agar is shown by the figures (arbitrary units).

agar block — shoot tip

25.8

tips kept dark

A

light

25.6

tips illuminated unilaterally

B

a) What can you conclude from comparing the results of treatments A and B? *(1)*

b) Suggest the purpose of the thin glass cover slip. *(1)*

c) Using the results from treatments C and D, describe the response of the shoot tip to unilateral light. *(2)*

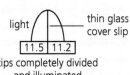

light — thin glass cover slip

11.5 | 11.2

tips completely divided and illuminated unilaterally

C

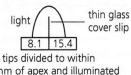

light — thin glass cover slip

8.1 | 15.4

tips divided to within 0.5 mm of apex and illuminated unilaterally

D

d) If the shoot tip from treatment D were on an intact shoot, describe and explain the response you would see. *(3)*

3 Describe the role of the SAN, AVN and Purkyne tissue in coordinating the heartbeat. (6)

4 The number of rods and cones in the field of view of an optical microscope was counted at frequent intervals along a horizontal line across the retina of a human eye. The graph shows the results.

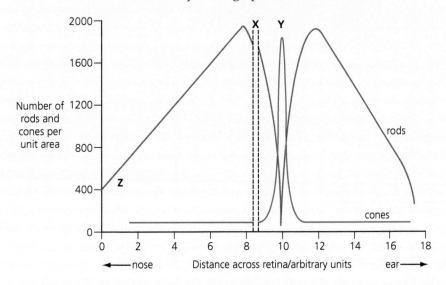

a) Identify the parts of the retina at points X and Y, giving reasons for each answer. (2)

b) Use the graph to find the ratio of rods to cones at 6 arbitrary units across the retina. (1)

c) At which distance across the retina would sensitivity to colour be highest? Explain your answer. (3)

d) i) Describe the sensitivity to light of receptors at Y compared with those at Z. (1)

ii) What is the reason for the difference in sensitivity of the receptors at Y and Z? (2)

e) Describe the connections made in the optic nerve by receptors at Z and explain how this would affect the visual acuity at this point on the retina compared to Y. (3)

Stretch and challenge

5 Evaluate the main theories to explain how the human eye is able to distinguish between so many different colours when cones are sensitive to just three specific wavelengths of light. To what extent are these theories supported by evidence?

6 To what extent are plant growth regulators important in horticulture? You may wish to discuss their use in selective weedkillers, producing seedless fruit and in plant propagation, among other applications.

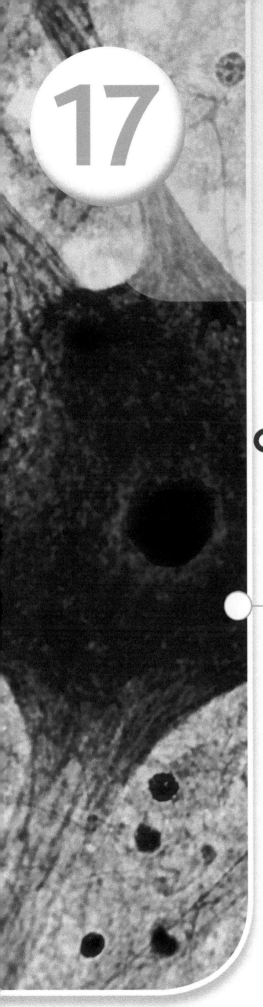

17 Nervous coordination

Introduction

The human brain is estimated to have about 100 billion (10^{11}) neurones. If we gave equal portions of a single brain to every person in England, he or she would get about 2000 cells. Moreover, each neurone has synapses connecting it to as many as 10 000 other neurones. This creates a neural network that makes the average computer seem like a child's toy.

The brain is the organ responsible for coordinating the great majority of activities in the body. It synchronises most of the automatic activities such as heartbeat and breathing, as well as such complex movements as walking or playing the piano. The brain receives and processes a constant flow of information from sensory receptors. This information may stimulate appropriate responses, or be stored or ignored. The brain is also responsible for what we consider to be the higher human activities, such as thinking, emotions, memory and consciousness.

So, how is it possible that tiny pulses of electrical activity and the transfer of chemicals across synapses and membranes can result in the complex mass of thoughts, feelings, actions, memories and emotions of which humans are capable? Neuroscientists are still a long way from full understanding, but slowly they are discovering the functions of different regions of the brain and how they interact. In this chapter we shall study the basic processes necessary to understand how the brain works and how the nervous system coordinates our activities. This will enable us to explain how chemicals such as alcohol and other drugs can disrupt the system.

Neurones

First we need to look at the structure of the nerve cells, or neurones, that form the conducting tissue in the nervous system (Figures 17.1 and 17.2).

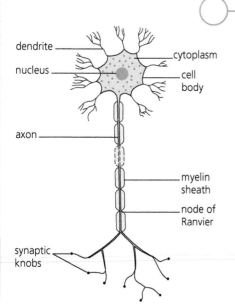

dendrite
nucleus
cytoplasm
cell body
axon
myelin sheath
node of Ranvier
synaptic knobs

Figure 17.1 The structure of a myelinated motor neurone.

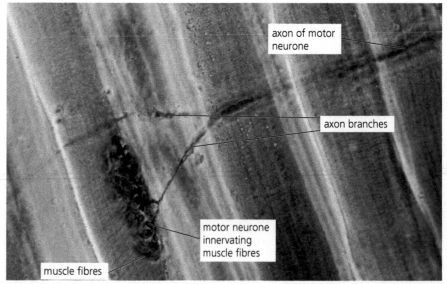

axon of motor neurone
axon branches
motor neurone innervating muscle fibres
muscle fibres

Figure 17.2 Light micrograph of a motor neurone and its effector, a group of muscle cells.

In Chapter 16 you learned about a simple reflex arc that involved three different types of neurone.

- A **sensory neurone** conducts impulses from a receptor to the spinal cord.
- One or more **relay neurones** act as links between the sensory and motor neurones.
- A **motor neurone** conducts impulses from the spinal cord to a muscle.

Look at Figure 17.1, which shows the structure of a motor neurone. You can see that the cell body contains cytoplasm and a nucleus and looks similar in structure to other animal cells. The cytoplasm also contains mitochondria and ribosomes.

In other respects a motor neurone is highly **specialised**. Its cell body is situated inside the spinal cord. It has large numbers of short branches, called **dendrites**. A motor neurone may have over a hundred of these branches, each with one or more synapses to a relay neurone. There is also one very long branch that extends from the spinal cord to a muscle. This axon can be as much as a metre in length and less than a micrometre in diameter.

At the far end of the axon are short branches that terminate in tiny **synaptic knobs**. These knobs are located close to the cell-surface membrane of muscle cells at the neuromuscular junction. Large numbers of neurones are bundled together into **nerves**. The sciatic nerve, for example, originates from the spinal cord in the lower back. Branches from it pass all the way down the leg to muscles in the foot. The sciatic nerve also contains sensory neurones conducting impulses in the opposite direction. Damage to the lower back can compress this nerve where it passes between the vertebrae, causing a painful condition called sciatica.

You can see from Figure 17.3 that the axon is surrounded by the **myelin sheath**. This sheath is not strictly part of the neurone. It is made from highly specialised cells, called **Schwann cells**, that lie alongside the axon. As the axon grows these cells wrap round and round the axon until there may be up to a hundred layers of lipid and protein membranes. This makes a fatty 'bandage' around the axon, shielding it from surrounding tissue fluid and electrically insulating it from other neurones.

Between each Schwann cell is a tiny gap where the axon is exposed, called a **node of Ranvier** (see Figure 17.1). These nodes are the only places that ions can pass between the tissue fluid and the axon through the cell-surface membrane. The nodes play an important part in speeding up the conduction of impulses along the axon, as we shall see later.

Axon The long fibre in a neurone that conducts impulses away from the cell body.

Neuromuscular junction A synapse between a neurone and a muscle cell.

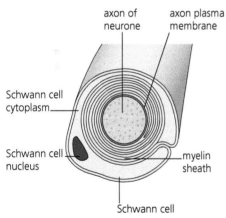

Figure 17.3 A section across an axon and Schwann cell.

Impulses

Impulses are waves of electrical activity passing along a neurone. The process is not the same as the conduction of electricity through a wire. It is much slower, although it is still quite fast. When a neurone is stimulated at a synapse, there is a brief change in the cell-surface membrane of the neurone. This allows ions to pass through rapidly. As soon as this happens it causes the next small section of the membrane to change, so ions can pass through here. Meanwhile the first section of neurone changes back as the distribution of ions is restored. This process carries on down the neurone. It is rather like a long line of standing dominoes that are knocked down one after another once the first one is pushed over, except that in this case the domino is immediately set upright again as soon as the next one falls over.

So, what causes the movement of ions? Before we explain, it might be a good idea to remind yourself of the ways in which ions can pass across a cell-surface membrane. Previously you learned about the ways that ions can pass through **channel proteins** and **carrier proteins** in the membrane. It may be helpful to refer to Chapter 3, page 46.

The resting potential

The cell-surface membranes of neurones contain carrier proteins called **sodium-potassium** pumps. These pumps are special carrier proteins that move sodium ions out of the axon and potassium ions in. The pumps use ATP to actively transport three sodium ions out for every two potassium ions they actively transport in.

There are also proteins in the membrane that allow ions to move back down their concentration gradients by facilitated diffusion. These proteins are called channels and are always open. However, the membrane is more permeable to potassium ions than to sodium ions. This means that the potassium ions diffuse back down their concentration gradient more rapidly than the sodium ions. This is called **differential permeability**.

The result of these two factors is that the outside of the axon membrane always has a slight excess of positive ions. This gives a **potential difference** of about 70 mV (millivolts) between the inside and the outside of the membrane. Because the inside of the axon is more negative and an electric current is a flow of negative electrons from a more negative to a more positive potential this is written as −70 mV. This is called the **resting potential**.

EXAMPLE

Investigating the action potential

Look at Table 17.1, which shows the concentration of sodium and potassium ions inside and outside the axon of a motor neurone.

	Concentration/mmol dm⁻³	
Ion	Inside axon of motor neurone	Outside axon of motor neurone
Sodium (Na⁺)	18.0	145.0
Potassium (K⁺)	135.0	3.0

Table 17.1 The concentration of sodium and potassium inside and outside the axon of a motor neurone.

1 Describe the differences in concentration inside and outside the axon for each of the ions in Table 17.1.
The concentration of sodium ions outside the axon is much greater than their concentration inside, whereas the opposite is true for potassium ions.

2 There are channel proteins in the membrane that allow ions to diffuse through. These channels are always open. From Table 17.1, what would you expect to happen to the concentrations of the sodium and potassium ions inside and outside the axon?
You would expect the concentrations to be equalised by sodium ions diffusing in and potassium ions diffusing out.

3 Sodium and potassium ions continuously diffuse through the open channel proteins in the axon membrane. However, the concentrations are not equalised. Explain why.
The sodium-potassium pumps operate all the time actively transporting sodium ions out of the axon and potassium ions in. For every two potassium ions moved in, three sodium ions are moved out.

The action potential

So far we have only considered an axon that is not conducting an impulse. What happens when an impulse passes along an axon?

You may recall from Chapter 3 that some channel proteins in membranes have **gates** that prevent facilitated diffusion when they are closed. Gated **ion channels** also play an important role in the action potential.

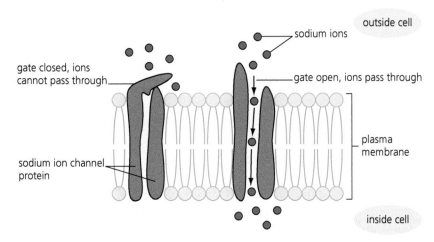

Figure 17.4 Gated sodium ion channels in the cell-surface membrane.

When an axon is inactive, the charges are in an unbalanced state all the time, with an excess of negative charges inside the axon and positive charges outside. As we have explained previously, this is the resting potential. ATP is used to maintain this uneven balance. However, this resting potential means that the axon is poised to conduct an impulse at any time, because of the steep concentration gradient of ions across the membrane.

Stimulation of a neurone at a synapse causes a small change in the potential difference in the membrane close to the synapse. When the change in the potential difference is sufficient (this is called the **threshold potential**), the gates in the sodium ion channels change their shape to 'open' for a brief period. This allows sodium ions to diffuse in. It is estimated that each open channel allows about 20 000 Na+ ions to diffuse through. This causes a sudden increase in positively charged ions inside the neurone. Instead of having a slight excess of negative ions, the inner surface of the membrane now acquires a positive charge of about 30 mV compared with the outside (so is represented as +30 mV). The change in the potential difference across the membrane is called depolarisation.

The sodium ion channels close as soon as the membrane potential reaches about +30 mV. At this point depolarisation is complete and the diffusion of sodium ions stops. Depolarisation causes potassium ion channels in the membrane to open, allowing potassium ions to diffuse out. They, however, diffuse out more slowly than the sodium ions diffuse in. The membrane **repolarises** because potassium ions continue to diffuse out. In practice, rather more potassium ions diffuse out so the drop in potential difference overshoots a bit, as you can see from Figure 17.6. This is called **hyperpolarisation**.

Repolarisation A return to the resting membrane potential.

Repolarisation does not immediately restore the concentration of ions inside the axon to their original state. The sodium-potassium pumps restore the balance by active transport. This maintains the resting potential and keeps the axon ready for another impulse.

TIP

The actual proportion of ions that move in and out during the passage of an impulse is tiny. Don't get the impression that all the sodium ions diffuse into the axon, or all the potassium ions diffuse out.

TEST YOURSELF

1 In the motor neurone in Figure 17.1, in which direction are impulses conducted?
2 What maintains the resting potential in an inactive neurone?
3 Describe the sequence of events during depolarisation.
4 How does repolarisation occur?

ACTIVITY

Analysing an action potential

Figure 17.5 shows the changes in the permeability of the membrane of an axon to sodium and potassium ions as an impulse passes a particular point. The graph also shows the changes in potential difference compared with the resting potential. These results were obtained from the neurone of a squid, which has large, unmyelinated axons where it is easier to take measurements than in a mammalian neurone.

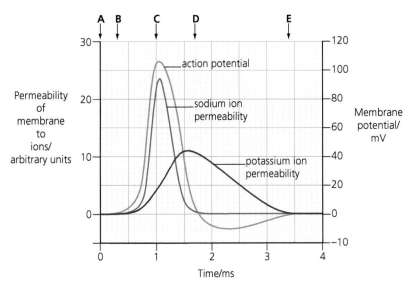

Figure 17.5 Graph of changes in permeability to ions of a tiny part of an axon membrane during an action potential.

1 Use the letters shown in Figure 17.5 to identify each of the following stages: resting potential, depolarisation, repolarisation, hyperpolarisation.
2 Explain how the change in permeability to sodium ions results in an action potential.
3 Why does the increase in sodium ion permeability start later than the change from the resting potential?
4 Explain the effect of the change in permeability to potassium ions.
5 Look at the curves on the graph showing the changes in the permeability of the membrane to sodium and potassium ions. How do these changes compare?
6 Suggest the advantage of differences in the permeability to sodium and potassium ions.
7 The resting potential in this axon was $-70\,mV$. What was the maximum value of the potential inside the axon?
8 For how long was the potential above $-70\,mV$?
9 For how long did hyperpolarisation last?

TIP

Remember that 1 ms is one-thousandth of a second.

How do impulses move along a neurone?

The depolarisation in one small section of the membrane sets off depolarisation of the next section of the neurone because the change in voltage in the membrane stimulates adjacent sodium ion channels to open. Therefore the action potential travels like a wave along the neurone Figures 17.6 and 17.7. This wave is what we commonly call a **nerve impulse**. Once started the impulse travels all the way to the next synapse.

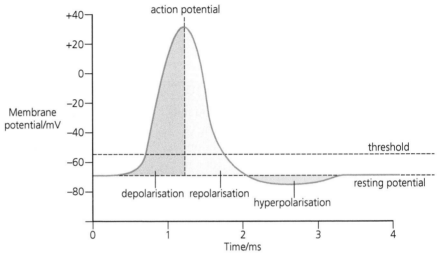

Figure 17.6 Changes in membrane potential as an impulse passes a particular point in a neurone.

The strength of the stimulus does not affect the impulse, as long as the stimulus is strong enough to get above the threshold value. This is known as the **all-or-nothing principle**. It is rather like squeezing the trigger on a rifle. As long as the trigger is pulled back far enough the bullet is fired. Pulling harder will not make the bullet go any further or faster.

However, a stronger stimulus does result in an increased frequency of impulses. This is how nerve impulses carry information. If receptors are stimulated more, they establish a larger generator potential (page 301), which in turn triggers more rapid impulses in the sensory neurone to which they are connected.

(page 301)

There is always a time gap between impulses. This means that impulses are always **discrete**, in other words they never merge together. This is because after the action potential has reached its peak the ionic balance has to be restored during repolarisation. Only after repolarisation can another action potential be generated. The minimum interval between action potentials, and therefore between impulses, is the refractory period. The refractory period means that there is a maximum frequency at which nerve impulses can be conducted along the axon. This means that there is a limit to the strength of a stimulus that can be detected.

When the sodium ion channels close at the peak of the action potential there is a short period of about 0.5 ms when it is impossible for the channels to re-open. Therefore no stimulus can generate an impulse during this period. The full refractory period lasts until repolarisation is complete. In practice most neurones can only conduct about 300 impulses per second.

Refractory period The time following an action potential during which another action potential cannot take place, regardless of the strength of the stimulus.

313

Figure 17.7 A change in charge at an axon membrane during the passage of impulses.

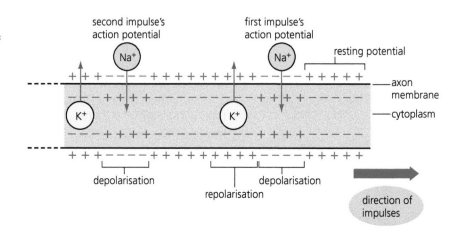

The myelin sheath

You may have noticed that so far we have talked about an axon as though the cell-surface membrane has a continuous and unobstructed outer boundary. But if you look back to Figure 17.1 you will recall that the axon of a motor neurone is myelinated. It is surrounded by a fatty myelin sheath that only has gaps at the nodes of Ranvier. Not all neurones in humans have myelin sheaths. **Non-myelinated neurones** conduct impulses in the way described above.

In myelinated neurones, ions can only pass through the cell-surface membrane at the nodes of Ranvier. Therefore action potentials can only occur at these nodes. The effects of the depolarisation cause almost immediate depolarisation/action potentials at the next node. In effect action potentials jump rapidly from node to node, as shown in Figure 17.8. This is called **saltatory conduction**.

At each node there are large numbers of sodium and potassium channels. Since the nodes are about 1 mm apart, saltatory conduction has the great advantage of being much faster.

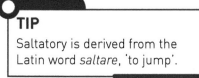

Myelinated neurones Neurones with axons surrounded by a series of Schwann cells.

TIP

Saltatory is derived from the Latin word *saltare*, 'to jump'.

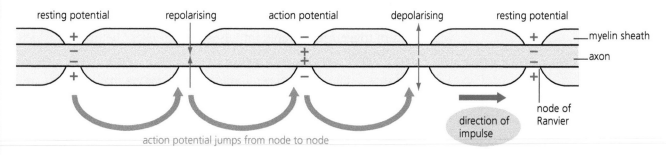

action potential jumps from node to node

Figure 17.8 Saltatory conduction in myelinated neurones. Conduction of impulses along these neurones is approximately 50 times faster than in non-myelinated neurones.

Two other factors affect the rate at which neurones conduct impulses – temperature and the diameter of the axon. The rate of conduction is slower at low temperatures and in narrow axons.

In mammals, temperature rarely makes a significant difference since they normally maintain a fairly stable temperature. In animals whose temperature varies with the environmental temperature, cold conditions can considerably slow reaction times because the facilitated diffusion of sodium and potassium ions during an action potential will be slower.

The diameter of an axon makes a difference because the surface area to volume ratio of an axon is related to its diameter. Axons with a greater diameter have a larger volume of cytoplasm, containing more ions, which reduces their electrical resistance. This means that an action potential in one part of the axon pushes the next section to threshold more quickly. Nerve impulses are therefore conducted faster in larger diameter axons.

Apart from vertebrates, many animal species possess only non-myelinated neurones. These usually have a small diameter and are relatively slow conductors. Some animal species have evolved exceptionally large-diameter axons. Much of the early research on nerve conduction was done using giant axons from squid, marine animals closely related to octopuses. We saw some of the results of these investigations in Figure 17.7. These giant axons have a diameter of nearly 1 mm, over a hundred times greater than other axons in the squid. They can conduct impulses about 10 times as fast as normal motor neurones.

The giant axons extend the full length of the main part of the body, with branches to the muscles along the way. When the squid is startled, impulses are conducted at about 35 m per second along the giant axons. The muscles contract almost instantly, providing the sudden force that squirts water out backwards and jet-propels the squid away from danger. Earthworms also have large-diameter axons running the length of their body. These allow them to contract muscles rapidly and withdraw quickly into a burrow when threatened by a bird.

> **TIP**
> You do not need to recall details of the giant axons in squid and earthworms.

> **TEST YOURSELF**
> 5 Explain the importance of the refractory period during impulse conduction.
> 6 Describe the all-or-nothing principle.
> 7 Give **three** factors that affect the speed of impulse conduction.
> 8 Explain what is meant by saltatory conduction.

Synapses

> **TIP**
> Remember that 1 μm is 1000 nm.

Synapses are the junctions between neurones. At a synapse, there is a small gap called the **synaptic cleft**, which is usually about 20 nm wide. This gap prevents electrical impulses passing directly from one neurone to another. Communication between neurones is by chemical **neurotransmitters** that diffuse across the cleft from one neurone to another. This is a slower process than the passage of an impulse along a neurone. Figure 17.9 shows a synaptic cleft.

> **TIPS**
> ☒ Note that you should be careful not to refer to impulses or action potentials crossing synapses. It is information that crosses a synapse.
> ☒ Communication between neurones is by chemical neurotransmitter.

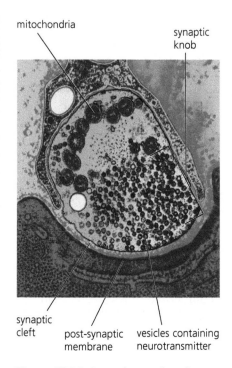

mitochondria
synaptic knob
synaptic cleft
post-synaptic membrane
vesicles containing neurotransmitter

Figure 17.9 Coloured scanning electron micrograph of a synaptic cleft. The synaptic knob containing vesicles of neurotransmitter is clearly visible.

A synapse is a gap between the tiny branching end of a neurone and either a dendrite or the cell body of the next neurone. The branching end is always slightly swollen into a **synaptic knob** (Figure 17.10). A motor neurone may have approximately 8000 synapses on its dendrites and another 2000 directly on the cell body. It has been estimated that neurones in the brain have an average of approximately 40 000 synapses and that some have as many as 200 000. The number of possible pathways between neurones is phenomenal, which helps to explain the amazing complexity of the brain (see the introduction to this chapter).

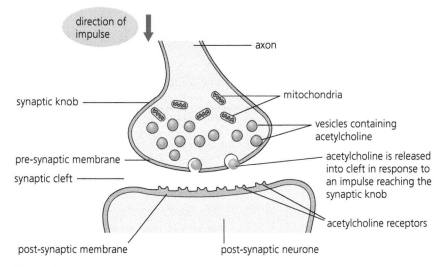

direction of impulse

axon

synaptic knob

mitochondria

vesicles containing acetylcholine

pre-synaptic membrane

synaptic cleft

acetylcholine is released into cleft in response to an impulse reaching the synaptic knob

acetylcholine receptors

post-synaptic membrane

post-synaptic neurone

Figure 17.10 Structure of a synapse.

Cholinergic synapse A synapse that uses acetylcholine as a neurotransmitter.

So, what happens when an impulse reaches a synapse at the end of an axon? We will describe here what happens in a cholinergic synapse, which uses **acetylcholine** as a neurotransmitter. Acetylcholine (ACh) is the main neurotransmitter in synapses in the nerves outside the central nervous system, as well as in some parts of the brain. It is also used at the junctions between motor neurones and muscles. These **neuromuscular junctions** work in much the same way as synapses, as we shall see in Chapter 18. Many other neurotransmitters are used in the nervous system, but the principles of synaptic transmission are similar in all cases.

When an action potential arrives at the **pre-synaptic membrane of an axon** at the synaptic knob it causes calcium ion channels to open. The concentration of calcium ions (Ca^{2+}) in the fluid of the synaptic cleft is higher than the concentration inside the synaptic knob. Therefore calcium ions rapidly move by facilitated diffusion into the knob.

As you can see from Figure 17.10, inside the synaptic knob there are many vesicles containing acetylcholine. These vesicles are tiny droplets of acetylcholine surrounded by a membrane. When calcium ions enter, some of these vesicles move to the pre-synaptic membrane and fuse with it. As a result, the acetylcholine is secreted out into the cleft and diffuses across a very short distance to the **post-synaptic membrane** on the other side. This post-synaptic membrane has receptor proteins on its surface to which the acetylcholine molecules attach. Since the diffusion distance is so short, the acetylcholine diffuses quickly across the synaptic cleft (see Chapter 3).

TIP

Don't confuse receptor molecules with receptor cells, such as the rods and cones that we studied in Chapter 16.

TIP

Don't refer to the binding site of a receptor protein as an active site. Although it has a complementary shape to the neurotransmitter, it is not an enzyme.

The receptor proteins have **binding sites** that are complementary in shape to acetylcholine molecules. Each receptor protein is located next to a sodium channel. When acetylcholine binds to a receptor protein, it changes shape and pushes the neighbouring sodium channel open. You will recall that the excess of sodium ions entering the neurone depolarises the membrane. If enough sodium ions enter, and the threshold is exceeded, an action potential is triggered (page 311). Once an action potential does start it travels all the way along the neurone as an impulse until it reaches the synapses at the other end. Figure 17.11 illustrates the sequence of the events that occur at a synapse.

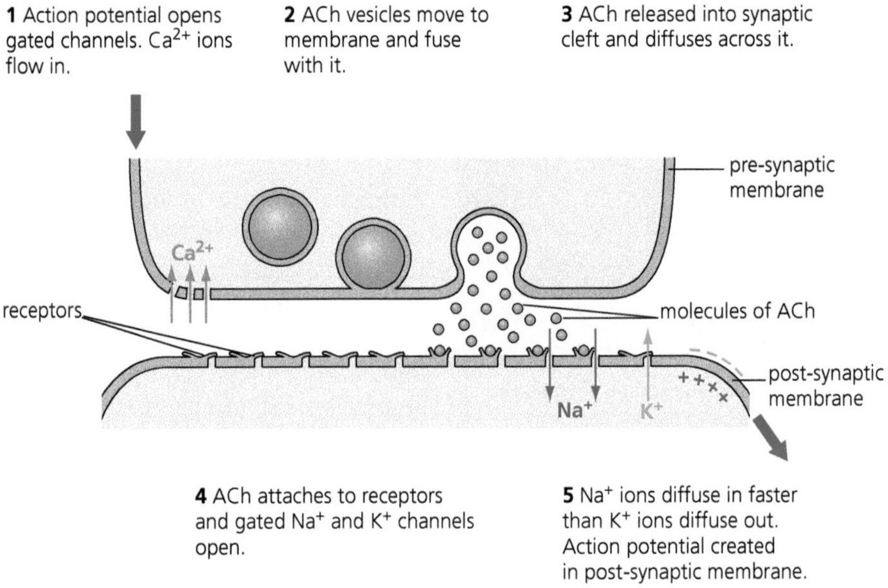

Figure 17.11 Sequence of events during an impulse transmission at a cholinergic synapse.

1 Action potential opens gated channels. Ca^{2+} ions flow in.

2 ACh vesicles move to membrane and fuse with it.

3 ACh released into synaptic cleft and diffuses across it.

pre-synaptic membrane

receptors

molecules of ACh

post-synaptic membrane

4 ACh attaches to receptors and gated Na^+ and K^+ channels open.

5 Na^+ ions diffuse in faster than K^+ ions diffuse out. Action potential created in post-synaptic membrane.

You will no doubt have realised that during transmission across a synapse changes take place that must be reversed if it is to keep on transmitting. Calcium ions are **actively transported back into the synaptic cleft**, ensuring that the concentration outside the membrane is always higher. The acetylcholine is removed from the receptors by an enzyme called **acetylcholinesterase** that hydrolyses it to acetate and choline. These are actively transported back into the synaptic knob where they are synthesised into acetylcholine once again. Vesicles are refilled, so the acetylcholine is recycled continuously.

The role of synapses

It takes much longer to describe what happens at a synapse than for the transmission to occur. But synapses do slow down the rate at which impulses pass through the nervous system. There is a delay of about 0.5 ms at each synapse. This is because the rate of diffusion is much slower than the rate at which a nerve impulse is propagated along a neurone, especially by saltatory conduction.

This does not mean that synapses are a bad thing. Synapses are vitally important as the points in the nervous system where the passage of impulses is controlled. Remember the all-or-nothing principle. If impulses could cross straight from one neurone to another without any control, stimulation at any point in the system would spread to all neurones.

Specialisation of functions in different parts would be impossible. Some of the control mechanisms involved are complex and there is still much to learn about them, especially about the processes involved in the brain.

One simple control method is that transmission across a synapse can only be in one direction. Only the synaptic knob contains the vesicles of neurotransmitter. Therefore neurotransmitter can only be released into the cleft from the pre-synaptic membrane. Once an action potential is established in the post-synaptic neurone it can only travel as an impulse in one direction because depolarisation always occurs in front of the action potential. This is called **unidirectionality**.

Summation

Impulses arriving at a synapse do not always result in impulses being generated in the next neurone. The electron micrograph in Figure 17.12 shows the synapses on the cell body of a motor neurone. As you can see there are many synaptic knobs all linking to this motor neurone.

A single impulse arriving at one synaptic knob is not likely to lead to the generation of an action potential in the post-synaptic neurone. It may release only one or a small number of vesicles of acetylcholine. If, as a result, only a few of the gated ion channels in the post-synaptic membrane are opened, not enough sodium ions will pass through the membrane to depolarise it and reach the threshold value. The acetylcholine only remains attached to the receptors for a very brief time since the acetylcholinesterase breaks it down within a couple of milliseconds. If several impulses reach the synaptic knob in quick succession, enough acetylcholine is released for the membrane potential to reach the threshold value. This causes the sodium ion channels to open and produces an action potential. Because the effects of several impulses are added together in a short time this process is called **temporal** (meaning 'time') **summation** (Figure 17.13).

A second way in which an action potential can be generated is when several impulses arrive simultaneously at different synaptic knobs stimulating the same cell body. The cell body may then be depolarised enough for an action potential to be generated in the axon. This is called **spatial summation**.

One advantage of summation is that the effect of a stimulus can be magnified. We have, for example, already come across summation in the retina (see page 304). It is also a way for different stimuli, or a combination of different stimuli, to trigger a response. Another effect of summation is that it avoids the system being swamped by impulses. A synapse acts as a barrier that only allows impulses to pass on if there is a significant input from receptors or other neurones, which may relate to the strength of the stimulus.

Inhibition

So far we have only described situations in which transmission across a synapse results in an action potential being generated in the post-synaptic neurone. Just as important are synapses where impulses are stopped altogether. Some neurotransmitters prevent action potentials being generated in the post-synaptic neurone. This is called **inhibition**. One way in which this occurs is that the inhibitory neurotransmitter

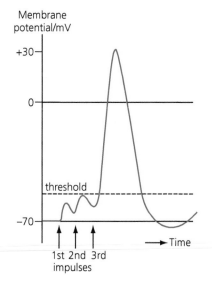

Figure 17.12 An electron micrograph showing a mass of synaptic knobs spread over the surface of the cell body of a neurone.

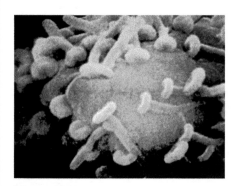

Figure 17.13 Temporal summation. You can see from the graph that the first impulse reaching the synapse fails to raise the membrane potential above the threshold for an action potential. The three impulses arriving in quick succession do produce an action potential.

stimulates gated potassium ion channels to open. Potassium ions therefore diffuse out of the cell body. If a cell body is affected by both excitatory and inhibitory synapses, the effect of sodium ions entering following stimulation by an excitatory synapse may be cancelled out by potassium ions leaving following stimulation by an inhibitory synapse. Therefore the membrane potential does not reach the threshold for an action potential to be produced.

Inhibition is very important in nervous circuits. It enables specific pathways to be stimulated, while preventing random impulses all over the body. For example in reflex actions it is desirable for neural pathways to produce a response only where it is needed. If your hand is on a hotplate you need your arm muscles to pull it away, but not your leg muscles to contract. The development of inhibitory pathways can be very important in learning specific skills, such as drawing and writing. If you watch a young child learning to draw you will see that at first the movements are uncontrolled. As the skill develops the movements become much more coordinated as the appropriate pathways are refined by inhibitory circuits.

Look at Figure 17.14. This shows in simplified form the cell body of a neurone and an axon with two synapses on it. Both are excitatory synapses.

If an action potential arrives at synapse A the membrane potential becomes more positive but not enough to reach the threshold value for an action potential. If an action potential arrives at synapse B at more or less the same time, the two effects are added together and the combined effect is now above the threshold value for an action potential to be generated in the axon.

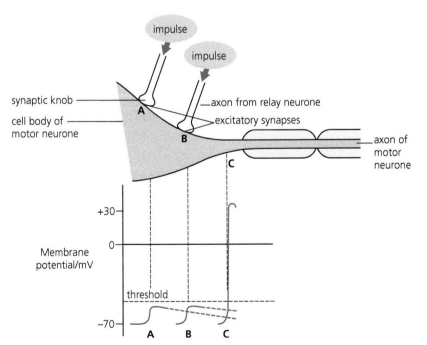

Figure 17.14 The cell body and an axon of a neurone with two excitatory synapses on it and a graph of the membrane potential that results.

Figure 17.15 shows a similar arrangement, but in this case synapse B is an inhibitory synapse. At an inhibitory synapse the effect is that the membrane potential actually decreases. In this example the two effects balance out the EPSP so the threshold value for an action potential is not reached.

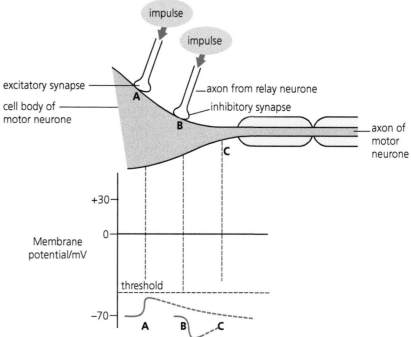

Figure 17.15 The cell body and an axon of a neurone with one inhibitory synapse and one excitatory synapse on it. The threshold value for an action potential is not reached.

TEST YOURSELF

9 Explain why synaptic transmission can only work in one direction.

10 How would inhibition of acetylcholinesterase affect synaptic transmission?

11 Explain the difference between temporal summation and spatial summation.

12 What does an inhibitory neurotransmitter do?

TIP

You do not need to recall the names or modes of action of any specific drugs on synpases. The curare example illustrates the sort of information you might be asked to explain.

The effects of drugs

We mentioned at the beginning of this chapter that chemicals may have major effects on the working of the brain and other parts of the nervous system. Many of these substances are chemically quite simple. We often refer to these substances as **drugs**.

The term drugs, however, is quite vague and is applied both to substances that are used as cures and to mind-altering substances that may be damaging or potential poisons. Studies of drugs and poisons have revealed that many have quite specific effects on transmission at synapses or on neural circuits within the brain. In some cases, research has helped scientists to understand more about how the nervous system works and to offer ways of using certain drugs to treat nervous disorders.

An example of a poison that affects synaptic transmission is curare. Curare is a naturally occurring poison that is found in the bark of trees in the Amazon forest. It has been used for centuries by South American Indians to coat the tips of arrows used for hunting. Curare molecules attach to the acetylcholine receptors on post-synaptic membranes, especially the membranes of neuromuscular junctions at the end of motor neurones. The curare molecules block the receptors and prevent sodium channels opening. Therefore impulses that pass down the motor neurones fail to stimulate muscle contraction, so any animals shot with such an arrow are paralysed.

EXAMPLE

The effect of drugs on synapses

Read the following passage carefully.

The brain has many different neurotransmitters. Over 50 have been identified. Many work in a similar way to acetylcholine. Many occur only in certain parts of the brain and are therefore restricted to particular functions. Some are inhibitory and some excitatory. Each binds to receptor molecules that are specific to the neurotransmitter. For example dopamine is a neurotransmitter that is important in parts of the brain concerned with control of muscle action. Parkinson's disease is a fairly common condition in elderly people in whom the neurones that produce the dopamine degenerate. The muscles become too tense, making delicate muscular movements difficult so that, for example, it is hard to pour a cup of tea without trembling. As the disease develops, walking and other everyday activities become harder. Two types of drug can be used in treatment. One is a substance from which dopamine is synthesised in the neurones. Another is an agonist, something that mimics the effect of dopamine by binding to the same receptors.

Another condition in which research suggests that dopamine plays a part is schizophrenia. This is a condition that affects about 1% of people at some time in their lives but most frequently as young adults. Schizophrenia causes people to behave untypically and often to suffer from delusions and hallucinations.

It appears to have a variety of causes but one feature that has been discovered is an excess of dopamine receptors in the brain.

It is not easy to develop drugs that have specific effects in the brain because many substances are blocked from entering the brain by the blood–brain barrier. However, a number of drugs have been produced that help in schizophrenia by blocking the dopamine receptors. Dosage is very important and there is, as with many drugs, a danger of side effects.

1 Two types of drug can be used to treat Parkinson's disease. Suggest how each type of drug mentioned in the passage helps to treat Parkinson's disease.
Providing the brain with more of the precursor for dopamine helps the neurones that have not degenerated to manufacture more dopamine. The agonist attaches to the same receptors and has the same effect as dopamine.

2 Use information in the passage to suggest:
a) one type of drug that might help to reduce the symptoms of schizophrenia
Drugs that block the dopamine receptors should help.
b) one side effect that could result from the use of this drug.
A possible side effect is loss of muscle control, as found in Parkinson's disease.

Practice questions

1 The graph shows some changes in the permeability of an axon membrane during an action potential.

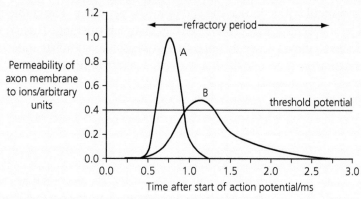

Time after start of action potential/ms

a) Name

   i) ion A

   ii) ion B. *(1)*

b) Explain the changes in permeability to ion A that take place between 0.5 and 1.0 ms. *(2)*

c) Explain how changes in permeability to ion B contribute to repolarisation of the axon membrane. *(2)*

d) Use the graph to explain what is meant by

   i) the refractory period

   ii) the all-or-nothing principle. *(2)*

2 The drawing shows a cholinergic synapse.

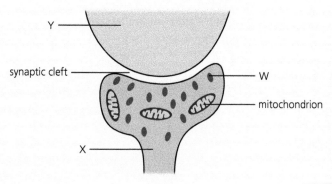

a) Name

   i) organelle W

   ii) the neurotransmitter used in this synapse. *(1)*

b) In which direction does synaptic transmission take place, X–Y or Y–X? Give a reason for your answer. *(1)*

c) Describe the sequence of events that occur at this synapse during synaptic transmission. *(4)*

d) Atropine is a chemical substance found in the deadly nightshade plant (*Atropa belladona*). It binds to and blocks acetylcholine receptors. Describe the effect of atropine on the sequence of events at a cholinergic synapse. (2)

3 Explain how a resting potential is established and maintained in a neurone. (5)

4 The diagram shows three different neurones synapsing with a motor neurone.

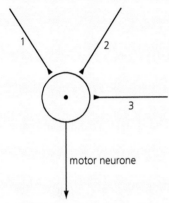

motor neurone

a) The motor neurone is myelinated. Describe and explain how this affects the conduction of action potentials. (3)

b) If an action potential arrives at synapse 1, no action potential occurs in the motor neurone. If action potentials arrive together at synapses 1 and 2, an action potential occurs in the motor neurone. Explain why. (3)

c) Synapse 3 releases an inhibitory neurotransmitter. Suggest what would happen if action potentials arrived at all three synapses at the same time. (2)

Stretch and challenge

5 Botox is a widely available but controversial cosmetic treatment. Discuss its mode of action and evaluate its use as a cosmetic treatment.

6 Examine the role of myelin in nerve tissue, by discussing the symptoms of multiple sclerosis. To what extent may it be regarded as an autoimmune condition?

18

Muscles and movement

PRIOR KNOWLEDGE

● Skeletal muscles work in antagonistic pairs.
● Muscles and skeleton interact to produce movement.
● Muscles are examples of effectors, and are stimulated by nerve impulses from motor neurones.
● An action potential results from depolarisation of the cell-surface membrane of a neurone.
● Action potentials spread across the cell-surface membrane of a neurone.
● Cholinergic synapses are those that use acetylcholine as a neurotransmitter.
● ATP hydrolase hydrolyses ATP to ADP and inorganic phosphate.

TEST YOURSELF ON PRIOR KNOWLEDGE

1 What is an antagonistic pair of muscles?
2 Explain why effective movement requires both muscle and an incompressible skeleton.
3 Describe the structure of a myelinated motor neurone.

Have you ever wondered why the meat of chicken breast is almost white and seemingly more or less bloodless, whereas the meat on a chicken leg is much darker and looks more like meat from mammals? After all, both are muscle. The obvious suggestion might be that farmed chickens can't fly, so they don't use their breast muscles for flying as a wild bird does. This, however, is not the complete answer. Chickens have been bred from wild junglefowl that live in the forests of South-east Asia. These birds spend most of their time scratching around for food on the ground, but when danger threatens they quickly fly to safety in a tree. Unlike most wild birds they never fly long distances. The breast muscles that flap their wings are only used in short bursts but have to produce enough force to lift quite a heavy bird off the ground at speed. On the other hand their leg muscles are used more or less all the time for wandering about the forest floor, scratching away at the leaf litter to find food and stopping them falling off their perch.

The junglefowl is adapted to its way of life by having different proportions of two types of muscle fibre in its muscles. The wing muscles mainly consist of fast fibres. These fibres contract rapidly but quickly become tired. They are well suited to a fast and short escape flight into a tree, lasting just a few seconds, but would not be suitable for sustained flight over a long distance. The muscle looks pale for two reasons. It has fewer capillaries containing blood with red blood cells than red muscle and therefore obtains its supplies of oxygen and glucose quite slowly. It also has very little of the red pigment myoglobin. Myoglobin is similar to haemoglobin and acts as a store of oxygen in muscles. Myoglobin also speeds up the absorption of oxygen from the capillaries.

The junglefowl's leg muscles contain mostly slow fibres. These have larger amounts of myoglobin and more mitochondria than the fast fibres. The muscles also have a denser network of capillaries. The combination of larger quantities of haemoglobin and myoglobin make the muscles appear a much darker red. Slow fibres contract more slowly than fast fibres but they do not fatigue anything like as quickly. They are therefore well suited for the walking, scratching and perching activities required of the legs.

It is not just junglefowl that have both fast and slow fibres. The skeletal muscles of vertebrates in general include a mixture of both muscle types. The proportions in different muscles vary according to the activity for which the muscles are adapted. For example, cheetahs and wolves pursue their prey in quite different ways (Figure 18.1).

Figure 18.1 a) Cheetahs depend on taking their prey by surprise with bursts of very high speed over short distances. They have a high proportion of fast fibres in their leg muscles. These muscles fatigue very quickly, so if a cheetah does not catch its prey within a few metres it has to give up.

b) Wolves work as a pack and may follow their prey, which is often much larger than themselves, over long distances before singling out a weak member of a herd. Their leg muscles are adapted for endurance by having a high proportion of slow fibres.

In this chapter we will find out how muscles contract and how they use the ATP that is needed for contraction. We will also see how their activity is coordinated by the nervous system.

Skeletal muscles

The human body has over 600 skeletal muscles. About 40% of the mass of an average man is skeletal muscle. Humans have both fast and slow fibres in their muscles. Some of our muscles have particularly high proportions of one type. For example the muscles that move our eyes and eyelids have mostly fast fibres. These operate rapidly but do not need to keep up their movements for long periods. The muscles in our back have a large proportion of slow fibres. They are required to remain contracted for long periods so that we can stand or sit upright but they rarely have to move quickly. Most people have roughly equal proportions of each type of fibre in their main leg and arm muscles, although some individuals have distinctly more of one sort. Studies of highly trained sportspeople and athletes have shown that the most successful often have a bias that benefits their type of activity. The table shows some results from one study of trained athletes.

Table 18.1 Results of a study of muscle fibre type in the leg/arm muscles of athletes trained in different disciplines.

Sport/discipline	Mean percentage of slow fibres in leg or arm muscles	Mean maximum oxygen consumption/cm³ per kg body mass per minute
Cross-country skiing	79	82
Marathon running	65	79
Swimming	56	63
Fit students	53	55
Weightlifting	51	48
Running 800 m	48	65
Sprinting 100–200 m	42	55

You can see from Table 18.1 that athletes in endurance events, such as cross-country skiing and marathon running, have high percentages of slow fibres. These enable them to maintain activity for long periods without fatigue. These muscle fibres are well supplied with blood capillaries that provide a good oxygen supply. In contrast, athletes trained for short bursts of activity, such as weightlifting and sprinting, have lower proportions of slow fibres, which, of course, means that they have more fast fibres. These contract rapidly and can carry out intense activity for a few seconds. However, the rate at which the muscles are re-supplied with glucose and oxygen is low, because there are relatively few capillaries, so it is only possible to keep up this rate of muscle contraction for a short time. The average running speed of an Olympic 100 m sprinter is almost double that of a marathon runner.

An obvious question is whether athletes are born or bred. There is clear evidence that some people are born with unusually high proportions of one type of muscle fibre and that this tendency is inherited. However, training can make one type of fibre develop much more than the other. For example, weightlifting routines can increase the size of fast fibres. The number of fibres in a muscle does not increase but the number and length of the contractile units inside the fibres can increase considerably. Training can also increase the numbers of mitochondria and capillaries in the muscles. So, although a person may be born with muscles that are more suited for sprinting or for long-distance running, appropriate training can make a big difference to their chances of success (Figure 18.2).

Figure 18.2 Inheritance of desired muscle fibre types and training methods can make an athlete more successful at their chosen discipline. a) For a short race at maximum speed the sprinter benefits from a high proportion of fast fibres in his leg muscles. b) It is an advantage for a long-distance runner to have a high proportion of slow muscle fibres that fatigue less easily.

Muscle mechanics

The muscular system is very complex: many different muscles cross each other in many directions. Figure 18.3 shows some of the muscles on the front (anterior) of the body.

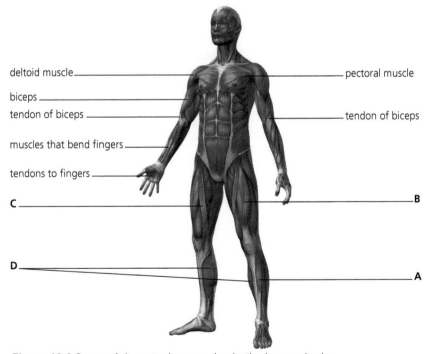

Figure 18.3 Some of the anterior muscles in the human body.

You will probably be familiar with the biceps and the triceps muscles in the arm from your GCSE course. These are used to flex and extend the arm at the elbow, respectively. But as you can see in Figure 18.3 there are several other muscles that enable us to raise and twist our arms as well as making delicate movements with our hands. The biceps and triceps are attached to

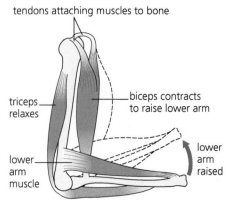

tendons attaching muscles to bone

triceps relaxes

biceps contracts to raise lower arm

lower arm muscle

lower arm raised

Figure 18.4 The triceps and biceps in the arm work antagonistically; here the lower arm is being raised.

bones by thick tendons made of strong fibrous connective tissue. A tendon has to be flexible, but not elastic. It must not stretch when a muscle is contracting and pulling on a bone. You can see in Figure 18.4 the tendon that attaches the lower end of the biceps to the radius bone in the lower arm. The upper end of the biceps has two tendons attaching it to the shoulder blade. These cannot be seen in Figure 18.3 as the deltoid muscle hides them.

Some muscles have very long tendons. The muscles that bend the fingers are situated in the arms just below the elbows. Their tendons stretch across the wrists and the palms of the hands. There are also muscles that attach directly to a bone instead of having tendons. For example, if you look at the pectoral muscle in Figure 18.3 you will see that it is attached all the way down the breastbone. The muscles generate force as they contract and pull on the bones of the skeleton, and this produces movement.

Muscles can only contract to move parts of the body. They can only *pull*; they cannot *push*. For this reason muscles in general operate as pairs, with one pulling in one direction at a joint and a partner that pulls the opposite way. This is called **antagonistic muscle action**. The biceps and triceps act as an antagonistic pair of muscles (Figure 18.4). When the biceps contracts the triceps relaxes so the lower arm is pulled upwards. When the biceps contracts it also pulls the triceps back out to its starting length so that it is ready to contract again. To straighten the arm again the triceps contracts and the biceps relaxes. When the biceps or triceps contracts, it pulls against the bones in the lower arm and because these bones can't be stretched, the arm is flexed around the elbow joint.

Extension

In practice, when you lift a heavy object in your hand, many muscles are involved. For example, another muscle connects from the lower end of the bone in the upper arm to the lower arm near the wrist. Several muscles are used to enable the hand to grip and the wrist to rotate. Others help to keep the shoulder steady. As the biceps contracts, the triceps slowly relaxes, so the movement takes place smoothly. This involves the brain, and in particular the cerebellum, in a highly complex process of coordination. Receptors in the muscles, tendons and joints constantly send impulses to the brain indicating the degree of stretching of muscles and the positions of tendons and joints. Reflex adjustments ensure that the operation is a continuously smooth process. Damage to the cerebellum, for example a stroke or tumour, can result in clumsy and jerky movements.

TEST YOURSELF

1 Which of the other muscles in the leg labelled B to D is antagonistic to muscle A in Figure 18.3? Explain your answer.
2 What are tendons and what is their function?
3 What sort of muscle fibres are mainly found in the eye muscles? What is the advantage of having this type of fibre here?
4 Describe and explain the relationship between mean percentage of slow fibres and mean maximum oxygen concentration in Table 18.1.
5 Why might people with a higher proportion of slow fibres in their muscles be more suited to long-distance running?

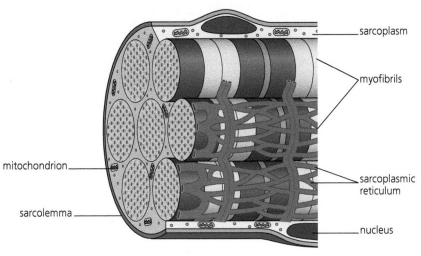

The structure of a muscle

The 'cells' that make up skeletal muscles are so different from most cells that they are referred to as **muscle fibres**. Each fibre is long, on average about 30 mm, but in some cases they can be as much as 300 mm. Part of a fibre is shown in Figure 18.5. You can see that it is surrounded by a cell-surface membrane, called the sarcolemma. Beneath the sarcolemma is cytoplasm, containing many mitochondria. There is also a specialised network of tubules, the sarcoplasmic reticulum, that store calcium ions. We shall see later how the sarcoplasmic reticulum is important in muscle contraction. Dotted within the sarcoplasm along the length of the fibre are many nuclei. Most of the fibre is filled with much thinner fibres, called **myofibrils** (Figure 18.5).

TIP

The names of many parts of muscles begin with the prefix 'sarco'. Examples include sarcolemma, sarcoplasm and sarcoplasmic reticulum.

Figure 18.5 Section through a skeletal muscle fibre.

The muscle fibres are packed together in bundles surrounded by a sheath of connective tissue. It is these bundles that can get stuck in your teeth when eating tough meat. Each bundle is well supplied with blood capillaries and branches of motor neurones. In larger muscles, such as the biceps, many bundles are wrapped together in a thick and tough connective tissue layer. This connective tissue is continuous with the tendon that attaches it to the bone (Figure 18.6).

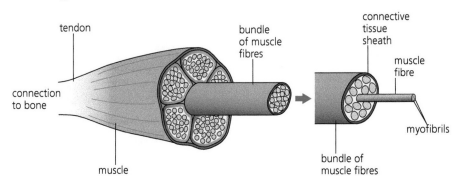

Figure 18.6 The arrangement of skeletal muscle fibres in a muscle.

Myofibrils

With an optical microscope very little detail of muscle fibres can be seen. Dark bands are just visible, as you can see in Figure 18.7a, and for this reason skeletal muscle is often called striated muscle. An electron microscope enables the structure of the myofibrils inside a muscle fibre to be seen. Figure 18.7b shows an electron micrograph of part of a muscle fibre showing short sections of five myofibrils.

a)   **b)**

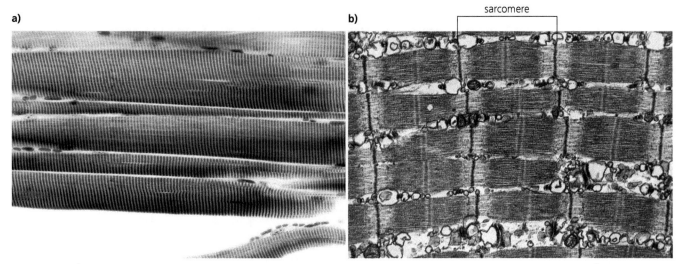

Figure 18.7 a) Skeletal muscle fibres viewed through an optical microscope, magnification approximately ×600; b) myofibrils in a skeletal muscle fibre viewed with an electron microscope with magnification approximately ×10 000.

> **TIP**
> Electron micrographs are black and white images. Figure 18.7b is a false-colour electron micrograph, where colour has been added using image-processing software.

Sarcomere The basic unit of contraction in a myofibril. The sarcomere is defined by the distance between two Z lines.

> **TIP**
> The decreasing size of these structures follows their alphabetical sequence: fibre, fibril, filament.

Look carefully at Figure 18.7b. You will see that each myofibril has dark bands across it. The section between a pair of dark bands is called a sarcomere. In the centre of each sarcomere is a paler (blue) band, and to either side of this are broad darkish (green) areas. You will notice that there appear to be many thin lines running along the length of the sarcomere. The explanation for this appearance is shown in Figure 18.8.

Each sarcomere contains thin filaments of **actin** and thick filaments of **myosin**. The different bands and zones are known by the letters shown in Figure 18.8.

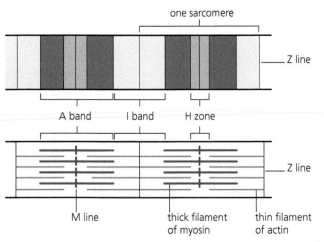

Figure 18.8 The arrangement of actin and myosin filaments in two sarcomeres.

You can see that the thin actin filaments are attached to the dark Z lines at the ends of each sarcomere. The thick myosin filaments are attached at the centre of the sarcomere to the M line and the tails of myosin molecules facing in opposite directions are bound to each other. Where there are myosin filaments the sarcomere looks darker than at the ends where there are only actin filaments. The darkest zones are where the actin and myosin filaments overlap. If you were to cut across a fibril at different points, you would see the mixtures of thin and thick filaments shown in Figure 18.9.

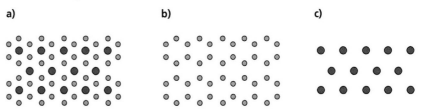

Figure 18.9 Actin and myosin filaments in section at different points along a sarcomere: a) where actin and myosin overlap within the A band, b) in the I band and c) in the H zone.

How do the muscle fibres contract?

Not until the development of electron microscopes was there any clear idea of how muscle contraction works. Then two scientists at London University, called Jean Hanson and Hugh Huxley, came up with the sliding filament hypothesis. They suggested that when a muscle fibre contracts the thin actin filaments slide between the myosin filaments. This is shown in Figure 18.10. They obtained evidence for their hypothesis by examining electron micrographs of the light and dark bands in contracted and relaxed myofibrils. If the actin filaments are sliding between the myosin filaments you would expect the light I bands to get smaller as the sarcomere contracts. This is exactly what Hanson and Huxley found. One feature that can be seen by electron microscopes is that there are 'cross-bridges' connecting the myosin and actin filaments. These can be seen in the scanning electron micrograph in Figure 18.11.

TIP

During contraction, neither the actin nor the myosin filaments get shorter. Instead, the actin slides between the myosin and the region of overlap becomes larger.

Relaxed

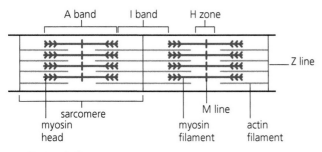

Figure 18.10 Uncontracted (relaxed) and contracted sarcomeres.

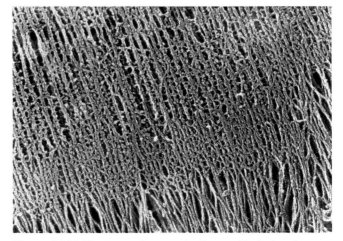

Figure 18.11 Electron micrograph of a myofibril showing the cross-bridges between myosin and actin filaments.

In simple terms the bridges are like tiny hands extending from the myosin towards the actin. These 'hands' can attach to the actin and then bend over. As a hand bends the actin is pulled along for a short distance before the

331

hand lets go and straightens up again. It is then able to attach to the actin filament a bit further along and give it another pull. As you can see from the diagram in Figure 18.12, these 'hands' are referred to as myosin heads. The summary below describes the process in rather more scientific terms.

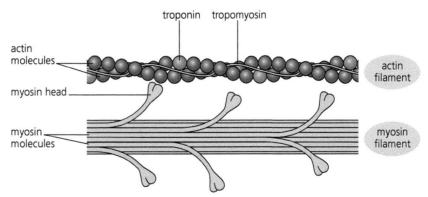

Figure 18.12 Actin and myosin filaments are quaternary proteins, each consisting of many individual actin and myosin molecules.

Actin and myosin filaments are quaternary proteins (see Chapter 1, page 13) which means that they each consist of many individual actin and myosin molecules held together (Figure 18.12). This is why the myosin filaments shown in Figure 18.10 have many heads and why actin filaments have many binding sites. You can see that the heads at each end of the myosin filaments in Figure 18.10 point in opposite directions. Figure 18.12 only shows one end of a myosin filament close up. When a myofibril is in a relaxed state the myosin heads protruding from the myosin filaments are detached from the actin filaments, although still very close, as shown in Figure 18.12. As well as molecules of actin the actin filaments have two other proteins, **tropomyosin** and **troponin**. In the relaxed state the tropomyosin covers the sites on the actin filaments to which the myosin heads can attach.

When nerve impulses reach a neuromuscular junction, the following steps occur, resulting in contraction of the muscle. These are illustrated in Figure 18.13.

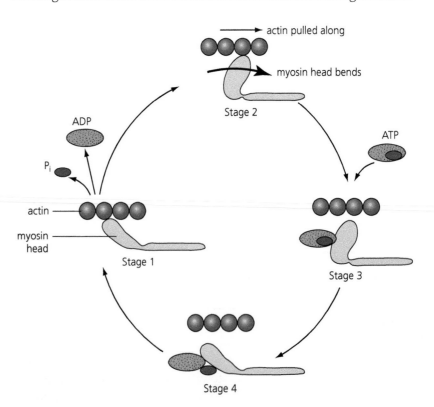

Figure 18.13 Stages in the cycle of actinomyosin bridge formation.

Stages in the cycle of actinomyosin bridge formation.

- A wave of depolarisation travels across the muscle cell-surface membrane, or sarcolemma, and calcium ion channels open.
- Calcium ions (Ca^{2+}) diffuse from the sarcoplasmic reticulum and also diffuse into the muscle cell across the sarcolemma.
- Calcium ions attach to the troponin and cause a change of shape, making the tropomyosin twist away from myosin-binding sites on the actin.
- The myosin heads attach to the binding sites forming **actinomyosin bridges**.
- Actinomyosin bridge formation causes the myosin molecules to spontaneously bend, releasing ADP and inorganic phosphate (P_i) and pulling the actin molecule along for short distance.
- Another ATP molecule attaches to each myosin head, leading to a change in shape, and it separates from the actin (but always in the same direction).
- ATP hydrolase hydrolyses the ATP to ADP and inorganic phosphate (P_i), resulting in straightening of the myosin molecule. This is called the recovery stroke.
- Each head is now able to repeat the process by attaching to another binding site and sliding the actin along a bit more.

The speed of this process is hard to imagine. Each myosin head can bind and detach up to 100 times a second. Each myosin filament is made up of many myosin molecules held together by their tails, and during contraction their many heads are all flicking to and fro. They do not all bind to actin filaments at the same time, which would make the sliding action jerky. However, the most important thing to understand is that the myosin heads at each end of the myosin filaments point in opposite directions, so when they bind they pull both sets of actin filaments in the sarcomere towards one another at the same time. This explains why it is the actin filaments and not the myosin filaments that slide, and why sarcomeres shorten when muscle contracts (see Figure 18.10). The combined effect of all the sarcomeres and therefore all the myofibrils contracting together can produce remarkably large forces in muscles at considerable speed.

While the calcium ions remain in the cytoplasm, the myosin heads can continue to bind to and detach from the same site, so the fibre stays contracted. This process requires a continuous supply of ATP. When nervous stimulation ceases the calcium ions are reabsorbed into the sarcoplasmic reticulum by active transport. The tropomyosin moves back, covering the binding sites, and the myofibrils relax. However, the sarcomeres remain the same length until an outside force such as an antagonistic muscle pulls the actin filaments back out from between the myosin filaments.

TIP

You would think that ATP hydrolysis would be used to drive the power stroke. However, it is actually used to drive the recovery stroke and allow the myosin head to reposition for the next cycle.

TEST YOURSELF

6 Explain why the myofibrils of skeletal muscle appear striated.
7 Use Figure 18.13 to describe the structure of an actin filament.
8 What do you expect to happen to the A band and the I band during contraction? Explain your answer.
9 What is ATP used for during the cycle of actinomyosin bridge formation?
10 What is the role of the sarcoplasmic reticulum in a muscle fibre?

How are skeletal muscle fibres stimulated to contract?

A skeletal muscle contracts when stimulated by nerve impulses reaching the end of a motor neurones. Figure 18.14 shows the branches of a motor neurone connecting with muscle fibres. The branches spread across the surface of the

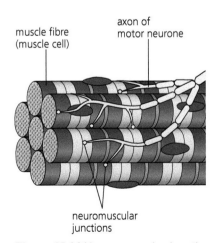

muscle fibre
(muscle cell)

axon of
motor neurone

neuromuscular
junctions

Figure 18.14 Neuromuscular junctions between a motor neurone and a group of muscle fibres.

333

muscle fibres and at various points terminate at a flattened structure called an endplate close to the sarcolemma of a muscle fibre. These points are called **neuromuscular junctions**. Neuromuscular junctions are very similar to synapses and work in much the same way. It might be useful to look back at the section on synapses in Chapter 17, particularly at Figure 17.10.

As you can see in Figure 18.15, the synaptic knob contains vesicles of acetylcholine. All neuromuscular junctions in skeletal muscles are cholinergic. When action potentials reach the pre-synaptic membrane, calcium ion channels open and calcium ions diffuse in across the membrane. This leads to vesicles containing acetylcholine fusing with the membrane. The acetylcholine diffuses into the cleft. This is stimulated by calcium ions that enter the pre-synaptic membrane from the cleft through channel proteins. The acetylcholine diffuses across the cleft to the sarcolemma of the muscle fibre, where it binds to protein receptors. This causes the sarcolemma to be depolarised in exactly the same way as a post-synaptic membrane is depolarised (see Chapter 17 page 317). Assuming the threshold value is reached, action potentials spread over the muscle cell-surface membrane and into the sarcoplasmic reticulum. This causes calcium ions to be released from the sarcoplasmic reticulum and to enter the muscle cell across the sarcolemma. As described earlier, the calcium ions attach to troponin and allow the binding sites on actin to be exposed.

Figure 18.15 A neuromuscular junction.

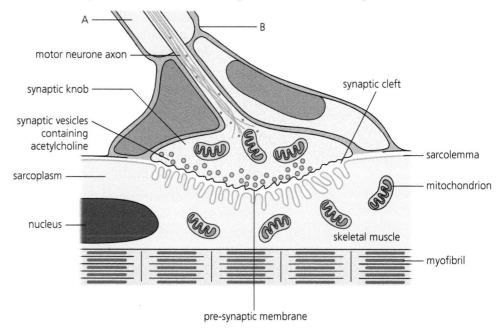

ATP hydrolysis is required for muscle contraction

A supply of ATP is required both for the return movements of the myosin heads that make the actin filaments slide and for the return of calcium ions into the reticulum by active transport. Resting muscle fibres only store enough ATP to maintain contraction for a very short time. The supply will only last for 3 or 4 seconds of intensive exercise. The muscle fibres contain mitochondria that can generate ATP by respiration of glucose. However, full aerobic respiration is relatively slow and even anaerobic respiration takes about 10 seconds to begin to produce some ATP.

But muscle fibres also contain stores of **phosphocreatine**. This can be used to produce ATP very rapidly by transferring a phosphate ion to ADP and thus replacing the ATP that has been hydrolysed.

$$\text{ADP} + \text{phosphocreatine} \longrightarrow \text{ATP} + \text{creatine}$$

TIP

Summation can occur at neuromuscular junctions as well as at ordinary synapses.

The amount of phosphocreatine available is limited and depends on the type of fibre but enables the muscles to keep going until more ATP can be supplied from the mitochondria. Nevertheless this means that really intense muscle activity can only be kept up for quite a short time. This may be enough for a trained athlete to sprint 100 m in 10 seconds, but such a level of activity could not be sustained for a much longer event. For prolonged activity the rate of muscle contraction has to be matched by the rate at which ATP can be provided from a combination of aerobic and anaerobic respiration.

Extension

How are muscle movements controlled?

When you pick up a heavy bag of books you may use much the same muscles as when you pick up a pen. But obviously you do not need to pick up the pen with the same force as the bag. If you carry the bag home you will have to hold it in the same position by keeping your muscles contracted. Individual muscle fibres are like neurones in that they work on an all-or-nothing principle. They either contract fully or, if the stimulus is below the threshold value, they do not contract at all. The strength of a muscle

contraction therefore depends on how many fibres contract and how often the stimulus is repeated. In skeletal muscles each motor neurone has branches to a group of muscle fibres called a **motor group.** This group may have several hundred fibres. When an action potential is conducted along the neurone, all the muscle fibres in the group contract. Different motor neurones in a motor nerve have slightly different thresholds, so how many groups of fibres are activated depends on the strength of the initial stimulus.

ACTIVITY

Muscle contraction

Figure 18.16 shows the effect of applying a *single* stimulus to a motor nerve connecting to a small muscle. The stimulus produced single action potentials in a few of the motor neurones in the nerve.

1 What fraction of a second is 100 ms?
2 For how long was the muscle contracting?
3 Before the muscle started to contract there was a short delay. Suggest what was happening during this period.
4 What stops the muscle contracting?

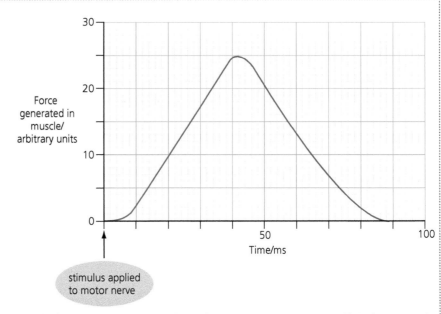

Figure 18.16 Graph showing the effect of a single stimulus on a small skeletal muscle.

Figure 18.17 shows the effect of **repeated** stimulation of the same motor nerve. The nerve is stimulated at the times shown by the arrows A, B, C and D.

5 Describe what happens when the muscle is stimulated several times in quick succession. Suggest a name for this effect.
6 Suggest how the repeated stimulation causes the muscle to contract with a greater force.
7 After the sarcolemma of a skeletal muscle fibre has been depolarised and an action potential has been produced, there is a delay of about 5 milliseconds before it can be depolarised again. This is similar to the period of delay in a neurone. What is this period called?

8 Skeletal muscle can start to contract again when relaxation is only about one-third complete. Cardiac muscle has a much longer delay of about 300 milliseconds. Use your knowledge of the cardiac cycle from Chapter 9 to suggest the advantage of this longer delay.

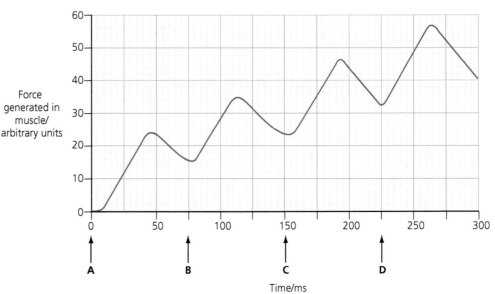

Force generated in muscle/ arbitrary units

Time/ms

Figure 18.17 Graph showing the effect of a repeated stimulus on a small skeletal muscle.

Fast and slow skeletal muscle fibres

The chapter introduction about junglefowl explained that muscle fibres are not all the same. As the name implies, fast fibres contract much more quickly than slow fibres. During contraction their myosin heads connect and disconnect from the binding sites on the actin up to five times as fast. Now that we've studied the detailed structure of muscles we can explain how the two types of fibre function. Table 18.2 shows the key differences.

Slow muscle fibres	Fast muscle fibres
Long contraction–relaxation cycle	Short contraction–relaxation cycle
Smaller store of calcium ions in sarcoplasmic reticulum	Large store of calcium ions and more sarcoplasmic reticulum
Dense network of blood capillaries around fibres for supply of oxygen and glucose for aerobic respiration	Fewer blood capillaries around fibres
ATP largely obtained from aerobic respiration	ATP largely obtained from anaerobic respiration
Many, large mitochondria, nearer the surface of the fibres	Fewer, smaller mitochondria, more evenly distributed
Small amount of glycogen	Larger amount of glycogen and phosphocreatine
Slower rate of ATP hydrolysis in myosin heads	Higher rate of ATP hydrolysis in myosin heads, so more actinomyosin bridges formed per second
Resistant to fatigue, since less lactate is formed	Quickly become fatigued, since more lactate is formed

Table 18.2 Comparing the properties of slow muscle fibres and fast muscle fibres.

Slow fibres are able to keep working for long periods. They have plenty of mitochondria, so they are able to obtain most of their ATP from aerobic respiration, even though it produces ATP relatively slowly. They have a dense capillary network, and their mitochondria are near the surface, so diffusion distances for oxygen are shorter. This maximises the delivery of oxygen, so slow fibres are more likely to remain aerobic. Because they contract more slowly they require a lower concentration of calcium ions to initiate contraction. They tend not to fatigue easily because lactate is less likely to build up if they are respiring aerobically.

Fast fibres have fewer capillaries to deliver oxygen and glucose and fewer mitochondria, and they obtain much of their ATP from glycolysis, or anaerobic respiration, which is faster. They work in rapid bursts for short periods of time so require a higher concentration of calcium ions to initiate more contraction. Larger amounts of phosphocreatine help to supply ATP for the first few seconds until more can be produced by respiration. Fast fibres have larger glycogen stores to act as an internal source of glucose, or respiratory substrate, for glycolysis. However, they tend to fatigue more quickly than slow fibres because anaerobic respiration produces lactate, which is acidic and lowers their pH.

EXAMPLE

Comparing fast and slow muscle fibres

1 Explain the importance of mitochondria in muscle contraction.
   As you learned in Chapter 14, page 255, the mitochondria are where the Krebs cycle takes place during aerobic respiration and where the majority of ATP is produced. This is a significantly slower process than anaerobic respiration but produces much more ATP per molecule of glucose than does anaerobic respiration.

2 Explain the importance of the dense network of capillaries in slow muscle fibres.
   Slow fibres depend on aerobic respiration for ATP. This requires a constant supply of oxygen and glucose, which has to be obtained from the haemoglobin in the red cells of the blood. The dense network ensures that all fibres are close to this source of oxygen and glucose.

3 How does the availability of calcium ions in the sarcoplasmic reticulum affect the rate of contraction?
   Calcium ions are released from the sarcoplasmic reticulum when a muscle fibre is stimulated. The ions diffuse into the myofibrils and attach to troponin, causing the tropomyosin to move away from the binding sites on actin. The larger network of sarcoplasmic reticulum in fast muscle fibres reduces the distance for diffusion and the higher concentration of calcium ions increases the diffusion gradient. Both factors help to speed up the rate at which the ions reach the troponin in the sarcomeres.

4 Explain the importance of ATP hydrolysis in the myosin heads.
   The hydrolysis of ATP into ADP and inorganic phosphate drives the recovery stroke. The myosin molecule straightens and becomes ready to attach to take part in another cycle of actinomyosin bridge formation. The faster hydrolysis occurs the faster the fibre contracts.

5 Explain how glycolysis provides ATP more rapidly than aerobic respiration.
   Glycolysis takes place in the cytoplasm whereas the rest of aerobic respiration occurs in the mitochondria. In the absence of oxygen the pyruvate produced by glycolysis is converted to lactate. This process occurs at a much faster rate than the diffusion of pyruvate into the mitochondria, so anaerobic respiration can occur very rapidly for a short period, making ATP available more quickly.

6 Suggest what causes fast muscle fibres to fatigue rapidly.
   Anaerobic respiration produces lactate and this increases the acidity inside the fibre. One effect is to disrupt the ATP hydrolase in the myosin head that hydrolyses ATP. In practice, the effects are complex and not fully understood. Other factors are interference with the release of and uptake of calcium ions and depletion of the reserves of phosphocreatine and glucose.

7 A top-class 100 m sprinter can complete a race without taking a breath. Suggest an explanation for this.
   All the ATP for muscle contraction during the race is derived from phosphocreatine and glycolysis. There is no need to replenish the supplies of oxygen through breathing. However, following the race, deep breathing is required to overcome the oxygen deficit. Oxygen is required for the oxidation of the accumulated lactate.

TEST YOURSELF

11 Name the parts labelled A and B in the neuromuscular junction shown in Figure 18.15.
12 Describe how a sarcolemma is depolarised by acetylcholine.
13 Explain how the resting potential of the sarcolemma is restored.
14 What is the role of phosphocreatine in muscle contraction?
15 Give three differences between fast and slow muscle fibres.

Practice questions

1 The figure shows a diagram of the structure of two filaments in a myofibril.

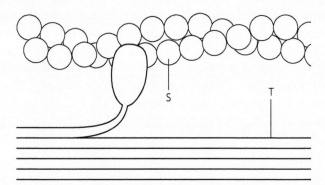

a) i) Identify filament S and filament T. *(1)*

   ii) Mark with an X on the diagram the location of ATP hydrolase. *(1)*

   iii) Indicate with an arrow on the diagram the direction in which filament S would move during myofibril contraction. *(1)*

b) Identify the role of ATP in myofibril contraction. *(1)*

c) Describe the sequence of events in the cycle of actinomyosin bridge formation. *(5)*

2 The figure is an electron micrograph of part of a skeletal muscle myofibril showing a sarcomere.

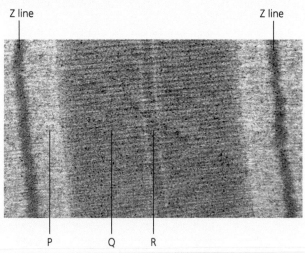

a) Which type of filaments would be present in a cross-section cut at

   i) R?

   ii) Q? *(1)*

b) If this myofibril were to contract, describe and explain what would happen to the distance between the Z lines. *(2)*

c) What would happen to the zone labelled P during contraction? Explain your answer. *(2)*

d) In a muscle cell, muscle fibrils are surrounded by sarcoplasmic reticulum. Describe the role of sarcoplasmic reticulum in myofibril contraction. (3)

3 The diagram shows a neuromuscular junction.

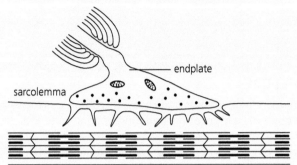

a) Name the neurotransmitter released at a neuromuscular junction. (1)

b) Suggest the advantage of a neuromuscular junction having a flattened endplate rather than a synaptic knob. (1)

c) Describe the events that take place when action potentials arrive at this neuromuscular junction. (5)

d) Explain how one action potential arriving at this neuromuscular junction may not cause contraction, but several arriving one after another might. (3)

4 The slow loris is a tree-dwelling mammal that shows very slow, deliberate movements along branches and cannot jump from tree to tree. The skeletal muscles in its limbs contain mainly slow fibres.

a) Explain why slow fibres are surrounded by very dense networks of blood capillaries. (2)

b) Give two other differences between slow and fast muscle fibres. (1)

c) The ATP hydrolase on the myosin heads of slow fibres has a slow rate of ATP hydrolysis. Explain why this prevents the slow loris from jumping from one tree to another. (3)

d) The slow loris is only active at night. Suggest why having mainly slow fibres in its limb muscles might be an advantage for the slow loris. (1)

Stretch and challenge

5 Cardiac muscle is similar to skeletal muscle in some ways but very different in others. Compare cardiac muscle fibres with skeletal muscle fibres and discuss reasons for the differences.

6 Examine the role of dystrophin in muscle tissue. Discuss its importance by contrasting muscle tissue in a healthy person with muscle tissue in a person with Duchenne muscular dystrophy.

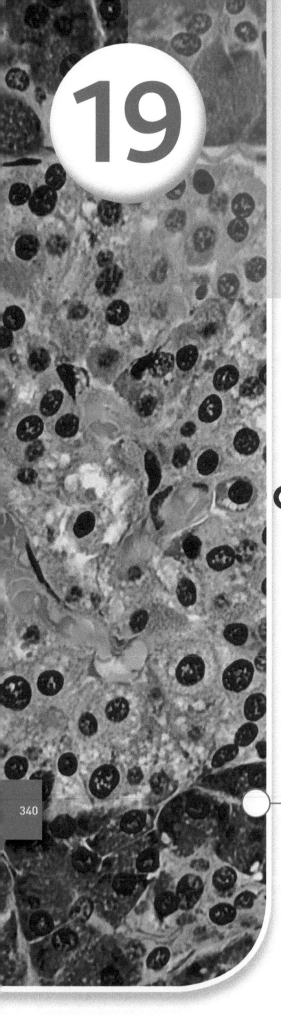

Introduction

It was only in the 20th century that biologists discovered that chemical 'messengers' or hormones are transported round the body in the blood and that these play an essential part in the body's communication system. Compared to a nervous response, mammalian hormones produce responses that are usually slow, long-lasting and widespread. The term 'hormone' was not coined until 1906.

In this chapter we shall look at some examples of the ways in which hormones control physiological processes in humans and other mammals.

One important role of hormones is maintaining more or less stable conditions in the body. Hormones ensure that the glucose concentration in the blood doesn't rise too high or fall too low. They keep the pH and water potential of the blood fairly constant.

What is a hormone?

A hormone is a substance that is secreted directly into the blood from the tissue where it is made. The blood carries it to all parts of the body but the hormone only has an effect on certain **target organs** or **target cells**. Many hormones work in a similar way to neurotransmitters. They attach to **receptor proteins** on the cell-surface membranes of target cells. This sets off reactions that activate enzymes in the cells, as we shall see later. Some hormones that are soluble in lipids can pass through the cell-surface membrane into the cell. Here they may combine with receptor proteins in the cytoplasm, then enter the nucleus and cause genes to be expressed, leading to the production of enzymes.

Hormones are synthesised in **endocrine** tissues. Endocrine glands are organs whose function is to release hormones. These include the pituitary gland situated below the brain and the adrenal glands, one at the top of each kidney. Other hormones are made by groups of cells in other organs, such as the cells that produce the sex hormones in the testes and ovaries, and the insulin-making cells in the pancreas.

Homeostasis in mammals

Homeostasis in mammals involves physiological control systems that maintain the internal environment within restricted limits. For example, it is obviously advantageous to control

- the temperature and pH of tissue fluid around cells since they both affect the rate of enzyme-controlled reactions
- the concentration of glucose in the blood since this affects both the water potential of the blood and the availability of a respiratory substrate to cells.

Negative and positive feedback

You will remember from Chapter 16 that receptors detect stimuli. Deviations from the normal range are stimuli, detected by receptors. This leads to a corrective mechanism to bring the factor back to the normal range. This is regulated by **negative feedback**. Usually, there is one corrective mechanism when the factor becomes too high and another corrective mechanism when the factor falls too low.

Figure 19.1 shows how, when a factor changes from its normal level, the body detects the change. It then causes a corrective mechanism to change the factor back to the normal level. This corrective mechanism may involve nervous mechanisms and/or hormones. The amount of correction needed to bring the factor back to the normal level is regulated by **negative feedback**. This makes sure that the corrective mechanism is reduced as the factor returns to its normal level.

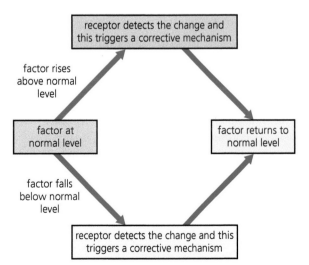

Figure 19.1 Negative feedback.

341

As you can see, with negative feedback the corrective mechanism reduces the original effect of the stimulus. In a positive feedback system, the stimulus brings about a response that changes the factor even further from the normal level.

Control of blood glucose concentration

The control of the concentration of glucose in the blood involves two homeostatic mechanisms. The possession of two separate mechanisms, controlling departures in different directions from the norm, provides a greater degree of control than would a single mechanism.

The normal value for the concentration in human blood is approximately 90 mg per 100 cm^3. It is important that there is always some glucose circulating in the blood to enable cell respiration to continue. The brain cells in particular will very soon die if their supply is cut off. On the other hand, if the concentration of glucose rises too high it has a major effect on the water potential of the blood.

Glucose enters the blood from three main sources

- absorption from the gut following the digestion of carbohydrates
- hydrolysis of stored glycogen
- conversion of non-carbohydrates such as lactate, fats and amino acids.

The amount of glucose absorbed into the blood from digestion can fluctuate greatly. The control systems remove excess glucose entering the blood after a carbohydrate-rich meal and release glucose rapidly from storage compounds when muscles are depleting the content of the blood at a fast rate during exercise.

What happens when blood glucose concentration rises?

A rise in the blood glucose concentration above the 'norm' is detected by the **beta cells (β cells)** in the **pancreas**. The β cells are situated in little groups of cells dotted around the pancreas called **islets of Langerhans** (Figure 19.2). The β cells synthesise a hormone called **insulin**. You will probably recall that the pancreas secretes digestive enzymes that pour through a duct into the upper part of the small intestine. Since they produce hormones and not enzymes, the islets are like mini-endocrine glands.

When blood with a high concentration of glucose reaches the islets, glucose is absorbed into the β cells. The cell-surface membrane of a β cell contains carrier protein molecules that transport glucose into the cells by facilitated diffusion. This stimulates vesicles containing insulin to move to the cell-surface membrane and release insulin into the surrounding capillaries. This is similar to the process at synapses where acetylcholine is released into the synaptic cleft (see page 317).

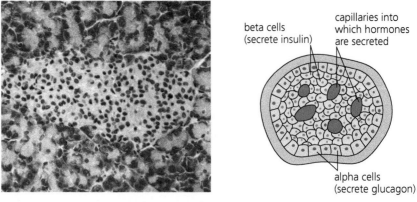

Figure 19.2 a) Coloured photomicrograph of islets of Langerhans in the pancreas, magnification ×100; b) diagram of an islet of Langerhans.

Insulin then circulates round the body in the bloodstream. Its main effects are to stimulate uptake of glucose by cells in muscles, adipose (fat storage) tissue and the liver. Insulin attaches to receptor proteins in the cell-surface membranes of the cells in these tissues. Glucose cannot diffuse into cells through the phospholipid bilayer. Cells have channel proteins, called glucose transporter proteins, that allow glucose to enter by facilitated diffusion; but the rate of uptake is limited by the number of these channel proteins. The glucose transporter molecules in muscles and adipose tissue (called GLUT4) are insulin-sensitive. Insulin causes additional transporter molecules to join the cell-surface membranes of muscle and adipose cells (Figure 19.3). By adding many more channel proteins the rate of uptake of glucose from the blood by facilitated diffusion is greatly increased.

Glycogenesis The formation of glycogen from glucose. This occurs in the liver.

Liver cells already have large numbers of glucose transporter molecules in their membranes. In the liver, insulin leads to an increase in glucose uptake in a different way. After the glucose has entered the liver cells it activates an enzyme that rapidly converts the glucose to glucose phosphate. This lowers the glucose concentration inside the cells and this maintains a steep **diffusion gradient** between the blood and the liver cells. Other enzymes then synthesise the glucose phosphate to glycogen, a process known as glycogenesis. Glycogenesis also occurs in the muscles, replenishing the stores there. In fat storage tissue, insulin activates enzymes that manufacture fatty acids and glycerol, which are then stored as fat.

1 Insulin molecule binds to receptor

cell-surface membrane

2 Chemical signal

glucose carrier protein

3 Vesicle moves to membrane, fuses and adds carrier proteins to it

343

Figure 19.3 In the cells of muscles and adipose tissue, insulin causes additional glucose transporter molecules to join the cell-surface membrane.

To summarise, insulin stimulates removal of glucose from the blood by:

● increasing the rate of facilitated diffusion into muscle and fat storage cells by stimulating these cells to add more carrier proteins (glucose transporter molecules) to their cell-surface membranes
● increasing the rate of glucose uptake in the liver by stimulating glycogen synthesis
● increasing the rate of glucose uptake in adipose tissue by stimulating fat synthesis.

What happens when blood glucose concentration falls?

When the glucose concentration in the blood drops below approximately 90 mg per 100 cm^3, insulin secretion declines. As blood concentration falls, this is detected by the beta cells, which respond by producing less insulin. This is a negative feedback mechanism by which insulin regulates its own secretion. Furthermore, the lower glucose concentration stimulates the **alpha cells (α cells)** in the islets of Langerhans to secrete another hormone, called **glucagon**. Glucagon's main effects are in the liver. It activates enzymes that break down the stored glycogen to glucose, which is then released into the blood. This process is called glycogenolysis. It can also activate enzymes that convert other substances to glucose, notably lactate and amino acids. This is a process called gluconeogenesis, which roughly means 'making glucose from new substances'. Both of these effects increase the blood glucose concentration again so that it quickly returns to its normal value. In practice, there are constant adjustments to the amounts of insulin and glucagon being secreted. Much of the time both hormones are secreted in small quantities with the proportions adjusted to maintain the glucose concentration at a fairly constant level.

Figure 19.4 summarises how negative feedback mechanisms operate in the control of the blood glucose concentration.

Glycogenolysis The breakdown of glycogen to release glucose.

Gluconeogenesis The synthesis of glucose from molecules that are not carbohydrates, such as amino acids and fatty acids.

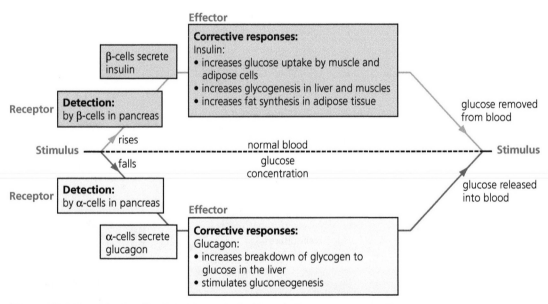

Figure 19.4 Negative feedback in the control of blood glucose concentration, showing the effects of insulin and glucagon.

The second messenger model

As often happens in scientific research, continuing investigations have revealed that the simple story does not give the full picture and that systems are more complex than at first thought. The breakdown of glycogen in the liver is not only stimulated by glucagon. **Adrenaline**, the 'fight-or-flight' hormone, has a similar effect. Adrenaline attaches to specific receptor proteins in the cell-surface membranes of liver cells, which, in turn, leads to the activation of enzymes involved in the conversion of glycogen to glucose.

The second messenger model

Unlike insulin, adrenaline and glucagon do not have a direct effect on the liver cells. Instead, when adrenaline or glucagon attaches to a specific receptor protein that spans the cell-surface membrane of a liver cell, it causes a cascade of reactions within the cell. This is called the **second messenger model**, where glucagon is the first messenger.

Figure 19.5 shows what happens in this model, using adrenaline as an example; the process is similar when glucagon attaches to its specific receptor protein. Binding of adrenaline causes a change in the shape of an enzyme also in the cell-surface membrane, called **adenyl cyclase**. This makes the enzyme active. This enzyme acts on ATP, which has three phosphate groups. Instead of producing ADP, adenyl cyclase removes two of the phosphate groups. This makes **cyclic adenosine monophosphate**, usually written as **cAMP**. cAMP is the second messenger. The cAMP then binds to another enzyme in the cytoplasm, called **protein kinase**. This exposes its active site, which leads to the hydrolysis of glycogen to produce glucose phosphate. Hormones, and indeed chemicals in general, often get a bad press. Hormones are blamed for erratic behaviour in teenagers and for a variety of eccentricities in men and women. Hormones are frequently thought to be minor players in the body's communication system compared with nerves. As you have seen in the previous chapters, the nervous system depends on chemicals, for example at synapses and neuromuscular junctions. Many of the neurotransmitters are now known to be chemically very similar to hormones.

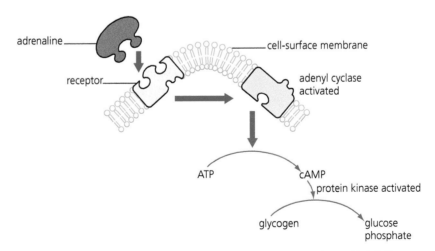

Figure 19.5 The second messenger model of hormone action, which follows stimulation of liver cells by adrenaline or glucagon.

The advantage of this process is that each molecule of hormone can stimulate many molecules of cAMP, and these in turn activate large numbers of enzyme molecules.

TEST YOURSELF

1 Explain the effect of high glucose concentration on the water potential of the blood.
2 What is meant by facilitated diffusion? (If you can't remember you may need to refer back to Chapter 3).
3 How does facilitated diffusion differ from active transport?
4 After its secretion, glucagon breaks down within a few minutes. Suggest the advantage of this.
5 The veins from the small intestine and the pancreas connect to the hepatic portal vein. This transports blood directly to the liver. Therefore the blood flows through the liver before returning to the heart and entering the general circulation. Suggest the advantages of this arrangement in relation to the control of the blood glucose concentration.
6 Suggest why it is an advantage for adrenaline to have its amplification effect through a second messenger.

Diabetes

Diabetes is a condition in which homeostatic control of the blood glucose concentration breaks down. In 2017, there were 3 689 509 adults in the UK living with diabetes, and its incidence is growing. The increase is attributed to the trend towards over-eating and obesity. Insulin function is disrupted and the concentration of glucose in the blood rises above the normal range after meals. One effect is that the kidneys are unable to reabsorb all the glucose into the blood as they normally do. Therefore glucose appears in the urine. In response the kidneys tend to produce excess quantities of urine. This dehydrates the body and makes the sufferer very thirsty.

There are two main types of diabetes, called **Type I** and **Type II**. Table 19.2 shows the proportion of these two types of diabetes in 2011 in the UK.

Table 19.2 The incidence of diagnosed Type I and Type II diabetes in the UK in 2011.

Age group	Percentage of cases	
	Type I diabetes	Type II diabetes
Children	98	2
Adult	10	90
All	15	85

Type I diabetes

Type I diabetes is caused by an inability to make enough insulin. It usually develops quite early in life, often in childhood (see Table 19.2). It is an auto-immune condition in which the immune system destroys some of the body's own cells. In Chapter 6 we described how T cells attack infected cells by

Auto-immune condition A condition in which the immune system destroys some of the body's own cells.

attaching to antigens on their cell-surface membranes. For some unexplained reason, in Type I diabetes T cells start to attack and destroy the β cells in the islets of Langerhans. The shortage of insulin causes fatigue because not enough glycogen is stored in the liver. The concentration of glucose in the blood can rise to dangerously high levels after a meal. This can lead to serious organ damage.

This type of diabetes can be controlled by eating appropriate food and taking regular blood tests and injections of insulin.

The food recommended for people with diabetes by health authorities and by Diabetes UK differs little from that recommended for everyone wishing to have a healthy diet. The recommendations include the consumption of food containing polysaccharides, rather than monosaccharides and disaccharides, at least five portions of fresh fruit and vegetables each day, and few processed foods.

Before the isolation and use of insulin, Type I diabetes was always fatal. For many years insulin was extracted from pancreas tissue of cattle and pigs. In the 1980s, synthesis of human insulin by recombinant DNA technology was one of the early successes of genetic engineering (see Chapter 12). Treatment has been transformed by the development of fast-acting soluble and slow-release insulin preparations. These allow either for rapid control at a mealtime or control that can last for many hours. Methods of injection have been made much more convenient, for example by the use of injector pens. As a result people with Type I diabetes can live a normal and full life. Nevertheless most have to inject themselves with insulin twice a day in order to maintain a stable blood glucose concentration. There is hope that transplants of islets of Langerhans will provide more permanent treatment, but so far this has been much less successful than, for example, kidney transplantation.

Type II diabetes

As you can see from Table 19.2, this form of diabetes is much more common than Type I. Usually it develops in people over 40 years old but it is increasingly being found at younger ages. It is caused, not by failure to produce insulin, but by reduced sensitivity of the liver and fat storage tissue to insulin. The receptors either fail to respond or are reduced in number so glucose uptake is much reduced. The exact effects are still unclear because they are often associated with a range of other factors, often resulting from obesity and unbalanced diets. One response is that the β cells are stimulated to produce larger quantities of insulin. If the condition persists the β cells are damaged and the condition becomes more like Type I.

In the early stages Type II diabetes can be treated simply by careful attention to the diet. Reducing the intake of refined carbohydrates, such as sugars and forms of starch that are rapidly digested to sugars, avoids large surges in blood sugar concentration. The fat content of the diet also needs to be kept low and exercise helps to improve the body's sugar metabolism.

TEST YOURSELF

7 Insulin is a protein. Explain why it has to be injected instead of being taken orally.

8 Suggest how Type II diabetes could be controlled by exercise and diet.

9 One effect of poorly controlled diabetes is high blood pressure. Use your knowledge of osmosis to suggest why.

EXAMPLE

The glucose tolerance test

One of the tests used in diagnosing diabetes is the **glucose tolerance test**. After a period of several hours without food the person being tested has a drink of water containing 75 g of glucose. The concentration of glucose in the blood plasma is measured at 30 minute intervals. The graph in Figure 19.6 shows the results of this test from three adults. One had no symptoms of diabetes. Both of the others had tested positive for glucose in the urine.

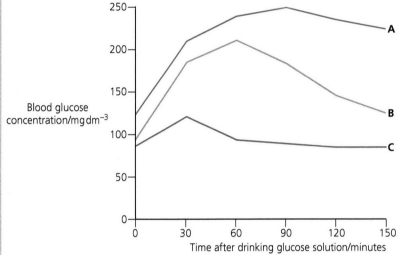

Figure 19.6 Results of a glucose tolerance test taken by three adults, labelled A, B and C.

1 Curve C shows the results for the person without diabetes. Suggest why the glucose concentration rose in the first 30 minutes.
   The glucose from the drink was absorbed into the blood.

2 Describe and explain the results for curve C after 30 minutes.
   The β cells of the pancreas release insulin, so glucose is removed from the blood and stored as glycogen in liver and muscle cells.

3 Describe the differences between curves A and B.
   Both curves show a large increase in glucose concentration in the blood, but while curve B peaks and starts to fall after an hour, curve A does not peak until 90 minutes and then falls more slowly than curve B.

4 Of the people A and B, one was diagnosed as having Type I diabetes and the other Type II. Which was which? Explain your reasoning.
   A has Type I diabetes as the glucose concentration rises in the blood for 90 minutes, which indicates that insulin is not produced. The concentration of glucose falls very slowly, which again indicates that there is no insulin produced and the glucose only gradually decreases as it is used in respiration. B has Type II diabetes as the curve peaks earlier than A and starts to fall more rapidly. This implies that insulin is produced and some of the gucose is converted to glycogen, but the body is not as responsive to insulin as in a healthy person.

REQUIRED PRACTICAL 11

Production of a dilution series of a glucose solution and use of colorimetric techniques to produce a calibration curve with which to identify the concentration of glucose in an unknown 'urine' sample

This is just one example of how you might tackle this required practical.

A colorimeter is a piece of equipment that passes light of a particular wavelength through a sample and measures the amount of light transmitted through the sample. You learned about this in Chapter 1 page 16.

In using a colorimeter, you need to produce a calibration curve. This is done using solutions of a known concentration.

> **TIP**
> Refresh your memory of colorimetry by reading the section about it in Chapter 15.

> **SAFETY**
> Wear eye protection when carrying out this practical.

A student was given a $1\,mol\,dm^{-3}$ solution of glucose and then made a series of dilutions. She made $20\,cm^3$ of each dilution.

1 Complete the table to give the volumes of water and $1\,mol\,dm^{-3}$ sucrose needed for each solution.

Table 19.3

Concentration of sucrose/ mol dm^{-3}	Volume of 1 mol dm^{-3} required/cm^3	Volume of distilled water required/cm^3
0.9		
0.8		
0.7		
0.6		
0.5		
0.4		
0.3		
0.2		
0.1		

An alternative version of Benedict's reagent for quantitative testing contains potassium thiocyanate and does not form red copper oxide. Instead the presence of reducing sugar is measured by the loss of the blue colour of copper sulfate and a white precipitate is formed. This will settle out or can be removed by filtering. Then the **filtrate** is placed in a cuvette in a colorimeter. The intensity of the blue colour is measured by the amount of light that is able to pass through the solution. This method can give an accurate measurement of the concentration of reducing sugar in a solution, and it is much more sensitive that the qualitative Benedict's test.

> **Filtrate** The precipitate that is left in the filter paper after filtering has taken place.

The student put $4\,cm^3$ of each solution into separate labelled test tubes.

Next she added $2\,cm^3$ of quantitative Benedict's reagent to each tube and placed the tubes in a boiling water bath for 5 minutes. After this time she filtered each solution to remove the precipitate.

The student set the wavelength on the colorimeter to red.

First of all she filled a cuvette with distilled water and put it into the colorimeter. This is called a 'blank'. She set the transmission of light through the tube to 100%. This means that she could compare the transmission of light through the test solutions to the blank.

The student put a sample of each test solution into cuvettes in turn, and measured the percentage transmission of light through each tube. Next she plotted a graph with concentration of glucose on the x-axis and percentage transmission of light through the solution on the y-axis.

After this, she used the same method to identify the concentration of glucose in two urine samples.

2 Why did the student use a red light in the colorimeter?

3 What is the purpose of the blank?

4 How could the student use her graph to find the concentration of glucose in an unknown solution?

Control of blood water potential (osmoregulation)

The body cells of mammals are bathed by tissue fluid. Exchange of substances, including water, occurs constantly between this tissue fluid and the cells. Unless this tissue fluid has the same water potential as their cytoplasm, the cells will lose or gain water by osmosis, which could prove fatal. You learned in Chapter 9 how tissue fluid is formed from capillary blood. Control of the water potential of tissue fluid is achieved by controlling the water potential of the blood from which it is formed.

> **TIP**
> Remember to use water potential terminology when discussing this topic.

The homeostatic control of water potential is called **osmoregulation**. Like the homeostatic control of blood glucose concentration, osmoregulation involves negative feedback loops. Unlike the homeostatic control of blood glucose concentration, the control of blood water potential involves nervous as well as hormonal coordination. In this section we will look at how the hypothalamus, the posterior lobe of the pituitary gland and antidiuretic hormone (ADH) regulate the amount of water that is lost from the body in urine. Look at Figure 19.7 to remind yourself where the hypothalamus and pituitary gland are located. Mammalian urine is produced by the kidneys. The two kidneys form part of the human urinary system shown in Figure 19.8. They are supplied with blood via renal arteries. The kidneys filter the blood brought in these arteries. Initially, they remove some ions and molecules including amino acids, glucose, urea and water. Later, they reabsorb ions and molecules that are useful but not molecules like urea, which is a breakdown product of amino acids that is excreted in the urine. As we shall see later, it is the control of water reabsorption that is at the heart of osmoregulation. The functional unit of the kidney that removes ions and small molecules from the blood and then selectively reabsorbs them is the kidney tubule, or **nephron**.

> **TIP**
> You will soon realise that two of the processes involved in urine production are fundamentally similar to the production of tissue fluid and the countercurrent flow of blood that you learned about in your first year of study.

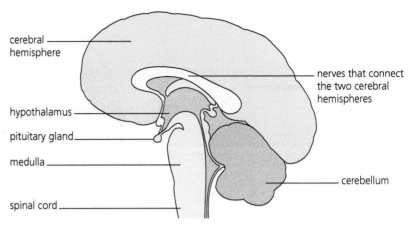

cerebral hemisphere

nerves that connect the two cerebral hemispheres

hypothalamus

pituitary gland

medulla

cerebellum

spinal cord

Figure 19.7 This shows a section through the centre of the brain, separating the two cerebral hemispheres. The pituitary gland and the hypothalamus have important roles in several different control systems. Although you do not need to learn the parts of the brain shown here, we shall be referring to them several times in this chapter so it will be useful to know where in the brain they are situated.

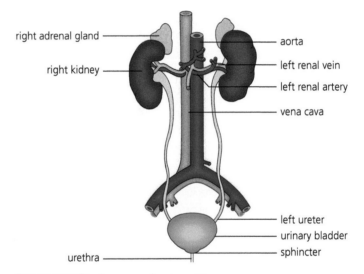

right adrenal gland

aorta

right kidney

left renal vein

left renal artery

vena cava

left ureter

urinary bladder

sphincter

urethra

Figure 19.8 The human urinary system.

Ultrafiltration and selective reabsorption by nephrons

Figure 19.9 (overleaf) shows the location of one nephron within a kidney. You can see that it begins and ends in a blood-rich outer layer, called the cortex, passing in the meantime into the central less-blood-rich region, called the medulla. You will soon understand what causes the appearance of the cortex and medulla.

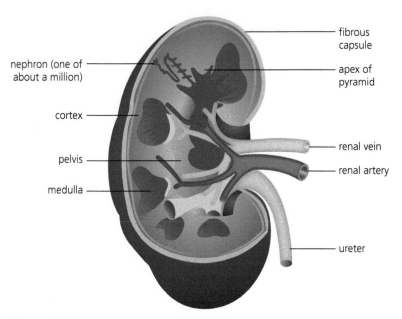

Figure 19.9 A section through a kidney with one nephron shown.

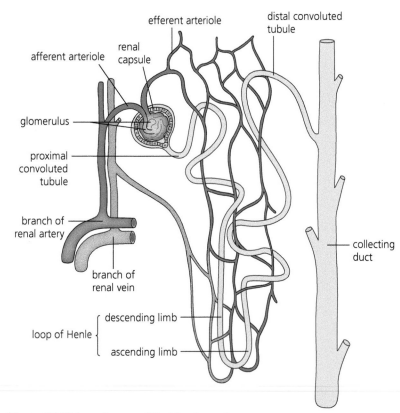

Figure 19.10 A nephron and its blood supply.

A human kidney contains an estimated 1 million nephrons; Figure 19.10 shows one of them. You can see that it has several sections: the renal capsule, proximal convoluted tubule and distal convoluted tubule are all present in the cortex of the kidney; the loop of Henle and the collecting duct are in the medulla. You can also see that within each renal capsule (see Figure 19.11) is a capillary network, called the **glomerulus**. Each glomerulus is supplied with blood by an afferent arteriole that branches from a renal artery. Blood leaves the glomerulus via an efferent arteriole that forms a second capillary network that wraps around the rest of the nephron. This is much narrower than the afferent arteriole, so this creates a high pressure in the capillaries of the glomerulus. As the blood flows through the glomerulus, the pressure forces water from the blood plasma, along with some molecules and ions, across a filtering system into the renal space. This process is called **ultrafiltration**.

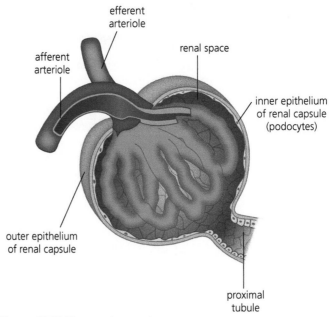

Figure 19.11 The renal capsule.

TIP
Ultrafiltration is similar to the process by which tissue fluid is formed. You learned about this in Chapter 9.

Ultrafiltration

Blood enters a glomerulus via an afferent arteriole that is wider than the efferent arteriole through which it leaves. As a result, a high pressure is created in the glomerulus, which forces water, ions and other small molecules into the renal space. This is the process of ultrafiltration. The fluid that is forced into the renal space is called the **glomerular filtrate**.

The filtering system of the renal capsule is shown in Figure 19.12. It has three layers.

- The capillary endothelium has large gaps that allow blood plasma through but not blood cells.
- The basement membrane, which is a mesh of protein molecules that supports the capillary endothelium, acts as a fine filter. It allows plasma proteins with a molecular mass of 68 000 daltons or less to pass through, but not larger ones.
- The **podocytes** forming the lining of the renal capsule have large gaps between them that allow the glomerular filtrate through into the renal space.

Ultrafiltration in the renal capsule is non-selective except by size. Any substance that is small enough (i.e. below the renal threshold) can be filtered out of the blood whether it is useful or not. This means that useful substances such as glucose and amino acids are filtered out, along with inorganic ions, water and urea. Many of the useful substances are reabsorbed in the proximal convoluted tubule.

353

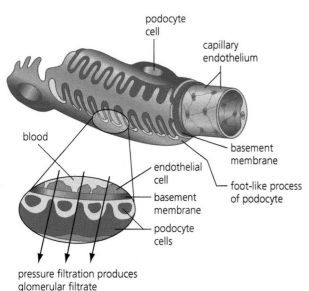

Figure 19.12 The filtering system of kidney renal capsule.

Reabsorption in the proximal convoluted tubule

Many of the substances in the glomerular filtrate are reabsorbed by active transport into the cells lining the proximal convoluted tubule. These normally include glucose, amino acids and some inorganic ions. Some of the urea is reabsorbed by diffusion. Figure 19.13 shows how the structure of an epithelial cell lining the proximal convoluted tubule is adapted for reabsorption.

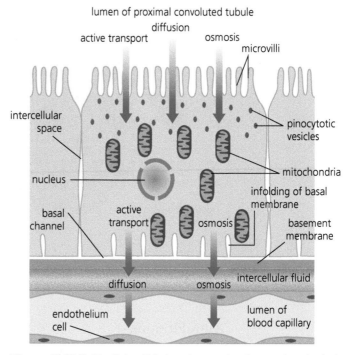

Figure 19.13 Epithelial cell lining the proximal convoluted tubule.

Water is also absorbed by cells lining the proximal convoluted tubule. The active transport of glucose, mineral ions and amino acids into the tubule cells lowers their water potential. As a result, water enters the tubule cells by osmosis. The glucose, amino acids and mineral ions absorbed by the tubule cells then diffuse into the blood capillaries around the proximal convoluted tubule and water moves into the blood capillaries by osmosis. Usually all of the glucose and amino acids in the filtrate are reabsorbed in the proximal convoluted tubule. However, in a person with diabetes, the glucose carrier proteins in the membrane cannot reabsorb all the glucose from the filtrate, so some is lost in the urine.

TEST YOURSELF

10 A person with a damaged basement membrane in the renal capsule has large proteins in their urine. Explain why.

11 Use information from Figure 19.13 to explain two ways in which an epithelial cell of the proximal convoluted tubule is adapted for the reabsorption of glucose and other essential nutrients.

12 Describe in terms of water potential how water is reabsorbed in the proximal convoluted tubule by osmosis.

13 Haemoglobin has a molecular mass of 68 000 daltons but it is not normally found in the urine. Suggest why.

The role of the loop of Henle

The ability of humans to produce urine that is more concentrated than blood plasma is due to the activity of the loop of Henle. The loop of Henle creates a high concentration of inorganic ions deep in the medulla of the kidney. It has two parts, the descending limb and the ascending limb.

- The descending limb of the loop of Henle is permeable to water but not very permeable to mineral ions such as sodium and chloride.
- The ascending limb of the loop of Henle has thick walls, which are impermeable to water. The narrow part of the ascending limb allows mineral ions to move passively into the medulla, but the wide ascending limb actively transports sodium chloride into the medulla.

The fluid in the descending limb of the loop of Henle flows in the opposite direction to the fluid in the ascending limb. The result is that the concentration gradient between the two limbs is maintained all the way along the loop.

You can see in Figure 19.14 that the loop of Henle creates a water potential gradient that allows water to be reabsorbed from the glomerular filtrate by osmosis. Some water has already been absorbed by osmosis in the proximal convoluted tubule, so the filtrate in the wide part of the descending limb (A) has the same water potential as the fluid in the surrounding tissues. However, because the surrounding tissue fluid in the medulla of the kidney has a high concentration of mineral ions, this creates a water potential gradient so that water is drawn out of the narrow part of the descending limb (B) by osmosis as the filtrate passes down the loop of Henle. The mineral ion concentration in the medulla increases towards the tip of the loop, so water can pass out along the whole length of the descending limb. This water is then carried away in the surrounding capillaries. The filtrate is now reduced in volume and contains a higher concentration of salts. The ascending limb is permeable to sodium chloride but impermeable to water. As the filtrate passes up the thin part of the ascending limb (C), sodium and chloride ions diffuse into the surrounding tissue fluid. Higher in the ascending limb (D), chloride ions are actively transported out of the limb and sodium ions follow into the tissue fluid. These processes maintain the high sodium and chloride ion gradient in the surrounding tissue fluid that is needed for water reabsorption in the collecting ducts.

> **TIP**
> You will remember that you learned about countercurrent exchange in Year 1, when you studied gas exchange in fish gills (see Chapter 7).

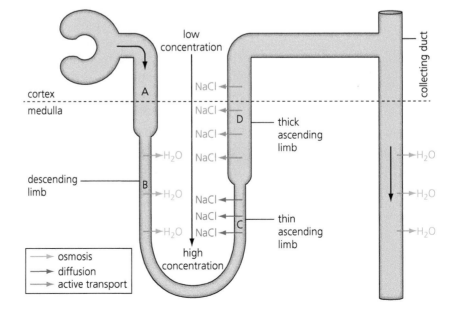

Figure 19.14 A single loop of Henle.

Control of water reabsorption by the distal convoluted tubule

Most of the useful substances, such as glucose and amino acids, have already been absorbed in the proximal convoluted tubule. However, some mineral ions and water remain. In the distal convoluted tubule some of the remaining mineral ions and water are reabsorbed. Almost all of the water that is required has already been reabsorbed, but this is the part of the nephron where water reabsorption is controlled.

Hormones control the amount of reabsorption of water by affecting the permeability of the distal convoluted tubule and collecting duct to water. Figure 19.15 shows how.

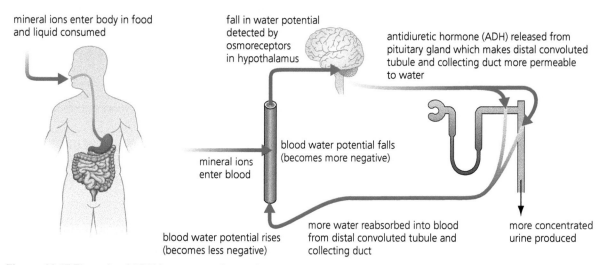

Figure 19.15 The role of ADH in water reabsorption.

Special receptor cells in the hypothalamus of the brain, called **osmoreceptors**, are sensitive to the stimulus of changes in the water potential of the blood. When mineral ions and other solutes in food and drink or loss of water from the body cause the water potential of the blood to fall (i.e. become more negative), these osmoreceptors send impulses to the **posterior pituitary** to release antidiuretic hormone (**ADH**) into the blood. ADH increases the permeability of both the distal convoluted tubule and the collecting duct. This happens because ADH binds to specific receptor molecules on the cells lining the distal convoluted tubule and collecting duct causing protein channels, called **aquaporins**, to move into their cell-surface membranes. The increase in the number of aquaporins allows more water to pass through the membrane by osmosis. As the water potential in the tubule is higher than the water potential of the blood, water is reabsorbed from the distal convoluted tubule and collecting duct. As some water has been reabsorbed from the tubule by osmosis, a smaller volume of more concentrated urine is produced.

When high fluid intake causes the water potential of the blood to rise, the osmoreceptors send impulses to the pituitary gland, which inhibits ADH release by the posterior pituitary. This fall in ADH release reduces the water permeability of both the distal convoluted tubule and the collecting duct, so less water is reabsorbed from the urine. This is another example of negative feedback. Secretion of ADH leads to an increase in water potential in the blood, which leads to a reduction in ADH secretion. You will probably have worked out that the lack of ADH causes the aquaporins to leave the cell-surface membranes of cells lining the ducts and move to the interior of these cells. Therefore a larger volume of more dilute urine is produced. This is summarised in Figure 19.16.

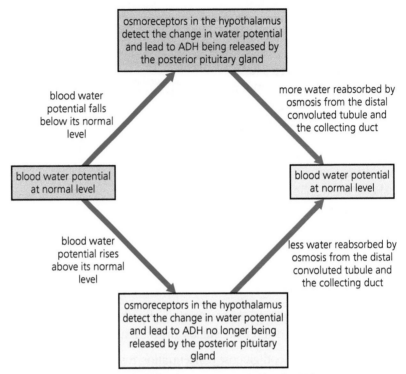

Figure 19.16 The control of blood water potential by the kidney.

TEST YOURSELF

14 Filtrate entering the loop of Henle has the same water potential as the blood. Explain why.

15 Explain how the high concentration of salts in the tissue fluid around the loop of Henle causes water to be drawn out of the thin part of the descending limb.

16 Suggest why the active transport of chloride ions out of the wide ascending limb causes sodium ions to follow.

17 Alcohol inhibits the release of ADH. Explain why a person working outdoors on a hot day should avoid drinking alcohol.

18 Use Figure 19.16 to explain the meaning of negative feedback.

Practice questions

1 One test for diabetes is called the glucose tolerance test. The person taking the test has nothing to eat, and only water to drink, for several hours before the test. Their blood glucose concentration is tested. They are given a sugary drink, and then their blood glucose concentration is measured regularly over the next few hours. The graph shows the results obtained for two patients.

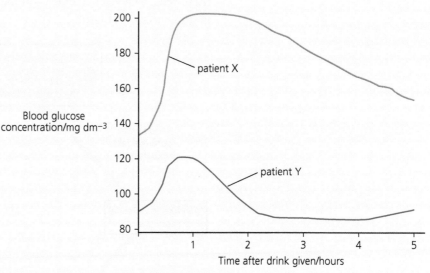

a) The patients had nothing to eat, and only water to drink, for several hours before the test. Explain why. *(1)*

b) Which of patients X and Y has diabetes? Explain your answer. *(2)*

c) i) The blood glucose concentration of patient Y fell between 1 and 2 hours after the sugary drink. Explain why. *(3)*

ii) Patient Y's blood glucose concentration rose slightly between 4 and 5 hours after the sugary drink even though the patient did not have anything to eat and drank only water. Explain how. *(2)*

2 The table shows some substances present in food and their effect on blood glucose concentration.

Substance present in food	Component(s) of substance	Effect on blood glucose concentration
Glucose	Glucose	Large increase
Starch	Glucose	Large increase
Sucrose		Moderate increase
Lactose	Glucose and galactose	Moderate increase
Cellulose	Glucose	No increase

a) Complete the table to show the component(s) of sucrose. *(1)*

b) i) Cellulose and starch have different effects on blood glucose concentration. Explain why. *(3)*

ii) Glucose and lactose have different effects on blood glucose concentration. Explain why. *(3)*

c) Glycogen and glucagon are both compounds that are involved in regulating blood glucose concentration. Explain their different roles. *(5)*

a) i) Describe how glomerular filtrate is formed in the kidney nephron. *(4)*

ii) Normally glucose is not present in the urine. However, people with diabetes may have glucose in their urine. Explain why. *(3)*

b) Furosemide inhibits carrier proteins that reabsorb sodium ions in the ascending limb of the loop of Henle. Explain the effect that this drug will have on water reabsorption in the kidney. (3)

4 The graph shows the effect of the loss of liver and kidney function on the concentration of urea in the blood.

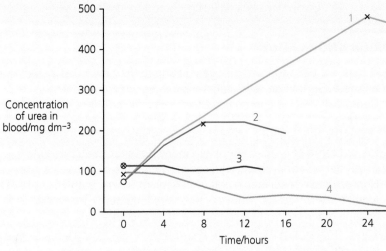

Key:

— kidney function lost at time 0, liver function lost 24 hours later   — liver function lost at time 0, kidney function not lost

— kidney function lost at time 0, liver function lost 8 hours later   × time when liver function was lost

— kidney and liver function both lost at time 0   o time when the kidney function was lost

a) What is the evidence from patient 1 and 2 that urea is produced in the liver? (3)

b) Explain why

   i) the concentration of urea stays approximately the same for 13 hours in patient 3 (2)

   ii) the concentration of urea in the blood falls steadily in patient 4. (2)

Stretch and challenge

5 To what extent does the length of the loop of Henle relate to the environment the animal lives in? Use reference materials to research the relationship between the length of the loop of Henle and water availability in the animal's environment. A good starting point would be desert animals such as the kangaroo rat.

6 Examine the role of the distal convoluted tubule in regulating blood pH. Contrast its role in regulating blood pH with the lungs and circulatory system.

7 Contrast negative feedback and positive feedback. Research hypothermia and hyperthermia, and explain how these are examples of positive feedback. Use your knowledge of physiology to justify the first aid treatments for these conditions.

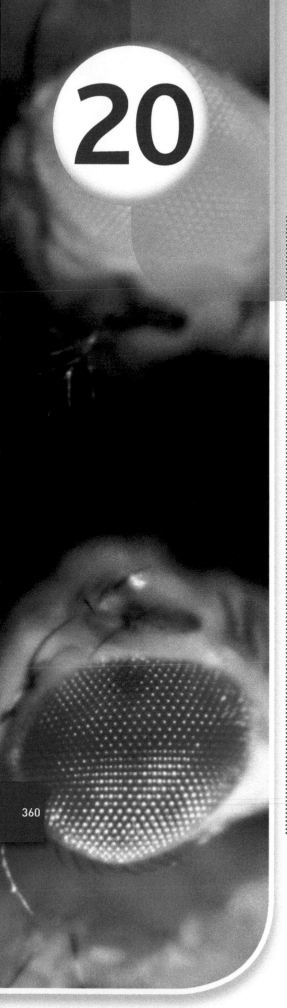

20 Genes, alleles and inheritance

PRIOR KNOWLEDGE

- Genes are passed from one generation to the next in both plants and animals. Simple genetic diagrams can be used to show this.
- Genetic diagrams can be used to predict the outcomes of monohybrid crosses.
- Using these diagrams it is possible to predict and/or explain the outcome of crosses between individuals for each possible combination of dominant and recessive alleles of the same gene.
- Homozygous describes a cell or organism where both variants of a gene are the same, whereas heterozygous describes a cell or organism where both variants of a gene are different.
- The term phenotype refers to the appearance of an organism and the term genotype refers to its genetic makeup.
- In human body cells, one of the 23 pairs of chromosomes carries the genes that determine sex. In human females the sex chromosomes are the same (XX); in human males the sex chromosomes are different (XY).
- Some characteristics are controlled by a single gene. Each gene may have different forms called alleles.
- An allele that controls the development of a characteristic when it is present on only one of the chromosomes is a dominant allele.
- An allele that controls the development of characteristics only if the dominant allele is not present is a recessive allele.
- A gene occupies a fixed position, called a locus, on a particular chromosome.
- A gene is a base sequence of DNA that codes for the amino acid sequence of a polypeptide.
- Random mutation can result in new alleles of a gene.
- Meiosis produces daughter cells that are genetically different from each other.
- The process of meiosis consists of two nuclear divisions, resulting usually in the formation of four haploid daughter cells from a single diploid parent cell.
- Genetically different daughter cells result from the independent segregation of homologous chromosomes.
- Crossing over between homologous chromosomes results in further genetic variation among daughter cells.

Introduction

TEST YOURSELF ON PRIOR KNOWLEDGE

1 Match each word with its definition.

Table 20.1

Word	Definition
1 Heterozygous	**A** An allele that controls the development of a characteristic when it is present on only one of the chromosomes
2 Mutation	**B** A cell in which the two alleles of the same gene are different
3 Genotype	**C** A different form of a gene
4 Phenotype	**D** A sequence of DNA bases that codes for one polypeptide
5 Homozygous	**E** A change in the base sequence of DNA resulting in a new allele
6 Haploid	**F** An allele that controls the development of a characteristic only when the dominant allele is not present
7 Diploid	**G** A cell that contains one set of chromosomes
8 Dominant	**H** The genetic makeup of an individual
9 Recessive	**I** The characteristics of an organism resulting from its alleles plus environment
10 Allele	**J** A cell in which both alleles of a gene are the same
11 Gene	**K** A cell that contains two sets of chromosomes

2 Draw a diagram to show how sex is determined at fertilisation in a human.

3 Describe what happens during crossing over and explain how it results in variation.

4 What is independent segregation?

5 A tall pea plant with genotype Tt is crossed with a dwarf pea plant with genotype tt. What are the possible genotypes and phenotypes in their offspring?

Introduction

Each gene occupies a position on a particular chromosome, called its **locus**. Ever since this was confirmed, less than 100 years ago, geneticists have tried to locate the position of genes on chromosomes.

The first gene maps were based on the results of genetic crosses, like the ones you will read about in this chapter. The reasons that the fruit fly has been used so often in genetics investigations are that it has a short life cycle, is easy to culture in large numbers (which gives statistically reliable results) and has only four pairs of homologous chromosomes.

Very early on, geneticists noticed that fruit flies sometimes inherited the same combination of alleles of different genes from generation to generation. For example, they inherited the alleles for black body colour, vestigial wings and cinnabar eyes as if they were in a single block of genes. The geneticists reasoned that this was because the genes were located very close together on the same chromosome. As a result, the combination of alleles in one parent would be inherited as a single unit, unless they were separated by crossing over in the first prophase of meiosis during gamete formation.

Extension

Cross-over values

By performing large numbers of crosses, geneticists estimated the frequency of crossing over between genes on sister chromatids of homologous pairs during meiosis. They called this the cross-over value. They reasoned that genes with a small cross-over value were closer together and thus less likely to be separated by crossing-over events than those with a larger cross-over value. Thus, cross-over values allowed geneticists to judge the relative positions of genes on chromosomes. Figure 20.1 shows the type of chromosome map they produced for the black body, cinnabar eyes and vestigial wing alleles on chromosome 3 of the fruit fly, *Drosophila melanogaster*.

A second type of chromosome map resulted from studies using optical microscopes. When chromosomes are stained with chemical dyes, bright-coloured and dark-coloured bands appear along their length. A gene's cytogenetic location can be related to these bands on the chromosome. Figure 20.2 shows human chromosome 19. The bands, caused by Giemsa stain, are very obvious.

Figure 20.1 A map showing the position of three alleles (b), (cn) and (vg) on chromosome 3 of *Drosophila*, based on cross-over values.

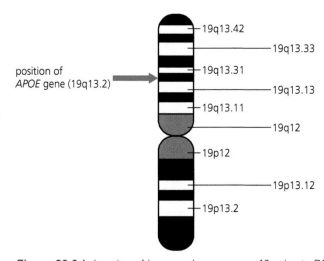

Figure 20.2 A drawing of human chromosome 19 prior to DNA replication. The restriction near the middle of the chromosome is called the centromere and it divides the chromosome into two 'arms'. The position of one gene – the *APOE* gene – is shown.

Extension

The Human Genome Project

The Human Genome Project (HGP) enabled biologists to produce a molecular map of DNA that gives more detailed information than the cytogenetic method above. The HGP was an international project completed in 2003. During the project a number of research teams around the world used different techniques to find the base sequence of each human chromosome. As a result the position of a human gene can be described in terms of the base pairs it occupies along a DNA molecule.

Phenotype and genotype

When we study an organism, we notice a number of distinctive features that it possesses. These might be observable features, such as flower colour or length of beak, or they might be chemical differences, such as the inability to produce lactase in lactose-intolerant people. These observable or measurable features make up the phenotype of an organism.

Figure 20.3 summarises how an organism's phenotype results from an interaction between its genes and the environment in which it lives. An organism's genetic constitution is called its genotype. We can refer to the genotype meaning all of an organism's genes or we can refer to the genotype controlling a single characteristic.

Phenotype The features of an organism that result from an interaction between the expression of its genes and their interaction with the environment.

Genotype The alleles of a gene (genetic constitution), or all of the alleles of all of the genes, that an individual inherits.

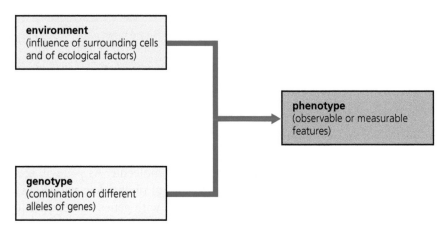

Figure 20.3 An organism's phenotype results from an interaction between its genotype and its environment.

The Himalayan rabbit is an intriguing example of the interaction of genotype and the environment (see Figure 20.4). This rabbit has a gene that encodes an enzyme that results in black fur. As you can see only a few parts of the rabbit's body have black fur; the rest of its fur is white. The enzyme encoded by the black-fur gene has a low optimum temperature. At temperatures above 34 °C this enzyme is denatured. Over most of the rabbit's body the enzyme is inactivated by body heat, resulting in white fur. Only its extremities – the nose, ears, tail and paws – are cold enough for the enzyme to be active.

Many non-scientists seem to believe that an organism's phenotype is controlled entirely by its genotype. For example, certain newspapers tell their readers that scientists have found a gene that 'causes' coronary heart disease (CHD), giving people hope that a genetic cure for CHD can be found. Readers of these newspapers seem to find it convenient to blame their genes (their genotype) for a high risk of CHD and to ignore the environmental risk factors over which they have control, such as a diet rich in saturated fats, smoking tobacco or failure to take exercise.

Figure 20.4 This Himalayan rabbit has inherited genes that result in the production of black fur. Only the fur on its nose, ears, tail and paws is black. Why is the rest of its fur white? The answer lies in the interaction of genotype and environment, in this case temperature, in producing the phenotype.

Scientists locate gene that causes some breast cancers

OBESITY GENE FOUND

Figure 20.5 Some newspaper headlines suggest that there are simple genetic causes of diseases or conditions. Although possession of a particular allele of a gene might *predispose* someone to a disease such as cancer, environmental factors, such as diet or smoking, are also involved.

Genes and alleles

A gene is a sequence of bases in a DNA molecule that encodes another functional molecule. As you will see in Chapter 23, some genes encode molecules of ribonucleic acid (RNA). However, in this chapter we are only concerned with genes that encode functional polypeptides. Thus, in this chapter, we are using the term **gene** to mean a sequence of bases in a DNA molecule that encodes a functional polypeptide. How many bases there are in the DNA sequence determines how many amino acids are in the encoded polypeptide.

Often a gene can have more than one form, each with a slightly different base sequence. An allele is one of two or more different forms of a gene. Look back to page 362; base pairs 50 100 901 to 50 104 488 on chromosome 19 carry the genetic code for apolipoprotein E: it is the *APOE* gene. However, this gene can have at least three different base sequences, called **e2**, **e3** and **e4**. These different sequences are alleles of the *APOE* gene. Figure 20.6 shows how different alleles of a gene result in the formation of polypeptide chains with slightly different amino acid sequences.

In Western Europe, the most common allele of the *APOE* gene is e3. The base sequence of this allele results in the production of a functional lipoprotein that carries excess cholesterol from the blood to the liver; that is, it is fully functional apolipoprotein E.

People with a copy of the e4 allele of the *APOE* gene produce a different polypeptide that is not as efficient at carrying cholesterol to the liver and which results in an increased risk of atherosclerosis (an accumulation of fatty deposits and scar-like tissue in the lining of the arteries). The progressive narrowing of the arteries that can result from atherosclerosis increases the risk of heart attacks and strokes. People with a copy of the e4 allele are also at increased risk of developing clumps of proteins, called amyloid plaques, in their brains. As a result, these people have an increased

Allele An alternative form of a gene.

Key to amino acid

Ala	=	alanine
Arg	=	arginine
Asn	=	asparagine
Cys	=	cysteine
Gln	=	glutamine
Glu	=	glutamic acid
His	=	histidine
Leu	=	leucine
Phe	=	phenylalanine
Ser	=	serine
Thr	=	threonine
Trp	=	tryptophan
Val	=	valine

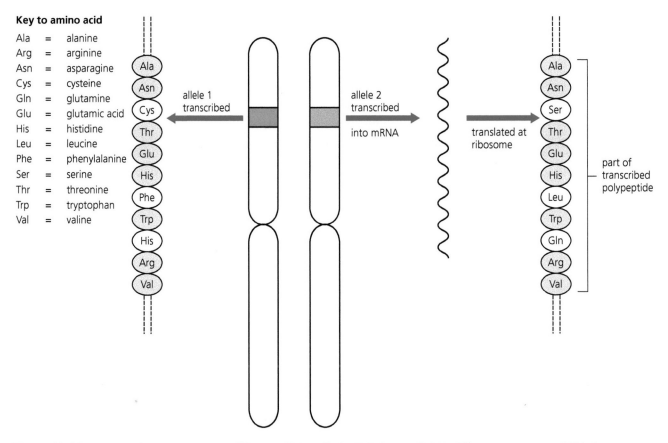

Figure 20.6 A gene may have one or more different alleles. Each allele has a slightly different sequence of DNA bases, resulting in the formation of a polypeptide chain with a different sequence of amino acids.

risk of developing a form of Alzheimer's disease. This risk is even greater if they possess two copies of the e4 allele.

People with a copy of the e2 allele of the *APOE* gene produce yet another polypeptide that increases their risk of developing hyperlipoproteinaemia type III. This is a condition associated with a very high concentration of cholesterol and triglycerides in the blood.

TIPS
- Make sure you know the difference between a gene and an allele, and that you use both terms correctly.
- There are no specific examples of inheritance that you need to learn, but make sure you understand the principles involved.

TEST YOURSELF

1 What is the source of the heat that denatures the enzyme that causes the Himalayan rabbit in Figure 20.4 to produce black fur?
2 Explain the difference between a gene and an allele.
3 Explain why it is wrong to say that the e4 allele causes heart attacks and strokes.

Homozygous and heterozygous genotypes

Look at Figure 20.7 to remind yourself of something you learned in the first year of your A-level Biology course. In sexual reproduction in humans, diploid cells divide by meiosis to produce haploid gametes: egg cells and sperm cells. Each haploid cell contains only one of the homologous chromosomes from each pair. Since these chromosomes carry the same genes in the same order, a haploid cell carries only one copy of each gene. At fertilisation a zygote is formed, which is diploid. This zygote, and every cell formed from it by mitosis, carries two copies of each gene.

In diploid organisms, such as humans, every **somatic** (body) cell has the same genotype. We can refer to the genotype of a somatic cell and the genotype of an individual interchangeably, since they are the same.

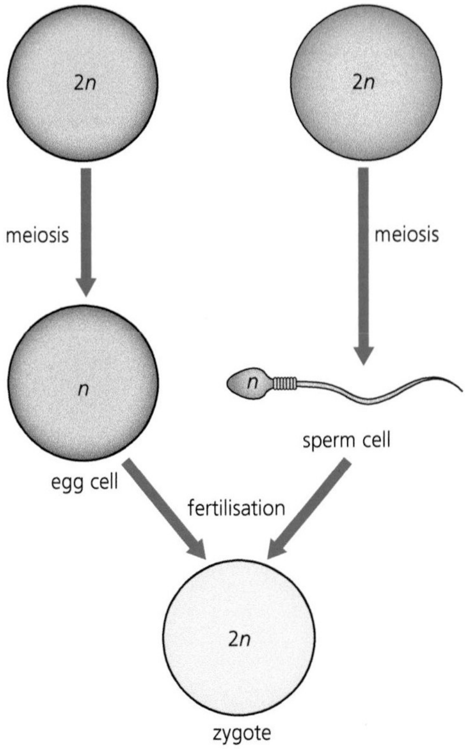

Figure 20.7 Egg cells and sperm cells have only one copy of each chromosome and thus only one copy of each gene carried on a chromosome. At fertilisation, the zygote cell becomes diploid: it has two copies of each gene.

- If an individual possesses two identical alleles of the same gene, we say the genotype is homozygous for that gene. We represent a homozygous genotype using only one type of symbol, such as AA, aa, $C^A C^A$ or $C^a C^a$.
- If an individual possesses two different alleles of the same gene, we say the genotype is heterozygous. We represent a heterozygous genotype using two different symbols, such as Aa or $C^A C^a$.

Homozygous Possessing the same alleles of genes at the same locus on homologous chromosomes.

Heterozygous Possessing different alleles of genes at the same locus on homologous chromosomes.

Dominant, recessive and codominant alleles

We have seen that a gene is a sequence of DNA bases that encodes a functional polypeptide. However, if a gene has two alleles, often one of them will be **dominant** and one will be **recessive**. This means that one (the dominant allele) will code for the functional polypeptide, whereas one (the recessive allele) will code for a different polypeptide, which may be non-functional. As a result, the second allele has an effect on the phenotype that is different from the normal allele of this gene.

Imagine an organism that is heterozygous for one gene. The dominant allele will contribute to the phenotype, but the recessive allele will contribute to the phenotype only if the dominant allele is not present. Look at

Table 20.2. It shows the effect of the three alleles of the human gene encoding the ABO blood group.

Table 20.2 The gene controlling the ABO blood group in humans has three alleles: I^A, I^B and I^O. The symbol 'I' stands for immunoglobulin, a type of globular protein.

Allele of gene controlling ABO blood group	Polypeptide in the surface membranes of red blood cells encoded by allele
I^A	antigen A
I^B	antigen B
I^O	neither antigen A nor antigen B

Notice from Table 20.2 that the I^A and I^B alleles encode functional polypeptides that are located in the cell-surface membrane of red blood cells. These polypeptides act as antigens, meaning that they will cause an immune response in the blood of another mammal that lacks the same polypeptide. In contrast, the I^O allele does not encode a functional polypeptide in the cell-surface membrane.

Table 20.3 shows the genotypes that are possible for the human blood group gene. Remember that, although there are three alleles of this gene, only two of them can be present in any one cell.

Table 20.3 The possible genotypes for the human ABO blood group and the phenotypes associated with them. Here, blood group is the phenotype.

Genotype	Antigen present in surface membrane of all red blood cells	Name of blood group
$I^A I^A$	antigen A	group A
$I^A I^O$	antigen A	group A
$I^B I^B$	antigen B	group B
$I^B I^O$	antigen B	group B
$I^A I^B$	antigen A and antigen B	group AB
$I^O I^O$	neither antigen A nor antigen B	group O

What can we learn from Table 20.3? First, notice that the name of the blood group derives from the polypeptides (antigens) in the surface membranes of the red blood cells. Now look at the first row of the table. This tells us that someone who is homozygous for the I^A allele has only antigen A in the surface membranes of their red blood cells. You might have expected this, but now look at the second row. This tells us that, like the homozygous $I^A I^A$ person, a heterozygous $I^A I^O$ person has only antigen A in the surface membranes of their red blood cells and is also blood group A. From this we conclude that the I^A allele is dominant and the I^O allele is recessive.

The third and fourth rows of the table tell a similar story about the I^B and I^O alleles. We conclude that the I^B allele is dominant and the I^O allele is recessive. Not surprisingly, only when the genotype is homozygous $I^O I^O$ does the effect of the I^O allele show in the phenotype.

Now look at the fifth row of the table. In the heterozygous genotype $I^A I^B$, both antigen A and antigen B are present in the cell-surface membranes of red blood cells. When both the alleles of a gene in a heterozygote show their effect in the phenotype we call them **codominant** alleles.

Monohybrid inheritance involving only dominant and recessive alleles

In studying **monohybrid inheritance**, we follow the inheritance of a single character that is controlled by one gene. As we learn more about gene action, we find that few characteristics are actually controlled by only a single gene. Figure 20.8 shows examples of Fast Plants®, a form of *Brassica rapa* that was bred for research activities and is often used in school and college investigations because it has a very short life cycle. The plants on the right produce a purple pigment, called anthocyanin, in all parts of the plant. The plants on the left do not produce anthocyanin.

Figure 20.8 Fast Plants® normally produce a purple pigment like those on the right of the photograph. These plants have a dominant allele that enables them to produce anthocyanin. The plants on the left of the photograph are homozygous for a recessive allele of this gene. They do not produce anthocyanin.

Constructing a genetic diagram to explain a monohybrid cross

Figure 20.9 shows what happens when a homozygous Fast Plant® that produces anthocyanin is crossed with a plant that does not produce anthocyanin. The dominant allele for anthocyanin production is given the symbol A; the recessive allele is given the symbol a. The **genetic diagram** has been laid out in a particular way.

The first row in the diagram shows the phenotypes of the parents; in this case one can produce anthocyanin and one cannot. The next line shows the genotypes of the parents.

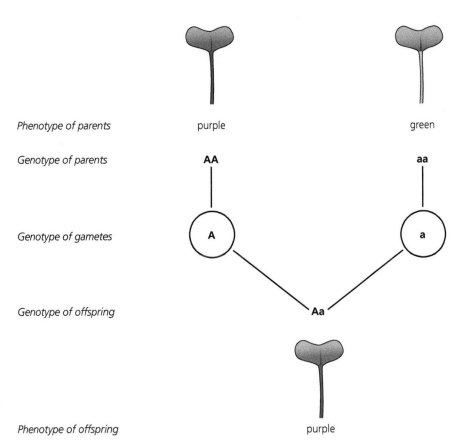

Phenotype of parents   purple   green

Genotype of parents   **AA**   **aa**

Genotype of gametes   **A**   **a**

Genotype of offspring   **Aa**

Phenotype of offspring   purple

Figure 20.9 A genetic diagram showing monohybrid inheritance of anthocyanin production when a purple Fast Plant® is crossed with a green Fast Plant®. You should use the layout of this genetic diagram when answering questions involving monohybrid inheritance.

We know that the parent that cannot produce anthocyanin must be homozygous, since the allele that results in no anthocyanin being produced is recessive (a). If the dominant allele (A) was present in the genotype, the plant would be able to produce anthocyanin. The next row of the genetic diagram shows the genotype of the gametes that each parent can produce. It helps to make the diagram clear if we put these genotypes in a circle.

The genotypes of the gametes contain only one allele, A or a, because gametes are always haploid. Therefore, the gametes produced during monohybrid inheritance will always contain only one of the alleles controlling the character we are investigating. The next row in Figure 20.9 shows the genotype of the offspring. Since all the gametes of one parent contain the A allele and all the gametes of the other parent contain the a allele, all the offspring must have the genotype Aa: they are heterozygotes. Their phenotype shows the effect of the dominant A allele, so they are all able to produce anthocyanin.

Figure 20.10 shows what happens if these heterozygotes are allowed to interbreed. It uses the same layout as Figure 20.9.

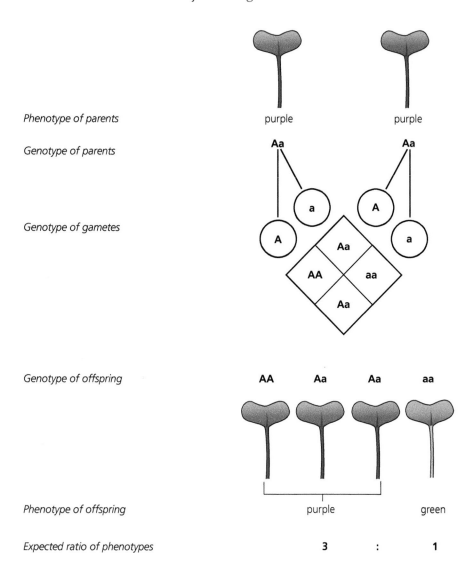

Phenotype of parents	purple purple
Genotype of parents	Aa Aa
Genotype of gametes	a A A a
	Aa
	AA aa
	Aa
Genotype of offspring	AA Aa Aa aa
Phenotype of offspring	purple green
Expected ratio of phenotypes	3 : 1

Figure 20.10 A genetic diagram showing the result of a cross between two heterozygous Fast Plants®. In a monohybrid cross involving a dominant and a recessive allele, a ratio of 3:1 is always expected in the offspring of heterozygous parents.

During the first division of meiosis, homologous chromosomes pair together. They are then pulled apart, one to each end of the spindle, when spindle fibres contract. One chromosome carries the A allele and the other carries the a allele, so the gametes with A and a will be produced in equal numbers.

Look at the fourth row in Figure 20.10. Instead of using the notation we used in Figure 20.9 to show fertilisation, we have used a Punnett square. If we represented fertilisation by drawing lines between the two gametes from one parent and the two gametes from the other, we would end up with a network of crossing lines. Not only does this look untidy, it can confuse us as we complete the genetic diagram. We are much less likely to make a mistake, and so get our answer wrong, by using a Punnett square.

In a monohybrid cross between two heterozygous parents, there are three possible genotypes among their offspring. In this case, they are AA, Aa and aa. Since the A allele is dominant, plants with genotypes AA and Aa can produce anthocyanin, so there are only two phenotypes, purple and green.

The expected ratio of phenotypes in Figure 20.10 is shown as 3 : 1. To make this prediction we have assumed that fertilisation between the gametes of the parents occurs at random. If there are a large number of fertilisations, we expect gametes containing an A allele from one parent to fuse with equal frequency with gametes containing the A allele or a allele from the other parent. This gives us the expected ratio of phenotypes in the offspring.

EXAMPLE

Monohybrid inheritance in *Drosophila*

The first photo at the beginning of this chapter (page 360) shows several adult fruit flies (*Drosophila melanogaster*). These particular flies have long wings, but other adults have small, stunted wings, called vestigial wings. Wing length is controlled by a single gene that has two alleles, one resulting in long wings and one resulting in vestigial wings.

Two heterozygous, long-winged *Drosophila* were mated together. There were 100 flies in the offspring generation. About three-quarters of the offspring had long wings and about one-quarter had vestigial wings.

1 Which of the two alleles for wing length is dominant and which is recessive? Explain your answer.
Long wings (wild type) is dominant because some of the offspring have vestigial wings but neither parent does. Therefore vestigial wings is recessive. We know this

because when you breed long wings and vestigial together you get long-winged offspring. However, few vestigial-winged flies would survive in the wild; therefore, if the gene was dominant the species might become extinct.

2 How many of the offspring generation would you expect to have vestigial wings? Use a genetic diagram to explain your answer.
The fact that some of the offspring are vestigial winged when neither parent is tells you that the allele for vestigial wings is recessive. However, vestigial winged flies have two recessive alleles, one from each parent. Therefore each parent must be heterozygous for wing length.
Let L be the symbol for long wings and l the symbol for vestigial wings.
The parent flies must both be heterozygous.
So of 100 flies, 75 would be expected to have long wings and 25 would be expected to have vestigial wings.

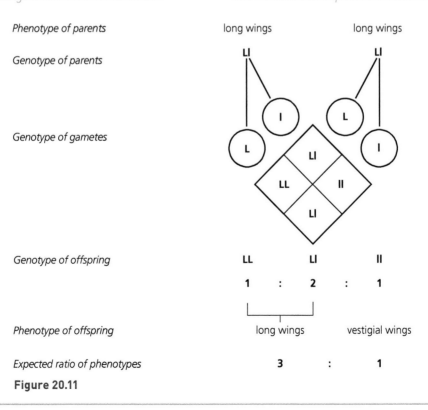

Phenotype of parents	long wings	long wings
Genotype of parents	Ll	Ll
Genotype of gametes		
Genotype of offspring	LL	Ll	ll
	1 : 2 : 1
Phenotype of offspring	long wings	vestigial wings
Expected ratio of phenotypes	3 : 1

Figure 20.11

20 GENES, ALLELES AND INHERITANCE

Monohybrid inheritance involving codominant alleles

Snapdragons are commonly grown in gardens. Some snapdragon plants produce red flowers and some produce white flowers. Flower colour is controlled by a single gene that has two alleles, one for red flowers and one for white flowers. In this example, the two alleles are codominant. Figure 20.12 shows what happens when a snapdragon with red flowers is crossed with a snapdragon with white flowers.

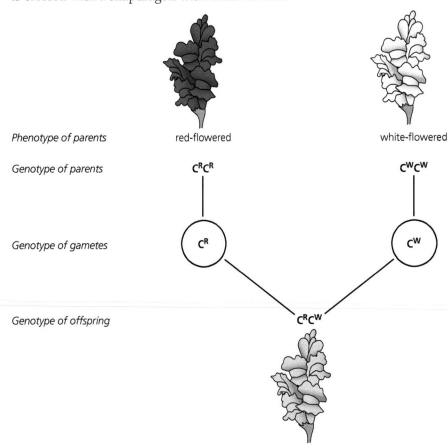

Phenotype of parents	red-flowered	white-flowered
Genotype of parents	$C^R C^R$	$C^W C^W$
Genotype of gametes	C^R	C^W
Genotype of offspring	$C^R C^W$	
Phenotype of offspring	pink-flowered	

Figure 20.12 Flower colour in snapdragons is controlled by one gene that has two codominant alleles. The offspring of a red-flowered plant and a white-flowered plant all have pink flowers.

TIP

Note that when alleles are codominant we use a different kind of symbol for the alleles. In this example we have used the letter C (to represent the gene for colour) with 'W' for white or 'R' for red in superscript. The W and R are capital letters because neither of them is recessive or dominant.

When the red-flowered plant is crossed with the white-flowered plant, all the offspring are heterozygous. If the allele for red flowers was dominant, all these heterozygotes would have red flowers; if the allele for white flowers was dominant, all these heterozygotes would have white flowers. As you can see in Figure 20.12, the heterozygotes produce flowers that are neither red nor white; they are pink. Why are they pink? Because the allele for red flowers and the allele for white flowers are codominant: they both show their effect in the phenotype of a heterozygote. Some red and some white pigment is produced, making the flowers pink.

You can see the effect of codominant alleles in Figure 20.13. This diagram shows two different crosses, one between two plants with pink flowers and another between a plant with pink flowers and a plant with white flowers.

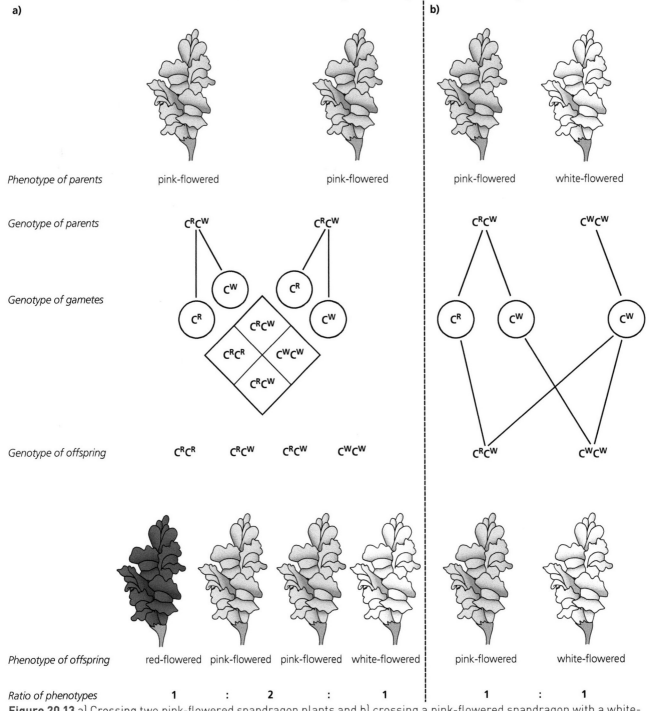

Figure 20.13 a) Crossing two pink-flowered snapdragon plants and b) crossing a pink-flowered snapdragon with a white-flowered snapdragon.

You can see that, instead of the 3 : 1 ratio we expected in monohybrid inheritance with a dominant and recessive allele (Figure 20.10), we get a ratio of 1 : 2 : 1 (Figure 20.13a).

EXAMPLE

Inheritance of coat colour in horses

Like all mammals, horses have hair. We refer to a horse's hair as its coat. Some horses have a reddish-brown-coloured coat, called bay, and some have a white-coloured coat. One gene for coat colour is involved, but it has two alleles, C^R and C^W. If a horse is homozygous $C^R C^R$ all the hairs in its coat will be reddish-brown and the horse will have a bay coat. If a horse is homozygous $C^W C^W$ all the hairs in its coat will be white and the horse will have a white-coloured coat. However, if a horse is heterozygous, $C^R C^W$, half the hairs in its coat will be reddish-brown and half will be white. The result is a horse with a pinkish-colour coat. This coat colour is referred to as red-roan. Coat colour in horses is an example of monohybrid inheritance with codominant alleles of a single gene. Figure 20.14 shows horses with these three coat colours.

Phenotype	bay coat	white coat	red-roan coat
Genotype	$C^R C^R$	$C^W C^W$	$C^R C^W$

Figure 20.14 Horses with a reddish-brown coloured coat (bay) are homozygous $C^R C^R$; horses with a white-coloured coat are homozygous $C^W C^W$. Half the hairs in the coat of heterozygotes, $C^R C^W$, are reddish-brown and half are white, producing a coat colour called red-roan.

1 In the coat of a red-roan horse, half the hairs are reddish-brown and half are white. What does this suggest about the alleles of the coat-colour gene in the hair-producing cells?
It suggests the alleles are codominant.

When a stallion with a bay coat was mated with a mare with a white coat, all the offspring were red-roan.

2 Why, in the above statement about Figure 20.14, was it unnecessary to explain whether the parent horses were homozygous or heterozygous?
A horse with bay coat or white coat must be homozygous, since the heterozygotes are red-roan.

3 What ratio of coat colour would you expect in the offspring of red-roan parents? Use a genetic diagram to explain your answer.
You would expect offspring that were red-roan, bay and white in a ratio of 2 : 1 : 1. Figure 20.15 shows the genetic diagram.

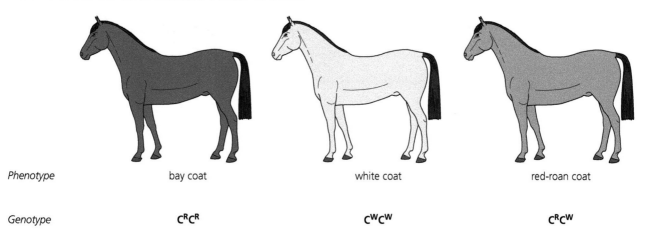

Parent phenotype	red-roan	×	red-roan
Parent genotype	$C^R C^W$	×	$C^R C^W$

Offspring genotypes	$C^R C^W$	$C^R C^R$	$C^W C^W$
	2 :	1 :	1
Offspring phenotypes	red-roan :	bay :	white
Expected ratio of phenotypes	2 :	1 :	1

Figure 20.15 Genetic diagram for coat colour in horses.

Monohybrid inheritance involving multiple alleles and codominance

We have looked at monohybrid inheritance involving genes with only two alleles. Many genes have more than two alleles; we refer to them as **multiple alleles**. We saw in Table 20.3 (page 367) the genotypes and phenotypes involved in the inheritance of the human ABO blood group. This feature involves multiple alleles (three alleles of one gene) and it involves codominance. Figure 20.16 shows a cross involving the ABO blood group. One parent is blood group AB and the other is blood group O.

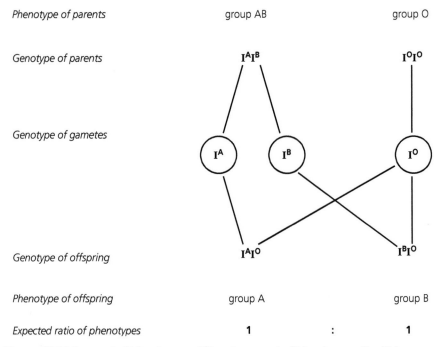

Phenotype of parents	group AB	group O
Genotype of parents	$I^A I^B$	$I^O I^O$
Genotype of gametes	I^A I^B	I^O
Genotype of offspring	$I^A I^O$	$I^B I^O$
Phenotype of offspring	group A	group B
Expected ratio of phenotypes	1 :	1

Figure 20.16 A parent of blood group AB and a parent of blood group O will have children of blood group A or blood group B, but none with the same blood group as the parents.

Because data from genetic crosses is in discrete categories, the best way to present the totals for each phenotype would be on a bar graph. You would use a chi-squared (χ^2) test to find out whether the number of each phenotype matched the expected ratio.

TIP

Please look at Chapter 26 on maths skills, to find out more about statistical tests and to see a worked example of this test.

TEST YOURSELF

11 What ratio of blood groups would you expect among the offspring of a mother with the genotype $I^A I^O$ and a father with the genotype $I^B I^O$? Explain your answer.

12 In Andalusian fowl, F^B is the allele for black plumage and F^W is the allele for white plumage. These alleles show codominance. The heterozygous condition results in blue plumage. List the genotypic and phenotypic ratios expected from the crosses:
   a) black × blue
   b) blue × blue
   c) blue × white

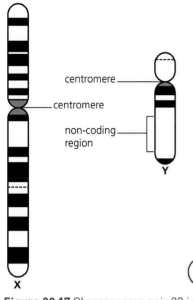

Figure 20.17 Chromosome pair 23 in a human male. The two homologous chromosomes are different sizes and only share a small number of genes. The shorter Y chromosome does not have a copy of many of the genes present in the X chromosome.

Inheritance of sex

In humans, the members of a homologous pair of chromosomes are usually the same size and carry the same genes in the same order. This is not, however, true of chromosome pair 23. Figure 20.17 shows chromosome pair 23 from a human male. One of the chromosomes (the Y chromosome) is very short and carries few genes; the other (the X chromosome) is long and carries many more genes. Although the short Y chromosome carries very few genes, one of them is crucial in determining sex. This gene, called the testis-determining gene (or *SRY*), turns the developing sex organ of a 7-week-old embryo into a testis, and the embryo develops into a male. Because the X chromosome carries some genes which the Y chromosome does not, a male is effectively haploid for these genes.

A human male has one X chromosome and one Y chromosome; his genotype is XY and half his sperm cells will carry an X chromosome and half will carry a Y chromosome. A human female has two X chromosomes; her genotype is XX and all the eggs she ever produces will carry an X chromosome.

Monohybrid inheritance involving a sex-linked character

A sex-linked character is one that is controlled by a gene located on one of the sex chromosomes. A sex-linked gene is on the X chromosome. Examples of sex-linked characters in humans include haemophilia (a disorder in which the blood of sufferers clots only slowly), red-green colour blindness and Duchenne muscular dystrophy.

Figure 20.18 represents what might happen when a woman with blood that clots normally has children with her husband who has haemophilia. The symbols we are using in this diagram represent two things: the sex chromosome and the alleles of the gene for blood clotting. The allele for normal blood clotting is dominant, so we represent it as H. The allele for slow blood clotting is recessive, so we represent it as h. However, this gene is located on the X chromosome.

Figure 20.18 Possible children produced by a normal woman whose husband has haemophilia.

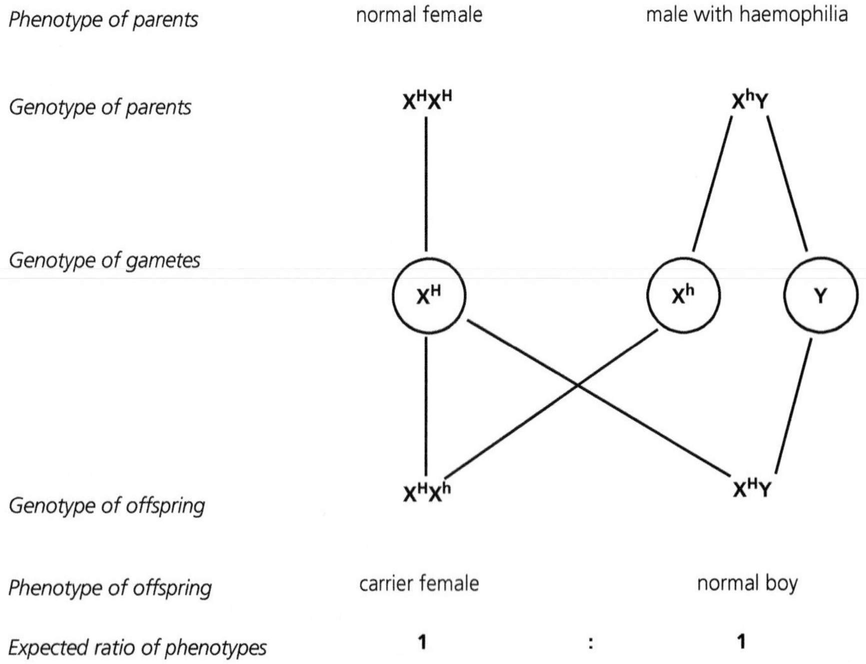

Phenotype of parents	normal female		male with haemophilia
Genotype of parents	$X^H X^H$		$X^h Y$
Genotype of gametes	X^H	X^h	Y
Genotype of offspring	$X^H X^h$		$X^H Y$
Phenotype of offspring	carrier female		normal boy
Expected ratio of phenotypes	1	:	1

We show this using the symbol X^H for an X chromosome with the H allele and X^h for an X chromosome with the h allele. Since the Y chromosome has no copy of the gene for blood clotting, it is shown simply as Y.

Notice in Figure 20.18 that any boy born to this couple will have blood that clots normally; he will not have haemophilia like his father. Any girl born to this couple will be heterozygous for the blood-clotting gene. We call her a **carrier** for haemophilia but, since the H allele is dominant, her blood clots normally. Figure 20.19 shows what might happen if this girl who is a carrier grows up, marries a man whose blood clots normally and has children.

Figure 20.19 Possible children produced by a woman who is a carrier for haemophilia and her normal husband.

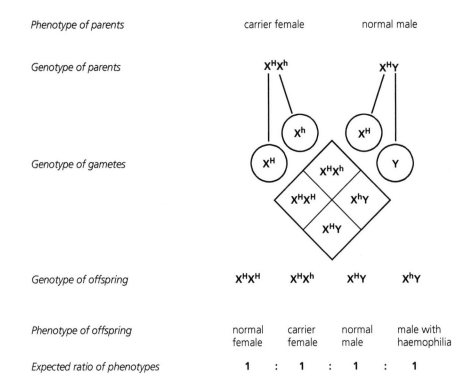

	carrier female	normal male
Phenotype of parents		
Genotype of parents	$X^H X^h$	$X^H Y$

Genotype of offspring	$X^H X^H$	$X^H X^h$	$X^H Y$	$X^h Y$
Phenotype of offspring	normal female	carrier female	normal male	male with haemophilia
Expected ratio of phenotypes	**1** :	**1** :	**1** :	**1**

TEST YOURSELF

13 Use a genetic diagram to explain why there is a 1 in 2 chance of a child being a girl. Use the symbols X and Y to denote X and Y chromosomes.

14 Suggest why sex-linked genes are usually on the X chromosome rather than on the Y chromosome.

15 Would it be possible for a girl to have haemophilia? Explain your answer.

Notice that the woman in Figure 20.19 can produce a son whose blood clots normally and a son who has haemophilia. The son with haemophilia inherits the recessive gene for haemophilia on the X chromosome from his mother, whose own father had the disorder (see Figure 20.18). This is typical of characteristics that are controlled by genes that are located on the X chromosome: the condition appears in males in alternate generations.

Interpreting pedigrees

So far, we have used genetic diagrams to explain a pattern of inheritance. An alternative method for presenting information about inheritance is shown in Figure 18.20 (overleaf). This diagram shows a **pedigree**. In pedigrees such as this, females are represented by circles and males by squares. Couples who have children are linked by a horizontal line and the children are represented in a hierarchical fashion below them. In the case of humans, we cannot breed them, so a pedigree is a record of the family tree of crosses in the past.

You can see in the figure that individual 1 is represented by a square, so he is male. The square contains the symbol A, so he has blood group A. Individual 2 is shown by a circle containing the symbol B, so she is a female with blood group B.

Figure 18.20 This pedigree shows the inheritance of ABO blood groups in three generations of a family.

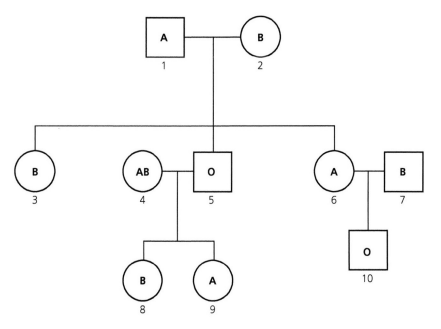

We can see who, out of individuals 3, 4, 5, 6 and 7, are the children of couple 1 and 2. The diagram shows a horizontal line between individuals 1 and 2. This indicates that they had children together. The vertical line from individuals 1 and 2 goes directly downwards to connect to each of their children, so individuals 3, 5 and 6 are the children of the parents 1 and 2. Individuals 4 and 7 are not connected by vertical lines to the parents 1 and 2, so are not their children. You can see from the horizontal lines that individual 4 is the partner of 5, and individual 7 is the partner of 6.

A pedigree only tells us phenotypes, it does not tell us genotypes. When we interpret a pedigree, we have to use clues provided by the phenotypes of some individuals to work out the genotypes of as many people in the pedigree as we can. The skill you need to develop is to identify which individuals provide these clues.

Individual 5 is blood group O, so we know his genotype is $I^O I^O$. Having found this, we can work backwards and forwards through the pedigree to work out other genotypes. Since individual 5 has the genotype $I^O I^O$, he must have inherited one I^O allele from each parent. However, his father (individual 1) has blood group A, so his father's genotype must be $I^A I^O$. The mother of individual 5 (individual 2) has blood group B, so her genotype must be $I^B I^O$.

Individual 3 is blood group B. She must have inherited the I^B allele from her mother (individual 2). If she had inherited the I^A allele from her father, she would be blood group AB. Since she is blood group B, she must have the genotype I^BI^O. Individual 6 is blood group A. She must have inherited the I^A allele from her father. She cannot have inherited an I^B allele from her mother or she would be blood group AB, so she must have the genotype I^AI^O.

Since individual 10 is blood group O, we know he must have the genotype I^OI^O. This means that his mother (6) and father (7) must carry the I^O allele, so they are I^AI^O and I^BI^O, respectively. Consequently, they are equally likely to produce offspring with the genotype I^AI^O, I^AI^B, I^BI^O or I^OI^O. If you are not sure why, draw a genetic diagram of $I^AI^O \times I^BI^O$ to see what offspring can result. This gives a probability of a child with blood group O of 0.25. The probability of a child being a boy is 0.5. So, the probability of a boy with blood group O is $0.25 \times 0.5 = 0.125$.

Dihybrid inheritance

Dihybrid inheritance involves a phenotype that is inherited as the result of two different genes. This involves the same principles you learned carrying out monohybrid crosses. However, the number of possible phenotypes increases because there are more different ways in which the alleles of two different genes can combine.

Dihybrid inheritance with no linkage

Let's look at the inheritance of two characters in pea plants: stem height and flower colour. Each character is controlled by a different gene. The two genes are located on different chromosomes; that is, they are unlinked.

- The gene for height has two alleles, T (tall) and t (dwarf).
- The gene for flower colour has two alleles, R (red) and r (white)

Figure 20.21 shows the results of crossing a homozygous tall, red-flowered pea plant with a dwarf, white-flowered plant.

Notice that the offspring of these two plants are all tall and red-flowered plants. This is because the parents can only produce one kind of gamete, so all the offspring have a dominant allele for height and a dominant allele for flower colour. However, when these mature offspring are interbred, each parent can produce four different kinds of gamete. This is because, during meiosis, either allele of the gene for height can end up with either allele of the flower colour gene. At fertilisation, these gametes give 16 different combinations in the offspring.

If you look carefully at the Punnett square in the diagram, you will see that the phenotypic ratio for each character is what you would expect from your previous work:
3 tall : 1 dwarf
3 red-flowered : 1 white-flowered.

However, due to independent assortment of non-homologous pairs of chromosomes, this dihybrid cross gives a phenotypic ratio of:
9 tall, red-flowered plants
3 tall, white-flowered plants
3 dwarf, red-flowered plants
1 dwarf, white-flowered plant.

Figure 20.21 Dihybrid inheritance in peas.

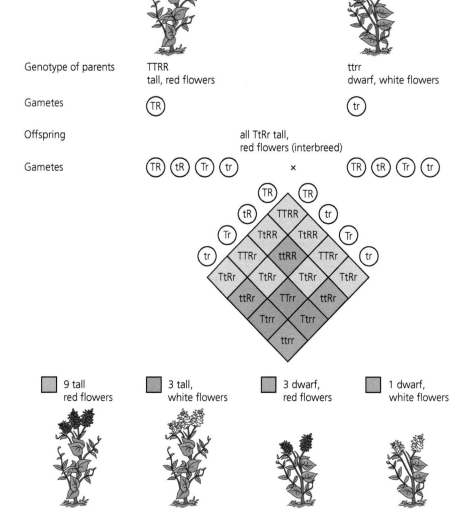

Genotype of parents	TTRR tall, red flowers		ttrr dwarf, white flowers

9 tall
red flowers

3 tall,
white flowers

3 dwarf,
red flowers

1 dwarf,
white flowers

TEST YOURSELF

16 In squash, the gene for fruit colour has two alleles. The allele for white fruit colour (W) is dominant over that yellow fruit colour (w). The gene for fruit shape also has two alleles. The allele for disc-shaped fruit (D) is dominant over that for sphere-shaped fruit (d).

A plant that produced white, disc-shaped fruit was crossed with a plant that produced yellow, sphere-shaped fruit. All the offspring produced white, disc-shaped fruit.

a) Give the genotypes of the two parent plants and of the offspring.

b) The offspring plants were interbred. Use a Punnett square to show the possible genotypes and phenotypes of their offspring, and the ratio in which you would expect them to be produced.

17 In shorthorn cattle, the gene for possession of horns has two alleles. The allele for the polled (hornless) condition, H, is dominant over that for horned (h). The gene controlling coat colour has two codominant alleles, C^R and C^W. The genotype C^RC^R results in a red coat and C^WC^W results in a white coat but the heterozygote, C^RC^W, is roan.

a) A homozygous polled white male is crossed with a horned, red female. What will be the appearance of the offspring?

b) If you bred two of the offspring together, what would be the possible genotypes and phenotypes of their offspring? Use a Punnett square.

18 In humans, red-green colour blindness is caused by a sex-linked recessive allele. A woman with blood group O who has normal vision but had a father with red-green colour blindness, marries a man with normal colour vision and blood group AB. Use a Punnett square to show the possible genotypes and phenotypes of their children.

Epistasis

In cells, enzymes control reactions. Many metabolic pathways, such as respiration and photosynthesis, consist of several steps, with each step controlled by a different enzyme. This means that each enzyme in the pathway depends on the previous enzyme to provide its substrate. If any of the enzymes in a pathway is non-functional, the pathway comes to a halt.

Figure 20.22 shows a metabolic pathway involving two enzymes. You can see that a different gene codes for each enzyme. The effect of enzyme B depends upon the action of enzyme A. Homozygous recessive individuals for gene A (aa) produce an inactive enzyme A, so they are unable to catalyse the reaction that results in a pale blue pigment. Even if this individual produces an active form of enzyme B, the individual will be colourless.

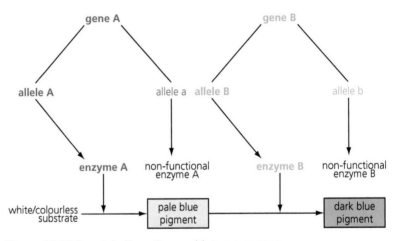

Figure 20.22 A metabolic pathway with two enzymes.

This is called **epistasis**, which means that the expression of one gene affects the expression of another. This occurs in metabolic pathways controlled by enzymes coded for by different genes. Epistasis reduces the number of possible phenotypes.

Figure 20.23, overleaf, shows inheritance of comb shape in chickens, which is another example of epistasis. Comb shape is controlled by two genes, each with two alleles, that interact to produce four different phenotypes:

the rose gene has two alleles, R and r
the pea gene has two alleles, P or p.

Table 20.4

Shape of comb	Genotype
Rose	RRpp or Rrpp
Pea	rrPP or rrPp
Walnut	RrPp or RRPp or RrPP
Single	Rrpp

Figure 20.23 Inheritance of comb shape in chickens.

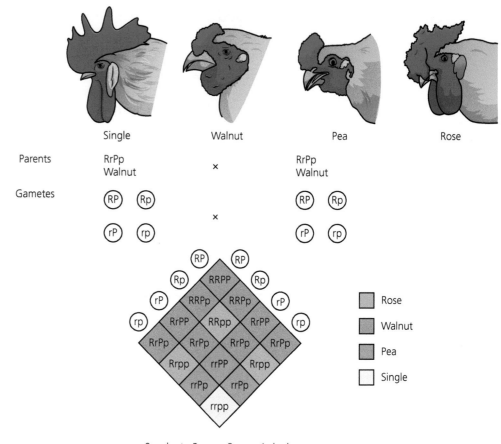

| | Single | Walnut | Pea | Rose |

Parents — RrPp Walnut × RrPp Walnut

Gametes — RP Rp rP rp × RP Rp rP rp

	Rose
	Walnut
	Pea
	Single

9 walnut : 3 rose : 3 pea : 1 single

TEST YOURSELF

19 One gene for fur colour in mice has three alleles, C^A, C^B and C^C.
The results of some crosses between mice are shown in Figure 20.24.

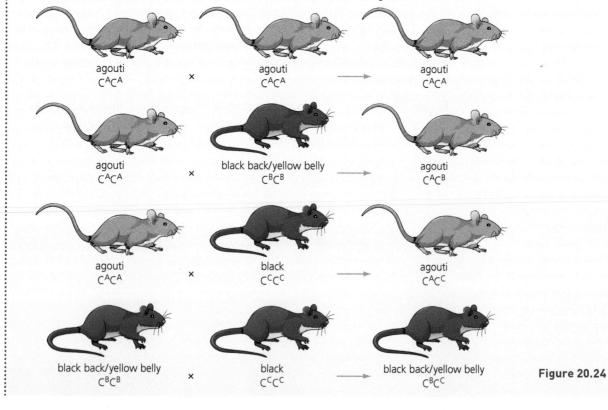

agouti C^AC^A	×	agouti C^AC^A	→	agouti C^AC^A
agouti C^AC^A	×	black back/yellow belly C^BC^B	→	agouti C^AC^B
agouti C^AC^A	×	black C^CC^C	→	agouti C^AC^C
black back/yellow belly C^BC^B	×	black C^CC^C	→	black back/yellow belly C^BC^C

Figure 20.24

a) Complete the table to show the possible genotype(s) for each phenotype

Phenotype	Possible genotypes
Agouti	
Black back yellow belly	
Black	

b) Use a Punnett square to show the possible offspring that would be produced from a cross between an agouti mouse of genotype C^AC^C and a mouse with a black back and yellow belly of genotype C^BC^B.

20 The diagram shows a metabolic pathway in sweet peas. Two plants with the genotype CcPp were interbred. Use a Punnett square to show the ratios of phenotypes and genotypes in their offspring.

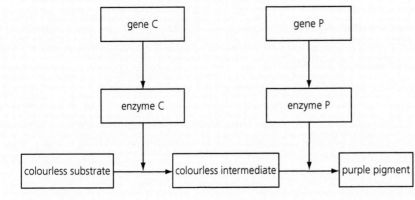

Figure 20.25

Dihybrid cross with autosomal linkage

Recombinant offspring An organism that contains a different combination of alleles from either of its parents.

In Figure 20.21 you saw that, when two tall, red-flowered (TtRr) pea plants were crossed, some of the offspring showed the same traits as their parents, i.e. tall and red-flowered, but others were dwarf and/or had white flowers. This is only possible because the gene for height in pea and the gene for flower colour are on different chromosomes. This allows independent assortment to occur so the parent plants could produce four different gametes. The T or t allele passed into the gametes independently of the R or r allele. As a result, the offspring included tall, white-flowered plants, dwarf, red-flowered plants and dwarf, white-flowered plants. These are called recombinant offspring because they are displaying new combinations of these traits.

In one kind of maize, there is a gene that determines seed colour. One allele of this gene (Y) codes for an enzyme that results in yellow seeds, while the recessive allele (y) codes for a faulty enzyme that results in colourless seeds. Another gene codes for an enzyme that controls the shape of the seeds. The dominant allele of this gene (S) results in smooth seeds whereas the recessive allele (s) results in wrinkled seeds.

Table 20.5 shows the result of a cross between a maize plant of genotype YySs and a plant of genotype yyss.

If these two genes were located on different chromosomes, independent assortment would occur. As we have seen earlier, this cross would then result in four different phenotypes in roughly equal proportions in the offspring. Instead, as you can see in Table 20.5, the results show a very strong tendency for the yellow allele, Y, and smooth allele, S, to be inherited together and for the colourless allele, y, and wrinkled allele, s, to be inherited together.

Table 20.5 Inheritance of seed colour and shape in one variety of maize.

Gametes			Gametes produced by YySs parent			
			YS	Ys	yS	ys
	Gametes produced by yyss parent	ys	YySs	Yyss	yySs	yyss
Phenotype			Yellow smooth	Yellow wrinkled	Colourless smooth	Colourless wrinkled
Proportion expected with independent assortment (%)			25	25	25	25
Actual results (%)			48.2	48.2	1.8	1.8

The explanation is that the gene loci for the seed colour gene is on the **same chromosome** as the seed shape gene. As they are on the same chromosome, alleles of the two genes cannot be separated by independent assortment and are inherited together,

Nevertheless, you can see in Table 20.5 that there are a few recombinants produced in this cross. Recombinant offspring have a different combination of alleles from both parents. This can be explained by crossing over during meiosis (see Chapter 11, page 195). If a chiasma is formed between the seed colour gene and the seed shape gene, the chromatids of homologous chromosomes exchange pieces and a new combination of alleles is produced.

Therefore the genes for seed colour and seed shape in this variety of maize are said to be **linked**; that is, they are located on the same chromosome. As this chromosome is not a sex chromosome – an **autosome** – we refer to this type of linkage as **autosomal linkage**.

Geneticists have studied many characteristics in organisms such as maize. They have found many genes that are linked. Some of these give a higher proportion of recombinants than the example we have just studied, while others give a lower proportion. Geneticists assume that the more recombinants produced, the further apart two gene loci are. From this information they have discovered linkage groups of genes that tend to be inherited together, and they have worked out the relative positions of the genes on the chromosomes. This is what you learned about at the start of this chapter.

Linkage group Sets of genes on the same chromosome which tend to be inherited together.

TEST YOURSELF

21 When tall tomatoes with red fruit are crossed with dwarf tomatoes with yellow fruit, all the offspring are tall with red fruit. When these offspring are interbred, they also produce offspring that are tall with red fruit, except for a very few dwarf plants with yellow fruit. What does this tell you about the inheritance of height and fruit colour in tomatoes?

22 In sweet peas, the allele of the gene for flower colour that results in purple flowers (R) is dominant over that for red flowers (r) and the allele of the gene for pollen shape that results in long pollen (L) is dominant over that for round pollen (l).

Two plants heterozygous for flower colour and pollen shape were interbred.

a) Use a Punnett square to show the ratios of genotypes and phenotypes that you would expect in the offspring.

b) The actual results were 296 plants with purple flowers and long pollen, 19 with purple flowers and round pollen, 27 with red flowers and long pollen and 85 with red flowers and round pollen. Explain these results.

Practice questions

1 Tay–Sachs disease is a rare and usually fatal genetic disorder that causes progressive damage to the nervous system. The diagram shows a family in which some individuals have Tay–Sachs disease.

  a) Give the evidence from the diagram that

   i) Tay–Sachs disease is caused by a recessive allele *(2)*

   ii) Tay–Sachs disease is not sex-linked. *(2)*

  b) Give the possible genotype(s) for

   i) individual 4

   ii) individual 6

   iii) individual 10. *(3)*

(from a family with no history of Tay–Sachs)

Key:

☐ = unaffected male   ▨ = Tay–Sachs male

○ = unaffected female   ◕ = Tay–Sachs female

2 a) What is a sex-linked gene? *(2)*

  b) In cats, the X chromosome carries a gene for coat colour. The gene has two alleles that show codominance: X^G results in ginger fur, X^B results in black fur. Cats with the genotype X^GX^B are covered in patches of ginger fur and patches of black fur, referred to as tortoiseshell. A tortoiseshell female was mated with a black male. Use a genetic diagram to show the possible genotypes and phenotypes among the offspring resulting from this cross. *(4)*

  c) Tortoiseshell male cats are very rare. Explain why. *(2)*

3 Some forms of clover are cyanogenic. This means that they produce cyanide gas when their leaves are damaged. The production of cyanide gas occurs using a metabolic pathway controlled by two enzymes. A different gene encodes each of these enzymes.

  a) Explain why the following genotypes of clover cannot produce cyanide gas (i.e. they are acyanogenic): aaBB, aaBb, AAbb, Aabb. *(1)*

  b) Two clover plants, each with the genotype AaBb, were crossed.

   i) Give all the possible genotypes among the gametes that each parent could produce. *(1)*

   ii) Use a genetic diagram to show the genotypes of the offspring of this cross. *(2)*

   iii) What proportion of the offspring are likely to be acyanogenic? *(1)*

4 The ABO blood group in humans is determined by a single gene with three alleles: I^A, I^B and I^O.

    a) i) Distinguish between the terms gene and allele. *(1)*

    ii) What is meant by codominance? *(1)*

    b) The diagram shows the inheritance of ABO blood groups in one family. The blood group of some individuals is given in the pedigree.

Using information in the diagram and the symbols I^A, I^B and I^O, give the phenotype and genotype of each of the following individuals in the pedigree. Give reasons for your answers. *(4)*

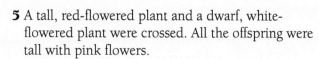

Individual	Phenotype	Genotype	Explanation
2			
5			
4			
9			

5 A tall, red-flowered plant and a dwarf, white-flowered plant were crossed. All the offspring were tall with pink flowers.

    a) What was the genotype of the tall, pink-flowered plants? *(1)*

    b) These tall, pink-flowered plants were interbred to produce a large number of offspring. Use a genetic diagram to show the ratio of phenotypes and genotypes in their offspring. *(3)*

    c) What results would you expect if you crossed a tall, red-flowered plant with a dwarf, white-flowered plant, and the genes for height and flower colour were on the same chromosome? Explain your answer. *(2)*

    d) A plant breeder wanted to find out whether a particular tall plant was homozygous or heterozygous for height. Explain how the breeder could carry out a genetic cross to solve this problem. *(3)*

Stretch and challenge

6 Geneticists can produce a chromosome map from the results of crosses involving autosomal linkage. Explain how they would do this.

7 Explain how sex is determined in humans. Turner's syndrome, Klinefelter's syndrome and Swyer syndrome are all abnormalities of sex chromosomes in humans, for example. Compare and contrast how these abnormalities occur.

8 Explain how sex is determined in birds, insects, reptiles and in some species of animal that are able to change sex.

20 GENES, ALLELES AND INHERITANCE

21 Gene pools, selection and speciation

PRIOR KNOWLEDGE

- Two organisms belong to the same species if they are able to produce fertile offspring.
- Genetic diversity can be defined as the number of different alleles of genes in a population and is a factor enabling natural selection to occur.
- Natural selection in the evolution of populations involves four principles.
- First, random mutation can result in new alleles of a gene.
- Second, many mutations are harmful but in certain environments the new allele of a gene might benefit its possessor, leading to increased reproductive success.
- Third, advantageous alleles are inherited by members of the next generation.
- Finally, as a result advantageous alleles usually increase in frequency in the population over many generations.

TEST YOURSELF ON PRIOR KNOWLEDGE

1 What is a mutation?
2 Scientists carrying out captive breeding of zoo populations of endangered species aim to interbreed animals that are as unrelated to each other as possible. Explain why.
3 Orangutans from the island of Borneo, *Pongo pygmaeus*, and orangutans from the island of Sumatra, *Pongo abelii*, are classified as separate species. However, in zoos it is possible for them to interbreed and produce fertile offspring.
   a) Suggest why they were originally classified as separate species.
   b) Responsible zoos do not allow these two species of orangutan to interbreed. Suggest why.

Introduction

In the 20th century, the rabbit population in the UK grew so large that rabbits became a serious agricultural pest. They also became a pest across mainland Europe and in Australia, where they had been introduced to provide a food source for early settlers.

In the late 19th century, laboratories in Uruguay had discovered a virus that caused skin lesions in the local rabbits but which killed European rabbits. This virus was the myxomatosis virus and the disease that killed European rabbits was myxomatosis. During the 1940s, a number of research organisations studied the use of myxomatosis as a potential biological control method to reduce rabbit populations.

Figure 21.1 You might have seen our common rabbit species, the European rabbit, feeding in fields or parks early in the morning or during the late evening. It might surprise you to learn that rabbits were introduced by the Normans from mainland Europe about 800 years ago and were originally bred for food and fur. With changes in agricultural practices, rabbits became serious agricultural pests in the 20th century.

Initially, governments were unwilling to sanction this method of biological control. One reason for the UK government's reluctance was public opinion: although farmers wanted to be rid of rabbits, most people in the UK had been brought up on cultural traditions that were sympathetic to rabbits. Perhaps you had a fluffy toy rabbit, or read stories involving rabbit characters, as a child. Another reason was that scientists at the time feared the virus might 'jump the species barrier' and infect animals other than rabbits. If this idea seems familiar, exactly the same concern has been expressed more recently about the avian (bird) influenza virus.

Before governments decided to sanction the use of myxomatosis to control rabbits, events took their course. A retired physician, who owned a rabbit-infested farm, released the virus in France in mid-1952. From France, the virus reached England, at Edenbridge in Kent, in 1953 and spread rapidly throughout the UK, killing almost 90% of the rabbit population. In Australia, the virus 'escaped' from a government laboratory in the Murray Valley sometime during the Christmas–New Year holiday in 1950–51. It spread quickly throughout the Murray-Darwin basin, killing millions of rabbits.

Although myxomatosis initially devastated rabbit populations, they have since recovered. Rabbit populations are nothing like as large as they were, but they are now in balance with the myxomatosis virus. How has this balance happened? One reason is that rabbits have become more resistant to the myxomatosis virus. This has occurred through a process called natural selection, in which rabbits with alleles giving resistance to myxomatosis had greater reproductive success than rabbits that were susceptible to myxomatosis.

It might surprise you to learn that natural selection has also changed the virus population. In England, the myxomatosis virus is transmitted from rabbit to rabbit by fleas, when they bite rabbits to feed on their blood. However, in Australia, the virus is transmitted by mosquitoes. A mosquito will only bite a living rabbit, so any virus that kills its host too quickly is unlikely to be passed on to another host by a mosquito. In Australia, natural selection has favoured myxomatosis viruses that are less virulent, that have hosts that live longer and provide greater opportunity for the virus to be passed to another host via a mosquito bite.

Gene pools

Members of a species do not live alone; they live in populations. A **population** is a group of organisms of the same species occupying a particular space at a particular time and potentially able to interbreed. In Chapter 20 we looked at the inheritance of genes by individuals. Now we will look at the alleles of a gene controlling a single characteristic in a population.

Figure 21.2 shows two adults of a single species of the banded snail, *Cepaea nemoralis*. One of the snails has a yellow shell and the other has a pink shell. This shell colour is controlled by two alleles of a single gene. The pink allele (C^P) is dominant over the yellow allele (C^y). Imagine there are 1000 snails in a population living in a beech woodland. Since every snail has two alleles in its genotype for shell colour (C^PC^P, C^PC^y or C^yC^y), we can say that there are 2000 alleles in the shell-colour genotypes of snails in this population. This represents the gene pool for this characteristic in this population. In general, we can define the **gene pool** as all of the alleles of all of the genes that are present in a population at any given time.

Figure 21.2 The banded snail is common in woodlands and grasslands in the UK. The difference in shell colour is controlled by two alleles of a single gene.

Allele frequencies in the gene pool of a population

We saw above that a population of 1000 individuals of banded snail has a gene pool of 2000 alleles of the gene for shell colour. If all the snails in this population had the genotype C^PC^P, all the alleles in the gene pool would be C^P. In other words, the **frequency** of the C^P allele in this gene pool would be 1. If all the snails had the genotype C^PC^y, the frequency of the C^P allele in the gene pool of this population would be 0.5 and the frequency of the

C^y allele would also be 0.5. Therefore, an allele frequency of 1 is equivalent to 100% of the alleles of that gene in the population, and an allele frequency of 0.5 is equivalent to 50% of the alleles in that population.

TIP

Note that C^PC^y, where colour pink is dominant over colour yellow, is the standard notation used by the vast number of researchers working on populations of *Cepaea nemoralis*. However, at A-level you would use the notation given in the question, which might be different.

Hardy–Weinberg: the principle and the equation

Hardy and Weinberg were two scientists who looked at **allele frequencies** within populations. The principle that carries their names, the **Hardy–Weinberg principle**, predicts that the frequency of the alleles of one gene in a particular population will stay the same from generation to generation. In other words, there will be no genetic change in the population over time. The principle is based on mathematical modelling and makes the following five important assumptions.

1 **The population is large**. In small populations, chance events can cause large swings in allele frequencies.

2 **There is no movement of organisms into the population (immigration) or out of the population (emigration).** Any such movement of organisms would result in new alleles entering the gene pool or existing alleles leaving the gene pool.

3 **There is random mating between individuals in the population.** This ensures that there is an equal probability of any allele of a gene being passed on to the next generation.

4 **All genotypes must have the same reproductive success.** This also ensures that there is an equal probability of any allele of a gene being passed on to the next generation.

5 **There is no gene mutation.** Any mutation of genes would cause some alleles in the gene pool to change to different alleles of the same gene.

Using the Hardy–Weinberg equation

We cannot always tell the genotype of an individual from its phenotype. For example, you learned in Chapter 20 that an organism showing the effect of a dominant allele could be homozygous for that allele or it could be heterozygous. This is where the Hardy–Weinberg equation is helpful. It allows us to calculate the frequencies of alleles in the gene pool and of genotypes and phenotypes in the population. Therefore we can use the Hardy–Weinberg equation to find the probability of certain phenotypes in a future generation of a population.

We can always tell which phenotype has the homozygous recessive genotype because the recessive allele only shows its effect in a homozygote. In a heterozygote, the effect of the recessive allele is masked by that of the dominant allele.

The Hardy–Weinberg equation is applied to two alleles of a gene. The symbol A represents the dominant allele, which will always show its effect in the phenotype, and the symbol a represents the recessive allele. Since we are dealing with allele frequencies, we need symbols to represent the frequencies of the two alleles in the gene pool. We use the following symbols

p = frequency of the A allele in the gene pool
q = frequency of the a allele in the gene pool.

The gene pool for this gene consists of only two alleles, A and a. The frequency of the gene itself is 1.0, since all organisms have the gene. Consequently, the frequencies of the two alleles added together must also be 1, i.e. $p + q = 1$.

To summarise so far, we have a population with a gene pool for a particular gene that has two alleles. The frequency of the dominant allele, A, is p and the frequency of the recessive allele, a, is q.

So, what is the frequency of genotype AA in the population of individuals? The frequency of the A allele is p, so the frequency of the AA genotype must be $p \times p$, or p^2. We can work out the frequency in the population of each of the three possible genotypes, AA, Aa and aa.

The Hardy–Weinberg equation is:

$$p^2 + 2pq + q^2 = 1$$

In a single population

the frequency of the AA genotype is p^2
the frequency of the aa genotype is q^2
the frequency of the Aa genotype is $2pq$.

If you do not feel confident about the mathematics, do not worry. So long as you remember the above frequencies, you will find the calculations involving the Hardy–Weinberg equation are very easy sums to do.

Let's go back to our beechwood population of 1000 banded snails. When a group of students investigated this population, they found that 160 of the snails had yellow shells and 840 had pink shells. We always start with the phenotype which gives away its genotype, i.e. the homozygous recessive. In this case, yellow-shelled snails are homozygous recessive (C^yC^y).

So, there were 160 snails with yellow shells in the population of 1000 snails. To work out the frequency of yellow-shelled snails in this population you must remember that frequencies are always given as a decimal value. There were 160 yellow-shelled snails in the population of 1000 snails. The frequency of these yellow-shelled snails is 160/1000 = 0.16.

Now we have worked out that the frequency of homozygous recessive snails is 0.16, we need to calculate the frequency of the recessive allele in this population. The Hardy–Weinberg equation tells us that the frequency of the homozygous recessive individuals is q^2. So, if we know the frequency, we simply need to use a suitable scientific calculator to find its square root. In this case the frequency of the yellow-shelled snails $q^2 = 0.16$. Therefore, the frequency of the allele for yellow shells $q = \sqrt{0.16} = 0.4$.

We can now calculate the frequency of the A allele in the gene pool. The Hardy–Weinberg equation tells us that $p + q = 1$. We have just calculated the value of q, so we find the value of p as $1 - q = 1 - 0.4 = 0.6$.

If you did these calculations correctly, you can pat yourself on the back. We are now in a position to work out the frequency of the three genotypes controlling snail shell colour in the beechwood population.

- The frequency of the genotype $C^P C^P = p^2 = 0.6 \times 0.6 = 0.36$.
- The frequency of the genotype $C^P C^y = 2pq = 2 \times 0.6 \times 0.4 = 0.48$.
- The frequency of the genotype $C^y C^y = q^2 = 0.4 \times 0.4 = 0.16$.

We know that our calculation must be correct because $p^2 + 2pq + q^2 = 0.36 + 0.48 + 0.16 = 1$.

We know from the data above that the students found 160 yellow-shelled snails in the woodland population. From this we can work out how many snails in the population were homozygous with pink shells and how many were heterozygous with pink shells. Hopefully, you will now find the last part of the calculation very easy; all we need to do is to multiply the frequency by the number of snails in the population.

- The frequency of homozygous pink-shelled snails is 0.36. There were 1000 snails in the population, so the number of homozygous pink-shelled snails is $0.36 \times 1000 = 360$.
- The frequency of the heterozygous snails is 0.48, so the number of heterozygous pink-shelled snails is $0.48 \times 1000 = 480$.

We can always check that we are correct, because if we add up the number of different snails we have calculated, we should find they give us the total in the population. In this case, $160 + 360 + 480 = 1000$.

TIP

Always check you have the right answers by adding your values for p^2, q^2 and $2pq$. If your arithmetic is right they will add up to 1.

TEST YOURSELF

1 Would you expect the Hardy–Weinberg principle to have held true over the past 20 years in the human population of the town in which you live? Explain your answer.

2 In humans, the ability to roll the tongue is determined by the dominant allele T. Non-rollers are homozygous recessive (tt).

In a school, the following data were obtained:

Tongue-rollers	Non-rollers	Total
490	210	700

How many of the tongue-rollers were heterozygous for tongue-rolling?

3 For one of the genes controlling coat colour in mice, the allele for agouti coat (A) is dominant over that for non-agouti (a). In a sample, 16% were found to have a non-agouti coat.

a) What are the frequencies of the agouti and non-agouti alleles in the population?

b) What percentage of the population would you expect to be homozygous for A, and what percentage would you expect to be heterozygous?

4 'Woolly hair' is common among Norwegian families. People with this condition have hair that is tightly kinked and easily broken. The gene controlling this condition has two alleles. The allele for woolly hair (H) is dominant over that for normal hair (h). In a population of 1200 people, 1092 individuals had woolly hair. Use the Hardy–Weinberg formula to calculate the frequency of each of the genotypes HH, Hh and hh.

5 In a population of 20000, 16800 people were found to have the Rhesus positive blood group. The allele for Rhesus positive is dominant over Rhesus negative. How many people in this population are heterozygous for the Rhesus blood group?

Natural selection resulting in differential reproductive success

The Hardy–Weinberg principle predicts that allele frequencies in a large population remain stable from generation to generation. One of the assumptions on which the Hardy–Weinberg principle is based is that all genotypes in the population have equal reproductive success. This will not be true if organisms with one particular genotype have a phenotype that makes them:

- more likely to die before reproducing
- unable to grow sufficiently well to reproduce successfully
- unable to attract a mate.

In all these cases, organisms with one particular phenotype are less likely to reproduce successfully than others in the population with different phenotypes and will leave fewer, or no, offspring. We say there is **differential reproductive success** between the phenotypes in the population.

Differential reproductive success is common in populations. In beech woodlands, the yellow-banded snails shown in Figure 21.2 are very conspicuous against the pink leaf litter lying on the ground. The snails with pink shells are better camouflaged and so are more difficult to find. Song thrushes are birds that eat banded snails. Like us, song thrushes have colour vision. A large number of investigations have shown that song thrushes find more of the conspicuous yellow-shelled snails than they do pink-shelled snails in beech woodlands. As a result, fewer yellow-shelled snails survive to reproduce. This means that fewer of the C^y alleles are passed on to the next generation. The process by which the frequency of alleles in a population changes, because different genotypes have differential reproductive success, is called **natural selection**. Table 21.1 summarises the process of natural selection.

Table 21.1 An explanation of natural selection in a beech woodland population of banded snails. The events in the left-hand column can be applied to any example of natural selection.

Sequence of events leading to natural selection	Application of these events to selection of yellow-banded snails in beech woodlands
Within a population there is variation in phenotypes. These phenotypes result from genetic variation in the genotypes controlling this characteristic and environmental factors.	The C^P and C^y alleles of the shell colour gene result in snails with pink shells and snails with yellow shells.
There is differential reproductive success between the different phenotypes. In other words, some phenotypes are better adapted to the environment, so they survive longer and therefore are more likely to reproduce more.	Yellow-shelled snails are more conspicuous than pink-shelled snails among beech litter. Song thrushes find yellow-shelled snails more easily than they find pink-shelled snails. Fewer yellow-shelled snails survive to reproduce.
Organisms with greater reproductive success leave more offspring than those with less reproductive success.	In a beech woodland, pink-shelled snails have more offspring than yellow-shelled snails.
Organisms with greater reproductive success will be more likely to pass their combination of alleles on to their offspring. As a result, the frequency of particular alleles will increase in the population; i.e. natural selection has occurred. **Evolution is a change in the allele frequencies in a population**.	In a beech woodland, the frequency of the pink allele will increase and the frequency of the yellow allele will decrease. Pink-shelled snails are at a selective advantage in beech woodlands.

Directional, stabilising and disruptive selection

Natural selection operates only on phenotypes that organisms possess and results in populations being better adapted to their environment. It can result in a change in a population from one phenotype to another or it can result in a reduction of variation of a particular phenotypic character about an optimum modal value.

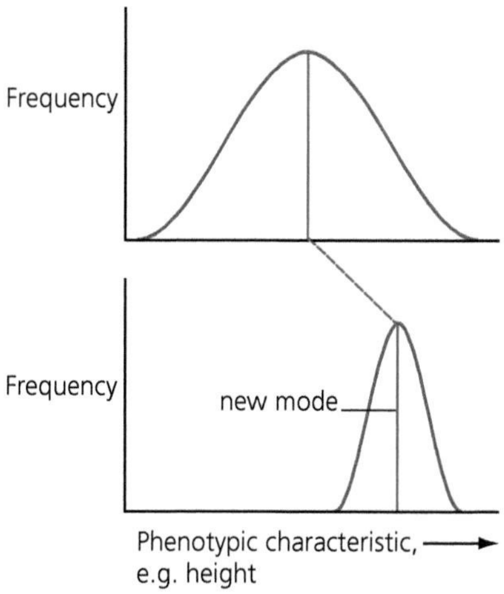

Directional selection

Directional selection is usually associated with a change in the environment. It acts against one of the extremes and/or in favour of the other extreme in a range of phenotypes, e.g. height. As a result, one or more phenotypes becomes rarer and alternative phenotypes become more common (Figure 21.3).

Look first at the upper of the two graphs. It represents a frequency distribution of phenotypes in a population. The graph is a normal distribution with a fairly large standard deviation. The mode (most frequent value) is marked in red. The graph represents the frequency distribution of this population before natural selection has occurred. The lower graph is a frequency distribution of the population after directional selection has occurred. The standard deviation of this curve is less than the upper curve and its mode has shifted to the right on the x-axis. Natural selection has caused a change in the allele frequencies in the population, favouring organisms with a characteristic towards the upper end of the range of the distribution.

Stabilising selection

Stabilising selection is normally associated with a stable environment in which populations have already become well adapted to the environment. It acts against both extremes in a range of phenotypes. As a result, variation about the mode is reduced (Figure 21.4). Again, the upper graph shows the frequency distribution of this population before natural selection. The lower graph shows the frequency distribution of the same population after natural selection has occurred. This time, the mode is in the same position: this is the most advantageous phenotype. Stabilising selection has reduced the variation about this modal value.

Stabilising selection occurs on birth mass in humans. Babies with very low or very high birth masses have a higher infant mortality rate than those with a birth mass near the mode of the range.

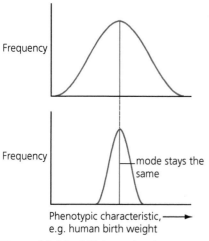

Figure 21.3 The graphs show variation in one characteristic of a population. The upper graph shows the range of phenotypes before natural selection; the lower graph shows the range of phenotypes after directional selection has occurred.

Figure 21.4 Stabilising selection reduces the variation of one phenotype in a population. The upper graph shows the range of phenotypes before natural selection; the lower graph shows the range of phenotypes after stabilising selection has occurred.

Disruptive selection

This is where both extreme phenotypes are selected for. You can see this in Figure 21.5.

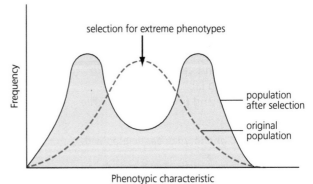

Figure 21.5 Disruptive selection. In disruptive selection, both extremes of the phenotype are selected for, and the intermediate phenotype is selected against.

An example of this would be where pale-coloured snails and dark-coloured snails are both camouflaged in an environment, but intermediate-coloured snails are not. Therefore intermediate-coloured snails are more susceptible to predators, meaning that snails that are either pale or dark are more likely to survive than the intermediate-coloured snails.

TEST YOURSELF

6 Populations of banded snails also live in grassland, where they are preyed on by song thrushes. Which shell colour – yellow or pink – would you expect to be more common in grasslands? Explain your answer.

7 Look back to Figure 21.3 on the previous page. How can you tell that the lower curve has a smaller standard deviation than the upper curve?

8 The National Health Service (NHS) was introduced to the UK in July 1948. Suggest why natural selection on human birth mass has had less of an effect in this country since 1948 than before this date.

EXAMPLE

Natural selection and sickle-cell anaemia

Many large towns and cities in the UK have sickle-cell anaemia clinics. Sickle-cell anaemia is caused by a mutant allele of the gene controlling the production of β-globulin, one of the polypeptides in a haemoglobin molecule (Figure 21.6). Although the mutant allele changes only one amino acid in the β-globulin chain, its effect is striking.

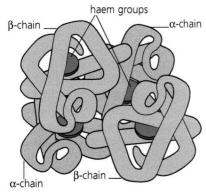

Figure 21.6 A haemoglobin molecule consists of four polypeptides, two α-globulin chains and two β-globulin chains, and four haem groups that bind to oxygen.

When the concentration of oxygen is low, haemoglobin molecules with the mutant β-globulin chains have abnormally low solubility and form fibres within red blood cells. This causes the red blood cells to change from disc-shaped cells to sickle-shaped cells. You can see these differences in Figure 21.7. The sickle shape reduces the surface area of red blood cells.

Sickle cells are also targeted for destruction by the immune system, so have a much shorter life span than normal red cells (about 15 days compared with the normal 120 days). Both these differences result in anaemia. Sickle cells are also liable to get stuck in capillaries, so that nearby tissues become starved of oxygen.

1 Sickle cells have a smaller surface area and shorter life span than normal red blood cells. Anaemia occurs when the blood cannot carry enough oxygen. Explain why both these differences result in anaemia.

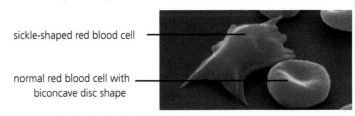

Figure 21.7 Red blood cells from a person suffering sickle-cell anaemia. Sickle cells have haemoglobin molecules with β-globulin chains that differ by one amino acid from those in normal disc-shaped cells.

Having a smaller surface area reduces the volume of oxygen that can diffuse into the cells, so the cells carry less oxygen than normal cells, leading to anaemia. Also, if the cells have shorter life span than normal cells, this means that, at any time, there will be fewer red blood cells in the blood. Therefore less oxygen is transported in the blood, causing anaemia.

The production of β-globulin is controlled by a single gene that has two alleles. We will call the β-globulin gene Hb and its alleles Hb^A and Hb^S. Table 21.2 shows the possible genotypes and phenotypes for this characteristic.

Table 21.2 The genotypes and phenotypes associated with sickle-cell anaemia.

Genotype	Phenotype
Hb^AHb^A	All red blood cells have normal haemoglobin and are disc-shaped.
Hb^AHb^S	The red blood cells contain a mixture of normal and abnormal haemoglobin. People with this genotype have the sickle-cell trait, but are healthy and usually show no symptoms of sickle-cell anaemia.
Hb^SHb^S	All red blood cells contain abnormal haemoglobin, which causes all red blood cells to become sickled in low oxygen concentrations. People with this genotype have sickle-cell anaemia, the complications of which can shorten their lives.

2 What does Table 21.2 tell you about the Hb^A and Hb^S alleles?
They are codominant. You can tell this because the heterozygote has a different phenotype from both the homozygotes. Both alleles produce a functional protein.

3 Sickle-cell anaemia can shorten people's lives. Explain why you might expect natural selection to reduce the frequency of this allele in human populations.
People with sickle-cell anaemia, without medical treatment, are likely to die in childhood. Therefore they are unlikely to pass on their alleles, which will reduce the frequency of the sickle-cell allele in the population.

Despite its apparent disadvantages, sickle-cell anaemia is common among people of African, Middle Eastern or southern Mediterranean descent. To explain this, we need to look at how malaria is spread. Malaria is common in Africa and in parts of the Middle East and the southern Mediterranean. Malaria is caused by a single-celled organism, called *Plasmodium falciparum*. This parasite is transmitted when a female *Anopheles* mosquito bites and sucks blood from an infected person and then injects saliva, containing several parasites, as she bites another person. Once in the blood system of a human host, the parasites enter the host's red blood cells where they use oxygen in the red blood cells for their own respiration.

4 Suggest how the behaviour of a red blood cell that is infected by *P. falciparum* will differ in someone who suffers from sickle-cell anaemia and someone who does not.
The P. falciparum parasite will respire inside the cell, using up oxygen. Therefore infected red blood cells will carry less oxygen. The reduction in oxygen concentration caused by respiration of P. falciparum causes the red blood cells of a sickle-cell anaemia sufferer to become sickled. This causes them to be destroyed by the body's immune system within about 15 days. As they are destroyed, the P. falciparum within them is also destroyed. This makes people with the sickle-cell trait less susceptible to malaria than people with no sickle-cell condition.

5 In an area where malaria is endemic, which genotype, Hb^AHb^A, Hb^AHb^S or Hb^SHb^S, will be at a selective advantage? Explain your answer.
Hb^AHb^S will be at an advantage. This is because Hb^SHb^S are likely to die prematurely from sickle-cell anaemia, whereas people with Hb^AHb^A are more likely to die of malaria. The heterozygotes have enough normal red blood cells to avoid the worst symptoms of sickle-cell anaemia, and they have resistance to malaria.

6 What is the probability that two parents, both of whom have the sickle-cell trait, will have a child that suffers from sickle-cell anaemia? Use a genetic diagram to justify your answer.

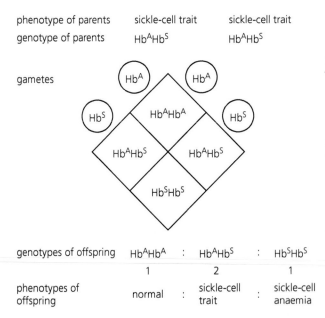

Figure 21.8 Genetic diagram to show the probability that two parents, both of whom have the sickle-cell trait, will have a child that suffers from sickle-cell anaemia.

Figure 21.8 shows the genetic diagram for working out the probability that two parents with the sickle-cell trait will have a child who has the condition. The probability of two parents who have sickle-cell trait having a child with sickle-cell anaemia is 1 in 4, or 25%.

Speciation

Natural selection brings about changes in the allele frequencies in a population (in other words, evolution). This can lead to the appearance of new species.

Figure 21.9 shows four closely related species in the taxonomic family Canidae. The two dogs belong to different breeds but are both the same species called *Canis familiaris*. Different breeds of dog originated when humans bred ancestral dogs that had useful characteristics. Jack Russell terriers (Figure 21.9c) were originally bred for hunting foxes, and bullmastiffs (Figure 21.9d) were bred to find and immobilise poachers. When humans breed animals or plants for their useful characteristics, we call this **artificial selection, or selective breeding**. The other animals in Figure 21.9, the wolf (*Canis lupus;* Figure 21.9a) and the jackal (*Canis aureus;* Figure 21.9b), are naturally occurring species. They arose by evolution, of which natural selection is a part. The formation of new species by natural selection is called **speciation**.

Figure 21.9 These animals belong to the same taxonomic family, Canidae. a) A wolf; b) a jackal; c) a Jack Russell terrier; d) a bullmastiff. The different species resulted from natural selection and the different breeds of dogs resulted from artificial selection by humans.

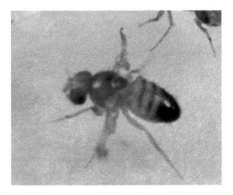

Figure 21.10 This male fruit fly has very short (vestigial) wings, which make him less successful in courtship than males that have long wings.

You learned in the first year of your Biology A-level course that organisms belong to the same **species** if they breed together in their natural habitat and produce fertile young. So how can one species give rise to another during speciation?

The accumulation of genetic differences is vital for speciation to occur by natural selection. Figure 21.10 shows a fruit fly, belonging to the genus *Drosophila*. You might have seen fruit flies like this on over-ripe fruit in your home or around your dustbin in summer. There are many different species of fruit fly. The differences between them are so small that you would probably not be able to tell one from another. However, the flies can tell. Before mating, male and female fruit flies undergo a courtship ritual. The male performs a dance, during which he vibrates his wings, changes his body position and licks the female. The whole sequence, including the response of the female, is controlled by many genes and is species-specific. If a male does not have the correct courtship dance, the female does not allow him to mate. The male in Figure 21.10 is homozygous for a mutant allele of the gene for wing length. As a result, he has very short (vestigial) wings. This will not kill him, but it means that his courtship dance is less likely to attract a female to mate with him. He is less likely to pass on his alleles of the gene for wing length.

As a result of gene mutation, the male fruit fly in Figure 21.10 will probably die without leaving any offspring. His mutated genes will not be passed on. However, suppose these genetic changes occurred not in an individual but in an entire population that was isolated from another population of the same species. We now have conditions in which speciation can occur.

For two populations to become two different species, they have to become reproductively isolated. The accumulation of differences in their separate gene pools might eventually result in these two populations not being able to reproduce and produce fertile offspring. Reproductive isolation may occur if a mutation causes one population to reproduce at a different time of year from the other population. This would prevent the two populations from being able to interbreed, so they would then become different species.

Allopatric speciation

Allopatric speciation is the formation of new species from populations of the same species that are geographically isolated. To understand this, let's imagine a population that has become isolated. This might happen, for example, if a pregnant rabbit was washed on a piece of driftwood from the mainland to an offshore island where there were no other rabbits. Here, our female would give birth to a group of offspring. Rabbits do not have the same taboos as humans about mating with their brothers and sisters, so these rabbits would mature and breed with each other. In a short space of time, there would be a population of rabbits on the island. Because rabbits are not strong swimmers, there would be no interbreeding between the rabbits in the island population and those in the mainland population. We can say that there is unlikely to be **gene flow** between the island population's gene pool and the mainland population's gene pool. Each population of a species lives in a different, though often similar, environment. Therefore we would expect different selection pressures to act on local populations.

TIP
You will find out about how genes mutate in Chapter 23.

TIP
Remember that mutations occur at random.

Gene mutations occur at a constant, low rate in any population. Some gene mutations are beneficial, resulting in organisms with phenotypes due to the beneficial allele of a gene reproducing more successfully than others. As a result, the frequency of these beneficial alleles in the gene pool of this population will increase. In our imaginary example, this process will happen separately in the mainland and island populations of rabbits. Over a period of time, differences in their gene pools may accumulate. Look back to the definition of species that we used earlier. If any of the accumulated differences in the two gene pools prevent rabbits from the mainland and island populations mating, then, by our definition, we now have two species of rabbit. If any of the accumulated differences in the two gene pools result in hybrids between the mainland and island populations being infertile then, again according to our definition, we have two species of rabbit, because they are reproductively isolated.

TEST YOURSELF
9 Canidae is the name of a taxonomic group, called a family. Name the taxonomic group represented by a) *Canis* and b) *familiaris*.
10 Islands in an archipelago often have species of a single genus that are unique to each island and all different from the species on the mainland. Explain why.
11 Explain the difference between the terms haploid, diploid and polyploidy.
12 What are the genetic causes of variation between the different members of a population?
13 Give examples of factors that might lead to some individuals surviving and others not surviving.

TIP
Remember that individuals cannot adapt to a change in the environment. Individuals that by chance have phenotypes that are, well adapted to the environment are more likely to survive and pass on their alleles. You should never write things like 'The species needed to adapt, so...'.

Sympatric speciation

Sometimes a new species can arise without two populations being geographically isolated. An example of how this could happen is the evolution of copper tolerance in certain plants.

The evolution of copper tolerance

The mine waste that formed the hill shown in Figure 21.11 contains high concentrations of copper ions. The fields around the copper waste contained populations of a species of grass called bent grass but for many years none grew on the hill.

1 Seeds from bent grass plants are dispersed great distances by the wind. Suggest why no bent grass plants grew on the hill in Figure 21.11.
There were copper ions present in the soil on the hill. These are toxic to bent grass, so it is unable to grow there.

After several years, bent grass plants began to grow on the hill. Experiments found that these plants were tolerant of copper in the soil.

2 Suggest how copper tolerance had originated in the copper-tolerant plants.
A mutation occurred in a plant that made it able to grow in the presence of copper. This plant survived, and passed on its allele for copper tolerance to its offspring, so this population of bent grass plants were able to grow on the hill.

3 The genes that enabled bent grass to become copper-tolerant spread through the population as a result of natural selection. However, the populations growing in the fields around the hill are not copper-tolerant. Suggest why.
Where copper is present in the soil, it is an advantage to be copper-tolerant. However, the allele for copper tolerance does not confer an advantage on plants growing where copper is not present, and in these areas copper tolerance may actually be a disadvantage.

Figure 21.11 This hill was formed when waste was tipped from a nearby copper mine.

4 Suggest how you could test whether these two populations have become separate species.
You could interbreed plants from the two different populations to see whether they produce fertile offspring. If they do, then they are the same species. In this example, the copper-tolerant grass is not yet a separate species from the copper-intolerant grass. However, over time it is possible that a mutation might occur in one of the populations that affects their reproduction. For example, one population might develop a mutation that causes them to flower at a different time of the year from the other population. If this occurs, the two populations will no longer be able to interbreed. At this point we can say that they are separate species.

Genetic drift

Genetic drift is an effect that can occur in small populations. It occurs when a few individuals, just by chance, either fail to reproduce or have more offspring than others. This then changes the allele frequency in the next generation. A good way to explain this is to imagine that you have a bag of jelly babies, with 100 red and 100 green sweets in it. Without looking, you remove 10 sweets from the bag. Next you pass the bag to a friend, who does the same. If you repeat this a few times, and then look at the sweets each person has, it is likely that the samples will be very different. It might be that you have five red and five green sweets, but it is even possible that you have 10 sweets of one colour and none of the other colour. If the jelly babies you have taken out were real organisms, this process would have separated them from the rest of the population and prevented them from breeding. You can imagine that the allele frequencies in the small populations could be very different from the allele frequencies in the original population, just by chance.

Genetic drift may occur because of genetic bottlenecks or the founder effect, both resulting from chance events in small populations.

Northern elephant seals provide an example of a **genetic bottleneck**. These seals were hunted by humans so much that, by the end of the 19th century, only about 20 seals were left. Their population is much bigger now – hundreds of thousands – but their genetic diversity (in terms of the variety of alleles) is still very limited because all the present-day seals have arisen from a small number of individuals.

The occurrence of polydactyly among the Old Amish population of Pennsylvania in the USA provides an example of the **founder effect**. Polydactyly is a genetic condition in which people have additional fingers and sometimes toes. In most human populations this condition is very rare, but in this Amish population it is much more common than in other groups of humans. The reason for this is that the Amish population arose from about 200 German immigrants in the 18th century, one or two of whom possessed the allele for polydactyly. This is a high frequency of the allele in a very small population. Since then, the Amish people have tended to marry people within their own population, so a relatively small variety of alleles has remained in the gene pool.

TIP

You do not need to recall the terms 'genetic bottleneck' or 'founder effect'.

TEST YOURSELF

14 Explain why genetic drift is important in small populations but not in large populations.

15 In a small region of north-west Venezuela there is an unusually high frequency of a severe inherited neuromuscular disorder known as **Huntington's disease**. Approximately 150 people in the area during the 1990s had this rare fatal condition and many others were at high risk of developing it. This disease usually does not strike until early middle age, after most people have had their children. All of the victims of Huntington's disease in this region trace their ancestry to a single woman who moved into the area in the 19th century. She had a very large number of descendants and there is now a population of about 20 000 people with a high risk of having this genetically inherited condition.

a) Explain why the disease is unusually common in this area.

b) Huntington's disease does not normally show any symptoms until after people have had their children. What effect will this have on the frequency of the Huntington's allele in this population?

Practice questions

1 In mice, the allele for black fur colour (B) is dominant to the allele
for white fur colour (b). Two black and two white mice were kept
in a large cage in the laboratory under ideal breeding conditions.
After a few months, there were 149 black mice and 84 white mice
in the cage.

   a) Use the Hardy–Weinberg equation to calculate the expected percentage
of heterozygous black mice in this population.   (3)

   b) Give two assumptions that are made when using the
Hardy–Weinberg equation.   (2)

2 **a)** Explain the meaning of the term gene pool.   (1)

In a type of stag beetle, the allele for gold body colour, G, is dominant over
that for black body colour, g.

   b) In a stag beetle with black body colour, how many copies of the
g allele would you expect to find in a nucleus taken from
a muscle cell in the beetle's leg? Explain your answer.   (1)

   c) In a population of stag beetles, 182 had gold body colour and
18 had black body colour. Showing your working in each case, calculate
the frequency of:

      i) the g allele   (1)

      ii) the G allele.   (1)

3 **a)** *Howea forsteriana* and *Howea belmoreana* are both kinds of palm found
on Lord Howe Island in Australia. It is believed that these
two species evolved sympatrically. Explain how sympatric
speciation occurs.   (4)

   b) Name the taxonomic group represented by:

      i) *Howea*

      ii) *forsteriana.*   (1)

4 Scientists have found a new species of river dolphin in the Araguaia
and Tocantins rivers. The scientists have studied its DNA. They say
that its genes suggest that the species formed 2.08 million years ago when
the Araguaia-Tocantins basin was cut off from the rest of the Amazon river
system by huge rapids and waterfalls, isolating the dolphins from the rest of
the population.

   a) What is a species?   (1)

   b) Suggest how this new species of river dolphin has developed from
ancestral dolphins in the Amazon.   (4)

   c) Suggest how studying DNA provided the scientists with
information about how long ago this new species formed.   (3)

5 Scientists carried out an investigation on a dozen tiny islands in the Bahamas. They measured the leg length of a large number of tiny *Anolis sagrei* lizards. In half of the islands, they introduced a larger lizard species, which preys on *A. sagrei*, but cannot climb trees. When they returned to the islands, they found that on the islands where the larger lizard had been introduced the leg length of the *A. sagrei* lizards had changed. Some had longer legs, enabling them to run faster, while others had shorter legs, enabling them to climb trees faster.

a) i) Why did the scientists introduce the larger lizard to only half of the islands? (2)

    ii) What kind of selection was shown in this example? Explain your answer. (1)

b) The scientists released the larger species on to the islands because they knew that, every few years, the islands are subjected to flooding which wipes out all ground-living species. Explain why this was important. (3)

c) In the few months after flooding, would you expect the leg length variation in *A. sagrei* to change? Explain your answer. (2)

Stretch and challenge

6 Suggest how you could adapt the Hardy–Weinberg equation to account for selection pressure against one phenotype.

7 Suggest why the cystic fibrosis allele is relatively common among white Europeans when homozygous recessive infants would have died in childhood until modern medicine was available.

Populations in ecosystems

TEST YOURSELF ON PRIOR KNOWLEDGE

1 How can you sample organisms using a line transect?
2 Insert the right word in the table to match its definition.

Term	Definition
	The middle value in a series of numbers.
	A calculated 'central' value of a set of numbers. To calculate it, you add up all the numbers, then divide by how many numbers there are.
	The number which appears most often in a set of numbers.

3 Give two ways in which farming reduces biodiversity.

Introduction

The World Wildlife Fund (WWF) published its latest Living Planet Index in October 2014. It publishes this index every 2 years. The report immediately hit the headlines because it showed that the world's populations of wildlife, including mammals, birds, reptiles, amphibians and fish, have dropped by more than a half over the past 40 years.

The WWF arrived at this index by measuring over 10 000 representative animal populations and found that these populations had fallen by 52% between 1970 and 2010. The previous report showed a drop of 28%

between 1970 and 2009. The populations that had fallen the most were those of freshwater fish, down by 76% over this period.

The Director General of the WWF International explained that this issue needs to be dealt with urgently over the next few decades. He said that humans must stop using the natural resources of the planet and destroying natural habitats as if they can be replenished, as these actions are jeopardising our future.

Sustainability The ability to use resources to meet the needs of the present without compromising the ability of future generations to meet their own needs.

In the report, Kuwait was named as the country with the worst record over the past four decades, consuming most resources per head of any country in the world. They were closely followed by Qatar and the United Arab Emirates. The USA was also named as having a bad track record, alongside other countries including Denmark, Belgium, Trinidad and Tobago, Singapore, Bahrain and Sweden. On the other hand, some of the poorer countries had better records for sustainability, including India, Indonesia and the Democratic Republic of Congo. The UK came 28th of all the countries mentioned in the report, but it still uses far more resources per head of population than most countries.

Figure 22.1 Green turtles are found in Kuwait. Hatchlings like these find it increasingly difficult to reach maturity and breed successfully. Global sea turtle numbers in general have fallen by 80% since 1970.

405

Some people may think that worrying about wildlife populations is a luxury while there are human populations suffering poverty and disease. However, as biologists we need to explain that human populations are dependent on the stability of populations of other organisms. We need to find ways of coexisting with the wildlife on our planet and using resources more sustainably.

Organisms and their environment

Nettles are common British plants. You can find them growing in woods, on grazing land, by the sides of ponds and streams, and in areas of wasteland in towns and cities. Figure 22.2 shows a nettle plant and some of the animals that can be found on it.

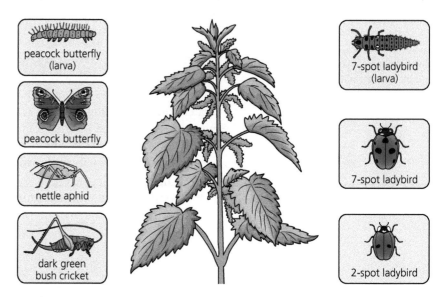

peacock butterfly (larva)

peacock butterfly

nettle aphid

dark green bush cricket

7-spot ladybird (larva)

7-spot ladybird

2-spot ladybird

Figure 22.2 Some common animals associated with nettles.

A **population** is a group of organisms of one species, occupying a particular space at a particular time, that can potentially interbreed. All the two-spot ladybirds in a nettle patch are a population. All the nettle plants are another population. A **community** is all the populations of different species living in the same place at the same time. When we talk about the community in a nettle patch, we mean the population of nettles, the population of aphids, the populations of ladybirds and all the populations of other organisms that we have not mentioned, such as the bacteria and mycorrhizal fungi that live in the soil surrounding the roots of the nettles.

The community, together with the non-living components of its environment such as soil, water mud and rock, form an **ecosystem.** Ecosystems can vary in scale, from a patch of nettles or a pond to a mangrove forest or an entire coral reef.

The **habitat** of an organism is the place where it lives. The two-spot ladybird lives on nettle leaves.

The set of conditions that surrounds all of the insects in a patch of nettles consists of abiotic factors and biotic factors.

- **Abiotic (physicochemical) factors** make up the non-living part of the environment. Nettles grow well, for example, where there is a high concentration of phosphate ions in the soil. Warm, humid conditions result in large numbers of aphids on the leaves. The concentration of phosphate ions in the soil and the air temperature and humidity are all abiotic factors. Abiotic factors affect population size by affecting intraspecific competition. For example, if water availability is limited, the population will compete for water, and some of the organisms might die.

- **Biotic factors** are those relating to the other populations in the environment. The numbers of aphids on the nettles will be affected by the numbers of ladybirds because ladybirds feed on aphids. Predation by ladybirds is a biotic factor.

Within a habitat, a species occupies a **niche** that is defined by a combination of abiotic and biotic factors. Its niche describes not only where it is found but what it does there. We can describe the niche of the two-spot ladybird, for example, in terms of the abiotic features of its habitat, such as the temperature range it can tolerate, and the position on the nettle leaves where it is found. We can also include in our definition of its niche the size and species of aphids that it eats. The total combination of tolerances and requirements describes the multiple dimensions of the niche of any species. Each species occupies its own niche which is different from that of any other species in that habitat.

Populations within an ecosystem can only reach a certain size. The number of individuals in a population will be limited by abiotic factors or by biotic factors such as competition or predation. The maximum size that a population can remain sustainable in an ecosystem is called the **carrying capacity**. If the population of a species exceeds the carrying capacity, its number will be reduced until it is at or below the carrying capacity. This means that the number of organisms in a population fluctuates around the carrying capacity.

TEST YOURSELF

1 Suggest reasons why estimates of seabird numbers based on counting breeding birds may not be accurate.
2 Match the words to their definition.

Word		Definition	
1	Population	A	Factors relating to the living part of the environment, e.g. predation
2	Niche	B	All the populations of different species that live and interact together in the same area at the same time
3	Community	C	The set of conditions that surrounds an organism, consisting of abiotic factors and biotic factors
4	Environment	D	Factors relating to the non-living or physical and chemical parts of the environment
5	Abiotic factors	E	The habitat of an organism and its role within it
6	Biotic factors	F	A group of organisms belonging to the same species found in the same area at the same time

Competition

Farmers grow crops for profit. One of the factors they take into consideration is the number of seeds they sow. The number of seeds sown per unit area is called the sowing density. Agricultural scientists have carried out research on how sowing density affects crop yield and weed growth. In one investigation they sowed wheat at different densities and measured the mass of grain produced by each plant and the mass of weeds. Some of their results are shown in the graphs in Figure 22.3.

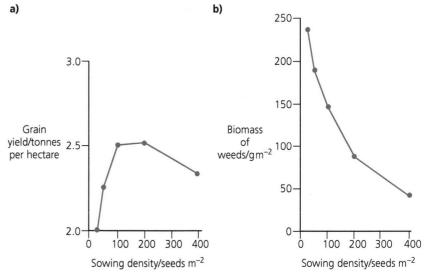

Figure 22.3 Graphs showing the effects of sowing density of wheat on a) grain yield and b) the mass of weeds per square metre.

Intraspecific competition Competition between organisms that belong to the *same* species.

Look first at Figure 22.3a. You can see that at sowing densities greater than 100 seeds m⁻² the grain yield decreased with increasing sowing density. This can be explained in terms of intraspecific competition, or competition between organisms that belong to the same species. The more wheat plants there are per square metre, the more competition there will be for resources such as light, water and mineral ions. The resources available to each plant will be fewer. This will result in smaller plants that produce less grain. Since they are the same species, with the same niche, they compete for the same things.

Interspecific competition Competition between *different* species.

Now look at Figure 22.3b. This is an example of interspecific competition, which means competition between different species. In this case, the wheat plants are competing with the weeds. The higher the density of wheat plants, the fewer the resources there are available for the weeds.

ACTIVITY

Intraspecific competition

We will look at some more data involving **sowing density**. Table 22.1 shows the results of a trial to find the effect of sowing density of cotton on the mean height of the plants.

Table 22.1 Effect of sowing density on the mean height of cotton plants.

Mean number of seeds sown per hectare	Mean height of cotton plants/m	Standard deviation
67 000	1.15	±0.18
33 000	1.27	±0.11
17 000	1.31	±0.11

1 The cotton seeds in the first row in the table were sown at a density of 67 000 seeds per hectare.

What was the mean number of seeds sown per square metre? (1 hectare = 10 000 square metres)

2 What do the figures in the first two columns of the table suggest about the effect of sowing density on the height of cotton plants? Suggest an explanation for your answer.

Plot the information in this table as a graph.

3 Use your graph to explain whether sowing density has a significant effect on the height of cotton plants.

4 What does the standard deviation tell you?

5 The scientists who carried out this investigation planned to repeat it the following year on a larger scale. Use your answer to question 4 to suggest why.

EXAMPLE

Squirrels and interspecific competition

There are two species of squirrel in the UK: the native red squirrel and the introduced grey squirrel. Red squirrels are smaller than grey squirrels. They store less body fat and they spend more time in the tree canopy. Scientists have suggested several hypotheses to explain why grey squirrels have replaced red squirrels in much of Britain. One hypothesis is that grey squirrels are better adapted to living in oak woods. Acorns are the fruit of oak trees. They mature in the autumn and fall to the ground where squirrels can collect them. Earlier research showed that grey squirrels are much better than red squirrels at digesting acorns.

The graph in Figure 22.4 shows the relationships between the numbers of the two species of squirrel and acorn numbers.

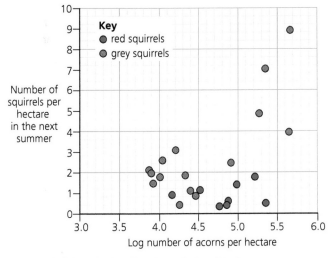

Figure 22.4 Graph showing the relationship between number of acorns produced per hectare and number of squirrels the following summer.

1 Why are the figures for the number of acorns given on a logarithmic scale?

A logarithmic, or log, scale increases in multiples of 10. On the scale shown here a log value of 3 represents 1000; a value of 4 represents 10 000; a value of 5 represents 100 000 and a value of 6 represents 1 000 000. Using a log scale lets us plot a much larger range of values on a single axis.

2 Suggest how the data relating to the number of acorns would have been collected.

The technique used would have to be based on a sampling method. Quadrats were probably used. This would allow the number of acorns in a given area to be found. This figure could then be converted to the number of acorns per hectare.

3 What is the relationship between the log of the number of acorns per hectare and the number of grey squirrels the next summer?

If you were to plot a curve of best fit you would get a line sloping upwards. This shows that there is a positive correlation: the greater the log of the number of acorns per hectare, the greater the number of grey squirrels the next summer.

4 What is the relationship between the log of the number of acorns per hectare and the number of red squirrels the next summer?

Plotting the curve of best fit this time gives a line parallel to the x-axis. This shows that there is no correlation.

5 Use the information in this section to suggest how interspecific competition could account for the absence of red squirrels in most of southern England.

Much of the native woodland in southern England is oak. Grey squirrels are more successful in oak woods because they spend more time on the ground where acorns fall. They can also digest the acorns better and put on more body fat as a result. A greater amount of body fat probably enables grey squirrels to survive over winter in oak woods better than red squirrels.

Predators and their prey

Predators are animals that kill and eat their prey. Figure 22.5 shows a spider that is feeding on a grasshopper. The spider is the predator and the grasshopper is its prey. Obviously, the number of predators influences the number of prey: as the number of predators increases, the number of prey decreases. Equally obviously, the number of prey influences the number of predators. Without enough food, fewer predators will survive.

Figure 22.5 A spider with a grasshopper in its web, an example of a predator and its prey.

EXAMPLE

Scientists use models to help them understand the relationship between the population of a predator and the population of its prey. We will look at one of these models, called the Lotka–Volterra model after the two scientists who originated it (Figure 22.6).

- Look at the part of the graph in Figure 22.6 between points A and B. It shows the prey population increasing. Therefore there will be more food available for the predator population. The predators will breed successfully and their numbers will increase.
- With more predators, however, more prey will be killed so we

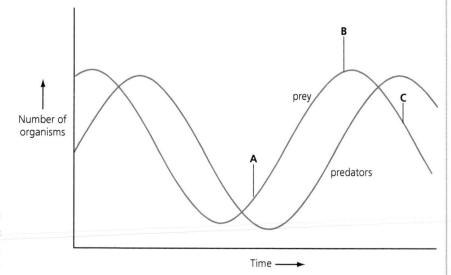

Figure 22.6 The Lotka–Volterra model predicts that the populations of a predator and its prey change over time.

see, between points B and C, a fall in prey numbers. There is now less food for the predators and their population starts to fall.

- Numbers of prey rise again and the pattern will be repeated. We will look at a piece of evidence that appears to support this model. Figure 22.7 shows a graph of changes in the numbers of skins of a predator, the lynx, and the number of skins of its prey, the snowshoe hare, bought from trappers by the Hudson Bay Company, Canada, for the years 1890 to 1920.

The curve for the number of lynx skins is different from the curve for the number of snowshoe hare skins in Figure 22.7. There are two features to note. The peak for the lynx skins always comes a little later than the peak for the snowshoe hare skins and the lynx numbers are always lower than the hare numbers. This is usually the case for predators and their prey, although there are some exceptions.

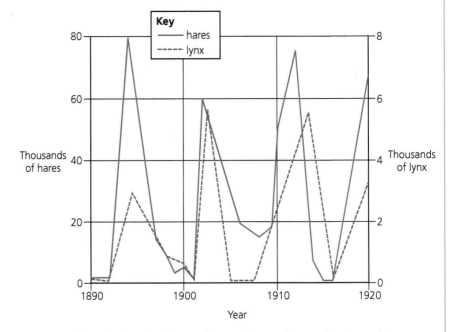

Figure 22.7 Graph showing the numbers of lynx and snowshoe hare skins bought by the Hudson Bay Company between 1890 and 1920.

The number of snowshoe hare skins purchased rises between 1908 and 1912. During this period lynx numbers are low, as shown by the number of skins bought by the Company. With few predators to kill them, snowshoe hare numbers therefore rise.

Note that, in one population cycle, lynx numbers rise after snowshoe hare numbers have risen. The rise in lynx numbers is the result of more prey being available. Therefore their numbers will only rise after snowshoe hare numbers have increased.

Between 1910 and 1914 the rise in the lynx numbers causes a fall in the numbers of snowshoe hares. There is therefore less food for the lynx and subsequently their numbers fall.

The changes in the numbers of snowshoe hare skins and lynx skins shown in Figure 22.7 seem to support the Lotka–Volterra model but more recent experimental work has shown that the situation is really much more complicated than this simple model suggests. Here is some information resulting from recent ecological investigations involving snowshoe hares and lynx.

- High populations of snowshoe hares are associated with a shortage of plant food on which the animals feed. When the snowshoe hare population is at its lowest, the plants start to grow again but, for the next 2 or 3 years, toxins in the young shoots are thought to delay the next rise in the population of snowshoe hares.
- When snowshoe hare numbers decrease, lynx eat other prey animals. There is very little evidence of lynx dying of starvation.
- Scientists attached radio collars to lynx. They showed that when the number of snowshoe hares was low, the lynx tended to move away. In one case, they travelled as far as 800 km.
- There are no lynx on Anticosti Island in Eastern Canada. The population of snowshoe hares on the island, however, still shows a regular population cycle.

What we can see from this information is that simple models, such as the Lotka–Volterra model, provide a very useful starting point for analysing population cycles. They cannot provide us with a complete explanation.

Counting and estimating

Every 10 years there has been a census of all the people living in the UK. It is not totally accurate because some people fail to fill in their census forms accurately, or don't fill them in at all, but it does give a good idea of the size of the UK human population.

There are other species where we can get a fairly accurate figure for the population by counting them. For example, seabirds nesting on the Farne Islands off the north-east coast of England can be counted because they nest in large colonies, but the data obtained are much less accurate than for human populations. In the Introduction to this chapter you learned that the WWF estimates world wildlife populations every 2 years, but again there are difficulties in ensuring reasonable accuracy.

We can only count all the animals or plants in a population in a few cases. The organism concerned needs to be large, conspicuous and confined to a relatively small area for a count to be accurate. If it is not possible to count every single organism, we need to take samples. Ecologists usually base their estimates of populations on samples. They must make certain, however, that the samples are representative of the population as a whole. In order to be sure of this, samples must be large enough to be representative and taken at random.

Sample size

The larger the size of a sample, the more reliable the data. Data from a very large sample, however, cannot usually be collected in a short period of time. When we sample an area we have to strike a balance. Look at the graph in Figure 22.8. It shows that a very small sample has very few species in it. As the sample size increases, the number of species that it contains increases. There comes a time when the number of species does not increase much more however much bigger the sample. This sample size is obviously representative of the community as a whole.

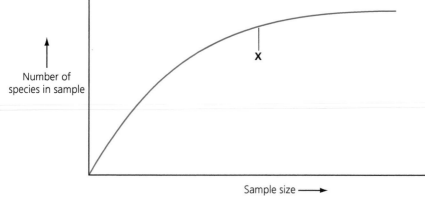

Figure 22.8 The effect of sample size on the number of species recorded. After point X the number of species does not increase much more with increasing sample size. The sample size at X is therefore representative of the community as a whole. A larger sample would simply mean more work.

Random sampling

When you studied the topic of variation in Chapter 11 you learned that, unless samples were taken at random the results may be biased. The same is true of ecological samples. They must be collected at random or, again, the results may be biased. Quadrats are often used to mark out areas to be sampled. When we use a quadrat for sampling, it is important that we use a method that will result in the quadrat being genuinely placed at random in the study area (Figure 22.9).

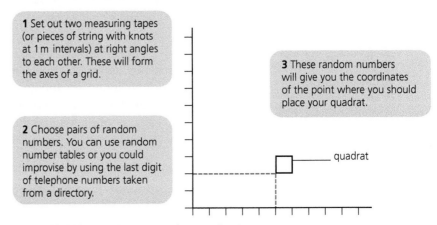

1 Set out two measuring tapes (or pieces of string with knots at 1 m intervals) at right angles to each other. These will form the axes of a grid.

2 Choose pairs of random numbers. You can use random number tables or you could improvise by using the last digit of telephone numbers taken from a directory.

3 These random numbers will give you the coordinates of the point where you should place your quadrat.

quadrat

Figure 22.9 How to place a quadrat randomly.

Quadrats and transects

Two important ways of sampling involve quadrats and transects. Both of these are in general used for plants, but they can also be used for organisms that do not move about very much, such as many of those that live on seashores. Table 22.2 shows when and how these techniques are commonly used. There are three different ways that we use to describe the distribution of organisms once the quadrat or transect is put in position. These are illustrated in Figure 22.10.

Table 22.2 Using quadrats and transects.

Sampling method	Method	What it is used for	How it is used
Quadrat	A frame, usually 0.5 m², or a rod 0.5 m long with 10 pins at 5 cm intervals, that can be placed on the ground to sample organisms	Studying the distribution of non-motile plants or animals in a fairly uniform area	Placed at random and used to find population density, frequency or percentage cover. In some investigations, such as those involving the effects of grazing, permanent quadrats may be used. They may remain in place for many years.
Transect	A line through a study area along which samples are taken	Usually used where one or more abiotic factors in the environment gradually vary of if there appears to be a change in communities from one place to another	Placed so that it follows the environmental gradient; for example, up a seashore or from sunlight into shade. This is called a **line transect**. Quadrats may be used continuously along a transect to sample in more detail. This is called a **belt transect**.

a) Population density

b) Frequency

c) Percentage cover

This quadrat measures 0.5 m × 0.5 m. It contains six dandelion plants. The **population density** of dandelions would be 24 plants per m². To get a reliable figure you would need to collect the results from a large number of quadrats. If a plant lies partly in and partly out of the quadrat, we normally count it if it overlaps the north or west side of the quadrat, and don't count it if it overlaps the south or east side.

This point quadrat frame is being used to measure **frequency**. The pins of the frame are lowered. Suppose three out of ten pins hit a dandelion plant. The frequency of dandelion plants will be three out of ten, or 30%.

Percentage cover measures the proportion of the ground in a quadrat occupied by a particular species. The percentage cover of the dandelions in this quadrat is approximately 40%.

Figure 22.10 Using a) population density, b) frequency and c) percentage cover to describe the distribution of dandelion plants.

Populations of the same species from two areas with different abiotic conditions can be compared by taking a random sample from both areas. Limpets are molluscs with a cone-shaped shell that attach to rocks with a strong foot. In an investigation, students sampled limpets by placing quadrats randomly on each section of the shore and measuring the height and base diameter of each limpet in the quadrats. After this they found the height : base diameter ratio for each limpet (Table 22.3).

Table 22.3 Height and base diameter of limpets sampled on a rocky shore

Sheltered shore			Exposed shore		
Height/ mm	Base diameter/mm	Height : base ratio	Height/ mm	Base diameter/mm	Height : base ratio
21.7	49.0	0.44	19.2	57.5	0.33
21.8	53.6	0.41	17.4	54.5	0.32
23.3	42.4	0.55	25.5	63.3	0.40
18.2	42.5	0.43	20.5	43.4	0.47
17.9	44.5	0.40	19.6	61.2	0.32
22.5	44.8	0.50	18.4	58.8	0.31
17.4	45.0	0.39	18.6	50.2	0.37
30.4	56.6	0.54	14.6	48.0	0.30
21.5	49.8	0.43	20.6	49.4	0.42
19.5	50.5	0.39	14.7	42.4	0.35
21.5	48.7	0.44	13.5	44.3	0.30
23.0	49.2	0.47	19.0	53.2	0.36
23.3	49.5	0.47	18.3	50.5	0.36
23.0	46.5	0.49	15.2	49.7	0.31
24.6	54.9	0.45	14.2	44.5	0.32

TIP

A ratio would normally be expressed as, for example, 0.44:1, but here we have expressed the ratios using the first figure alone, e.g. 0.44.

TIP

- In an exam you might be asked to recognise the type of data you have been given and to select an appropriate statistical test.
- Look at Chapter 26 on maths skills to find out more about statistical tests and to see a worked example of this test.
- You won't need to do statistical calculations in a written paper.

Because the students were dealing with measured data the best way to present it would be as mean height to base diameter ratios on a bar chart. The bars would be different heights indicating that limpets on the sheltered rocky shore had a higher mean height to base diameter ratio than those on the exposed shore; in other words, those on the exposed shore had grown wider and flatter than those on the sheltered shore. You would use a t-test to find out whether the difference between the means was significant.

REQUIRED PRACTICAL 12

Investigation into the effect of a named environmental factor on the distribution of a given species

This is just one example of how you might tackle this required practical.

Some students decided to investigate the effect of trampling on plant cover by placing a transect across a path. They placed a measuring tape at right angles to the path, including the vegetation at the side of the path.

The students placed a point quadrat at right angles to the measuring tape every 10 cm. The point quadrat had ten pins, and they recorded what each pin was touching. They wanted to convert this to percentage cover, so they multiplied the number of pins touching a specific plant or substrate by 10. Table 22.4 shows their results.

Table 22.4

Distance along quadrat/m	Percentage cover				
	Bare soil	Rock	Grass	Plantain	Dandelion
0	0	0	80	10	10
0.1	0	0	70	10	20
0.2	0	0	90	0	10
0.3	10	0	70	20	0
0.4	20	10	50	20	0
0.5	20	0	60	20	0
0.6	30	10	40	20	0
0.7	40	20	30	10	0
0.8	60	20	10	10	0
0.9	70	30	0	0	0
1.0	60	40	0	0	0
1.1	80	20	0	0	0
1.2	70	10	10	10	0
1.3	70	0	10	20	0
1.4	50	10	20	20	0
1.5	40	10	30	20	0
1.6	30	10	30	20	10
1.7	30	0	40	30	0
1.8	40	10	20	30	0
1.9	30	0	40	20	10
2.0	10	0	70	10	10
2.1	10	0	70	10	10
2.2	0	0	80	10	10
2.3	0	0	90	0	10
2.4	0	0	80	0	20
2.5	0	0	90	0	10

They used these data to plot a kite diagram, which displays the density and distribution of plant or animal species in a particular habitat. It shows the percentage of certain species spread over a certain distance. You can see how to plot a kite diagram in Figure 22.11.

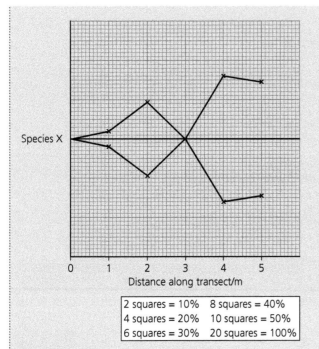

Species X

Distance along transect/m

2 squares = 10%	8 squares = 40%
4 squares = 20%	10 squares = 50%
6 squares = 30%	20 squares = 100%

Figure 22.11 How to plot a kite diagram.

TIP

You do not need to recall the details of how to plot a kite diagram.

For each species, you draw a straight line. Then you choose a scale for your kite, so that the width of the kite represents the percentage cover of the species. Crosses are placed each side of the line for each sampling point along the transect. The crosses are joined with a straight line and then the 'kite' is shaded in.

1 Plot a series of kite diagrams to display the data in the table. Set out the graphs as shown in Figure 22.12.

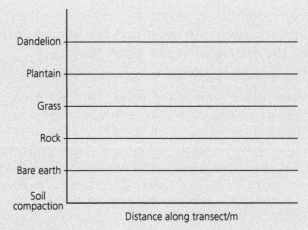

Distance along transect/m

Figure 22.12

2 Suggest how the students could measure soil compaction.

3 What are the advantages and disadvantages of using a point quadrat in this investigation, rather than a quadrat square?

4 What do these data suggest about the effects of trampling on plant cover?

5 Suggest reasons why trampling affects plant cover. Look at Figure 22.13 for some clues.

Figure 22.13 a) *Plantago major* (greater plantain). b) Dandelion

Mark-release-recapture

Most animals move around (are motile) so it is not usually possible to estimate the size of an animal population using a quadrat. Instead we use the mark-release-recapture method. This relies on capturing a number of animals in an area and marking them harmlessly so that they can be recognised again. The marked animals are then released. Some time later a second sample is trapped in the same area and the numbers of marked and unmarked animals in the second sample are recorded. From the data collected the size of the population can be estimated. The calculation relies on the assumption that the proportion of marked animals in a sample (the second sample) is the same as the proportion of marked animals in the whole population.

This equation makes several assumptions:

● the animals all come from the same population
● marking does not harm the animal or make it more likely to be seen by a predator
● there is no migration into or out of the population during the period of the investigation

---

EXAMPLE

Estimating the size of a population of grasshoppers

Here is how we would estimate the size of a population of grasshoppers:

Number of grasshoppers caught, marked and released = 56

Number of marked grasshoppers in second sample = 16

Total number of grasshoppers in second sample = 48

$$\frac{\text{Number of marked grasshoppers in sample}}{\text{Total number of grasshoppers in sample}} = \frac{\text{Number of marked grasshoppers in population}}{\text{Total number of grasshoppers in population}}$$

$$\frac{16}{48} = \frac{56}{\text{Total number}}$$

Rearranging this equation:

$$\text{Total number} = 56 \times \frac{48}{16}$$

$$= 168$$

● there are no births or deaths during the period of the investigation.
  Scientists have modified the mark-release-recapture technique to estimate the size of whale populations.
● Whales cannot be trapped and they cannot easily be marked.
● Scientists can use a small boat to approach a surfaced whale. They can remove a tiny piece of skin from the whale.
● The scientists are able to analyse the DNA from this piece of skin and produce a DNA fingerprint. Each whale has its own unique DNA fingerprint.

Ecosystems are dynamic. This is reflected by fluctuations in populations, among other things. Numbers of a particular species vary around the carrying capacity from place to place and from one time of year to another. These variations may result from differences in abiotic factors. They may also result from changes in biotic factors.

Figure 22.14 Because whales spend most of their time below the surface of the sea they cannot be trapped or easily marked, so scientists needed to find a method other than observing and counting the whales in order to estimate the size of whale populations. They developed a DNA fingerprinting method.

ACTIVITY

Investigating the mark-release-recapture method of analysing population size

Materials:

4 or 5 packets of small chocolate sweets with a coloured sugar coating (e.g. Smarties)
paper bag
egg cup
plastic dish (e.g. margarine tub).

- Tip all the coloured sweets into the plastic dish. Take out the red sweets and count them.
- The coloured sweets represent a population of field mice and the red ones represent the mice trapped on the first evening that are marked and released.
- Tip all the sweets, including the red ones, into the paper bag and shake them gently.
- Use the egg cup to take a sample of sweets from the bag without looking. Tip the sample into the plastic dish. This represents the mice captured the second time.
- Copy Table 22.5 and use the results to calculate the population size.
- Repeat this nine more times so that you can find the mean of 10 population estimates.
- Now that you have estimated the population size, tip all the sweets into the plastic dish and count them.

1 How accurate was the first estimate that you calculated?
2 Was the mean estimate after 10 samples more or less accurate?
3 If this was a real population of mice, would it be feasible to sample the population 10 days in a row? Explain your answer.
4 Now think about a variable you might change. For example, does the number of traps set matter? What happens if one or two of the marked animals are eaten by a predator after they are released? What happens if two or three of the unmarked animals die, or if new mice enter the area?

Think about how you can simulate this change, and carry out the calculations again. For example, you can simulate fewer traps by half-filling the egg cup, or setting more traps by using a slightly bigger egg cup. You can remove a couple of red sweets to simulate marked animals being killed, but remember that you wouldn't know this had happened if you were investigating a real population, so your calculations would be based on the original number of marked animals.

5 Based on your investigation, evaluate the use of the mark-release-recapture method of estimating the size of real populations.

Table 22.5

Sample number	Total number in sample (a)	Number of marked mice in sample (b)	Number of marked mice in population (c)	Population size = $\frac{ac}{b}$
1				
2				
3				
4				
5				
6				
7				
8				
9				
10				
Mean				

TEST YOURSELF

7 Which of the following methods would allow a quadrat to be placed at random?
   A Closing your eyes, turning on the spot and throwing the quadrat over your shoulder.
   B Placing quadrats along a tape at 5 m intervals.
   C Picking random numbers from a hat to give coordinates on a grid.
8 A sample of 40 trout in a fish pond was caught in a net. Each fish was harmlessly marked on one of its fins. The fish were then released back into the pond.

One week later a second sample was caught. It contained 17 marked trout and 41 unmarked trout. Estimate the total number of trout in the pond.

9 Use the information on sampling whale populations, page 179, to suggest how scientists could modify the mark-release-recapture technique to estimate the size of a whale population.

10 For the mark-release-recapture method to be reliable, certain assumptions need to be made. Suggest what these are.

Succession

In some places, the populations change over time, creating a new community. This is called succession. It is an ecological process resulting from the activities of the organisms themselves. Over a period of time the organisms modify their environment. These modifications produce conditions better suited to the growth of other species (Figure 22.15).

Figure 22.15 Succession is an important ecological process associated with slow-flowing streams and rivers.

Some of the plants in the foreground of Figure 22.15a have leaves that float on the surface. Their roots trap particles of silt carried in the water and this forms mud. As the water becomes shallower, these floating plants are replaced by upward growing reeds and other plants.

The first organisms to colonise the bare rock in Figure 22.15b are lichens, but they are only able to grow when water is available. Lichens are involved in breaking the surface of rocks, creating a 'soil' that accumulates under them. This soil holds enough water and mineral ions to allow new species to colonise. As this soil is often washed away, succession is very slow. Ultimately though, the lichens will be out-competed by mosses. These, in turn, will be out-competed by ferns and species of flowering plant.

Succession in sand dunes

Sand dunes occur in many coastal areas and can provide a good example of succession. Succession is a gradual process and the time scale involved prevents us from sitting in one place and observing successional changes. If we walk, however, from the sea shore in to the sand dunes we will pass through different areas which represent different stages in succession (Figure 22.16).

We will start nearest the sea on a sandy shore, above high tide level. Some plants, such as sea couch grass and lyme grass, are able to colonise and survive in the bare sand. We call species such as sea couch grass and lyme grass **pioneer species** because they are the first plants to colonise the area. Their roots and shoots form a dense network that binds sand particles together and, as a result, sand starts to pile up and form the fore dunes. As the plants grow larger, their aerial parts trap more and more

TIP

Ecosystems are constantly changing because of changes in populations, for example as a result of predation or competition, and changes in abiotic factors. Abiotic factors such as temperature or water availability may vary with the season, or change gradually over a long period of time. This is particularly seen in the case of succession.

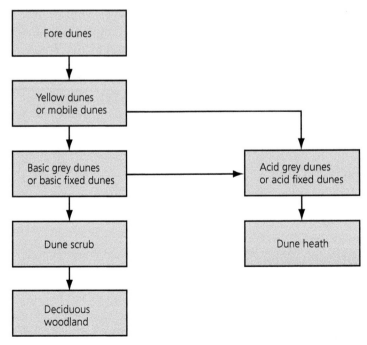

Figure 22.16 Succession in sand dunes.

sand. They cannot grow fast enough, however, to avoid being smothered by sand when it is piling up at a rate faster than 30 cm year⁻¹. Under these conditions they are replaced by marram grass.

Marram grass-dominated dunes are called yellow dunes or mobile dunes: yellow because there is very little humus in the sand and it shows yellow through the marram grass cover; mobile because they are continually changing shape as the wind scours the face and blows the sand. This is a very harsh environment and few plants can survive in these conditions. Look at Table 22.6. This table compares the abiotic conditions in these yellow dunes with the conditions in the grey or fixed dunes later in the succession.

Table 22.6 A comparison of some of the abiotic factors in yellow and grey dunes.

Abiotic factor	Yellow dunes	Grey dunes
Mean wind velocity 5 cm above the dune surface/km h⁻¹	12.1	2.4
Organic matter/%	0.3	1.0
Concentration of sodium ions/ppm	8.5	4.2
Concentration of calcium ions/ppm	637.0	297.0
Concentration of nitrate ions/ppm	48.0	380.0

The high wind speed has two important effects. It will pile sand on top of any plants growing there and it will lead to a high rate of water loss by transpiration. There is still very little organic matter, so the sand does not retain water very well and nutrients tend to leach out. Also, there is little material for decomposers to act on and recycle nitrogen-containing compounds. It will dry out rapidly after rain. Important soil nutrients such as nitrates are in short supply, but sea spray results in high concentrations of sodium and calcium ions, so this lowers the water potential of the sand dune, creating osmotic problems for plants. Very few plants can grow here but one that does is marram grass. Marram grass has a number of xerophytic adaptations that enable it to grow in these conditions (Figure 22.17).

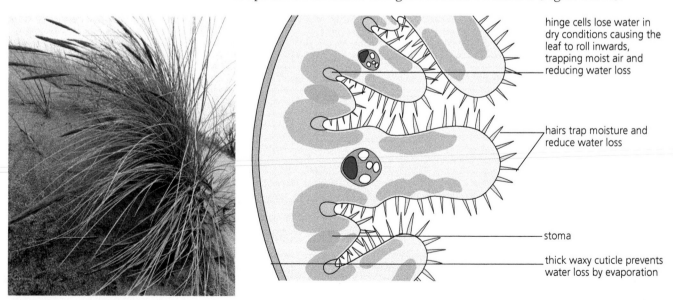

hinge cells lose water in dry conditions causing the leaf to roll inwards, trapping moist air and reducing water loss

hairs trap moisture and reduce water loss

stoma

thick waxy cuticle prevents water loss by evaporation

Figure 22.17 a) On sand dunes, marram grass is an important species. Its leaves grow from a vertical underground stem, so remain above the sand that the wind deposits on the plants. It also has adaptations that enable it to grow in areas where there is often little water available. Some of these adaptations can be seen in the section through a leaf shown in b).

Marram grass illustrates an important biological principle: the more hostile the environment, the fewer the species that are able to survive and the lower the species diversity. In hostile environments, it is generally abiotic factors that determine which species are present.

Now we will go inland to the area of grey dunes or fixed dunes. They are called grey dunes because humus in the sand colours them grey; they also are referred to as fixed dunes because the sand is no longer being blown about, so they are much more stable. Table 22.6 on the previous page shows how abiotic factors change and many of these changes are due to the activity of organisms on the yellow dunes. The roots of marram grass bind the particles of sand and the leaves act as a windbreak, reducing wind-chill and creating shade. The wind velocity is lower so less sand blows about. Dead material falls from the marram grass and is broken down by soil bacteria to form humus. The amount of humus is higher, so the developing soil retains moisture better and the concentration of important soil nutrients such as nitrates rises. Rain is also beginning to leach the soluble sodium and calcium ions from the surface layers of the soil. Other plants can now grow and they gradually replace marram grass because they compete better with marram grass when the soil has more nutrients. We are beginning to arrive at a situation where the environment is less harsh. There are more species and a greater species diversity, creating more complex and therefore more stable food chains and webs. In this environment it is biotic factors, such as competition, that determine whether particular species survive.

The process of succession continues. The species found on these grey dunes may change their environment in such a way that they are replaced by other species. Woody shrubs start to grow. Hawthorn and elder grow and shade out the shorter vegetation. Ultimately woodland develops. We have reached a stage that remains relatively unchanged over long periods of time. This is the climax community. In Britain, it is usually woodland of some sort. If you look again at Figure 22.16 on page 182 and you will see that not all succession on sand dunes follows this pattern. The sand particles that make up the dunes have different origins. Where they originate from rocks such as granite, the soils they produce are more acidic. They support different plants and gradually give way to dune heath dominated by plants such as bracken and heather. Basic soils containing high concentrations of calcium carbonate can also become more acidic as soluble basic ions are gradually leached from them.

The details of the processes of succession in slow-flowing streams and rivers and on bare rock shown in Figure 22.15 are different from those on sand dunes but the principles are very similar. They are summarised in Figure 22.18 overleaf.

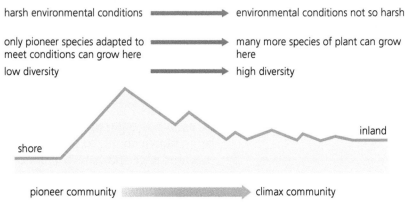

harsh environmental conditions ➡ environmental conditions not so harsh

only pioneer species adapted to ➡ many more species of plant can grow
meet conditions can grow here   here

low diversity ➡ high diversity

shore   inland

pioneer community ➡ climax community

Figure 22.18 A summary of the processes involved in succession.

Managing succession

Succession does not always proceed all the way to a climax community. It can be stopped by various factors often associated with human activity.

There is little, if any, of the landscape of Britain that has not been modified by human activity. Much of our current landscape is the result of centuries of agricultural practice. Chalk grassland is one example. It consists of a diverse mixture of grasses and herbs and can support up to 50 species of flowering plant per square metre (Figure 22.19). There is also a high diversity of invertebrates, such as insects and snails.

At the beginning of the nineteenth century more than 50% of the South Downs was chalk grassland. Today, the figure stands at just 3%. There are many reasons for this massive reduction. Understanding them requires not only a knowledge of ecology, but also an understanding of other aspects of biology.

Figure 22.19 Orchid-rich chalk grassland is identified as a European conservation priority.

- Chalk grassland resulted from sheep grazing the thin, nutrient-poor soils overlying chalk. A massive decline in the number of sheep and the resulting spread of scrub has reduced the amount of chalk grassland. The bushes that form this scrub also spread to remaining areas of chalk grassland. These bushes, however, are also part of the chalk ecosystem and are important in contributing to the overall biodiversity by providing a habitat for many species. A balance needs to be achieved by controlling this process of succession.
- The disease myxomatosis wiped out large numbers of rabbits in the 1950s. Since then rabbits have become resistant to the disease and numbers have increased dramatically. They have little effect in controlling scrub vegetation and their warrens and overgrazing cause serious erosion.
- In order to improve yield, many farmers have added nitrogen-containing fertilisers and selective herbicides to fields. Some species have benefited at the expense of others and there has been a resulting loss in biodiversity.
- Much of the existing chalk grassland is fragmented. This fragmentation causes isolation and makes populations more vulnerable to local extinction from disease and predation, because each population has a limited gene pool and there is limited variation for natural selection to operate on.

Conserving habitats such as chalk grassland requires careful management, to prevent succession taking place. By grazing sheep on chalk grassland, scrub cannot develop, and this maintains a suitable habitat for many different species of wildflowers and insects such as butterflies. Many areas of moorland are mowed to remove tree species and preserve the habitat for moorland birds and other wildlife.

Investigating succession when plant cover has been destroyed

In an investigation, ecologists compared control plots with plots on which the plant cover had been partly destroyed by allowing motor cycles to be ridden over it. The results are shown in Figure 22.20.

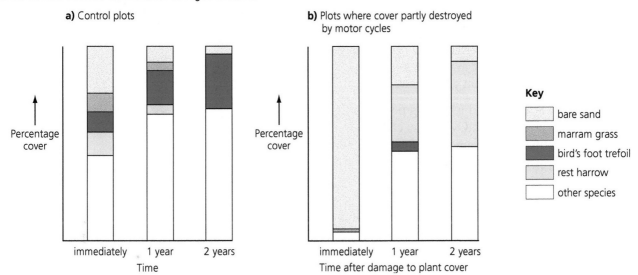

Figure 22.20 A comparison of changes in the vegetation of a) control plots and b) plots in which the plant cover had been partly destroyed by motor cycles.

1 Describe how the control plots should have been treated.

In this case, the important thing was to treat the control plots in the same way as the experimental plots in all aspects except in riding a motor cycle over them.

2 What was the purpose of the control plots?

They provide a comparison with the experimental plots by showing what would happen to similar areas of dune that were not damaged by being ridden over.

3 Describe the process of succession that takes place in the control plots over the period covered by this investigation.

There is a decrease in the amount of bare ground and a change in plant cover. The percentage cover of marram grass and rest harrow decreases and there is an increase in bird's foot trefoil and other species.

This investigation illustrates an important point. When we halt the process of succession, in this case by partly destroying plant cover with a motor cycle, it rarely returns to where it began.

4 Use the results in the bar chart to describe how the process of succession is different after the plant cover has been destroyed. Suggest an explanation for the difference.

In a natural succession, marram grass is present early in succession. In the investigation, we are starting from a position where species other than marram grass are present, although most of the plant cover has been destroyed. The sand in the areas from which the plant cover has been removed is quite different from the sand in which the marram grass became established initially. The sand with reduced plant cover contains more humus and it is not so likely to dry out. It also has a lower concentration of sodium and calcium ions and a higher concentration of nitrate ions. These are conditions in which rest harrow grows better than marram grass. You can see quite clearly how human activity has altered the path of succession.

---

11 Use your knowledge of competition to explain why the use of nitrogen-containing fertilisers on chalk grassland has resulted in a loss of biodiversity.

12 Explain two ways in which farming prevents succession occurring in fields.

13 Succession takes place after a fire has destroyed a large area of forest. Suggest how plant species are able to colonise the area.

Practice questions

1 a) What is species richness? *(1)*

b) The graph shows how species richness changes along a sand dune system.

i) Describe the graph. *(1)*

ii) Explain the results. *(3)*

c) How would you expect the animal species richness to change along the same transect from 0 to 400 m? Explain your answer. *(2)*

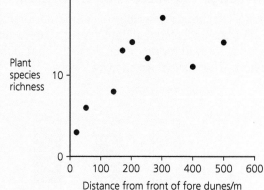

2 If a field is abandoned for a long period of time, succession will occur. Over time, changes in the composition of the plant and animal communities will occur until eventually no further change takes place.

a) What name is given to the final stable community at the end of a succession? *(1)*

b) One community is gradually replaced by another during succession. Explain how. *(2)*

c) Explain how the following farming activities interfere with the process of succession:

i) regular grazing by sheep

ii) ploughing fields each year. *(2)*

3 a) What is meant by an ecological niche? *(1)*

b) In a study of one population of badgers, 72 animals were trapped and marked with a harmless dye on their underside. They were then released. One month later, scientists trapped 120 badgers and found that 14 of these had been marked with dye. Use these figures to calculate the size of the badger population. Show your working. *(2)*

c) Suggest two reasons why this answer may not be completely reliable. *(2)*

4 a) Give the meaning of these ecological terms:

i) population

ii) community. *(2)*

b) Some students used the mark-release-recapture technique to estimate the size of a population of woodlice. They collected 97 woodlice and marked them before releasing them back into the same area. Later they collected 86 woodlice, 14 of which were marked. Calculate the number of woodlice in the area under investigation. Show your working. *(2)*

c) Describe how you would use a quadrat to estimate the number of buttercup plants in a field measuring 100 m by 50 m. *(3)*

5 *Paramecium aurelia* and *Paramecium caudatum* are single-celled protoctists that feed on algae. A scientist cultured the organisms in flasks containing a suitable culture medium. He grew them separately and together. The graph shows the results.

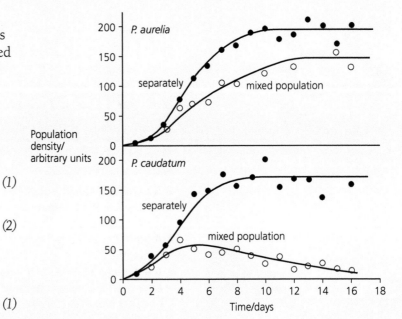

a) Explain why the two species of *Paramecium* were grown separately as well as together. *(1)*

b) Explain the results when both species are grown together. *(2)*

c) Suggest two conditions that the scientist would have kept the same for these results to be reliable. *(1)*

d) Evaluate the benefit of laboratory studies like this in understanding interactions between organisms in a natural environment. *(4)*

6 Scientists investigated the effect of a predatory mite on a herbaceous mite. They grew a population of a herbaceous mite that feeds on oranges in a container. They added the predatory mite that feeds on the herbaceous mite. The population sizes of both mites were measured at regular intervals for a period of just over a year.

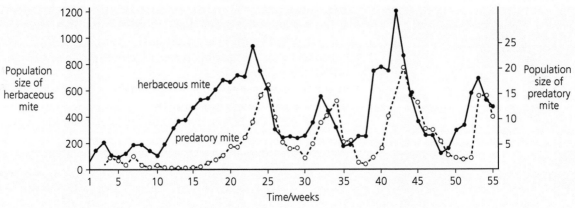

a) Describe and explain the results shown in the graph. *(3)*

b) Explain how you would calculate the percentage increase in the population of the herbaceous mite between weeks 10 and 24. *(1)*

c) Explain two advantages of studying these two mites in understanding the relationship between predators and prey. *(2)*

Stretch and challenge

7 Design an investigation into the effect of an abiotic factor on the distribution of a plant species. Plan your investigation to ensure reliable results, and choose a suitable statistical test to analyse your results.

23

The control of gene expression

PRIOR KNOWLEDGE

- A gene is a base sequence of DNA that codes for the amino acid sequence of a polypeptide or a functional RNA (including ribosomal RNA and transfer RNA).
- A gene occupies a fixed position, called a locus, on a particular DNA molecule.
- A sequence of three DNA bases, called a triplet, codes for a specific amino acid. The genetic code is universal, non-overlapping and degenerate. Also, three bases in mRNA or tRNA code for one amino acid.
- In eukaryotes, much of the nuclear DNA does not code for polypeptides. There are, for example, non-coding multiple repeats of base sequences between genes. Even within a gene only some sequences, called exons, code for amino acid sequences. Within the gene, these exons are separated by one or more non-coding sequences, called introns.
- The genome is the complete set of genes in a cell and the proteome is the full range of proteins that a cell is able to produce.
- Messenger RNA (mRNA) and transfer RNA (tRNA) have specific structures.
- Transcription is the production of mRNA from DNA. RNA polymerase is used in joining mRNA nucleotides.
- In prokaryotes, transcription results directly in the production of mRNA from DNA.
- In eukaryotes, transcription results in the production of pre-mRNA; this is then spliced to form mRNA.
- Translation is the production of polypeptides from the sequence of codons carried by mRNA.
- Ribosomes, tRNA and ATP are involved in translation.

TEST YOURSELF ON PRIOR KNOWLEDGE

1 Give two similarities and two differences between an RNA and a DNA nucleotide.
2 The coding strand of DNA has the following base sequence: CGGTACGA. What is the base sequence of the mRNA for which it codes?
3 Put the following sentences in order to describe the process of protein synthesis.
   A A tRNA molecule brings a specific amino acid to the ribosome.
   B RNA polymerase joins the nucleotides together to make a molecule of mRNA.
   C A section of DNA (a gene) unwinds and the hydrogen bonds break.
   D The amino acids join by a peptide bond.

E The mRNA leaves the nucleus and attaches to a ribosome in the cytoplasm.

F RNA nucleotides attach to the exposed DNA bases by complementary base pairing.

G One strand of the DNA becomes a template.

H The anticodon on the tRNA is complementary to the codon on the mRNA.

I Once the first tRNA has passed on its amino acid, it leaves the ribosome and the ribosome moves along the mRNA, three bases at a time.

J A second tRNA brings its specific amino acid to the ribosome, next to the first tRNA.

4 What is the difference between a polypeptide and a protein?

5 During transcription, the base sequence of only one of the strands of DNA is used to make a molecule of mRNA; the other DNA strand is not used. Give **one** advantage of a molecule of DNA having both strands.

Introduction

You might be familiar with animals that have different body forms during their life cycle. The cabbage white butterfly is a major economic pest in the UK. Its larval stage, the caterpillar, is a voracious leaf-eater and can destroy cabbage crops. The adult butterfly does not eat plant leaves; instead it feeds infrequently on the sugary nectar found in flowers.

Have you ever taken cuttings from a plant, such as the stem, and used them to grow new plants? If so, you have created a plant clone, a group of genetically identical plants. Humans have been doing this for hundreds, if not thousands, of years. Research botanists perform a more sophisticated version of the same process when they isolate individual cells from the root tip of a plant and grow them in tissue culture to produce large numbers of genetically identical plants.

In these examples, we can see different body forms within a single species. Body form is one aspect of an organism's phenotype. Although butterflies have different phenotypes during their life cycle, their genotype is the same at each stage. Similarly, the genotypes of the root and stem used to make plant clones are identical.

Making clones from animals is not as easy as for plants. As you will see in this chapter, one reason for this is that animal cells lose the ability to express many of their genes as they differentiate and mature. However, some animals have been cloned by removing the nucleus from one of their cells and inserting it into an egg cell whose nucleus has been removed. This technology is called somatic cell nuclear transfer (SCNT); we usually refer to it as animal cloning.

You are probably familiar with Dolly, shown in Figure 23.1. She was produced by a team of scientists in Scotland, using SCNT, and is thought to be the first mammal to have been cloned successfully. Dolly was put down when she was only 6 years old. Although she had given birth to a number

Clone An organism or cell, produced asexually from one ancestor, to which it is genetically identical.

Figure 23.1 Dolly the sheep was the first mammal to be produced by somatic cell nuclear transfer. Do you know why she was called Dolly?

of healthy lambs, at the time of her death Dolly showed symptoms seen more commonly in much older sheep. These included arthritis, lung disease and obesity. Some of these symptoms are thought to have resulted from problems associated with cell division and the control of gene expression in her body cells.

In this chapter you will learn how genes are controlled at a cellular level, how they are switched on or off in different cells within an organism. This knowledge will help to explain how different body forms within one life cycle and different organs within one body are possible.

Stem cells

Unlike plants, only a small number of cells in animals retain the ability to divide and give rise to new tissues. Most cells in animal that can divide are only able to produce cells of the same type. These are called unipotent cells. However, some cells retain the ability to divide and are called **stem cells** and can be found in embryonic tissue and in some adult tissues. Whatever their source, stem cells have three general properties.

- They can divide and renew themselves over long periods.
- They are unspecialised.
- They can develop into other specialised cell types.

Human embryonic stem cells are stem cells that exist in all human embryos. They are taken from an embryo that has developed following *in vitro* fertilisation of an egg (in an *in vitro* fertilisation clinic) and been donated for research purposes with the informed consent of the donors. About 30 cells are removed from the inside of a human embryo that is 4–5 days old. These cells are plated into dishes that contain a coating of embryonic mouse skin cells and an appropriate culture medium, where they divide.

Very small numbers of **human adult stem cells** are found in specific tissues, such as bone marrow and the brain. They lie dormant for many years until stimulated to begin to divide, following an injury. In addition to being difficult to isolate, they are also more difficult to grow in tissue culture than embryonic stem cells. Unlike human embryonic stem cells, adult stem cells can only give rise to a limited number of body tissues: they are **multipotent**. In contrast to human adult stem cells, embryonic stem cells are pluripotent; this means, if cultured in appropriate conditions, they can develop into most of the body's cell types. This type of stem call can divide in unlimited numbers and can be used to treat human disorders. Cells taken from a very early embryo, in the first 3 or 4 days, are totipotent and can develop into any kind of cell type.

The use of embryonic stem cells is controversial. In some countries, including the USA, it is currently illegal to use embryonic stem cells, even for research. In other countries, the use of embryonic stem cells is currently legal but is tightly regulated. In April 2008, members of the European Parliament (MEPs) voted to ban across the European Union any research involving embryonic stem cells. The following month, members of the UK parliament, which is a member of the European Union, voted to continue to allow such research.

Unipotent A cell that can divide to form only one kind of cell.

Pluripotent A cell that can mature into many different kinds of specialised cell.

Totipotent A cell that can mature into any kind of cell type.

More recently, scientists have developed **induced pluripotent stem cells** (iPS cells). These are made from somatic (body) cells which have already differentiated. All cells in an organism contain the same genes, but once a cell differentiates some of the genes are 'switched off'. The scientists who developed iPS cells found that there were four genes expressed in a mouse embryo that control pluripotency. They added these four genes to somatic cells and found that the transcription factors produced by the genes made the cells pluripotent. You will learn more about transcription factors later in the chapter. There is a great deal of evidence that these iPS cells are very similar, if not identical, to embryonic stem cells. It may be possible in the near future to use these cells rather than embryonic stem cells, to avoid the ethical issues mentioned above. However, at the moment there is still a great deal of testing to be done, as iPS cells can lead to tumour formation. Scientists need to understand the processes going on in the cells much better before they can be used therapeutically in humans.

Tumour A group of one type of cell that is dividing rapidly and uncontrollably. The formation of a tumour might result from one, or only a few, genetic changes in a cell.

TEST YOURSELF

1 Embryonic mouse skin cells are used in culturing human embryonic stem cells to provide a surface to which the human cells can attach. Suggest one potential danger of this use of mouse cells.
2 Scientists hope that transplanting cultures of stem cells might be used to repair damaged tissues or to replace malfunctioning tissues.
   a) How can transplanted stem cells repair or replace malfunctioning tissues?
   b) Suggest one biological advantage and one biological disadvantage of using embryonic stem cells in such transplants.

Gene mutation

Complementary base pairing is essential if transcription and translation are to work properly. Table 23.1 summarises the pairing relationship between a base in a DNA nucleotide and the bases in complementary nucleotides of mRNA and then of tRNA.

Table 23.1 Complementary base pairing is important during DNA replication, transcription and translation.

Base on DNA nucleotide	Complementary base in the codon of a nucleotide of mRNA	Complementary base in the anticodon of a nucleotide of tRNA
Adenine	Uracil	Adenine
Cytosine	Guanine	Cytosine
Guanine	Cytosine	Guanine
Thymine	Adenine	Uracil

You also learned in Chapter 4 that three bases, called a codon, code for one amino acid. Table 4.6 on page 70 shows the mRNA codons that code for each amino acid.

As you also learned in the first year of your A-level Biology course, DNA base sequences are copied during DNA replication. At a rate of about one in a million, spontaneous errors occur during this copying process.

Some of the errors that can occur are outlined below.

- **Base deletion**: a base is lost from the DNA sequence. As a result, the whole base sequence following the deleted base changes. This is called a frame-shift mutation and results in a new sequence of amino acids after the deletion.
- **Base addition**: a new base is added, which changes the whole base sequence following this addition. This also results in a frame shift and a new sequence of amino acids after the deletion.
- **Base substitution**: the 'wrong' base is included in the base sequence. This mutation might result in a different amino acid being included in the polypeptide chain. It does not cause a frame shift.
- **Base duplication**: an important source of evolutionary change. This happens when one gene, or part of a gene, is copied, so that there are two copies on one chromosome. The second copy can develop new functions by mutation, while the original copy continues to make the protein, so there is no harmful effect on the organism. This is how the different kinds of haemoglobin – alpha, beta and fetal – are believed to have evolved. Also, in ice fish, which survive sub-zero temperatures in the Arctic Ocean, duplication of a gene coding for a digestive enzyme is believed to have resulted in the second copy of the gene mutating into an antifreeze protein.
- **Base inversion**: this can occur when two breaks in the DNA of a single gene occur. The 'cut' portion can rotate 180° and then re-join the original DNA. As this type of mutation usually affects several amino acids, it generally results in a non-functional protein being produced. However, sometimes it results in a very different protein.
- **Base translocation**: this can occur when part of a gene breaks off and re-attaches to another gene. This usually results in the original gene with a missing section, and the translocated portion, coding for non-functional proteins. However, if an inversion or a translocation occurs in a proto-oncogene or a tumour suppressor gene, this may lead to cancer. For example, a faulty tumour suppressor gene may lead to a cell with faulty DNA continuing to replicate, and translocation of a proto-oncogene to another part of the DNA may cause the gene to be expressed more strongly, or to produce a protein that has oncogenic activity.

Table 23.2 shows the effects that can arise from base deletion, base substitution and base insertion.

Frame-shift mutation A mutation caused by the addition or deletion of bases so that all of the triplets are changed from the point of mutation.

Table 23.2 The first two rows of this table show the amino acid sequence encoded by part of a molecule of mRNA. The mRNA sequence is shown as individual codons to help you to read the code. The table demonstrates the effect on the encoded amino acid sequence of a deletion, substitution and insertion of the bases shown in red.

Original base sequence on mRNA	AG**A**	UAC	GCA	CAC	AUG	CGC
Encoded sequence of amino acids	Arg	Tyr	Ala	His	Met	Arg
mRNA base sequence after base deletion	AGU	ACG	CAC	ACA	UGC	GCX
Encoded sequence of amino acids	Ser	Thr	His	Thr	Cys	Ala
mRNA base sequence after base substitution	AGC	**U**AC	GCA	CAC	AUG	CGC
Encoded sequence of amino acids	Ser	Tyr	Ala	His	Met	Arg
mRNA base sequence after base insertion	AGG	**A**UA	CGC	ACA	CAU	GCG
Encoded sequence of amino acids	Arg	Ile	Arg	Thr	His	Ala

Mutagenic agents

Natural mechanisms occur within cells that identify and repair damage to DNA. Many environmental factors increase the rate of mutation. They are called **mutagenic agents** and include:

- toxic chemicals, for example bromine compounds, mustard gas (used in a large number of conflicts to kill soldiers and civilians) and peroxides
- ionising radiation, for example gamma rays and X rays
- high-energy radiation, for example ultraviolet light.

TEST YOURSELF

3 a) Look at the middle two rows of Table 23.2. The final codon in the sequence is incomplete (GCX). Why can we safely identify Ala (alanine) as the encoded amino acid?

b) Use the middle two rows to explain the term 'frame shift'.

c) Use the bottom two rows in Table 23.2 to explain why a base substitution does not cause a frame shift.

4 a) Mutations in some cells are not important. Suggest why a mutation in the gene for haemoglobin is unimportant in a white blood cell.

b) Explain why a gene mutation can have important effects if it is found in:

    i) a gamete

    ii) a gene controlling cell division.

5 Some people are concerned that the aerials used to transmit mobile phone messages emit radiation that increases the rate of mutation. For this reason they oppose the erection of these aerials in their neighbourhood. Suggest what you would need to measure to investigate whether these aerials increased the rate of mutation leading to cancer.

Control can occur at several stages of gene expression

Gene expression involves the following flow of genetic information: DNA → mRNA → polypeptide. Gene expression can be controlled at any stage in this flow of information, including transcription and translation.

- Control of transcription: only some genes are transcribed at any given time.
- Control of translation: mRNA might be destroyed or its translation by a ribosome blocked.

Control of transcription by specific transcription factors

Every gene has one or more DNA base sequences that control its expression. Figure 23.2 shows one of these, called a **promoter region**. A promoter region is located near the gene it controls, usually about 100 base pairs before the start of its gene. Figure 23.3 shows how a protein, called a **transcription factor**, binds to a gene's promoter region and, in doing so, enables RNA polymerase to attach to the start of the gene and begin its transcription.

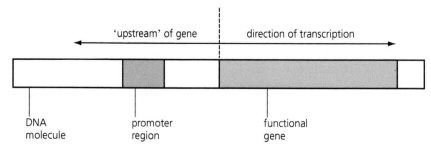

'upstream' of gene   direction of transcription

DNA
molecule

promoter
region

functional
gene

Figure 23.2 Every gene in eukaryotes is controlled by one or more promoter regions. The promoter region lies close to the gene it controls and is 'upstream' of it.

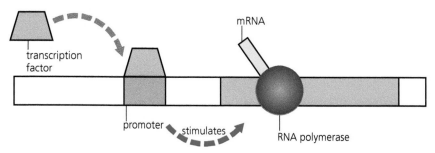

mRNA

transcription
factor

promoter   stimulates   RNA polymerase

Figure 23.3 The role of a transcription factor and a promoter region in stimulating transcription. The position of a gene encoding a polypeptide is shown in part of a molecule of DNA. In the DNA sequence preceding this gene (referred to as 'upstream' of the gene) is a promoter region. Only if an appropriate transcription factor attaches at this promoter region can RNA polymerase begin to transcribe the gene.

The role of oestrogen in initiating transcription

Oestrogen is a mammalian steroid hormone. It is involved in the control of the mammalian oestrous cycle; it also stimulates sperm production in males. Because it is a small, hydrophobic molecule, oestrogen can diffuse through the plasma membranes of cells. Once in the cytoplasm, oestrogen diffuses into the cell's nucleus, where it binds to a type of oestrogen receptor, called ER alpha (**ERα**).

The ERα oestrogen receptors are transcription factors that can bind to the promoter region of up to 100 different genes. Figure 23.4 shows one way in which these ERα oestrogen receptors work. In the cell, oestrogen receptors are normally held within a protein complex that inhibits their action. When oestrogen binds to an ERα oestrogen receptor it causes the oestrogen receptor to change shape and leave its protein complex. The oestrogen receptor can now attach to the promoter region of one of its target genes, stimulating RNA polymerase to transcribe that target gene.

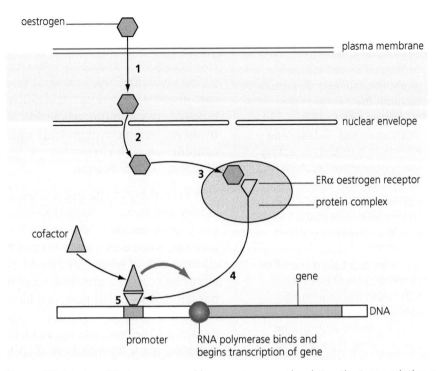

Figure 23.4 A simplified summary of how oestrogen stimulates the transcription of a target gene.
(1) Oestrogen diffuses through the plasma membrane of a target cell and then diffuses into its nucleus.
(2) Here it attaches to an ERα oestrogen receptor that is contained within a protein complex.
(3) This causes the oestrogen receptor to change its shape and leave the protein complex that inhibits its action.
(4) The oestrogen receptor can now attach to the promoter region of a target gene
(5) where it attracts other cofactors to bind with it. The oestrogen receptor, with combined cofactors, enables RNA polymerase to transcribe its target gene.

Oestrogen receptors, oestrogen-dependent breast tumours and hospital budgets

Many people in the UK develop breast tumours; the vast majority are women. About 35% of breast tumours are associated with over-stimulation of the gene encoding oestrogen. They are called **oestrogen-dependent breast tumours**.

For over 20 years the main treatment of breast tumours has been a drug called tamoxifen. This drug is effective because it has a chemical shape similar to that of oestrogen (Figure 23.5). This enables tamoxifen to bind permanently to oestrogen receptors in tumour cells. As a result, these tumour cells can no longer bind with oestrogen, which they need in order to grow.

More recently, new drugs have been developed that inhibit an enzyme, aromatase, which is required for the synthesis of oestrogen. Like tamoxifen, these drugs stop tumour cells responding to oestrogen, so stopping their growth. Large-scale clinical trials suggest that aromatase inhibitors, such as anastrozole and letrozole, can be even more effective in treating breast tumours than tamoxifen.

However, not all breast tumours are stimulated by oestrogen. Tamoxifen and aromatase inhibitors are ineffective against breast tumours that are not oestrogen-dependent. A relatively new drug, called herceptin,

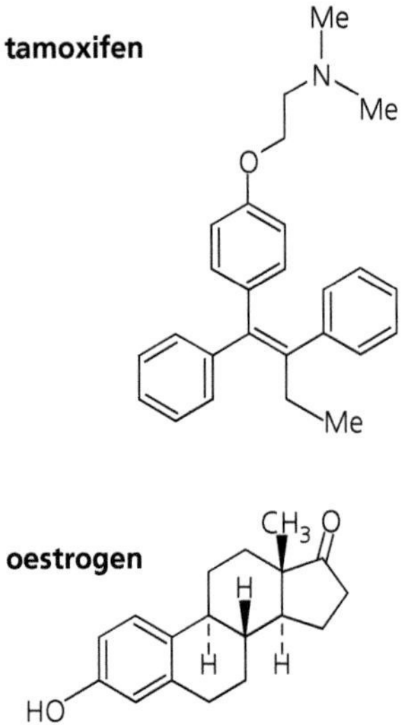

tamoxifen

oestrogen

Figure 23.5 Tamoxifen is a drug that is used to treat oestrogen-dependent breast tumours. Tamoxifen has a similar chemical shape to oestrogen.

433

TEST YOURSELF

6 Explain why having hydrophobic molecules enables oestrogen to diffuse through the cell-surface membrane of a cell.

7 Oestrogen has widespread effects in humans, affecting the reproductive system, the cardiovascular system, the immune system and bone tissue. Use information in the text to suggest how this is possible.

8 If you were a board member of an NHS trust involved in a decision about whether herceptin should be used to treat a patient with breast cancer, what type of evidence would you take into account?

is effective against some of these breast tumours because it controls them in a different way. Herceptin is a monoclonal antibody. You learned about monoclonal antibodies in the first year of your A-level Biology course. Herceptin works by binding to a growth factor receptor that is embedded in the surface membranes of some types of breast-tumour cells. As a result, it inhibits the growth of the tumour cells. The type of breast-tumour cell against which herceptin is effective is very invasive. Treatment of patients with this type of breast tumour, used in conjunction with chemotherapy, has been shown to lead to long periods of disease-free remission.

However, there is one major drawback with the use of herceptin. It is very expensive, costing the National Health Service about £22 000 per patient per year. NHS trusts have to consider the cost of treating patients, since they have a limited budget and have to balance the cost-effectiveness of what they can do. Some trusts have not been able to afford herceptin treatment for patients with invasive breast cancers. This has led to a so-called postcode lottery: depending on where a sufferer lives she might, or might not, be able to receive herceptin treatment. As is often the case, the use of biological advances is dependent on decisions made by members of society.

Control of translation by RNA interference

In eukaryotes and some prokaryotes translation of mRNA from target genes can be inhibited by RNA interference, or RNAi.

EXAMPLE

Gene regulation in petunias

TIP

The main idea that you need to remember from this is that interfering RNA binds to mRNa by complementary base pairing.

In 1990, a group of scientists reported their attempts to produce petunia flowers with a very deep purple colour. They used genetic engineering techniques to insert many copies of the gene encoding chalcone synthase into the cells of petunia plants with pale purple flowers.

Petunias are popular plants that are grown in hanging baskets in the UK. The purple colour of the flowers in Figure 23.6 on the next page results from a series of reactions in which a white pigment is converted to a purple pigment. One of the enzymes in this series of reactions is called chalcone synthase.

Figure 23.6 Petunia flowers show a wide range of colours.

1 Use your knowledge of how enzymes work to suggest why the scientists expected the plants with the extra genes for chalcone synthase to produce deep purple flowers.

Enzymes speed up a reaction by combining with molecules of substrate to form enzyme–substrate complexes, which break down to release molecules of the product. We can increase the rate at which the product is formed by adding more enzyme molecules. As a result, more enzyme–substrate complexes will be produced and so the product will be formed faster. The scientists predicted that if they added more genes for chalcone synthase to cells of petunia plants, more mRNA would be transcribed from them and so more molecules of enzyme would be made in the cells of the petunia.

Instead of deep purple flowers, the transformed petunia plants produced flowers that were mottled white. The scientists could not explain this result.

In 2004, in a totally unrelated investigation, a different team of scientists reported their discovery of enzymes called **RNA-dependent RNA polymerases (RDRs)**. These RDRs catalyse the production of **double-stranded RNA (dsRNA)**. They do this by using molecules of mRNA that are in the cytoplasm as a template to synthesise a complementary strand. The two RNA strands are held together by hydrogen bonds between complementary base pairs.

2 What is unusual about the RNA produced by RDRs?

You learned in Chapter 4 that RNA molecules are single-stranded, but the RNA molecules produced by RDRs are double-stranded (dsRNA).

Figure 23.7 shows what happens when RDRs produce a dsRNA molecule from a molecule of mRNA in the cytoplasm.

3 Look at Figure 23.7. What is the first thing that happens to the dsRNA molecule?

*The dsRNA molecule is cut into small fragments. These fragments are about 23 base pairs long and are called **small interfering RNA (siRNA)**.*

Figure 23.7 shows what happens to the siRNA. In a reaction requiring the hydrolysis of ATP, a protein complex in the cytoplasm takes up one of these siRNA fragments and separates its two RNA strands.

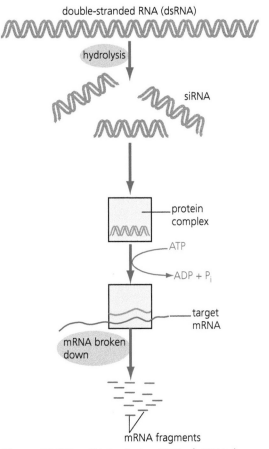

Figure 23.7 Small interfering RNAs (siRNAs) are small double-stranded RNA molecules. They are used by protein complexes in a cell's cytoplasm to break down mRNA. By breaking down the target mRNA, a cell can control the expression of the gene from which the mRNA was transcribed.

4 Suggest why the hydrolysis of ATP is involved in the reaction between the protein complex and the siRNA it takes up.

Like many reactions in metabolism, the reaction between the protein complex and the siRNA requires energy. This energy is released when ATP is hydrolysed to ADP and inorganic phosphate.

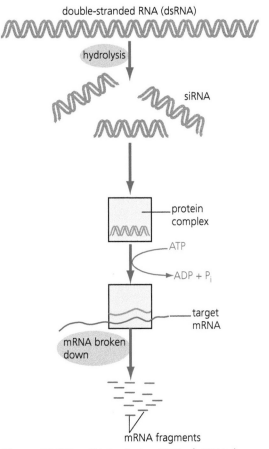

Look at the protein complex containing the single strand of RNA in Figure 23.7. The protein complex uses the RNA strand to bind to a molecule of mRNA in the cytoplasm. Once it has attached to the mRNA molecule, the protein complex breaks down the mRNA molecule. This stops the mRNA being translated by the cell's ribosomes.

5 To which type of mRNA will the protein complex attach?
The RDR in Figure 23.7 used a molecule of mRNA to make the dsRNA. One strand of the dsRNA must, therefore, have a complementary base sequence to the mRNA from which it was made. When the protein complex separates the two strands of the siRNA, it uses this complementary base sequence to attach to any of the original mRNA molecules in the cytoplasm. Thus, the protein complex specifically breaks down the mRNA from which the siRNA was made.

6 Will the base sequence of the siRNA be complementary to the whole of the mRNA in Figure 23.7?
The siRNA is a small fragment of the mRNA molecule, so it will only be complementary to part of the base sequence of the mRNA. However, this is enough to enable the protein complex to bind with, and destroy, the target mRNA.

7 The destruction of mRNA by siRNA is stimulated when the concentration in the cytoplasm of one type of mRNA becomes high. Can you use this information to suggest how the discovery of RNA-dependent RNA polymerases and small interfering RNA provided an explanation for the failure of the first team of scientists to produce petunia flowers with a deep purple colour?
The first team of scientists increased the concentration of mRNA encoding chalcone synthase by inserting into petunia cells many copies of the gene for the enzyme. A high concentration of mRNA encoding chalcone synthase stimulated RDRs to produce dsRNA from it. As a result many more siRNA molecules carrying part of the mRNA code for chalcone synthase were produced by the plants, so the mRNA was destroyed by the mechanism in Figure 23.7. Unwittingly, the scientists had stimulated destruction of mRNA encoding chalcone synthase instead of stimulating more of it to be transcribed to produce more enzyme molecules.

Epigenetic control of gene expression in eukaryotes

Epigenetics involves heritable changes in gene function without changes to the base sequence of DNA. Along with environmental influences, it explains why identical twins, whose DNA is identical at fertilisation, become increasingly different from each other as they get older. Figure 23.8 shows two identical twin sisters. They still look very much alike at 78, but the small differences between them are much easier to see now that they are older.

Figure 23.8 These are identical twin sisters, whose DNA base sequence is identical. In this photo they are aged 78. As twins age, more and more small differences can be seen between them as the result of epigenetic imprinting. (Note, however, that the difference in hair colour is a result of hair dye and not epigenetic imprinting!)

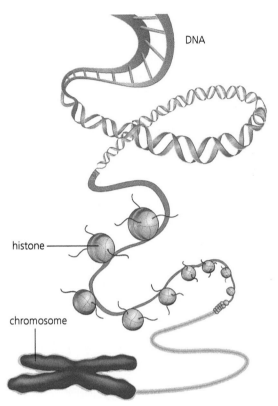

Figure 23.9 DNA winds around histone proteins to form chromosomes. Chemical 'tags' (see Figure 23.12) attached to the DNA and the histones affect how tightly the DNA winds around the histones.

Epigenome The sum of the chemical changes to the histone proteins and the DNA (but not including changes to the base sequence) of an organism.

Figure 23.10 Acetylation of the histone proteins makes the DNA less tightly wound round the histone proteins. This allows DNA transcription to occur.

In the cells of eukaryotes, the DNA is wrapped around proteins called histones. Chemical 'tags' can join on to the histone proteins and the DNA (see Figure 23.9). The chemical 'tags' affect how tightly the DNA is wound around the histone proteins. If the DNA is wound tightly round the histone proteins, the gene is effectively 'switched off' but if the DNA is loosely wound, the gene may be 'switched on'. This system is flexible, so the epigenome can be different in one cell type from another.

Acetylation of histones

Histone molecules contain the amino acid lysine. Acetyl groups ($COCH_3$) may be added to these lysine residues, replacing one of their hydrogen ions. This removal of positively charged ions causes the histone proteins to be less tightly wrapped around the DNA. As a result, the enzyme RNA polymerase and other factors needed for transcription can bind to the DNA more easily. You can see this in Figure 23.10. Therefore, in most cases

- adding acetyl groups to the histone proteins (**acetylation**) stimulates transcription
- removing acetyl groups from the histone proteins (**deacetylation**) suppresses transcription.

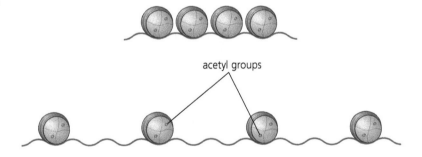

acetyl groups

Methylation of DNA

DNA methylation occurs when methyl groups (CH_3) are added to a DNA molecule, usually to a carbon atom of cytosine bases where they occur in a cytosine–guanine sequence. This methylation suppresses transcription of the affected gene, effectively switching the gene off.

Methyl and acetyl groups are examples of the chemical 'tags' that form part of the epigenome. They can be added to chromosomes as the result of environmental factors, such as diet, stress, smoking or exercise. They can also be the result of signals from neighbouring cells or even from within the same cell. Methylation is an important mechanism for 'switching' genes on and off during embryonic development, when cells are differentiating.

Epigenetic imprinting

You will remember that our body cells have two sets of chromosomes in them, one (paternal) set from our fathers and the other (maternal) set

437

Epigenetic Describing inherited changes in the DNA that do not involve a change in the DNA base sequence.

from our mothers. During the formation of oocytes and sperms, DNA methylation of certain genes occurs; this process is called epigenetic **imprinting**. This imprinting is reversible. For example, when a woman inherits a chromosome from her father, it will be epigenetically imprinted as 'paternal'. However, when the daughter passes the same chromosome on to her child, it will have become imprinted as 'maternal'.

Recently, scientists have found that the same allele or alleles can have a different effect, depending on which parent it was inherited from. For example, Prader–Willi syndrome is a genetic condition that affects one in every 12 000–15 000 people. It affects both sexes and all ethnic groups. It is caused by inactivation of some of the alleles on chromosome 15.

You can see a pedigree showing the inheritance of Prader–Willi syndrome in one family in Figure 23.11. You will notice that individuals who inherit the defective chromosome from their mother do not develop Prader–Willi syndrome.

Figure 23.11 Inheritance of Prader–Willi syndrome in one family.

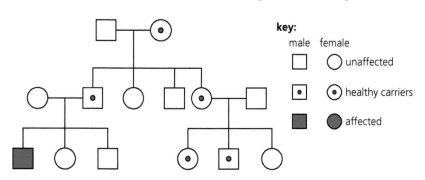

key:

male   female

☐ ○ unaffected

◻ ◉ healthy carriers

■ ● affected

Extension

Prader-Willi syndrome causes a wide range of effects, which may include:
- a constant feeling of hunger, which leads to obesity
- restricted growth
- reduced muscle tone
- no development of secondary sexual characteristics
- learning difficulties and behavioural problems.

Epigenetics is very important in cancer research. The DNA in human tumours shows changes in DNA methylation and histone protein modification. This can cause tumour suppressor genes to be silenced, or oncogenes to be activated. Scientists are developing drugs that treat cancer by reversing epigenetic changes, e.g. by removing acetyl 'tags' on histone proteins or removing methyl groups on DNA.

TEST YOURSELF

9 How does the pedigree in Figure 23.11 show that Prader–Willi syndrome results from a defective chromosome inherited only from the father?

10 The mule, born when a female horse and male donkey breed together, looks very different from a hinny, born to a female donkey and a male horse. Use your understanding of epigenetics to suggest why.

11 In the plant *Arabidopsis*, the *FT* gene causes flowering. This gene is inhibited when the FLC protein is present. The gene that codes for FLC is switched off when environmental temperatures are cold, but the gene is switched on again when the environmental temperature is higher. Suggest

a) the advantage of this process to *Arabidopsis*

b) how the *FT* gene might be switched on or off.

Gene mutations and cancer

Mitosis occurs during the cell cycle of eukaryotic organisms. The rate of mitosis is controlled by two groups of genes

- proto-oncogenes, which control cell division
- tumour-suppressor genes, which slow cell division. These genes also promote programmed cell death (apoptosis) in cells with DNA damage that the cell cannot repair.

Gene mutations can occur in both these two types of gene. A mutated proto-oncogene, called an oncogene, stimulates cells to divide too quickly. Proto-oncogenes often code for proteins (growth factors) that stimulate cell division by binding to receptors in the cell membrane. They may also code for receptors in the cell membrane that control cell division. Mutated proto-oncogenes may result in over-production of these growth factors, or protein receptors in the cell membrane that stimulate cell division even when the growth factor is not present. A mutated tumour-suppressor gene is inactivated, allowing the rate of cell division to increase. Another way in which tumour-suppressor genes may be inactivated is if they undergo epigenetic changes, such as becoming hypermethylated.

ACTIVITY

An investigation into tumour formation in transgenic mice

Clones of laboratory mice are often used in investigations into the causes and effects of tumours.

1 What is a clone?

Scientists genetically transformed mice from a single clone to contain different oncogenes, called *myc* and *ras*. One group of mice contained only the *myc* oncogene, a second contained only the *ras* oncogene and a third contained both the *myc* and the *ras* oncogenes. The scientists then recorded the age at which the mice in each group developed tumours.

2 What is an oncogene?
3 Explain why the scientists used clones in this experiment.

Figure 23.12 shows the percentage of mice in each clone that were free of tumours during the course of the experiment.

4 For how many days were all the mice in this experiment free of tumours?
5 Use the graph to compare the effects of the *myc* and *ras* oncogenes when present alone.
6 Suggest an explanation for the curve showing mice with both the *myc* and *ras* oncogenes.

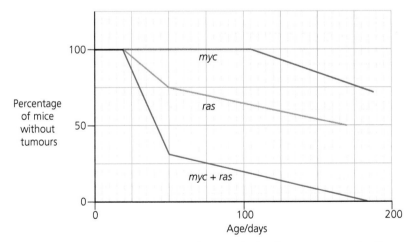

Figure 23.12 Three groups of mice from a single clone were genetically engineered to contain either *myc* oncogenes or *ras* oncogenes or both. The graph shows the percentage of mice in each clone that did *not* have tumours.

Figure 23.12 shows the results of an investigation involving the formation of tumours in mice. A tumour is a group of one type of cell that is dividing rapidly and uncontrollably. The formation of a tumour might result from one, or only a few, genetic changes in a cell. Tumours can be benign or cancerous. Benign tumours are tumours which grow in one place and do not spread. They are not cancerous tumours. However, depending on where they grow, they may still be harmful.

Cancerous tumours have cells that can break off and spread around the body, in a process called metastasis. The cancerous cells invade organs and tissues throughout the body and secondary tumours develop. These are called metastases.

Cancers are tumours in which some cells break away from the group and invade organs and tissues throughout the body. Figure 23.13 shows that the change from tumour cells to cancer cells requires many more genetic changes. Thus, cancer does not usually result from a single gene mutation.

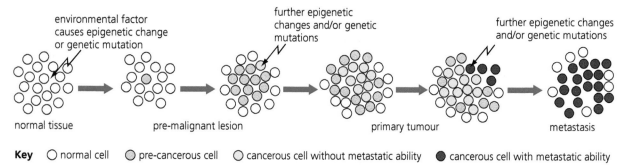

environmental factor causes epigenetic change or genetic mutation

further epigenetic changes and/or genetic mutations

further epigenetic changes and/or genetic mutations

normal tissue   pre-malignant lesion   primary tumour   metastasis

Key ○ normal cell   ◔ pre-cancerous cell   ○ cancerous cell without metastatic ability   ● cancerous cell with metastatic ability

Figure 23.13 Healthy cells become tumour cells as a result of mutation in one, or a few, genes controlling cell division. Many more mutations are required for a tumour cell to become a cancer cell.

EXAMPLE

Meat consumption and cancer

The graph in Figure 23.14 shows the results of a large-scale epidemiological study of Armstrong and Hill (1975). It shows there is a strong link between colon cancer incidence and meat consumption.

1 Describe the relationship shown by the graph.

It shows that the more meat is eaten per person per day, the higher the incidence of colon cancer in women. However, this is not a perfect correlation. For example, meat consumption in Sweden and Hungary is very similar, yet the incidence of colon cancer in women in Sweden is more than double the incidence in Hungary.

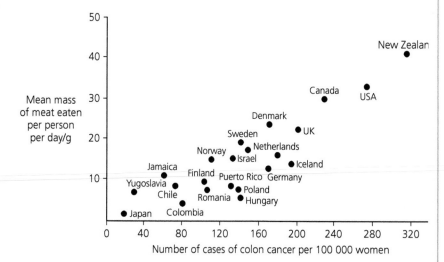

Figure 23.14 Meat consumption and incidence of colon cancer.

2 Explain why the incidence of colon cancer is given per 100 000 women.

Each country has a different size of population, so to give the incidence of colon cancer in each population would be misleading. For example, the population of New Zealand is very much lower than the population of the USA. Giving the incidence per 100 000 means you can compare countries even though their populations are different in size. It is also useful to look at rates of colon cancer among women as the incidence of many cancers varies depending on the sex of the individual.

3 A journalist writing for a vegetarian food magazine wrote an article about these data, under the heading 'Here's the proof that eating meat causes cancer'. Evaluate the journalist's conclusion.

There are many points that you can give here. The word 'evaluate' means 'give the arguments for and against'. You can certainly say that the graph gives evidence that meat-eating increases the incidence of colon cancer, as there is a positive correlation. However, this is a correlation and does not prove cause and effect. It is possible that there is another factor that causes cancer. People who don't eat meat usually eat more vegetables, so it might be that fibre and vegetables protect people from cancer. Countries where people eat a lot of meat tend to be wealthier than countries where people eat very little meat, so there might be a factor connected with affluence that increases the incidence of colon cancer, such as lack of exercise or drinking more alcohol. The rates of cigarette smoking are likely to be very different between these countries. The data does not distinguish between types of meat. It may be that red meat, or processed meat, is more likely to cause cancer than some other kind of meat. The problem with these data is that environmental factors will vary between these countries, so meat-eating is not the only variable.

TIP

Sir Richard Doll was the world's leading epidemiologist in the 20th century and proved the causal relationship between smoking and lung cancer. Look back at the section about smoking, cancer, correlations and causal relationships in Chapter 7.

Practice questions

1 Read the following passage.

Scientists think that some of the genetic changes associated with ageing may be the result of epigenetics, which suggests they could be reversed.

Molecules can attach to DNA, enhancing or preventing gene activation without changing the underlying genetic code. Such epigenetic changes are already suspected as factors in psychiatric disorders, diabetes and cancer.

They may also play a role in ageing. A group of scientists looked at the DNA of 86 sets of twin sisters aged 32 to 80, and discovered that 490 genes linked with ageing showed signs of epigenetic change through a process called methylation. They found that these genes were more likely to be methylated in the older twins than the younger twins. This can be triggered by environmental factors.

a) What is epigenetics (line 2)? (2)

b) What is methylation (line 10)? (3)

c) Name two environmental factors that might lead to methylation. (1)

d) Scientists think that some of the genetic changes associated with ageing might be reversed (line 2). Suggest how. (2)

2 Read the following passage.

A woman in Japan has received the first medical treatment based on induced pluripotent stem cells, 8 years after they were discovered. The iPS cells were made by taking skin cells from the woman's arm and reprogramming them, so they could be transformed into specialised eye cells. These were used to treat age-related macular degeneration (AMD), a condition that affects millions of elderly people worldwide and often results in blindness.

The woman had a patch of the cells measuring 1.3 × 3 millimetres grafted into her eye. She is one of six people who will receive this treatment, in an attempt to investigate the safety of stem cell treatments.

a) What is the meaning of pluripotent (line 2)? (1)

b) Explain why the skin cells must be reprogrammed before using them for this treatment. (2)

c) Explain the advantages of using stem cells produced from the woman's own skin cells, rather than cells from an embryo. (4)

d) Suggest reasons why it is important to investigate the safety of stem cell treatments such as this. (4)

3 Some kinds of breast cancer cells have HER-2 receptors on their cell-surface membrane. These receptors cause the cells to be stimulated by a growth factor so that they divide too much. Trastuzumab is a monoclonal antibody that binds to the HER-2 receptors and stops the cells from dividing.

    a) What are monoclonal antibodies? (2)

    b) Explain how trastuzumab stops these cells from dividing. (2)

    c) Explain why trastuzumab is not effective against all kinds of breast cancer. (2)

    d) Trastuzumab is administered by injection into the bloodstream. Suggest why it is not given in the form of a tablet. (2)

4 RNA interference (RNAi) has been proven effective against a human disease, respiratory syncytial virus (RSV), which is harmful to young children but relatively harmless in adults. Eighty-five healthy adults were given a nasal spray containing either a placebo or small interfering RNA (siRNA) designed to silence one of the genes of RSV. The adults used the spray daily for 5 days. On day 2, all the volunteers were infected with live RSV. By day 11, just 44% of those who received the RNAi nasal spray had RSV infections, compared with 71% of the placebo group. The scientists hope to use siRNA on lung transplant patients soon, and then they want to test it on infants.

    a) Describe how siRNA can silence a gene. (4)

    b) i) What would the placebo contain? (2)

        ii) Suggest why the trial was done on adults rather than children. (2)

    c) Suggest why siRNA might be useful for people who have had lung transplants. (2)

Stretch and challenge

5 Investigate Angelman syndrome and how it is inherited. Contrast the inheritance of Angelman syndrome with the inheritance of Prader–Willi syndrome.

6 Research the genetic inheritance of tortoiseshell coat colour in cats. Investigate Carbon Copy, the cloned tortoiseshell cat, and explain why she does not look like her clone.

24

Gene cloning and gene transfer

TEST YOURSELF ON PRIOR KNOWLEDGE

1 The DNA code is universal. Explain what this means, and how this makes genetic engineering possible.
2 What is a gene?
3 What is the name for three bases in DNA that code for an amino acid?

Do you intend to study biology at university? If so, are you planning to use your degree to make a career? Many graduate biologists do, for example your biology teachers. Some biologists are employed by large pharmaceutical companies, where they are employed to develop new drugs.

In the 1980s, pharmaceutical companies were attracted by new developments in DNA technology, including the genetic manipulation of organisms. Some molecular biologists even set up their own companies to capitalise on their research. The potential profits from this biotechnology industry attracted people to invest their money in these companies. They believed that buying shares in the companies would give them a better return on their money than other types of investment. This sudden growth in ownership of shares in the biotechnology industry was called the 'biotechnology bubble'. Biotechnology companies became very valuable in a short space of time and some molecular biologists who had set up their own companies became very wealthy on paper.

Although the potential benefits of biotechnology were promising, there were many difficulties in converting research methods into profit-earning products. Many of these problems were technical, but some resulted from negative public opinion. As a result, the price of shares in these companies fell dramatically in 1999; the biotechnology bubble burst and investors lost a good deal of the money they had invested.

Recombinant DNA technology

Entire molecules of DNA are, in general, too large to be used in gene technology. Instead, molecular biologists use fragments of DNA, which contain the gene or genes they are interested in. They can produce these fragments using three different methods. They use:

- a reverse transcriptase to convert mRNA to cDNA
- a restriction endonuclease to cut DNA fragments out of a large DNA molecule
- a 'gene machine' to synthesise the required piece of DNA.

We will look at each of these methods in turn.

Using a reverse transcriptase to convert mRNA to cDNA

You learned in Chapter 23 that most cells do not transcribe all their genes. Thus, the cytoplasm of a specialised cell contains only mRNA transcribed from some of the genes in its nucleus. This is because specialised cells make a lot of a limited number of different proteins. It is often easier to extract mRNA from the cytoplasm of a cell than to find the gene from which it was transcribed. For example, mRNA encoding the mammalian hormone insulin will be found in high concentrations in the cytoplasm of β-cells in the islets of Langerhans of the pancreas, but in no other cell.

Molecular biologists can extract this mRNA and make a DNA copy from it (Figure 24.1). Purified mRNA is mixed with free DNA nucleotides and an enzyme called **reverse transcriptase**. You learned about reverse transcriptase when you studied HIV in the first year of your course (see Chapter 6). This enzyme catalyses a process that is the reverse of transcription: DNA is made from mRNA. The DNA formed in this way is called **complementary DNA (cDNA)**. The cDNA is single stranded, but it can be made into double-stranded DNA using DNA polymerase.

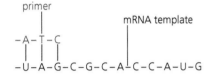

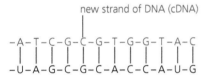

Figure 24.1 Complementary DNA (cDNA) is made using reverse transcriptase to copy the base sequence of mRNA.

Using a restriction endonuclease to cut fragments out of a large molecule of DNA

Nucleases are enzymes that break the bonds linking one nucleotide to the next in a DNA strand. An **exonuclease** removes nucleotides, one at a time, from the end of a DNA molecule. We are more interested in **endonucleases**. Figure 24.2 shows how an endonuclease hydrolyses bonds within the DNA molecule, producing fragments of DNA.

Restriction endonucleases (or restriction enzymes, for short) are enzymes that break bonds in the sugar–phosphate backbones of *both* strands in a DNA molecule. As a result, they produce double-stranded fragments of DNA, like the ones shown in Figure 24.2. Each restriction enzyme cuts DNA only at a particular sequence of bases, called the **recognition sequence** of the enzyme.

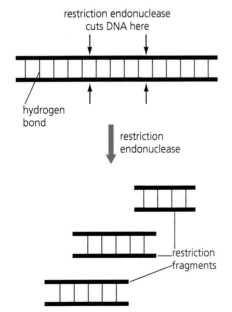

Figure 24.2 A restriction endonuclease cuts DNA into double-stranded fragments, called restriction fragments.

445

A restriction endonuclease can cut a DNA molecule in one of two ways. Look at Figure 24.3a. This shows how one restriction enzyme makes a simple cut right across the middle of its recognition sequence. You can see that this results in DNA fragments with **blunt ends**. Figure 24.3b shows how another restriction enzyme makes a staggered cut across its recognition sequence. This produces fragments with short single-stranded overhangs. Since these overhangs can form base pairs with other complementary sequences, they are called **sticky ends**.

a) Production of blunt ends by the restriction enzyme *Alu*1

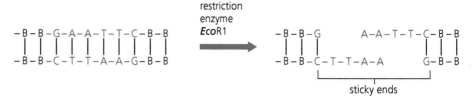

b) Production of sticky ends by the restriction enzyme *Eco*R1

Figure 24.3 A restriction endonuclease produces double-stranded fragments of DNA by breaking DNA molecules at a specific recognition sequence. The bases involved in the recognition sequence are shown in red. A is adenine, C is cytosine, G is guanine and T is thymine. The symbol B represents other organic bases that are not involved in the recognition sequence.

EXAMPLE

The properties of recognition sequences

1 What does Figure 24.3 tell us about the recognition sequence of the restriction enzymes from these two bacteria?

The recognition sequence of Alu1 is different from the recognition sequence of EcoR1. In fact, every restriction enzyme has a unique, specific recognition sequence where it cuts a DNA molecule.

2 Can you use your knowledge of how enzymes work to explain why each restriction enzyme has a specific recognition sequence?

You should recall that enzymes are specific. An enzyme is specific because only one type of substrate can fit into its active site. The active sites of restriction endonucleases are specific to particular sequences of bases; that is, to specific recognition sequences.

3 How long are the recognition sequences in Figure 24.3?

In Figure 24.3a, the recognition sequence is only four base pairs (4 bp) long. In Figure 24.3b, the sequence is only 6 bp long. Although some restriction enzymes have recognition sequences that are longer than this, most have recognition sequences between 4 bp and 8 bp long.

4 Look carefully at the sequence of bases in the upper and lower strands of each DNA fragment shown in Figure 24.3a and b. What is the common feature about their patterns of bases?

In Figure 24.3a, the upper strand of the recognition sequence reads AGCT. If you read the lower strand from right to left, it also reads AGCT. Look again at Figure 24.3b. The upper strand of the recognition sequence reads GAATTC, which is the same as the lower strand when read from right to left. The order of bases in one strand of a recognition sequence is always the same as the complementary strand when read backwards. Something that reads the same forwards and backwards is called a palindrome. A simple palindrome in English is 'never odd or even'. Figure 24.3 shows us that, in addition to being short, recognition sequences are palindromic. We can now define the recognition sequence of a restriction enzyme as a short, specific, palindromic base sequence.

5 How could molecular biologists ensure that they only cut a specific gene out of DNA?

6 Bacteria can be infected by viruses that inject their
nucleic acid into the host cell. Suggest one natural

Creating a gene in a 'gene machine'

If the primary structure of a protein is known, then it is possible to
synthesise the gene required to produce the protein using a 'gene machine'.
The amino acid sequence required is entered into a computer. The triplet
code for each amino acid is known (see Table 4.6 on page 70) so the DNA
sequence that will produce that protein is worked out. The computer
then controls the machine and the required DNA sequence is made. The
advantage of synthesising a gene in this way is that it does not contain
introns, so the gene can be transcribed and translated in prokaryotic cells.

Analysing restriction fragments

Since they are produced by restriction endonucleases, the DNA fragments
produced by these enzymes are called **restriction fragments**. DNA is
digested by restriction endonucleases into a number of restriction fragments
of different lengths. We can find out the number and size of the restriction
fragments using **gel electrophoresis** followed by visualisation of the DNA.
This is important in many genetic analyses, such as genetic fingerprinting
(see page 473) or analysing DNA to find out whether the DNA contains a
specific allele or not.

Separating restriction fragments using electrophoresis

Electrophoresis uses a slab of gel made of agarose or polyacrylamide. An
electrode is placed at each end of the gel and an electric current is passed
through it. A sample containing restriction fragments is placed in a well
that is cut in the gel near the negative electrode. Since DNA has a negative
charge, the restriction fragments will migrate through the pores in the gel
towards the positive electrode. The smaller fragments will move faster
through the pores in the gel than the larger molecules and so will move the
furthest from the well. Figure 24.4 shows a cross section through a gel with
restriction fragments migrating through it. The fragments produce bands of
restriction fragments of different sizes.

Figure 24.4 Restriction fragments migrate through a gel towards the positive electrode of an electric field. The smaller fragments move through pores in the gel faster than the larger fragments. Thus, bands of restriction fragments are formed in the gel.

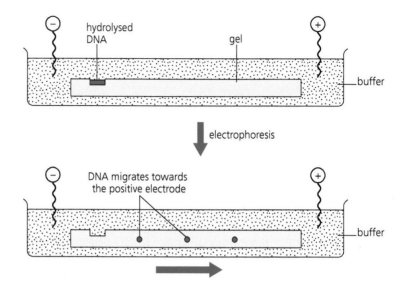

Visualisation of DNA bands

DNA is colourless, so we must treat the DNA in such a way that we can see the bands in the gel after electrophoresis. This can be done using a suitable stain. The result is a series of coloured bands in the gel. An alternative is to treat the DNA with a **radioactive marker** before it is digested by restriction endonucleases. The fragments can then be seen by placing the gel in contact with a photographic film that is sensitive to X-rays. The developed film, called an **autoradiograph**, shows the DNA bands. Figure 24.5 shows a developed film showing labelled restriction fragments. Alternatively, the DNA may be visualised by adding fluorescent probes that bind to the DNA (see page 469).

Figure 24.5 Radioactively labelled DNA has been separated using gel electrophoresis. After separation by electrophoresis, the gel has been placed in contact with a photographic film. After development, the different restriction fragments appear as dark bands on the film, called an autoradiograph.

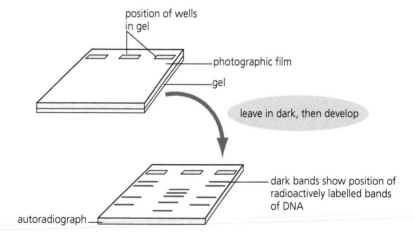

TEST YOURSELF

1 What will be the base sequence of cDNA made from a section of mRNA with the base sequence: ACG CGA UCA UGA?

2 Reverse transcriptase can only begin to copy mRNA in the presence of a primer. Use Figure 24.1 to suggest what a primer is.

3 Reverse transcriptase is found in some viruses such as HIV. What is its role in HIV?

4 The recognition sequence of a certain restriction endonuclease occurs six times along a molecule of DNA. How many fragments of DNA will be produced by the action of the restriction enzyme?

5 Why does DNA have a negative charge?

6 The agarose gel used in electrophoresis must be very pure and uniform. Suggest why.

Gene cloning

You learned in Chapter 23 that a clone is a group of cells, or organisms, that are genetically identical to each other. **Gene cloning** involves making identical copies of a gene. There are two major technologies that enable molecular biologists to clone genes (Figure 24.6).

One form of gene cloning or DNA amplification is the **polymerase chain reaction (PCR)**. Because the PCR occurs in a test tube, this method of gene cloning is called *in vitro* gene cloning (literally 'in glassware'). You are probably familiar with the term *in vitro* in the context of *in vitro* fertilisation (IVF). In an IVF clinic, the fertilisation of an egg cell by a sperm cell occurs in glassware.

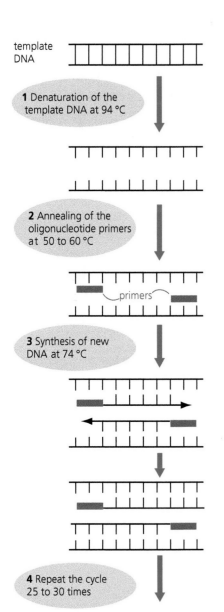

Figure 24.7 The basic steps in the polymerase chain reaction (PCR).

template DNA

1 Denaturation of the template DNA at 94 °C

2 Annealing of the oligonucleotide primers at 50 to 60 °C

primers

3 Synthesis of new DNA at 74 °C

4 Repeat the cycle 25 to 30 times

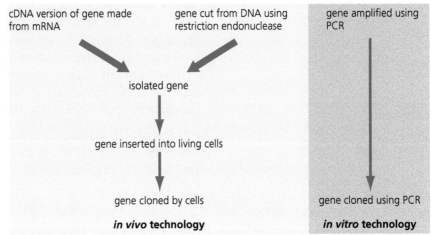

cDNA version of gene made from mRNA

gene cut from DNA using restriction endonuclease

gene amplified using PCR

isolated gene

gene inserted into living cells

gene cloned by cells

gene cloned using PCR

in vivo technology

in vitro technology

Figure 24.6 Gene cloning involves making identical copies of a gene. This can be done using two different technologies: *in vivo* and *in vitro*.

An alternative way of cloning a gene involves isolating the gene and inserting it into a suitable host cell. As the host cell divides, it replicates the inserted gene along with its own DNA. Since this method of gene cloning requires the use of living cells, it is called *in vivo* gene cloning (*in vivo* means 'in life').

Using the PCR to amplify part of a DNA molecule

According to scientific folklore, the idea of the PCR occurred as a brainwave to its inventor one evening as he drove along the coast of California. What is true is that the PCR has since become one of the most useful tools in molecular biology. It is highly likely that you have seen the PCR 'performed' in TV crime programmes. As long as molecular biologists can design primers that mark the beginning and end of a section of DNA, the PCR can be used to amplify a gene or any length of DNA or DNA fragment.

PCR involves mixing the DNA to be copied (**template DNA**) with a set of reagents in a test tube and placing the tube in a thermal cycler. The reagents include an enzyme, short lengths of single-stranded DNA (called **oligonucleotides**) and free DNA nucleotides, each with an adenine, cytosine, guanine or thymine base. The thermal cycler is an automated machine that changes the temperature at which the test tube is incubated in a pre-programmed sequence. Figure 24.7 shows the three basic steps in one PCR cycle, which we will now describe in more detail.

- **Step 1**: denaturing of the template DNA. The thermal cycler heats the mixture to 94 °C. This causes the hydrogen bonds that hold together the two strands of the template DNA to break down. As a result, the two individual DNA strands separate and can now act as templates for building complementary strands.
- **Step 2**: annealing the primers. The cycler cools the mixture to between 50 °C and 60 °C. This allows hydrogen bonds to re-form. In theory, the two strands of template DNA could re-join at this temperature. Most do not because the oligonucleotides bind to the template DNA strands instead. This binding is called **annealing** and occurs only at a site on the template DNA that has a base sequence complementary to that of the oligonucleotides. These oligonucleotides are called **primers** because the enzyme involved in the next step must attach to one of them before it can start to work.
- **Step 3**: synthesis of new DNA. The cycler raises the temperature to 74 °C. This is the optimum temperature of the **DNA polymerase** used in this step of the PCR. The enzyme attaches to one end of each primer and synthesises new strands that are complementary to the template DNA. At the end of this stage the original molecule of template DNA has been copied and two molecules are in the mixture. The cycler now raises the temperature back to 94 °C and the cycle occurs all over again.

The PCR is usually repeated about 25 times. As a result, over 50 million copies of the template DNA are formed. The result of a small number of cycles is shown in Figure 24.8.

Figure 24.8 For each cycle, the PCR produces two molecules of DNA from a template molecule of DNA. In a short time, many copies of the template DNA can be made.

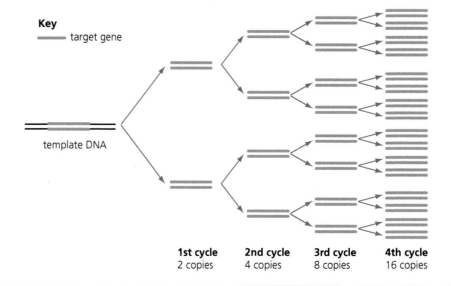

TEST YOURSELF

7 The enzyme most commonly used in the PCR is *Taq* DNA polymerase. This enzyme is extracted from a bacterium, *Thermus aquaticus*, that lives in hot-water springs. Many of this bacterium's enzymes, including *Taq* DNA polymerase, are thermostable.

a) What is the advantage to the bacterium of having thermostable enzymes?

b) Explain why it is important that the DNA polymerase used in the PCR is thermostable.

8 You have learned already about the importance of molecular collisions in explaining the action of competitive inhibitors. Use this understanding to suggest why it is important to the success of Step 2 that the PCR mixture contains a high concentration of primer.

Extension

How do scientists use the PCR to amplify specific fragments of DNA, including genes?

So far, our account of the PCR suggests that the whole of a DNA molecule will be copied in each cycle. In eukaryotes, this would involve copying entire chromosomes, which would not be helpful. A whole chromosome contains too much information to transfer or to store. Molecular biologists usually want to amplify only part of a DNA molecule, for example a gene that they wish to transfer from one organism to another or a gene that they wish to store in a gene library.

During the PCR, DNA polymerase starts replication where it binds to a primer. It will stop when it reaches another primer on the same strand of DNA. The correct design of primers is critical to the success of amplifying fragments of DNA.

The primer is a sequence of nucleotides that attaches to template DNA wherever it meets a base sequence that is complementary to its own. It would be possible to design a primer that contained only two nucleotides, but not very useful. A sequence of only two bases, for example CT, is likely to occur a large number of times within a short length of template DNA. DNA polymerase starts copying when it attaches to a primer, then moves along the template DNA but stops when it reaches another primer. With primers attached only short distances apart, this would result in a large number of very small sections of amplified DNA.

It is easy to estimate how often base sequences of different lengths are likely to occur along template DNA. For example, a particular sequence of four bases is likely to occur once in every $4^4 = 256$ nucleotides, a particular sequence of eight bases is likely to occur once in every $4^8 = 65\,536$ nucleotides and a particular sequence of 16 bases is likely to occur once in every $4^{16} = 4\,294\,967\,296$ nucleotides.

You will have spotted that a particular sequence of n bases is likely to occur every 4^n nucleotides in a DNA molecule. This means that a primer with only two bases is likely to attach to template DNA every $4^2 = 16$ nucleotides and our PCR would amplify fragments of DNA only 16 nucleotides long. These would be too short to be useful. Using primers that are too short would also make it almost impossible to amplify an entire gene during gene cloning.

As we have seen, short primers attach at short distances along a DNA molecule. Therefore, it is highly likely that short primers would attach to a DNA molecule at several points within a gene. As DNA polymerase copies DNA from one primer to the next, only fragments of the gene would be copied, not the whole gene.

By using a variety of primers, molecular biologists could collect an array of DNA fragments. It is much easier for them to find a gene in a DNA fragment than in the entire genome of an organism. It is also easier for them to sequence an organism's genome fragment by fragment. For an increasing number of organisms, molecular biologists have been able to determine the base sequence of genes and of the DNA at either end of genes.

If molecular biologists know the base sequences at each end of a gene, they can design primers that will attach to them. Provided the primers do not attach within the gene, DNA polymerase will amplify the entire gene plus any bases to either side. Genes are controlled by promoter regions that lie close to them. If they wished to transfer a human gene into a bacterium, molecular biologists would need to amplify the gene and its promoter region.

If molecular biologists wish to transfer a human gene successfully into a bacterial cell they would need to amplify the gene and its promoter and terminator regions. Bacteria are prokaryotic cells whereas human cells are eukaryotic. Prokaryotic cells have a mechanism for controlling expression of their genes that is different from the mechanism in eukaryotes. Without its promoter region, the human gene might not be expressed in the bacterial cell. Also, if the human gene is cut out of the human DNA using a restriction enzyme, it will still contain its introns. If this gene is transferred to bacteria, the correct protein will not be made as bacteria do not splice introns out of the pre-mRNA before translation. Therefore any human gene that is transferred to bacteria needs its introns removed.

An overview of *in vivo* gene cloning using bacteria

Because they are easy to culture and increase in number at a very fast rate, bacteria are often used to clone genes. This involves the following steps:

- **Step 1**: a fragment of DNA containing one or more genes is obtained by one of the three methods described on pages 452–454.
- **Step 2**: the DNA fragment is inserted into a vector which will transfer it into a bacterium. In this case, the vector is a small circular DNA molecule called a **plasmid**.
- **Step 3**: the vector is transported into a bacterial cell.
- **Step 4**: the bacterium is allowed to multiply.
- **Step 5**: in reality, only some bacteria will have successfully taken up the plasmid containing the target gene(s). In Step 5, these bacteria are identified so that they can be cultured. Any that have not taken up the target gene(s) are destroyed.

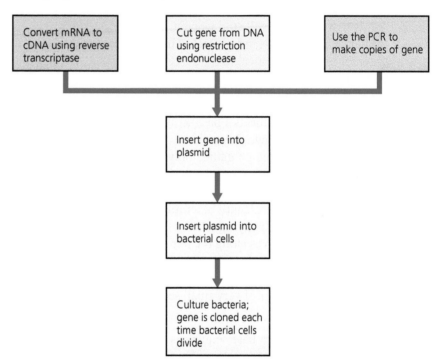

Figure 24.9 The basic steps in *in vivo* gene cloning. Although other cells can be used, bacteria are the most common host cells for this technology.

Inserting DNA fragments into vectors

Fragments of DNA can be produced by any of the three methods shown in Figure 24.9. Each method has been described in some detail earlier in this chapter. Having obtained an appropriate DNA fragment, a promoter and a terminator sequence need to be added. You learned about promoters in Chapter 23, and will realise that these are needed for transcription to start. A terminator is a sequence of DNA that acts as a signal for transcription to stop at the end of the gene sequence. The DNA fragment, with its promoter at one end and the terminator at the other end, must be inserted into a **vector** that will carry it into a target cell. Viruses, including those that infect bacteria (bacteriophages) can be used, but plasmids are the most commonly used vectors. **Plasmids** are small, circular DNA molecules that lead an independent existence in many bacterial cells. Each plasmid carries a small number of genes that are expressed in the phenotype of the bacterium. For example,

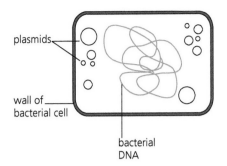

Figure 24.10 This bacterial cell contains many plasmids. They vary in size but they are smaller than the bacterial 'chromosome'.

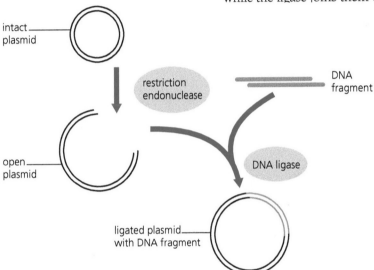

Figure 24.11 The cut ends of a plasmid and of a DNA fragment are joined together by a ligase enzyme to form recombinant DNA. This process is much more efficient if the plasmid and DNA fragments have sticky ends rather than blunt ends.

TIP

You need to be able to interpret information given to you about any method of identifying transformed bacteria but you do not need to recall any particular method.

the ability to survive toxic concentrations of antibiotics, such as ampicillin, chloramphenicol and kanamycin, is often due to the presence inside the bacterium of a plasmid carrying genes for antibiotic resistance. Figure 24.10 shows that a single bacterial cell can contain many plasmids, each of which is much smaller than the circular DNA that makes up the main bacterial DNA.

Before it can be used as a vector, a plasmid must be cut open. This is done using a restriction endonuclease. The DNA fragment can now be inserted into the cut plasmid. Figure 24.11 shows how, under the right conditions, the ends of the DNA fragment and of the plasmid join together. This process is called **ligation** and is catalysed by a **ligase** enzyme. Some ligases join together DNA fragments and plasmids that have blunt ends. However, ligation is much more efficient if the plasmid and the DNA fragment have sticky ends (page 446). This allows complementary base pairing to hold the two ends of DNA together while the ligase joins them up. If either the DNA fragment or the plasmid does not have sticky ends, they can be added before ligation occurs.

If successful, the result of this process is a plasmid that contains a fragment of foreign DNA. Because it contains DNA from two sources, the DNA of this plasmid is referred to as **recombinant DNA**.

Transferring vectors into host bacteria

Most species of bacteria are able to take up plasmids. There are several methods for getting vectors into bacterial cells. Most were found by trial and error and examiners will not expect you to recall any of them. One of the earliest methods involves soaking bacterial cells, together with plasmids, in an ice-cold solution of calcium chloride followed by a brief heat shock, during which the temperature is raised to 42 °C for 2 minutes. Quite why this works is poorly understood, but the treatment increases the uptake of plasmids into the bacteria. Bacteria that have taken up plasmids are said to be **transformed** and, because they contain a gene from another organism, they are also said to be **transgenic**.

Identification of transformed bacteria

When presented in a diagram, such as Figure 24.11, the process for producing transformed bacteria seems simple enough. In reality, biologists cannot see the bacterial cells or the plasmids. They carry out the procedures described above not knowing whether:

- any plasmids have recombinant DNA or whether they have just joined back on themselves (**self-ligated**)
- any bacterium has taken up a plasmid
- any bacterium that has taken up a plasmid has taken up a plasmid containing recombinant DNA. Only about 0.01% of them do.

Biologists need methods to identify which bacteria have been transformed, so that they can grow these, and only these, in a culture medium. There are many ways that they can identify transformed bacteria. Most involve the use of **marker genes** on the plasmids.

EXAMPLE

Finding transformed bacteria

Escherichia coli bacteria are often used in gene cloning experiments. This bacterium is able to hydrolyse lactose into its constituent monosaccharides.

1 Name the monosaccharides formed by the hydrolysis of lactose.
Glucose and galactose.

E. coli uses a series of enzyme-controlled reactions to hydrolyse lactose. One of the enzymes in this series is β-galactosidase. The enzyme is normally encoded by the bacterial gene *lacZ*. The normal *lacZ* gene contains segments that encode different peptide portions of the β-galactosidase molecule. Some strains of *E. coli* have a *lacZ* gene which lacks a segment, called *lacZ'*, that encodes the α-peptide portion of the enzyme.

2 Does β-galactosidase have a quaternary structure? Justify your answer.
Yes, because it consists of more than one chain of amino acids.

Figure 24.12a shows a plasmid, called pUC8, which is commonly used as a vector. This plasmid contains two genes that are of interest to us: one confers resistance to ampicillin and the other is the *lacZ'* gene. If a cell of *E. coli* that lacks the *lacZ'* gene contains a copy of the normal pUC8 plasmid shown in Figure 24.12a, it can produce normal β-galactosidase. Figure 24.12b shows where a foreign gene can be inserted into the pUC8 plasmid using a restriction enzyme called *Bam*HI. Note that the recognition sequence of *Bam*HI is in the middle of the *lacZ'* gene in the plasmid.

a) Normal vector molecule produces α-peptide

gene for resistance to ampicillin — pUC8 plasmid — recognition sequence of *Bam*HI — *lacZ'* gene

b) Recombinant vector molecule does not produce α-peptide

foreign DNA spliced into plasmid

lacZ' gene disrupted

Figure 24.12 The plasmid pUC8 is commonly used as a vector in gene cloning experiments. a) The normal plasmid with two of its genes indicated; b) a recombinant plasmid, containing a foreign gene.

3 Explain why a cell of *E. coli* that lacks the lacZ' gene is able to make β-galactosidase if it also contains a copy of the normal pUC8 plasmid, shown in Figure 24.12a.
The normal pUC8 plasmid contains a copy of the lacZ' gene.

Having followed the protocol to splice foreign DNA into pUC8 plasmids and then insert the plasmids into cells of *E. coli*, molecular biologists cultured the *E. coli* in a suitable medium.

4 What is a protocol?
A protocol is a detailed plan for a scientific procedure.

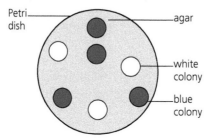

Figure 24.13 This agar plate was inoculated with bacteria. The circles represent colonies of bacteria that have grown on the agar plate. The agar contains the antibiotic ampicillin and X-gal. This helps us to identify which bacteria have taken up a transformed pUC8 plasmid.

After some time, the biologists inoculated the bacteria on to agar plates containing ampicillin and X-gal. Figure 24.13 shows the appearance of one agar plate after it had been incubated for 48 hours. The circles represent *E. coli* colonies that had grown on the agar. Each colony is a clone of a single *E. coli* cell that was able to grow on the agar.

5 All the colonies in Figure 24.13 are clones of cells that had taken up the pUC8 plasmid. Explain why they are clones.
Because they have all grown from one original cell by binary fission.

6 Before inoculating the bacteria on the agar plates, the scientists cultured the bacteria in a suitable medium. Use your knowledge of gene expression to suggest why.
Some of the bacteria would be unable to hydrolyse lactose so they need to be cultured in a medium that contains a suitable carbon source such as glucose.

In addition to ampicillin, the agar contained X-gal. This is a lactose analogue. It is white but is broken down by β-galactosidase to produce a blue-coloured product.

7 Use your knowledge of enzyme action to suggest the meaning of 'lactose analogue'.
This means it is similar in shape to lactose, so can fit into the active site of β-galactosidase and will break down to produce a blue product.

8 Which of the colonies shown in Figure 24.13 contain bacteria that can produce β-galactosidase?
The blue colonies.

9 Use Figure 24.13 and your answer to Question 8 above to explain how we can identify bacteria that have taken up plasmids containing the foreign gene.
These are the white colonies as they are resistant to ampicillin but they cannot break down X-gal so they do not have a functional lacZ' gene.

Relative advantages of *in vivo* and *in vitro* gene cloning

We have looked at two methods of cloning genes: the *in vivo* method, using bacteria, and the *in vitro* method, using the PCR.

Each cycle of the PCR takes between 3 and 5 minutes. Normally, 25–30 cycles are carried out in any one experiment. Consequently, a PCR experiment takes only a few hours to produce a large gene clone. In contrast, the *in vivo* method takes several weeks. This gives the PCR such a time advantage that you might wonder why scientists continue to use the *in vivo* method at all. Table 24.1 compares the relative advantages of the PCR and of the *in vivo* method and explains why the *in vivo* method continues to be used.

Table 24.1 The relative advantages of the PCR and cell-based methods of cloning genes.

In vitro method, using the PCR	*In vivo* method, using bacteria
The PCR can copy DNA that has been partly broken down. This makes it useful in forensic science.	Partly broken down DNA is not copied.
The PCR is very sensitive. Even minute amounts of DNA, such as that contained in a single cell, can be copied.	*In vivo* methods are less sensitive, so large amounts of sample DNA are needed. This makes *in vivo* methods less useful in forensic work.
DNA that is embedded in other material can be copied. This makes the PCR useful for analysing DNA in formalin-fixed tissues or in archaeological remains.	Unless DNA can be isolated from the medium in which it is embedded, it cannot be copied by *in vivo* methods.
The cloned genes are produced in solution. They cannot be used directly to manufacture the protein that they encode.	The cloned genes are already inside cells that will manufacture the protein that they encode.
DNA polymerase will only copy a gene if it is marked at each end by complementary primers. If we do not know the base sequences at each end of the gene, we cannot make appropriate primers. Thus, we cannot use the PCR to copy genes that have not been studied before.	Once incorporated into its host DNA, the gene will be copied by the host cell. This method can be used to copy genes that have not been studied before.
The PCR becomes unreliable when we use it to copy DNA fragments longer than about 1000 base pairs (1 kbp).	*In vivo* methods reliably copy genes up to about 2 Mbp long.
In vitro methods lack error-correcting mechanisms, so the error rate is higher than in cell-based methods.	Cells have mechanisms for correcting any errors that are made when copying genes. This reduces the error rate of gene copying.

TEST YOURSELF

9 Extracts from bacterial cells will contain the DNA of the bacterium as well as plasmids. Suggest one property of plasmids that will allow them to be separated from bacterial DNA for use later as vectors.

10 Complete the table with the names of the enzymes that carry out the processes described.

Enzyme	Process
	Cuts DNA at a specific base sequence
	Makes a single-stranded DNA copy of an RNA sequence
	Joins a new piece of DNA into a plasmid

Transgenic organisms can be useful to humans

A transgenic organism is one that contains a gene from another organism. You have seen how genes can be transferred into bacterial cells, producing transgenic bacteria. When a gene is successfully transferred into another organism, the transgenic organism will produce the protein encoded by the transferred gene. This happens because the genetic code is universal, so a gene from any organism can be transcribed and translated by any other organism.

Collecting human proteins from transgenic organisms

Many human diseases result from an inability to produce a vital protein. For example, Type I diabetes is caused by failure to produce insulin and haemophilia is caused by failure to produce an essential blood-clotting protein. If the gene encoding one of these proteins is successfully transferred into a transgenic organism, the protein it produces can be extracted and used to treat patients with the relevant disease. Table 24.2 shows a range of organisms that have been genetically modified to produce proteins useful to humans on a commercial scale.

Table 24.2 The range of transgenic organisms that have been used in the commercial production of human proteins. Many more are at the pre-clinical trial stage of production.

Transgenic organism	Human protein	Clinical use of protein
Bacteria	Alpha-1-antitrypsin	Treatment of emphysema
	Growth hormone	Treatment of dwarfism
	Insulin	Treatment of Type 1 diabetes
Yeast	Alpha-1-antitrypsin	Treatment of emphysema
Plants	Collagen	Reconstructive surgery
	Growth hormone	Treatment of dwarfism
	Interleukins	Treatment of cancer
	Vaccine	Prevention of measles
Cattle	Fibrinogen	Treatment of blood clotting disorders
Goats	Anti-thrombin	Treatment of deep-vein thrombosis
Pigs	Factor VIII	Treatment of haemophilia

Figure 24.14 These large industrial fermenters contain a culture of transgenic bacteria that produce a human protein. Since the bacteria secrete the human protein, it is relatively simple to extract it from the culture medium.

Extraction of the human protein is much easier if the transgenic organism secretes it into the fermentation medium in which it is growing, or in its milk if it is a mammal. This explains the presence of microorganisms and mammals listed in Table 24.2. Plants do not secrete large volumes of protein; extraction of a human protein from plants is easier if we know where in the plant it is produced. Scientists can ensure that a human protein is secreted in milk or deposited only in the seeds of plants by transferring the human gene into a specific part of the genome of the transgenic organism. By placing it downstream of a promoter controlling lactation in mammals, or of a promoter controlling seed production in plants, scientists control where in the transgenic organism the human protein is made. Producing pharmaceutical products using transgenic farm animals (sometimes called 'pharming') and transgenic plants is an increasingly important source of new drugs.

Transgenic plants produce genetically modified (GM) food

In addition to producing human proteins, research into gene transfer in plants has focused on improving plant productivity. In particular, research has focused on two plant characteristics: resistance to insect pests and resistance to herbicides. One species of soil bacterium, *Agrobacterium tumefaciens*, has been especially important in transferring new genes into crop plants. Figure 24.15 explains why this bacterium has been so important.

A. tumefaciens produces chemicals that make the bacterium resistant to attack by insects. The genes encoding the production of these **natural insecticides** have been successfully transferred from bacteria to crop plants, reducing the need to spray crops with insecticides. In 1995, maize was the first plant to be marketed that had been genetically modified in this way, with potato plants and cotton plants soon following them on to the market. Today, there are many competing agrochemical companies around the world, marketing many types of transgenic insecticide-producing crop plants.

No one has yet produced a plant that is resistant to weeds growing around it. Instead, transgenic plants have been produced that are resistant to the most commonly used herbicide, **glyphosate** (sold commercially as Roundup). Glyphosate inhibits an enzyme (EPSP synthase) involved in the pathway, which plants use to make essential amino acids. Without a functional EPSP synthase, weed plants cannot make some of the proteins they need and die. Unfortunately, the same happens to crop plants that have been sprayed to get rid of weeds. The solution was to transfer into crop plants a bacterial gene that confers resistance to glyphosate. Such herbicide-resistant crop plants (called Roundup Ready) are routinely grown in many countries.

The food produced by transgenic plants is commonly called genetically modified food (**GM food**). It was hoped that transgenic plants would increase crop production and so help to reduce starvation in the world. You would probably support this aim, to reduce world starvation, but are you in favour of genetically modified food? The issue has caused great debate in the UK. The growth of transgenic crops in the UK is restricted by government legislation and any food produced by transgenic plants must be clearly labelled as GM food. Many people refuse to buy GM food and are angry when they find that the constituents of processed food include GM products. Some people, like the activists in Figure 24.16, take extreme action and destroy experimental GM crops where they discover them.

People's concerns about GM foods tend to fall into one of three categories.

1 **Does eating GM foods harm us?** Some people are worried about eating GM foods, and the major UK supermarkets do not sell them. However, some people are concerned that meat and animal products, such as milk and eggs, can be sourced from animals that have been fed GM animal feed.

2 **Do GM plants adversely affect the environment?** Many people are concerned about the damage that pesticides might do to natural communities. Some people point out that the use of herbicides seems to have increased where herbicide-resistant GM crops have been grown.

Figure 24.15 The damage to this plant is called a gall. It is caused by a bacterium, *Agrobacterium tumefaciens*, which has infected the plant. When *A. tumefaciens* infects a plant cell, it injects its own plasmid, a T-plasmid, which the plant cell incorporates into its own DNA. By inserting genes into T-plasmids, biologists have been able to transfer foreign genes into plants.

Figure 24.16 Some people take extreme measures in support of their opinions. These activists are destroying what they believe to be a crop of transgenic plants (GM plants).

457

Others are concerned that **horizontal gene transfer**, where pollen from one species cross-pollinates another plant species, might result in plants other than the GM crop gaining the transferred genes for herbicide resistance.

3 **Does globalisation disrupt local enterprise?** These concerns are about the dominance of large, multinational companies and the effect of their behaviour on small farmers, rather than about the science of genetic modification. The seeds of genetically modified plants are produced by a small number of multinational companies. They sell these seeds as part of a package; the farmer must agree to buy seeds, fertiliser and pesticides from the same company. Tying in to a single provider increases the risk to the farmers of changes in company policy or of changes in political relations between countries.

You might like to discuss the GM issue in class, including why people in some countries appear to be less concerned about GM crops and GM foods than people in the UK.

TIP

You need to be able to discuss aspects of genetic engineering, but you do not need to recall any particular examples.

TEST YOURSELF

11 Unlike eukaryotic cells, bacterial DNA does not contain introns.
  a) Suggest the advantage of transforming bacteria by inserting complementary DNA (cDNA) rather than a copy of the original gene.
  b) Suggest one advantage of using transgenic yeast rather than transgenic bacteria to produce human proteins.

12 In addition to the DNA found in plant chromosomes, chloroplasts contain their own DNA. Genes conferring resistance to insecticides and herbicides are commonly inserted into the DNA of chloroplasts, rather than into the transgenic plant's chromosomes. Suggest why this technique reduces the risk of foreign genes passing from one plant species to another by cross-pollination.

Gene therapy may be used to overcome the effect of defective genes

Many people object to the concept of producing transgenic humans, yet this is what gene therapy does. The aim of gene therapy is to treat an inherited disease by providing the sufferer with a corrected copy of their defective gene. The process of putting a corrected gene into a cell is called **transfection** and the cell that has received the new gene is said to have been **transfected**. There are two approaches to gene therapy: somatic cell therapy and germ cell therapy.

Extension

Somatic cell therapy

During **somatic cell therapy**, copies of a functional gene are inserted directly into the body cells of sufferers. In some cases, body cells are removed, copies of the functional gene inserted into them and the transfected cells put back in the same patient's body. This works well with blood diseases, such as leukaemia, where cells from the patient's bone marrow are extracted, transfected and replaced. In other cases, where large numbers of body cells cannot be removed safely, the genes are inserted directly into the affected tissues. This technique has been used to treat lung diseases, such as cystic fibrosis.

Vectors are needed to transfer genes into the cells of a sufferer. **Retroviruses** are often used as vectors for bone marrow because they have been found to transfect a large number of stem cells successfully. Viruses, known as adenoviruses, have also been used as vectors for treating cystic fibrosis, as have **liposomes**: small droplets of lipid that fuse with the surface membranes of epithelial cells in the lungs (Figure 24.17).

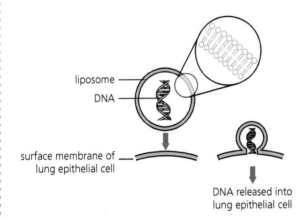

liposome

DNA

surface membrane of lung epithelial cell

DNA released into lung epithelial cell

Figure 24.17 The corrected gene for cystic fibrosis can be carried into the epithelial cells of the lungs by small droplets of lipid, called liposomes. A liposome is able to fuse with the surface membrane of a lung epithelial cell, releasing its DNA into the cell.

When stem cells from bone marrow are transfected and replaced in the marrow of patients, all the different types of white blood cells produced from them contain the added gene. This provides a long-term treatment

for blood diseases, such as leukaemia. Where stem cells cannot be used, the effects of the corrected gene last only as long as the transfected cells remain alive. This explains why gene therapy to treat cystic fibrosis must be repeated every few weeks.

Attempts have been made to use gene therapy to treat cancer. As you know from Chapter 23, cancers can result from activation of **oncogenes** or from inactivation of tumour-suppressor genes. Gene therapy has been used to insert corrected copies of active tumour-suppressor genes into cancer cells and to attempt to prevent the expression of an oncogene in cancer cells. The success of this research is currently hampered by a lack of vectors that will ensure the genes are taken up by cancer cells.

Germ cell therapy

In **germ cell therapy**, a corrected gene is inserted into an egg cell that has been fertilised using *in vitro* fertilisation (IVF) techniques. Once fertilised, the embryo is allowed to develop before being re-implanted into the mother's uterus. If gene transfer is successful, mitosis will ensure that every cell in the embryo contains a copy of the corrected gene. As an adult, the transfected individual will pass on copies of the corrected gene to her or his offspring. Consequently, germ cell therapy provides a treatment for an inherited condition that not only lasts the lifetime of the sufferer but also crosses generations. Germ cell therapy is currently illegal in humans.

Moral and ethical issues

As a biology student, you should be able to recognise and discuss some of the moral and ethical issues raised by DNA technology. It would be wrong for this book to suggest that there is a 'correct' response. You must make up your own mind.

However, you will be expected to discuss the controversial issues raised by gene cloning and DNA technology using appropriate scientific terminology and understanding. Three examples of how you could show your biological knowledge and understanding are given below.

You might express concern that producing transgenic animals causes the animals to suffer.

- As a biologist, you should be able distinguish between the effects of a foreign gene in an animal and the procedures involved in producing transgenic animals. Although the presence of a foreign gene does not appear to cause suffering, the manipulations involved in producing a transgenic animal might do so. For example, animals produced by somatic cell nuclear transfer (page 427) suffer a relatively high rate of birth defects and even the healthy animals appear to suffer from premature ageing.

- You might use Dolly the sheep as an example of premature ageing. If so, you should make it clear that she was *not* a transgenic animal. Her premature ageing was associated with the somatic cell nuclear transfer technique rather than any foreign gene.

You might express concern about the use of genetically modified plants to improve crop productivity.

- As a biologist, you should not justify this concern using statements such as, 'this harms the environment'. You can show knowledge and understanding of biological concepts and principles by giving a specific example of how a GM crop could cause damage. For example, you could say that horizontal gene transfer between bacteria could result in the foreign gene for herbicide resistance being transferred from the GM crop plants to weed plants and that this would cause the weeds to become an even greater pest. You might also recall that foreign genes are often inserted into the DNA of chloroplasts to reduce the likelihood of their horizontal transfer to other plants via pollen.

You might express concern about the use of gene therapy to cure human diseases. There is no easy answer to problems such as this, but you can give a balanced argument that includes biological knowledge. For example, you might argue that:

- it is unreasonable to object to the management of cystic fibrosis using liposomes in respiratory inhalers to insert functional versions of the gene into the lung cells of sufferers
- if bone marrow transplants are acceptable, it is difficult to reject the correction of blood disorders by inserting functional genes into blood stem cells
- germ cell therapy has been beneficial in producing cows that yield milk with less fat and more protein. However, since this type of genetic manipulation involves changing the genotype of this animal and of future generations of this animal in a directed way, it would be unacceptable to use it with humans.

TIP

It is *never* a good idea to use the expression 'playing at God', since this shows no biological knowledge or understanding. Remember that a good scientist uses evidence from reliable sources, and not unsubstantiated opinions, when formulating an argument.

Practice questions

1 Scientists have isolated a gene that codes for the protein that a type of spider uses for making a web. They have inserted the gene into a very early goat embryo. The modified goat produces the spider web protein in its milk. The protein is very useful for medical applications, such as treating nerve damage and repairing wounds.

a) Describe how the following are used in genetic engineering:

　i) a vector　　　　　　　　　　　　　　　　　　　　　(1)

　ii) restriction endonuclease.　　　　　　　　　　　　　(2)

b) i) Why is it important that the gene is inserted into a very early embryo?　　　　　　　　　　　　　　　　　　　　　(2)

　ii) Describe how the DNA inserted into the goat cells produces the spider web protein.　　　　　　　　　　　　　　(6)

c) One animal rights organisation described this example of genetic engineering as 'an unethical use of farm animals'. Do you agree with this statement? Give reasons for your answer.　　(4)

2 Genetically engineered insulin can be used to treat diabetes. The figure shows some of the stages in genetically engineering insulin.

a) Complete the table to give the names of the items being described.　　　　　　　　(4)

Item	Description
	The term used for the small circular piece of DNA
	The name for the structure labelled A
	The enzyme used to cut the DNA
	The enzyme used to attach the human gene to the circular DNA

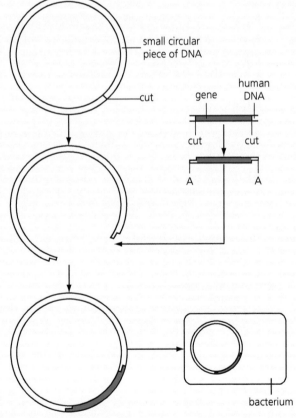

b) What is a marker gene?　　　　　　　　　　　(1)

c) Erythropoietin is a glycoprotein used to treat some medical conditions. It is produced by genetically modified animal cells.

　i) Give two differences between an animal cell and a bacterial cell.　　　　　　　　(2)

　ii) Explain why a glycoprotein, such as erythropoietin, cannot be produced using bacterial cells.　　　　　　　　　　　(2)

3 Some scientists wanted to carry out the PCR on a sample of DNA. They put the DNA they wanted to copy in a test tube together with two types of primer.

a) What else should be added to the tube for PCR to be successful?　(1)

b) Why are two different primers needed?　　　　　　　　(1)

c) Starting with just one DNA molecule, what is the maximum number of DNA molecules that would be present after six cycles of PCR? *(1)*

d) One cycle of PCR involves heating the mixture and then cooling it again. What happens during the heating stage of the cycle? *(2)*

4 Some scientists have devised a test to detect an animal parasite, *Sarcocystis*, in meat. This parasite can cause harm to humans if it is present in undercooked meat that is then eaten by a human. The test involves PCR, using primers that bind to certain genes in *Sarcocystis*.

a) Meat will contain DNA from the animal and may also contain bacterial DNA. Explain why this kit will detect *Sarcocystis* DNA, and not any other DNA present. *(1)*

b) i) The PCR uses DNA polymerase enzyme. Why is this enzyme needed? *(1)*

ii) The DNA polymerase enzyme is heat-stable. Explain the advantage of this. *(2)*

5 The Mediterranean fruit fly is a serious agricultural pest that causes extensive damage to crops. It can be controlled by insecticides and biological control but these are not always effective.

Now scientists have genetically engineered flies by adding a female-specific gene into the insects. This stops females developing to reproductive maturity. The male flies are released into the environment, where they mate with wild females. However, they pass on the female-specific gene to their offspring, ensuring that only males develop into adulthood.

a) Suggest how this new gene could be inserted into an insect zygote. *(4)*

b) If a chemical repressor is added to the food supply of the genetically modified fruit flies, the female-specific gene is inactivated. Suggest the advantage of this. *(2)*

c) Evaluate this use of genetically engineered Mediterranean fruit flies. *(4)*

Stretch and challenge

6 Discuss the concerns that many people in the UK have about genetic modification. To what extent are these concerns justified?

25 Using gene technology

TEST YOURSELF ON PRIOR KNOWLEDGE

1 Fragments of DNA move towards the positive electrode in gel electrophoresis. Explain why.
2 Name the enzyme used in the PCR.
3 What are primers used for in the PCR?

Introduction

Genome is the term for all the genes in an organism. The Human Genome Project (HGP) was an international, collaborative research programme that aimed to completely map and understand all the genes in a human being.

The Human Genome Project was launched in 1990. It started in the USA with collaborators from several other countries. In the UK, scientists in Cambridge had been working for several years on mapping the genome of a nematode worm that is widely used in research. Although it was not complete, they recognised that they had the technology to be able to achieve this. In the UK, the Sanger Institute near Cambridge provided the British contribution to the Human Genome Project. Laboratories all over the world were allocated different sections of different chromosomes to sequence.

One key decision made by these scientists was that all the data generated would be shared publicly before being published. This allowed data to be shared as quickly as possible. Because this was a publicly funded project, they wanted to make sure that all information discovered could lead to as many benefits for humans as possible without commercial interests being involved.

On 26 June 2000 a draft sequence of the human genome was published. In 2003, the International Human Genome Sequencing Consortium announced they had completed the detailed reference human genome with 99.99% accuracy. This work involved huge numbers of scientists from 20 institutions all over the world.

Figure 25.1 The Wellcome Trust funded the building of the Sanger Institute near Cambridge to provide the British contribution to the Human Genome Project.

Genome projects

The finished sequence of the human genome was over 3 billion letters long. Scientists were surprised to find that the human genome contains only about 25 000 genes, which was a great deal fewer than they had expected. However, the scientists still needed to find out the **proteome**. This is the full range of proteins that a cell is able to produce.

Applications of the Human Genome Project

Scientists also need to work out how to use this sequence to improve human health and cure diseases. In the human genome, scientists have identified a number of genes that can increase the likelihood of people developing certain types of cancer and other diseases, such as Alzheimer's disease. It is hoped that, in time, it will be possible to understand the causes of inherited diseases, and to find effective treatments.

Sequencing of genomes

Scientists at the Sanger Institute near Cambridge, as well as sequencing one-third of the human genome, have sequenced the genomes of many other organisms. These include organisms that are used in genetic research, such as zebra fish, roundworms and the plant *Arabidopsis*, and organisms that cause disease, including parasites such as *Plasmodium*, which causes malaria.

One of the most severe forms of malaria is caused by *Plasmodium falciparum*. Scientists have sequenced hundreds of different *P. falciparum* parasites. These genomes are then examined by computers to find differences between the sequences. Scientists can then identify the genes that show the most variation between parasites. This indicates the genes that are under the greatest selection pressure, so these can be investigated further to find out whether they code for proteins that function as antigens that could be targeted for a vaccine. Similarly, scientists are investigating whether specific gene variants are associated with drug or insecticide resistance. They are also trying to find human genes associated with protection against severe malaria.

TIP
Remember that you learned about the immune response in the first year of your course (see Chapter 6).

Identifying genes that might be a suitable target for a vaccine is just the first stage. Scientists need to work out the protein that the gene codes for. Then they need to inject that protein into people living in areas where malaria is common, to see whether they produce antibodies against the protein. Scientists at the Sanger Institute are working with centres in Africa and Asia to carry out these tests. It may be some years before an effective anti-malaria vaccine is readily available, but sequencing DNA enables scientists to focus their efforts on the most likely targets.

One reason why scientists can sequence so many genomes is that sequencing methods are improving rapidly. Most of the processes involved are automated. Data from sequences can be fed into huge banks of computers that are programmed to compare sequences from different organisms.

TEST YOURSELF

1 The scientists involved in the Human Genome Project decided that all the data generated would be shared publicly before being published. Evaluate this decision.
2 Explain why a gene for an antigen that shows a great deal of variation is likely to be under a high selection pressure.

Extension

DNA sequencing

DNA sequencing finds the base sequence of genes that have not been studied before. The methods for doing this are all based on the use of **dideoxyribonucleotides**. Look at Figure 25.2, which shows molecules that you might recognise. Molecule a) is ribose, the five-carbon sugar found in every RNA nucleotide. Molecule b) is deoxyribose, the five-carbon sugar found in every DNA nucleotide. Note that it has one oxygen atom less than ribose, hence the name *deoxy*ribose (see Chapter 4). Now look at molecule c). Can you see how this differs from ribose and deoxyribose? It has one less oxygen atom than deoxyribose and two less than ribose: this is *dideoxy*ribose.

Just like ribose and deoxyribose, a molecule of dideoxyribose can join up with phosphate and an

organic base to form a nucleotide. During DNA replication, a nucleotide containing dideoxyribose (a **dideoxynucleotide**) can pair with a deoxynucleotide on the template strand that has a complementary base. However, DNA polymerase stops replicating DNA when it encounters a nucleotide containing dideoxyribose on the developing strand. This is the basis of the **chain-termination technique** for sequencing DNA.

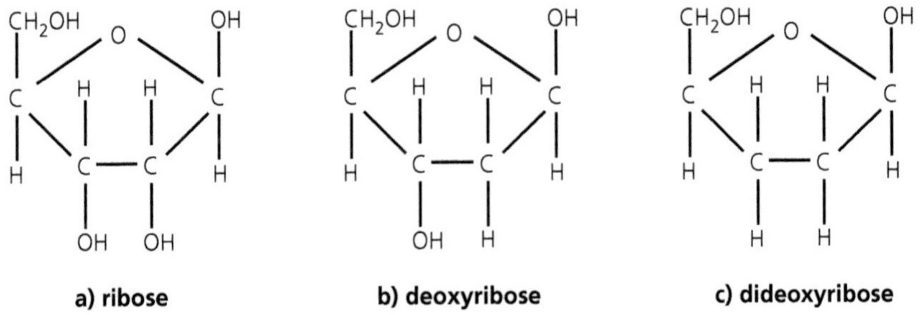

Figure 25.2 A molecule of a) ribose, b) deoxyribose and c) dideoxyribose.

Figure 25.3 The principles involved in the use of an automated DNA sequencer. a) The length of single-stranded DNA to be sequenced is cloned into a vector and a primer added. b) DNA polymerase attaches to the primer and begins DNA replication, adding nucleotides with bases that are complementary to the bases on the recombinant DNA. c) The DNA polymerase inserts a dideoxynucleotide and replication of the chain terminates. d) Provided the ratio of normal (deoxynucleotides) to dideoxynucleotides is high enough, chains of one to several hundred nucleotides are produced.

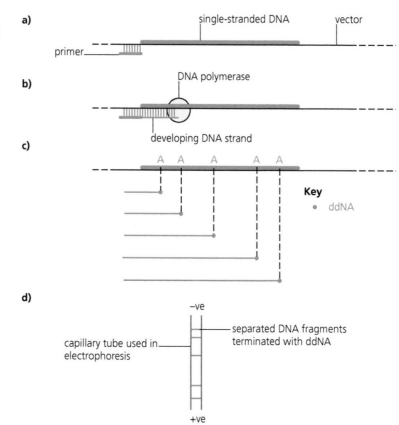

Automated DNA sequencing

Figure 25.3a shows a single-stranded version of the DNA we wish to sequence. It has been cloned into a vector, which in this technique is usually the DNA of a bacteriophage called M13. A short oligonucleotide has been annealed to the DNA to act as a primer.

DNA polymerase is added to the recombinant DNA, together with a mixture of deoxynucleotides containing each of the four bases: adenine, cytosine, guanine and thymine. Under normal circumstances, the DNA polymerase attaches to the primer and begins the process of replicating the recombinant DNA. The new strand starts to grow as free nucleotides form hydrogen bonds with complementary bases in the starting strand (Figure 25.3b).

However, also present in the mixture of nucleotides are some dideoxynucleotides. When, by chance, DNA polymerase inserts one of these instead of a deoxynucleotide, DNA replication stops. If the ratio of deoxynucleotides to dideoxynucleotides is high enough, we end up with complementary chains of DNA varying in length from a single nucleotide to several hundred nucleotides. Figure 25.3c shows how this is possible, using just one of the four types of dideoxynucleotide (ddNA). Whenever ddNA forms a complementary base pair with thymine on the recombinant DNA strand, replication stops.

Each type of dideoxynucleotide is labelled with a different fluorescent dye. The dideoxynucleotides carrying adenine (ddNA) are labelled with a green fluorescent dye, ddNC with a blue dye, ddNG with a yellow dye and ddNT with a red dye.

At the end of the incubation period, the newly formed DNA fragments are detached from the template DNA and these new single-stranded chains are separated by size, using a special type of electrophoresis that takes place inside a capillary tube (Figure 25.3d). The resolution of this technique is so high that it separates strands that are different in length by a single nucleotide.

When illuminated by a laser beam, each of the four dideoxynucleotides fluoresces. The colour and position of the fluorescence is read by a detector which then feeds this information into a computer, which either stores the data for future analysis or produces a printout, like the one in Figure 25.4, overleaf. An automated sequencer can read almost 100 different DNA sequences in a period of 2 hours.

You will remember that the genome is the DNA sequence of an organism. Scientists have found the DNA base sequence of many different organisms, including humans. You will also remember that genes code for proteins, and we already know the triplet sequences (codons) that code for each amino acid.

Figure 25.4 As the replicated fragments of DNA are separated, they pass through a laser and fluoresce. The fluorescence detector 'reads' the fluorescence and sends the data to a computer for analysis. The coloured printout shows the base sequence of the DNA sample.

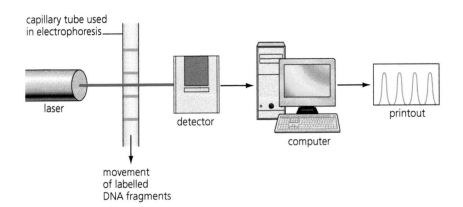

This means we can work out the primary structure of the proteins that the genes code for. The structure of the proteins that a cell's DNA codes for is called its **proteome**. Knowing the proteome can be very useful, for example in identifying potential antigens to use as targets for vaccines. Scientists trying to develop a vaccine against malaria have sequenced the genome of many different malaria parasites. They compare the results. The genes that show the greatest variation must be those that are subjected to the greatest selection pressure, so these are the best targets for a vaccine.

However, it is not easy to work out the entire proteome of an organism from sequencing the DNA. You will remember that complex organisms such as humans have a great deal of non-coding DNA and regulatory genes, and it is not always easy to identify these from the DNA that codes for proteins.

Interpreting the results of manual DNA sequencing

The manual method is similar to the automated method above, except that:

- a separate run is made for each type of dideoxynucleotide – ddNA, ddNC, ddNG and ddNT
- after incubation, the four mixtures are placed in separate wells of a gel and separated by gel electrophoresis
- the dideoxynucleotides are labelled with radioactivity, rather than with fluorescent dyes
- a Southern transfer of the gel is made and an autoradiograph made from it.

Reading the DNA sequence is relatively easy. Figure 25.5 shows part of an autoradiograph. For each dideoxynucleotide base, it shows bands that are DNA fragments for which replication was terminated. As you know, the band that has moved the furthest is the smallest fragment. This will be a fragment where replication stopped at the first base.

In Figure 25.5 the fragment that has moved furthest is in the ddNA track. This is the smallest fragment that was produced when a dideoxyribonucleotide stopped DNA polymerase forming a new strand of DNA. The smallest fragment of DNA that could be formed is a single nucleotide. By chance, a dideoxynucleotide was the first to be inserted into the new DNA strand and its presence stopped DNA polymerase inserting any more nucleotides. We have now found the first nucleotide that is incorporated into the new DNA strand. It carries adenine.

The next smallest fragment is in the ddNC track. This is a fragment of DNA that is only two nucleotides long and it carries cytosine. The first two bases in the test DNA are adenine and cytosine. The fragment which has moved the third furthest is in the ddNG track. This fragment stopped DNA replication because it carries the ddG nucleotide. So our third base is guanine and the base sequence of the test DNA becomes ACG. By repeating what we have done so far, you should find that the base sequence is ACGCATGTTC.

Figure 25.5 The wells at the top of this autoradiograph contained a sample of a reaction mixture that had been incubated with dideoxynucleotides containing adenine (ddNA), cytosine (ddNC), guanine (ddNG) or thymine (ddNT), respectively. Below each well is the track of the separated DNA fragments produced by DNA replication in the presence of each dideoxynucleotide. The DNA sequence of the newly synthesised DNA is read by identifying the distance each fragment has moved, starting with the one that has moved furthest.

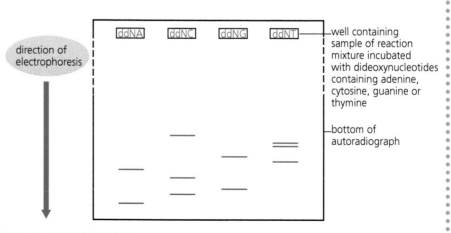

direction of electrophoresis

well containing sample of reaction mixture incubated with dideoxynucleotides containing adenine, cytosine, guanine or thymine

bottom of autoradiograph

Using labelled DNA probes and DNA hybridisation to locate specific alleles of genes

DNA hybridisation The process of combining two complementary single-stranded DNA molecules and allowing them to form a single double-stranded molecule through base pairing.

If we know at least part of the base sequence of a harmful allele of a gene, we can use DNA hybridisation to find out if this allele is present in DNA samples. Before hybridisation, the test DNA is treated using techniques you learned about in Chapter 24. Let's revise these techniques.

First, the DNA is extracted from the sample of cells taken from a patient. After purification, the test DNA is amplified using the polymerase chain reaction (PCR). The test sample will only contain a few cells, for example cheek cells, cells from the amniotic fluid or cells from the umbilical cord. These cells only yield a small amount of DNA. We need a large amount of DNA for analysis, so the PCR is used to produce this extra DNA. Of course, the PCR produces DNA that is identical to the test DNA.

The amplified test DNA is then digested using a restriction endonuclease. Whole DNA molecules are too long to be analysed successfully in one go. This is why the test DNA is digested using restriction endonuclease.

The restriction fragments are then separated using gel electrophoresis. Since DNA has a negative charge, DNA fragments migrate to the positive electrode of an electric field. The gel through which they move has fine pores. Small fragments of DNA move faster through these pores than large fragments. As a result, the fragments are separated, with the smallest nearer the positive electrode, forming bands in the gel.

In the final preparatory stage the DNA fragments on the nylon membrane are treated to break the hydrogen bonds holding the DNA together. Hydrogen bonds between complementary base pairs hold

DNA probe A piece of single-stranded DNA that has a specific base sequence that is complementary to a specific base sequence of a specific allele (of a gene).

together the two polynucleotide chains in a DNA molecule. When these bonds are broken, the nylon membrane contains bands of single-stranded DNA fragments. These fragments anneal with any strands of nucleotides that have complementary base sequences. This is where labelled DNA probes come in. A DNA probe is a piece of single-stranded DNA that is complementary to a specific base sequence. The DNA probe is attached to a fluorescent or radioactive label. This is called **labelling** the probe.

At least part of the base sequence of the harmful gene is known. This enables us to produce single-stranded DNA probes with a complementary sequence using a gene machine. These probes will anneal to any complementary DNA fragment on the nylon film. This is the process of DNA hybridisation. If the DNA probe is labelled, its position can be found. Figure 25.6 shows how a DNA probe will attach to a complementary base sequence if it is present in the test DNA and how, if labelled, we can find the hybridised DNA after electrophoresis of the restriction fragments.

Figure 25.6 A labelled DNA probe can be made which has a sequence that is complementary to part of the harmful allele but not to the normal allele of a gene. Once the DNA probe has attached to a fragment of the harmful allele we can find it in the separated restriction fragments.

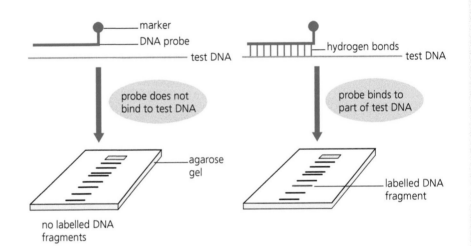

We can use Figure 25.6 to explain how DNA hybridisation enables biologists to locate the harmful allele of a gene in a sample of DNA taken from, for example, cells in amniotic fluid. Since the restriction fragments were made from the entire DNA in the sampled cells, one of them will contain all, or part, of the gene in question, for example the gene for β-globulin. The sequence of bases that is different in the harmful allele is known. The labelled DNA probe has a base sequence that is complementary to that part of the harmful allele. If the label attaches to any of the bands of restriction fragments on the nylon membrane, the DNA fragment in that position must be from the harmful allele. If no label attaches to any band, the DNA does not contain the harmful allele of the gene.

469

TEST YOURSELF

3 Use your knowledge of natural selection to suggest why the genes in a pathogen that show most variation are good targets for a vaccine.

4 Suggest how DNA probes are made.

5 Before a DNA probe is added to DNA on a nylon membrane, the DNA on the membrane is made single-stranded. Explain why.

6 If the DNA probe, complementary to the base sequence found in a harmful allele, does not bind to a person's DNA on a membrane, this is not 100% proof that the person does not carry a harmful allele. Explain why.

Medical screening and genetic counselling

DNA sequencing, DNA probes and DNA hybridisation are important techniques used in DNA analysis. Many human diseases result from mutations of genes, which give rise to alleles that are useful in one context but not in another. Sickle-cell anaemia results from an allele of the gene encoding β-globulin, one of the two types of polypeptide in a haemoglobin molecule. In areas of the world where malaria is endemic, heterozygotes for this gene are at a selective advantage because they are resistant to malaria. In countries where there is currently no malaria, for example the UK, heterozygotes are at a disadvantage because some of their children might be homozygous for the sickle-cell allele. These children will suffer uncomfortable and life-threatening symptoms.

Gene probes can also be used to test DNA from cancer cells to detect certain alleles. This can lead to targeted therapy for that specific mutation. Some cancer drugs only work on cancer caused by a specific mutation. Another use of gene probes is to find whether a person has a gene that makes them more likely to develop a particular kind of cancer. For example, women with the *BRCA1* and *BRCA2* alleles are at higher risk of developing breast cancer than other women.

A couple intending to have children might be concerned if they are aware that there is an inherited disease, such as sickle-cell anaemia, in their family. They might take medical advice about the risk of having a baby with the inherited condition. This advice is called **genetic counselling**. In the past, genetic counsellors were able to use the sort of information in Chapter 7 to work out the chances of an affected baby being born. For example, you can use your understanding of monohybrid inheritance to work out that the chances of two people who are both heterozygous for the sickle-cell allele having a baby with sickle-cell anaemia is 1 in 4 (a probability of 0.25).

Samples of the amniotic fluid that surrounds a developing fetus may be taken. This fluid contains cells from the fetus. Fetal DNA can be analysed using gene probes to find out whether the fetus has a specific genetic condition. A couple who are at risk of having a child with a genetic condition, such as cystic fibrosis, may opt for *in vitro* fertilisation (IVF). This means creating a number of embryos, which can be tested using gene probes. Healthy embryos can be selected and implanted into the mother's uterus.

ACTIVITY

Breast cancer and ovarian cancer

Each year in the UK about 200 000 cases of breast cancer and about 25 000 cases of ovarian cancer are diagnosed. Research has shown that about 10% of these cases are inherited. At least two genes are involved in the development of breast cancer: *BRCA1* (BReast CAncer 1) and *BRCA2*. Normally, the *BRCA1* and *BRCA2* genes control cell growth but mutations in these genes increase the risk of breast and ovarian cancers.

1 What type of gene is *BRCA1* and *BRCA2*?

We each have two copies of the *BRCA1* and *BRCA2* genes, inheriting one from each parent. Someone who carries one of the mutant alleles of either gene has an increased risk of developing breast cancer.

2 What does this tell you about the nature of the mutant alleles of the *BRCA1* and *BRCA2* genes?

Table 25.1 shows data about the risk of a woman developing breast cancer and ovarian cancer in the UK.

3 The risk of developing cancer in the middle column of Table 25.1 is shown as a range, whereas the risk in the right-hand column is shown as a single value. Suggest why.

4 Give two conclusions about the effect of carrying a mutant allele of a *BRCA* gene on the risk of developing breast cancer.

5 Suggest why the risk of developing breast cancer is higher than the risk of developing ovarian cancer.

Table 25.1 The risk of developing cancer among women who are carriers of one mutant allele of the *BRCA1* gene or *BRCA2* gene compared with that of the general population.

Risk of developing	Carrier of one mutant allele of *BRCA1* or *BRCA2* gene/%	General population/%
Breast cancer by age 50	33–50	2
Breast cancer by age 70	56–87	7
Ovarian cancer by age 70	27–44	<2

Genetic counsellors not only advise people about their chances of having a baby with an inherited disease or about their chances of developing an inherited disease, such as cancer, later in life. They also provide emotional and medical support to the affected families. This is needed because the results of screening tests provoke difficult decisions. For example, a woman who finds that she is carrying a child with an inherited disease might have to make a decision about carrying her pregnancy to full term or having an abortion. Similarly, a woman who finds she has an increased risk of developing breast cancer might have to decide to have her healthy breasts surgically removed so they cannot later develop cancer. None of these decisions is easy to make and affects not only the person making the decision but other members of their families as well.

It is likely that shortly after you were born, a drop of blood was taken from your heel. This 'pin prick' procedure collected blood that could be used to test for the presence of common inherited disorders, such as phenylketonuria. This represents a simple form of medical screening. Advances in DNA technology have resulted in much more sophisticated medical screening procedures. Using the techniques you learned about in Chapter 24 it is now possible to analyse DNA taken from scrapings of cells from inside your cheek, cells taken from the amniotic fluid surrounding an embryo in its mother's uterus or cells from the umbilical cord connecting the embryo to its placenta.

TEST YOURSELF

7 Is it always useful to be able to screen a person for alleles that might lead to health problems?

8 What are the arguments for and against screening men for the *BRCA1* and *BRCA2* genes if there is a family history of breast cancer?

Genetic fingerprints

You have learned that any one of our chromosomes carries the same genes in the same order. Consequently, you might expect that we would get the same pattern of restriction fragments if we used the same restriction endonucleases to digest copies of the same chromosome from two different people. In fact, we would not; we would get different patterns of restriction fragments. The reason for this is that, in the non-coding regions of our DNA, each of us has different repeated sequences of bases. These repeated base sequences include **variable number tandem repeats** (**VNTRs**). Because the length of these repeated base sequences is different in different people, we can use them to identify the source of DNA in tissue samples. This is the basis of classical **genetic fingerprinting**. VNTRs are chromosomal regions in which a short DNA sequence (such as AGCT) is repeated a variable number of times end-to-end at a single location (tandem repeat). Everyone has these tandem repeats, but a variable number of repeats. However, the number of tandem repeats that a person has is inherited, and so they could be used to identify biological parents.

Figure 25.7 shows how the distance between the recognition sites of a restriction endonuclease on two samples of DNA varies if the number of repeated sequences varies. As a result of these differences, DNA fragments of different lengths would be produced by digestion of these two DNA molecules with the restriction endonuclease. Figure 25.8 shows how these different lengths of DNA would produce different patterns of restriction fragments after electrophoresis. The important idea to grasp is that only identical twins have the same pattern of non-coding, repeated base sequences and so only they would produce the same pattern of restriction fragments.

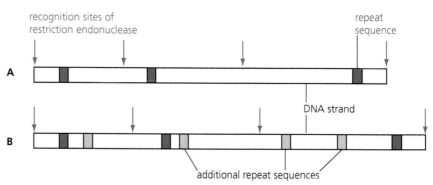

Figure 25.7 Each band represents part of a DNA molecule that has been taken from the same chromosome location from two different people, A and B. The recognition sites of an endonuclease are shown on each DNA molecule. The recognition sites are further apart in B because the non-coding regions of this DNA contain more repeated base sequences.

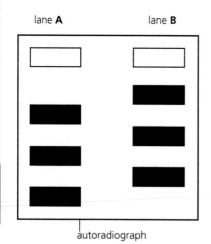

autoradiograph

Figure 25.8 The banding patterns produced after the two sets of DNA shown in Figure 12.7 have been digested using the same restriction endonuclease. Lane B contains longer DNA fragments than lane A because the DNA of B contained more repeated base sequences than the DNA of A.

Genetic fingerprinting by hybridisation with labelled probes

This is the classical method for genetic fingerprinting. You have come across the techniques involved earlier in this chapter. If the sample of DNA is very small, more copies are made using the PCR. A sample of DNA is digested into restriction fragments using a restriction endonuclease. The restriction fragments are of different lengths, depending on the number of repeat sequences within them, and are unique to each individual. These fragments are then separated using gel electrophoresis and a Southern blot is prepared from it. Radioactively labelled probes with the complementary base sequence to the repeat sequence are added to the blot. The probes reveal a labelled band wherever they hybridise with a restriction fragment containing the repeat sequences. Thus, each band in the fingerprint represents a DNA fragment that contains the repeat sequence. Figure 25.8 shows the genetic fingerprints of two people. The pattern of bands is different because the location of the repeat sequences was different in the two sets of DNA.

The very first repeated sequence to be used in classical genetic fingerprinting had the base sequence GGGCAGGABG, where B represents any base. This sequence was repeated a different number of times in the non-coding regions of the DNA samples used.

Figure 25.9 shows the results of the very first analysis that led to a criminal receiving the death sentence in the USA. A young couple had been murdered while they slept in their car. Their bodies were discovered the next day. A post mortem showed they had both died of gunshot wounds and that the woman had been raped. One man was later arrested driving the couple's stolen car. Under police questioning he identified a friend who had been with him on the night of the murders. DNA from the semen found in the woman's body and a DNA sample from each suspect was digested using the same restriction enzyme, and the restriction fragments separated using gel electrophoresis. If you look at Figure 25.9, you will see that the genetic fingerprint of the semen does not have the same pattern of restriction fragments as suspect 1 but does have the same pattern as suspect 2. On the basis of this evidence, suspect 2 was found guilty of rape and murder and given a double death sentence. The jury at the time was told that the chance of an innocent person showing the same pattern of restriction fragments was about 1 in 9 billion. At the time of the trial, the human population of the world was less than 6 billion.

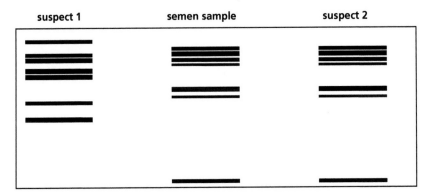

Figure 25.9 The results of the first repeated sequence analysis which led to a criminal receiving the death sentence. The genetic fingerprint of suspect 2 and of the semen taken from a murdered woman show that he was the rapist. He was convicted of rape and murder and received a double death sentence.

The Bains family

Mr Bains first settled in England in 1990. Two years later, he applied to bring his wife and four children to join him. The immigration authorities needed to be sure that the children were entitled to come into England and so they carried out genetic fingerprinting on the whole family. Figure 25.10 shows the results.

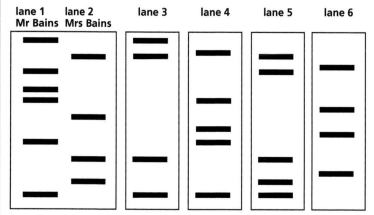

Figure 25.10 The genetic fingerprints of the Bains family members.

Mr Bains had been married before. Sadly, his first wife had died giving birth to their son. Some years later, Mr Bains remarried and he and his new wife had two daughters. They also adopted another son.

1 Look at lanes 1 and 2 in Figure 25.10. They show the separated restriction fragments of Mr and Mrs Bains. Explain why the two patterns of restriction fragments are different.
   This is because the different DNA samples have different numbers of repeats of the sequence.

2 Lane 6 of Figure 25.10 shows the restriction fragments from Mr and Mrs Bains' adopted son. Explain why this conclusion is valid.
   His bands don't match either of Mr or Mrs Bains's bands, so he cannot be their biological child.

3 Use Figure 25.10 to identify Mrs Bains' stepson. Explain your answer.
   All the bands in a child's DNA must match those from either the mother or the father. This is true for the bands in lanes 3 and 5 showing these contained the DNA from Mr and Mrs Bains' daughters. Some of the bands in lane 4 match those of Mr Bains, suggesting he is the biological father. The bands that do not match those of Mr Bains do not match those of Mrs Bains either. These bands must represent DNA from Mr Bains' first wife. So lane 4 is the DNA from Mrs Bains' stepson.

DNA profiling by PCR of variable number tandem repeats (VNTRs)

A more powerful technique has replaced the classical fingerprinting technique. This involves DNA profiling.

As we have seen, variable number tandem repeats (VNTRs) are found in the non-coding regions of our DNA. VNTRs are repeated blocks of base sequences such as GCAT or GC. In our own genome, the most common repeat is a sequence of two bases, cytosine and adenine, repeated between 5 and 20 times, for example CACACACACACACA. The number of repeats in any one VNTR is variable; in a single population, there might be as many as ten different versions of one VNTR. In DNA profiling, the different versions of a selected number of VNTRs are determined.

Using primers that anneal to the DNA to each side of a particular repeat sequence, the PCR is used to make many copies of the DNA containing these VNTRs. After amplification by the PCR, the DNA fragments are separated using gel electrophoresis. Normally, DNA profiling is carried out using several selected VNTRs, so that many more bands are present in the gel.

Just as with the automated sequencing technique described earlier, the VNTR fragments can be labelled with fluorescent dyes and separated using capillary electrophoresis. Scanners 'read' the sequences and feed the data into a computer which compares them with sample material. This technique is much more sensitive than the standard test and is now commonly used by teams of forensic scientists. Figure 25.11 shows the type of printout from an automated scanner.

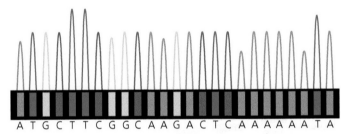

Figure 25.11 The printout from an automated DNA profiler. You might have seen this type of DNA fingerprint in a film or TV programme about crime scene investigators.

TEST YOURSELF

9 Why must the DNA probes used be radioactive or fluorescent?

1 Scientists in South Africa have used genetic engineering to create a vaccine for farm animals that protects them against several different diseases. They have taken a gene that codes for an antigen of Rift Valley virus and inserted it into a smallpox virus. Then they added a gene that codes for an antigen of lumpy skin virus and inserted that into the smallpox virus.

   a) i) What is a vaccine? (1)

   ii) This vaccine will produce immunity to three different diseases. Explain how. (5)

   b) Suggest the advantage of adding the genes for these antigens into a live virus, rather than injecting the animal with the antigens alone. (2)

2 a) Describe how a genetic fingerprint is produced. (6)

   Some blood was found on a broken window at the scene of a crime. The police found four suspects. They carried out genetic fingerprinting on the crime-scene blood, plus blood samples from each of four suspects. The results are shown in the diagram.

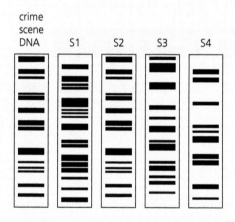

   b) Which suspect does the crime-scene DNA come from? Explain your answer. (2)

   c) It has been suggested that a blood sample should be taken from every person in the country, and the person's genetic fingerprint found. This could be stored on a national database. Any crime-scene genetic fingerprint could then be compared to this database. Do you think this is a good idea? Give reasons for your answer. (4)

3 A test can be carried out to find out whether a person's DNA carries a gene mutation that can cause tumour formation. This mutation is recessive.

   DNA is taken from the patient and the section where the gene is found is copied many times using the PCR. The DNA is incubated with a restriction enzyme. If the mutated allele is present it is cut by the restriction enzyme. If the normal allele is present it is not cut by the restriction enzyme. The diagram shows possible results.

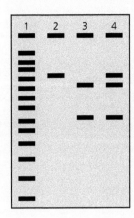

Lane 1 Standard ladder
Lane 2 Control or normal DNA
Lane 3 Mutated DNA for tumour cells
Lane 4 Carrier for the mutation (one
    normal DNA and one mutated DNA)

a) **i)** How is it possible to amplify only the section of DNA that carries the gene being investigated? (2)

   ii) Lane 1 contains DNA fragments of different known lengths. Suggest why this is used. (2)

   iii) Draw a circle around the shortest piece of DNA on the gel. (1)

b) Lane 2 contains DNA from a person who does not carry the mutated allele. Explain the results. (2)

c) Lane 3 contains DNA from a person who has a tumour caused by the mutated allele. Explain the results. (2)

d) Explain the results for lane 4. (2)

Stretch and challenge

4 It has been suggested that a child's genome should be sequenced at birth. This would enable the parents and the child to be aware of harmful alleles, for example those that predispose to cancer, and to make lifestyle choices to reduce exposure to risk factors. Evaluate this proposal, giving reasons to support your point of view.

26

Developing mathematical skills

Using units

In any measurements or calculations, using units correctly is critical. You need to be familiar with the common units used in biology and be confident converting between units, for example from millimetres to micrometres or from cubic centimetres to cubic millimetres.

There are a thousand millimetres (mm) in a metre (m), a thousand micrometres (μm) in a millimetre and a thousand nanometres (nm) in a micrometre. There are also a thousand cubic millimetres (mm^3) in a cubic centimetre (cm^3) and a thousand cubic centimetres in a cubic decimetre (dm^3).

Since these are the units you are most likely to need to convert between, the 'rule of thousands' is a helpful one. But there are some important exceptions such as the centimetre. There are, of course, only 100 cm in a metre and 10 mm in a centimetre.

> **TIP**
>
> Try to avoid using centimetres for measuring lengths. Use millimetres and then you are less likely to make a mistake if you need to convert them.

> **TIP**
>
> Avoid using mixed units such as minutes and seconds. Convert the whole time to seconds.

Sometimes units are combined, for example in a rate such as how fast oxygen is being produced ($mm^3\,minute^{-1}$) or how fast an animal is moving ($cm\,second^{-1}$). It is easy to forget to combine the units when you give an answer following a rate calculation. Sixty is the important number to remember for converting units of time. There are 60 minutes in an hour and 60 seconds in a minute.

For some rates, the first unit is a count such as the number of individuals in a quadrat or the number of beats in a heart rate. In this case, the correct units are number of beats $minute^{-1}$.

Sometimes three units are combined, such as when giving figures for gross and net primary productivity ($kJ\,m^{-2}\,day^{-1}$ or $kJ\,m^{-3}\,day^{-1}$ depending on whether it is a terrestrial or aquatic habitat).

Calculating areas, volumes and circumferences

You need to be confident with arithmetic calculations for finding the size and surface area of biological structures such as cells, organs and whole organisms. You also need to be able to give answers in the correct units and with the appropriate number of decimal places. It is also very useful to be able to recognise when your answer is outside of the expected range so that you realise when you may have made a mistake.

Calculating areas

Areas of rectangles are calculated by multiplying the lengths of the two sides whereas areas of circles are calculated by multiplying the square of the radius by π. Area calculations could be used to find the area of the field of view of a microscope, the area of a disc cut out from a leaf or the size of the area in which some random quadrat samples have been taken. Area calculations are often used to find the surface area of an organism or a cell such as a red blood cell.

If you assume that a red blood cell has a flat surface rather than a concave one (Figure 14.1), the approximate surface area of a red blood cell can be calculated. In Chapter 9 you saw that a red blood cell has a diameter of 7 μm. The radius is half the diameter, which is 3.5 μm. One side of a typical red blood cell therefore has an area of $3.5^2 \times \pi = 38.48\,\mu m^2$. Note that the units are micrometres squared because this is an area measurement.

But this is the area of just one side of the red blood cell. Since it has two sides, the approximate total surface area of a typical red blood cell is $38.48 \times 2 = 76.96\,\mu m^2$.

Decimal places and rounding

When you carry out calculations like finding the area of a disc cut out from a leaf, the answer shown on your calculator will often include a number of **decimal places**. This is the number of figures after the decimal point.

76.96 has two decimal places. In this case, giving the answer to the nearest whole micrometre is quite sufficient, because the original diameter of the red blood cell was only given as 7 μm. So the answer should be **rounded** and it becomes $77\,\mu m^2$ rather than $76.96\,\mu m^2$.

Using significant figures

Significant figures in a number include all the non-zero digits, any zeros between non-zero digits and, in numbers containing a decimal point, all zeros written to the right of the digits. The number of significant figures in a measurement gives an indication of its uncertainty.

Table 26.1 shows the mean diameters of the different types of blood vessel in a bat's wing. For each vessel, the cross-sectional area has been calculated by multiplying the square of the radius by π. You can see how much larger the cross-sectional area of a vein is compared to a capillary

red blood cell imagined as a flat disc

7.0 μm

top view

Figure 26.1 A red blood cell can be imagined as a flat two-sided disc.

TIP
Rounding should be done using the rule that below half is rounded down and half or more is rounded up, so 1.24 becomes 1.2 and 1.25 becomes 1.3.

TIP
When you do calculations, you should not round any numbers used in the calculation, only round the answer.

Table 26.1 Mean diameters and calculated cross-sectional areas of the blood vessels in a bat wing.

Vessel	Mean diameter/μm	Cross-sectional area/μm²
Artery	52.6	2170
Arteriole	19.0	284
Capillary	3.70	10.8
Venule	21.0	346
Vein	76.2	4560

479

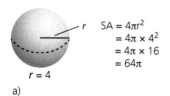

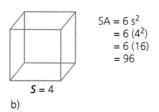

a)

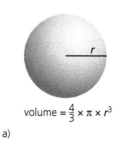

b)

Figure 26.2 Calculating the surface area of a) a sphere and b) a cube.

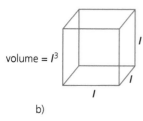

volume = $\frac{4}{3} \times \pi \times r^3$

a)

volume = l^3

b)

Figure 26.3 Calculating the volume of a) a sphere and b) a cube.

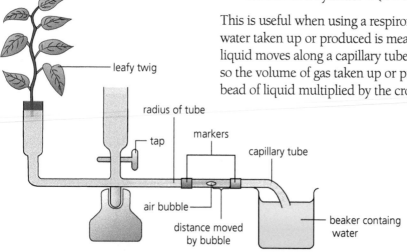

Figure 26.4 (a) Potometer showing the cylinder of water between the markers that has been taken up.

If you were calculating the area of part of a meadow in which you had carried out some quadrat sampling, you might measure two sides of the area with a tape measure. Let's assume you measured one side as 40.50 m and the other as 35.75 m. When you multiply the two to get 1447.875 m² on your calculator, you should only give the area as 1448 m² because the two original measurements were both to four significant figures.

Calculating areas of cubes and spheres

Different sized cubes are often used to compare imaginary organisms of different sizes. The surface area of a cube is calculated by finding the area of one side and multiplying that by six, because it has six sides. So a cube of side length 8 mm has a surface area of $(8 \times 8) \times 6 = 384$ mm². A cube of side length 16 mm has a surface area of $(16 \times 16) \times 6 = 1536$ mm². Notice that although the side length has been doubled, the surface area has increased four times. You should always estimate roughly what the answer should be for a calculation like this so that you can spot if your answer is different from what you are expecting and look for your mistake.

In the same way that cubes can represent different sized organisms, spheres can be used to represent cells of different sizes. You need to remember that

Area of a sphere = $4 \times \pi \times \text{radius}^2$

Again, this means that if the radius of a cell is doubled, the surface area increases by a factor of four.

Calculating volumes

Being able to calculate the volumes of cubes and spheres is also useful.

Volume of a cube = breadth × width × height

Volume of a sphere = $\frac{4}{3} \times \pi \times r^3$

Apart from calculating the volumes of cubes and spheres, you should also be able to calculate the volume of a cylinder. This is the cross-sectional area of the cylinder multiplied by its height.

Volume of a cylinder = $(\pi \times r^2) \times h$

This is useful when using a respirometer or potometer. The volume of gas or water taken up or produced is measured by finding how far a bead of coloured liquid moves along a capillary tube. The gas in the tube forms a cylinder shape, so the volume of gas taken up or produced is equal to the distance moved by the bead of liquid multiplied by the cross sectional area of the capillary tube **lumen**.

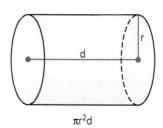

$\pi r^2 d$

(b) The volume of water in the cylinder can be calculated by knowing the radius of the capillary tube.

The cross-sectional area of the lumen is π × r^2. So to find the volume of gas or water taken up or produced you need to know the radius of the capillary tube (or its diameter so you can halve it). The radius of the capillary tube will most probably be in millimetres, and you should also measure the distance the bead of coloured liquid moves in millimetres so the answer will be in mm^3. If you have timed how long the bead of liquid took to move this distance, you could give the answer as a rate by dividing the distance by time in minutes. The appropriate units would then be $mm^3\,minute^{-1}$.

Calculating circumferences

Remember that the circumference of a circle can be found by multiplying the diameter by π.

It is difficult to measure the diameter of a tree trunk directly, but instruments called dendrographs can be used to measure the circumference. They have a band that is stretched tightly around the trunk, attached to a sensor. If the circumference increases or decreases, the sensor can measure the change. In rubber trees, mean daytime trunk shrinkage has been found to be 312 μm near to the base of the trunk and as much as 569 μm 2 m up the trunk.

If

Circumference = diameter × π

Then

$$Diameter = \frac{circumference}{\pi}$$

This means that near the ground the rubber tree trunk diameter decreases during the day by as much as $\frac{312}{\pi}$ = 99.3 μm. Further up the trunk it decreases by $\frac{569}{\pi}$ = 181 μm. Note that because the mean measured changes in circumference were given to three significant figures, the calculated diameters have only been given to three significant figures.

Using ratios and percentages

Standard form

Standard form is a way of using powers of 10 to describe very large or very small numbers. Standard form for small numbers moves the decimal place to the right. If you measured the length of a resting skeletal muscle sarcomere as 2.2 μm, that is 0.0022 mm. In standard form, this becomes 2.2×10^{-3} mm.

Standard form for large numbers moves the decimal place in the opposite direction. If you found that a growing population of yeast cells contained 240 000 cells per cubic centimetre, 240 000 in ordinary form becomes 2.4×10^5 in standard form; in other words, the decimal place is actually five places to the right.

Standard form avoids having to write out all the zeros in small or large numbers. However, you must be very careful to keep the same number of significant figures when converting numbers to and from standard form. A value of 0.0040 indicates that the value is exactly 0.004 rather than a rounded value. The last zero is a significant figure as well as the figure four, so 0.0040 has two significant figures.

TIP
You should keep all your units the same in calculations like this to avoid decimal place errors.

TIP
When using standard form, there can only be **one** digit to the left of the decimal point so 240 000 must be written as 2.4×10^5, **not** as 24×10^4 or 240×10^3.

If you were converting 0.0040 to standard form, you would need to include the last zero by writing it as 4.0×10^{-3}. Both significant figures are shown in the standard form.

Ratios

Ratios can be used to relate one attribute of something to another. For example, **C:N ratios** are helpful in describing the composition of organic material. Organisms that decompose organic material use carbon-containing biological molecules as their respiratory substrates. They use nitrogen-containing biological molecules for protein and nucleic acid synthesis.

Microorganisms use far more carbon-containing compounds as respiratory substrates than they use for protein and nucleic acid synthesis. Since they use about 30 parts of carbon for each part of nitrogen, a C:N ratio of 30:1 promotes rapid composting. If the C:N ratio is greater than this, it means there is more carbon and less nitrogen. If the C:N ratio is lower than this it means there is less carbon and more nitrogen so the soil microorganisms have more nitrogen than they can use.

Surface area to volume ratios are also helpful in understanding why larger organisms need specialised exchange surfaces. Surface area to volume ratios are also used when comparing the rate of heat loss to their surroundings in mammals of different sizes generally. Smaller mammals have a high surface area relative to their volume so lose heat more quickly.

A sphere of radius 10 mm has

Volume equal to $\frac{4}{3} \times \pi \times 10^3 = 4188.79 \, \text{mm}^3$

A surface area equal to $4 \times \pi \times 10^2 = 1256.64 \, \text{mm}^2$.

This means it has a surface area to volume ratio of 1257:4189, which simplifies to 1:3.33. The ratio is **simplified** by making the first number 1, so the second number is $\frac{1}{1257} \times 4189 = 3.33$.

A larger sphere of radius 20 mm has

Volume equal to $\frac{4}{3} \times \pi \times 20^3 = 33510.32 \, \text{mm}^3$

Surface area equal to $4 \times \pi \times 20^2 = 5026.54 \, \text{mm}^2$.

It has a surface area to volume ratio of 5027:33510, which simplifies to 1:6.67 by dividing both numbers in the ratio by the first one. So the larger sphere has a smaller surface area to volume ratio.

In genetics, you will come across **phenotypic ratios**. These relate the proportion of one phenotype to the proportion of another. For example, if a pair of fruit flies had 160 offspring and 120 had red eyes and 40 had white eyes, the ratio of red:white would be 3:1.

Fractions

Fractions are used to describe the portion of a value that fits into different categories. For example, you might find that $\frac{1}{2}$ of the sunlight hitting some leaves is reflected by the shiny surface or $\frac{1}{3}$ of the woodlice you are watching have moved to a more humid area of a choice chamber after 5 minutes. Fractions are useful for description, but they are often converted into decimals or a percentage to make them easier to work with. You probably already realise that $\frac{1}{3} = 0.33$ as a decimal, or 33% as a percentage. You may need to do this for genetic crosses. If $\frac{1}{4}$ of the offspring are a particular phenotype, that is the same as 25%.

You may come across a situation involving **fractions** when looking at cells undergoing mitosis. The **mitotic index** is a way of expressing the amount of cell division taking place in a tissue. In a sample of cells, the fraction undergoing mitosis gives an indication of how much the tissue is actively dividing.

$$\text{Mitotic index} = \frac{\text{the number of cells in mitosis}}{\text{total number of cells counted}}$$

Cells in any stage of mitosis can be identified because their chromosomes are visible (Chapter 5). Other cells are counted as being in interphase. So if a sample of 200 cells contains eight in prophase, 14 in metaphase, five in anaphase and nine in telophase, the mitotic index is:

$$= \frac{8 + 14 + 5 + 9}{200} = \frac{36}{200}$$

This fraction can also be given as a decimal answer $\left(\frac{36}{200} = 0.18\right)$ or converted to a percentage by multiplying by 100 ($0.18 \times 100 = 18\%$). Figure 26.5 shows the percentage of time a sample of cells spent in each phase of mitosis. In this case, the mitotic index would be:

$$= \frac{27 + 9 + 6 + 3}{100} = \frac{45}{100}$$

The mitotic index can be used to find how long mitosis takes. If the whole cell cycle takes 22 hours and the mitotic index is 0.18, then mitosis takes $22 \times 60 \times 0.18$ minutes, which is 237.6 minutes.

Percentages

You will come across **percentages** and especially the idea of a **percentage change**, quite often in biology. Converting data to percentage values allows a valid comparison of data from populations of different size. You will also come across the idea of percentage cover when sampling with quadrats (see pages 414 and 514).

Percentages should normally always add up to 100%. If you know the percentage of some of the bases in a length of DNA, you can calculate the percentage of their complementary bases. Table 26.2 shows an example where you have to calculate the missing values.

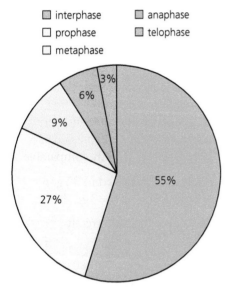

interphase anaphase
prophase telophase
metaphase

Figure 26.5 How much time some cells spend in each stage of mitosis.

Table 26.2 Percentages of bases in two complementary DNA strands.

	A	T	C	G
Strand 1	20	35		
Strand 2			30	

Adenine (A) is complementary to thymine (T), so if there is 20% A in one strand, there must be 20% T in the other. If there is 35% T in one, there must be 35% A in the other. However, finding the values for cytosine (C) and guanine (G) is slightly harder. If there is 30% C on one strand there must be 30% G on the other. Both rows must add up to 100%, so the remaining values for G must both be 15%. Table 26.3 shows the answers in bold.

Table 26.3 The missing percentages are shown in bold.

	A	T	C	G
Strand 1	20	35	**15**	**30**
Strand 2	**35**	**20**	30	**15**

Expressing information as a percentage sometimes allows a valid comparison. For example, if you wanted to compare the incidence of heart disease in Scotland with that of England, you could not directly compare the number of people with heart disease, because the population of Scotland is smaller than the population of England. Instead, you would need to calculate the percentage of people with heart disease in each country. These percentages would then be directly comparable.

Calculating percentage change

If you are collecting data in an experiment where the starting values for a measurement are all different, you can calculate percentage changes to make sure your results are comparable. When investigating the increase or decrease in mass of potato cylinders in different sucrose concentrations, the starting mass of all the cylinders will all be slightly different, even if you try and cut them the same size. So any increase or decrease in mass can be expressed as a percentage of the original mass, and all the results become directly comparable.

If a potato cylinder has an initial mass of 32 g which increased to 35 g in dilute sucrose solution, then the increase in mass is 35 − 32 = 3 g. The percentage increase is therefore $\frac{3}{32} \times 100 = 9.37\%$ which is correctly rounded up to 9.4%.

A quick estimate suggests that 3 is roughly a tenth of 32, so the percentage increase should be approximately 10%. If not, you may have made a mistake.

Percentage change is often used when considering changes in population size. An increase in the number of individuals in a population in a year may be expressed as a percentage increase. This is the additional individuals as a percentage of the number in the original population. If a population had 1300 individuals in one year and a year later contained 1576, the percentage increase is $\frac{1576 - 1300}{1300} \times 100 = 21.2\%$.

Remember that you should always try to estimate an answer to check that your calculated value is appropriate, rather than simply accepting it. In this case, the increase in the population is 276 compared to an original population of 1300 so an answer of roughly a fifth, or 20%, seems realistic.

Percentage error

When you make measurements in practical work, the uncertainty in the measurement can be expressed as a percentage error. The percentage error in the measurements obtained from different pieces of apparatus can be calculated by dividing the uncertainty by the measured value and multiplying by 100.

If a volume of 25 cm^3 measured with a measuring cylinder has an error of ±0.5 cm^3, then the percentage error for this piece of apparatus is $\frac{0.5}{25} \times 100 = 2\%$.

If a mass of 0.120 g measured using a balance has an error of ±0.001 g then the percentage error is $\frac{0.001}{0.120} \times 100 = 0.833\%$.

Dealing with orders of magnitude

Drawings and images of small structures such as cells and organelles are often larger than reality. Micrographs (photographs taken using a microscope) are magnified thousands of times. The magnification of the micrograph is usually given as a magnification factor such as ×10 000. This means that the image is 10 000 times larger than the object really is. You might be asked to calculate the actual size of something in the image, such as the width of a mitochondrion or the diameter of a cell. The method for doing this is always the same, regardless of the object or the magnification.

Measure the length of the structure in the image in millimetres. To convert it to micrometres, multiply by 1000. This is the image length. To find the real length of the structure, divide the image length you have found in micrometres by the magnification you have been given using the formula

$$\text{Size of real object} = \frac{\text{size of image}}{\text{magnification}}$$

This will give you a real length in micrometres. Try this for the length of the sarcomere in Figure 26.6. For microscopic structures, micrometres are the appropriate unit to use for the answer.

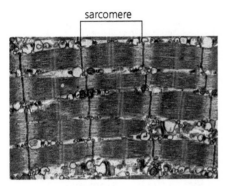

Figure 26.6 Electron micrograph of myofibrils (magnification approximately × 10 000).

> **TIP**
>
> Always convert your measured length in millimetres to micrometres by multiplying by a thousand before doing anything else. Do not measure in centimetres. Always measure the image in millimetres. This helps to avoid decimal place errors.

To calculate the magnification of a structure, use the formula

$$\text{Magnification} = \frac{\text{size of image}}{\text{size of real object}}$$

You should realise that magnification is how big something looks (its image length) compared with how big it really is (its real length). The answer should be given in the form ×2000 rather than just the number.

If there is a **scale bar** on the micrograph, this is the real length (see Figure 26.7). Measure it with your ruler in millimetres and convert to micrometres. Then divide your measurement by the real length, the value on the scale bar. This gives the magnification factor.

100 μm

Figure 26.7 Electron micrograph of a Pacinian corpuscle with scale bar.

Using logarithms to deal with orders of magnitude

Sometimes you may be faced with a very large range of numbers to plot on a graph. If you have a mixture of very small numbers and very large numbers, it is difficult to work out a suitable scale for your axis. To fit the large numbers on, you need a scale that means the small numbers are impossible to plot accurately.

In this sort of situation, using a logarithmic, or log, scale on the graph is a useful approach. You may see graphs like this used for plotting the numbers of yeast in a culture over several hours. The numbers increase very rapidly from a small number of yeast cells in the first place.

Figure 26.8 shows a graph that uses a log scale. You can see that the numbers on the y-axis are not evenly spaced. The number of cells at each time interval has been converted to a logarithm before being plotted. This means that the wide range of values more easily fit onto the same scale.

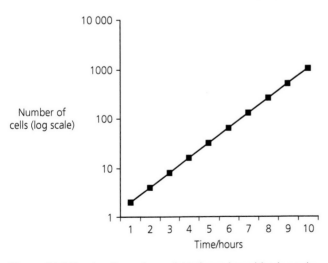

Figure 26.8 Yeast cell numbers plotted on a logarithmic scale.

Table 26.4 The meaning of some symbols used in equations.

Symbol	Meaning
=	is equal to
<	is less than
<<	is much less than
≤	is less than or equal to
>	is greater than
>>	is much greater than
≥	is more than or equal to
∝	is proportional to
≈	is roughly similar to

Using symbols and equations

You need to know the meaning of the mathematical symbols shown in Table 26.4 and when to use them.

Sometimes in questions you will need to rearrange an equation in order to find the answer. This was done with the magnification factor equation in the previous section. Magnification = size of image/size of real object can be rearranged as Size of real object = size of image/magnification depending on what you already know and what you need to find.

Two equations you will have seen in Chapter 14 and Chapter 21 are

$$N = I - (F + R)$$

or

Net production of consumers = chemical energy in ingested food − (chemical energy lost in faeces + heat lost to the environment)

and

$$p^2 + 2pq + q^2 = 1$$

or

Frequency of homozygous dominant phenotypes + frequency of heterozygous phenotypes + frequency of homozygous recessive phenotypes = 1

These can be both rearranged depending on which values you know and which you need to find out. If you know the net production, the chemical energy lost in faeces and the heat lost to the environment, you can find the chemical energy in ingested food (e.g. $I = N + F + R$).

Using the Hardy–Weinberg equation, if you are given the frequency of black rabbits where black is the dominant allele, you are being given $p^2 + 2pq$. The frequency of white rabbits is $q^2 = 1 - (p^2 + 2pq)$.

Often it is only necessary to put the known values into the equation to calculate an answer. An example of this is the index of diversity (see Chapter 13).

$$\text{Index of diversity, } d = \frac{N(N-1)}{\Sigma n(n-1)}$$

Where Σ means 'the sum of', N is the total number of organisms in the community and n is the number of organisms of each species in the community. All that has to be done here is to use the data for the species in the community to put the right numbers into the equation.

Another example is the mark-release-recapture equation (see Chapter 22 page 417), $N = n_1 \times n_2/n_3$ where n_1 is the number initially marked, n_2 is the number caught in the second sample and n_3 is the number re-caught.

Another equation you may need to use is the equation to calculate pH from the hydrogen ion concentration of a solution.

$$pH = -\log_{10}[H^+]$$

In this equation, square brackets mean concentration. You would need to find the logarithm of the value you are given for the concentration using a calculator. Reversing the sign would give you the pH. You may also have to use your calculator to calculate values such as x^y where y is a **power**. In this case, it might be to calculate how many bacteria would be present in a culture after a certain time. The number of bacteria doubles every generation. If there were 12 000 bacteria in a culture and they divide every 15 minutes, you could find the number of bacteria you would expect after 8 hours:

Number of generations = 8 × 60/15 = 32

Your calculator will have a power function, usually a key called x^y, which you can use to do these calculations. For this example you would type 2 x^y 32 which would return the answer of 4.29×10^9, which is the number of cells one cell would give after 32 generations. You would then multiply this by 12 000. The number of bacteria after 8 hours would be 5.15×10^{13}.

You may also have to solve expressions such as 4^n where n is a **power**. It might be to calculate how many combinations of three bases, or DNA triplets, are possible using four different bases. In this case, the expression is 4^3 and the answer is $4 \times 4 \times 4 = 64$. The same sort of expression can be used to find the number of possible combinations of chromosomes following meiosis. If a fruit fly has 16 chromosomes and there is a homologous pair of each, then the number of possible combinations is 16^2 or 16×16, which is 256.

Plotting and using graphs

Drawing a graph helps you to see the relationship between two variables much more clearly than looking at data in a table. The two types of graph you will see most frequently are scattergrams (also called scattergraphs) and line graphs.

Scattergrams

Scattergrams are used to look for a correlation between two variables. However, a scattergram can only indicate if there is a correlation. It cannot tell you if the correlation is significant or not. A Spearman's rank correlation test is required for that (see page 494). You will know that when you draw a scattergram you should ensure that you label the axes and include units. Someone else should be able to look at the scattergram and know exactly what it shows without any further explanation. Plot each point as a dot with circle round it (⊙) or a cross (×). Draw (or imagine) a line of best fit if asked to.

> **TIP**
> If you are drawing your own scattergram, you should try to use a similar scale on each axis so that the results are spread out equally in both directions.

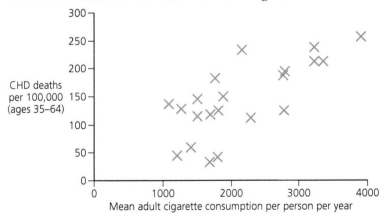

Figure 26.9 Scattergram showing smoking as a risk factor for coronary heart disease.

Correlation

If your line of best fit slopes upwards, you can say that there is a **positive correlation** between the two variables. In other words, as one variable increases, so does the other. If your line of best fit slopes downwards, you can say that there is a **negative correlation**. In this case, as one variable increases, the other decreases. Sometimes the line of best fit is horizontal, sometimes it is vertical and sometimes it is completely impossible to draw a line of best fit. In these cases, all you can say is that there is no correlation.

Line graphs

You use a line graph to show measured results for a dependent variable when you alter an independent variable in an experiment. You might draw line graphs for the growth rate of a population or for the rate of photosynthesis at different light intensities.

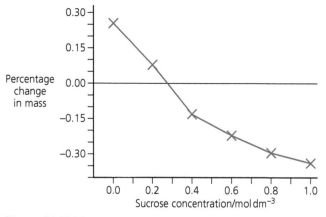

Figure 26.10 Line graph showing the results of putting potato tissue in different sucrose solutions.

When you draw a line graph the independent variable is plotted on the *x*-axis, and the dependent variable is plotted on the *y*-axis. The axes should be labelled and you should include units. Someone should be able to look at the graph and know exactly what it shows without any further explanation.

To choose a suitable scale, make sure all the points you need to plot will fit on the graph. Avoid a scale which involves parts of grid squares on the paper. Using whole squares is better. This makes it easier to plot points accurately. You should plot the individual points as clearly and accurately as possible. Use either a dot with a circle around it ⊙ or a cross (×). You can draw either a smooth curve or straight lines joining the points.

You should learn to recognise the other sorts of curves you usually see in biology. You will see some curves that are **linear**, others that reach a **plateau** and some that are **exponential**.

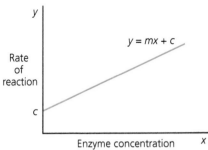

Figure 26.11 Relationship between the rate of reaction and substrate concentration when the enzyme is in excess.

Linear relationships

An example of a linear relationship is that between the rate of a reaction and substrate concentration when the enzyme is in excess. Provided the enzyme concentration remains in excess, the line will continue as a straight line. Graphs like this show a **linear relationship** which is described by the equation

$$y = mx + c$$

where m is the **gradient** and c is the **intercept** on the y-axis. Figure 26.11 shows a line graph with a linear relationship like this.

Sometimes the intercept value is zero. If it is not zero, you will need to read it off the graph. The intercept value is where the line reaches or crosses the y-axis.

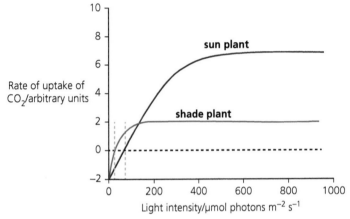

Figure 26.12 Rate of carbon dioxide uptake at different light intensities for sun and shade plants.

An example of a graph with intercepts is shown in Figure 26.12. The graph shows the effect of light intensity on the rate of carbon dioxide uptake by two different plants. You need to be able to see that the y intercept for the linear or straight part of each line is −2 arbitrary units.

On this graph there are also intercepts on the x-axis. When the light intensity becomes so low that the rates of photosynthesis and respiration are balanced, the plant is at the **compensation point**. There is no net uptake of carbon dioxide. You could find the light intensity at the intercept, in other words where the line crosses the x-axis.

Notice that the intercept is different for shade-tolerant species and species that prefer full sun.

Measuring rate of change

You will do some practical work involving enzymes. If the enzyme concentration becomes the limiting factor, the curve will reach a plateau. An example of this type of curve is shown in Figure 26.13.

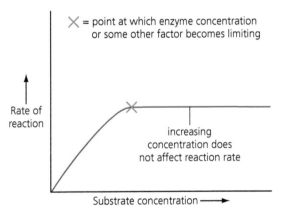

Figure 26.13 When the enzyme concentration becomes the limiting factor, the curve flattens off.

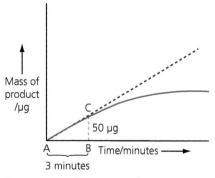

Figure 26.14 The progress of an enzyme-catalysed reaction.

Another line graph in which the curve flattens off is shown in Figure 26.14. Here, note that the *x*-axis shows time rather than substrate concentration and the *y*-axis shows the mass of the product rather than rate. Initially, after an enzyme is added to some substrate, the reaction is fast because there is plenty of substrate to form enzyme–substrate complexes. Once some of the substrate has been used, the reaction begins to slow down.

You can find the initial rate by drawing a **tangent** at the start of the curve. A tangent is a straight line that matches the gradient on the curve. The rate can be found by measuring AB and BC and then calculating BC/AB. This gives the rate at which product is formed per minute. So if 50 µg of product were formed in 3 minutes, the rate would be $\frac{50}{3} = 16.7\,\mu\text{g minute}^{-1}$.

TIP

Draw a tangent by placing a ruler on a short part of the start of the curve. Start from the origin and match the gradient. Make sure the ruler is positioned carefully. A slight change can alter the position of the tangent a lot.

Interpreting data from a sample of measured values

As you know, whenever a sample is taken from a larger group, the sample should be representative of the group as a whole. When these sorts of data are collected, a large number of observations or measurements is made and so summarising the data is often helpful. Biological data from samples can be summarised in a variety of ways, often called **descriptive statistics**.

Arithmetic means

One way of summarising sample data is by finding the **mean**. You will already have calculated mean values in practical work and will already know that the mean is found by adding up all the measurements and then dividing this total by the number of measurements.

$$\text{Mean} = \frac{\text{sum of all measurements}}{\text{number of measurements}}$$

Standard deviation

You will also know from last year that a mean becomes more useful if a standard deviation is also given with it. You will recall that standard deviation is found from the formula

$$SD = \frac{\Sigma(x - \bar{x})^2}{(n-1)}$$

where Σ means 'the sum of', $(x - \bar{x})$ is the difference between any measurement and the mean and *n* is the number of measurements. To do the calculation, you need to find the difference between each measurement and the mean, square it and then add all the squared values together. Then divide this by one less than the number of measurements.

Standard deviation can be calculated manually, by using a scientific calculator or by using spreadsheet software.

Range, median and mode

Other ways of summarising the data are the **range**, the **mode** and the **median**.

The range is the difference between the highest and lowest values. Range and standard deviation are both measures of dispersion for a given set of data, but they both have advantages and disadvantages. Range is easier and quicker to calculate, but it is heavily influenced by outlying values at the extremes of the range. An outlier can increase the range significantly but the outlier is just one result in the whole set of data. The range simply uses the highest and lowest values so takes no account of the rest of the data. In this situation, standard deviation can be more useful. It takes longer to calculate, but it takes account of all of the results in a set of data and so enables a better comparison of the dispersion of two sets of data.

The median is the middle value of the data arranged in rank order. The mode is the value that appears most often.

Frequency tables and histograms

One way of summarising data is a frequency table. A frequency table is made by dividing the observations or measurements into a number of **classes**. Classes are either categories or ranges of measured values to which observations can be allocated. The **class frequency** or number of observations or measurements in each class can then be tallied up.

Table 26.5 Frequency table for the mass of a sample of fish.

Mass/kg	Tally	Frequency
2.0–2.9	///// //	7
3.0–3.9	///// ///// ///// /	16
4.0–4.9	///// ///// ///// ////	19
5.0–5.9	///// ///// //	12
6.0–6.9	///// /	6
7.0–7.9	///	3
Total	63	63

A histogram is plotted when the frequency table shows measured data. Counted data in categories, such as the number of fish of different species, would have to be plotted as a **bar chart.** In a bar chart, the bars are each separated by a space because the data are in discrete categories.

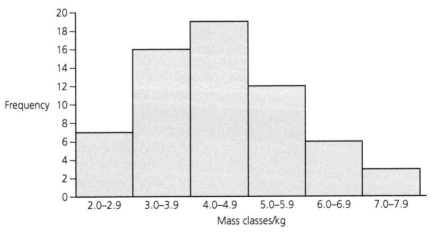

Figure 26.15 Histogram showing the frequencies of different mass classes for fish.

Statistical tests

During your A-level Biology course you have to choose one of three different statistical tests to analyse the data you collect in different practical situations. You will need to be able to recognise the type of data presented in a question, suggest how it could be presented on a graph and say which of the three different statistical tests best applies to the data. Using a statistical test in this way is called **inferential statistics**.

Why do scientists use statistics? Look at a roadside verge in early spring and you will almost certainly see dandelions in flower. Examine a particular length of verge more closely and you will see that most of the dandelions are growing near the road.

If you were to carry out a transect, you might find that the dandelions are less common further from the road and other species are more numerous. How can you explain this observation? Perhaps the road was treated with salt during the winter and dandelions can tolerate high salt concentrations better than other plants. There is another explanation, however, that has very little to do with biology. Maybe the differences in distribution of the various species is simply due to chance. It could be that you just picked an area where there were more dandelions growing closer to the road. A statistical test can help you judge how likely it is that there is a significant difference in the density of dandelions nearer or further away from the road.

As biologists, we accept that we can never completely rule out the effect of chance. There is still a small probability that the distributions you observed were due to chance. But it seems quite reasonable to say that the probability of different densities of dandelions nearer and further from the road arising due to chance is so low that we can safely reject chance as an explanation. We need a cut-off point and, for biological investigations, we normally accept this as a probability of 1 in 20 or 0.05 ($p = 0.05$). If there is a probability of less than 1 in 20 ($p < 0.05$) that the higher dandelion density near the road could have arisen by chance, there is another explanation. In this case we could say that it might be the salt from the road.

Figure 26.16 Dandelions growing near a roadside.

Different statistical tests for different purposes

A statistical test is a tool and, like all tools, you don't need to know how it works to use it. What you need to be able to do, however, is to select the right tool for the right job.

The practical work that you are likely to carry out could generate data that you might need to look at in different ways. It is likely that you will want to do one of the following.

● See if there is a significant difference between the means of two variables: suppose you were investigating the effect of temperature on the growth of duckweed. You could present the data as a line graph or a histogram. But you might also want to see if there was a significant difference between the mean numbers of duckweed plants when the plants were kept at two different temperatures.
● See if there is a significant correlation between two variables: if you investigated the number of stonefly larvae at different points in a stream, you could plot the data as a scattergram. But you might also want to see if there was a significant correlation between the number of stonefly larvae and the velocity of the water.

- Investigate data that fall into distinct categories. Suppose you were looking at whether maggots moved to light or dark conditions or offspring of a genetics cross matched a certain ratio. You could plot the data as a bar graph. But you might also want to know if there is a significant difference between the data you have collected and the data which your hypothesis predicts.

Each of these situations requires a different statistical test. The decision chart in Figure 26.17 should help you to choose which one to use.

Making the decision over which test relates to your data is very important. The way you plot your data on a graph and which statistical test you use are closely linked.

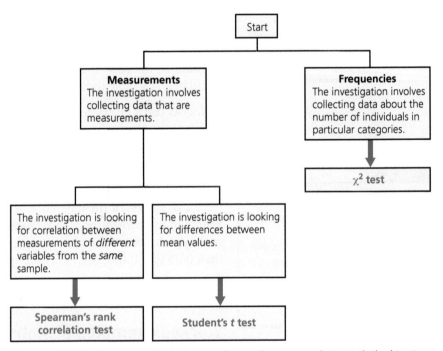

Figure 26.17 Decision chart to help you choose the appropriate statistical test.

The null hypothesis

Each of the statistical tests that you will use is based on what is called a null hypothesis. It is very important that you understand what a null hypothesis is and why it is used. Suppose you had a table of data showing the distribution of aphids on the upper surface and the lower surface of nettle leaves. In this example, your null hypothesis could be:

There is no difference between the numbers of aphids on the upper surface and on the lower surface of nettle leaves.

You can then use the results of a statistical test to come to a conclusion about whether or not to accept the null hypothesis.

- If the test indicates that you should accept the null hypothesis, then all you can say is that there is no significant difference in the numbers of aphids on the different surfaces and that the differences you found could be explained purely by chance.
- If the test indicates that you should reject the null hypothesis, then you can say that the probability that the differences in numbers of aphids on the upper and lower leaf surfaces are due to chance is low and you can suggest a biological explanation.

TIP

You will not be asked to carry out statistical calculations in written exams. Instead you could be asked to suggest what sort of graph or what sort of statistical test is most appropriate for the data and why.

TIP

'There is no difference in the result' is **not** an acceptable null hypothesis. You must mention both variables.

Calculating the test statistic

Once you have set up your null hypothesis, you can carry out the test you have selected. As a biologist, carrying out statistical tests is like using a recipe book to cook a meal. If you work methodically, and can use a standard scientific calculator, there is nothing to worry about. Remember that calculating test statistics will not be required in written exams but you may need to do a statistical calculation following practical work. If you look in the section at the end of this chapter, you will find worked examples of the three tests that you need to be able to use.

Interpreting the results of a statistical test

Calculating the test statistic gives you a number. What does this number mean? You always need to interpret this number in terms of probability. You can then make a decision as to whether to accept or reject your null hypothesis. In order to do this, you need to look up the value that you have calculated in a table to find the probability level, or 'p value'.

Biologists use a probability level of 0.05 (often written as $p = 0.05$) against which to interpret χ^2, t test or correlation coefficient values. The p value is like a hurdle, which is set to determine if the results might simply have been obtained by chance.

If the probability is less than 0.05 ($p < 0.05$), then the difference between the mean values, or the difference between the observed and expected results or the correlation, can be regarded as statistically significant. The null hypothesis can then be rejected. If the probability is 0.05 or greater than 0.05 ($p > 0.05$), then the null hypothesis must be accepted.

If you look at the following worked examples of each of the statistical tests, you will see how the test statistic has been interpreted in each case.

Worked examples of statistical tests

Using Spearman's rank correlation test

You should choose this test when you wish to find out if there is a significant correlation between two sets of variables. You need between 7 and 30 pairs of measurements, such as the nitrate data in Chapter 15 (page 280).

- Start by producing a null hypothesis.

Your null hypothesis is that there is no correlation between the total mass of nitrogen added to the surrounding fields and the mean concentration of nitrate in the stream.

- Calculate the correlation coefficient.

Start by ranking the total mass of nitrogen and the mean concentration of nitrate (Table 26.6). Note that when two or more values are of equal rank, each of the values is given the mean of the ranks that would otherwise have been allocated.

> **TIP**
> If you are given the result of a statistical test, or just the p value, you need to be able to interpret what this means.

> **TIP**
> It is the difference between results that is significant or non-significant. You should never just write that 'the results are significant'. You must say what it is that is significant and refer to the variables involved.

Table 26.6 Rank values of total mass of nitrogen added to the surrounding fields and the mean concentration of nitrate in the stream.

Total mass of nitrogen/kg ha⁻¹	Rank	Concentration of nitrate/mg dm−3	Rank	Difference in rank (D)	D2
41	1.5	1.2	1	0.5	0.25
41	1.5	1.3	2	0.5	0.25
51	3	1.5	3	0	0
56	4	1.8	5	1.0	1
63	5	1.6	4	1.0	1
69	6	1.9	6	0	0
72	7	2.0	7	0	0

Calculate the difference between the rank values (D) and square this difference (D^2).

Find the sum of the squares of the differences: $\Sigma D^2 = 2.5$.

Now calculate the value of the test statistic, R_s, from the equation:

$$R_s = 1 - \left(\frac{6 \Sigma D^2}{n(n^2 - 1)} \right)$$

where n is the number of pairs in the sample and D is the difference between ranks.

$$= 1 - 0.04$$
$$= 0.96$$

● Interpret the value of R_s.

The value of R_s will always be between 0 and +1 or −1. A positive value indicates a positive correlation between the variables concerned. A negative value shows a negative correlation.

Table 13.3 shows the values of R_s. Look in the table under the correct number of pairs of measurements for our data. You had seven pairs of values, so the critical value is 0.79.

Your calculated value of R_s is larger than the critical value so you reject your null hypothesis and say that there is a significant correlation between the total mass of nitrogen added to the surrounding fields and the mean concentration of nitrate in the stream.

Although in this case you have identified a significant correlation between the variables concerned, you still have to be careful about the conclusions you draw. A correlation between these variables does not necessarily mean that the application of fertiliser causes pollution in the stream.

Table 26.7 Critical values of R_s for different numbers of paired values.

Number of pairs of measurements (n)	Critical value of R_s
5	1.00
6	0.89
7	0.79
8	0.74
9	0.68
10	0.65
12	0.59
14	0.54
16	0.51
18	0.48

TIP
- In an exam you might be asked to interpret a given value of R_s.

Using the chi-squared test

You should choose this test to determine whether the data you obtain is significantly different from what you would have expected. The null hypothesis in the woodlice example from Chapter 16 (page 291) would be that there is no significant difference between the number of woodlice going to the dark side and the number going to the light side.

O represents the observed results. If you add together the numbers in the light and dark from the 10 trials you will find that there were 61 woodlice on the dark side and 39 in the light. If the null hypothesis is true, you would expect equal numbers on each side, i.e. 50. This is represented in the equation by E. Table 26.8 shows the results of the calculations.

Table 26.8 Using the chi-squared (χ^2) test to determine whether the results are statistically significant.

	Light side	Dark side
Observed results (O)	39	61
Expected results (E)	50	50
$(O - E)2$	121	121
$(O - E)2/E$	2.42	2.42

TIP
You may need to work out the number of degrees of freedom in a given investigation.

Now, taking the sum of the values for the light and dark sides, you will find that $\chi^2 = 4.84$. You now need to look up the values for χ^2 in a table of critical values (Table 26.10 on the next page). Since there are only two categories, light and dark, there is only one degree of freedom. The critical value for rejecting the null hypothesis with one degree of freedom at a probability level of 0.05 is 3.84, so you can have over 95% confidence that there is a significant difference in the number of woodlice moving to the dark. In other words, the probability of getting these results by chance is less than 5%.

EXAMPLE

Using the χ^2 test in examining genetic crosses

Data from genetic crosses are often categoric, for example data from the inheritance of the human ABO blood group. You should choose this test if you wish to check that the numbers of offspring obtained from a genetics cross are a close enough match to the expected ratio.

Geneticists investigated the ABO blood group phenotype of children born to a large number of parents. For each pair of parents, the blood group genotype of one was heterozygous $I^A I^O$ and the other was heterozygous $I^B I^O$. The geneticists expected the blood groups of children in the investigation to be A, B, AB and 0 in a ratio of $1:1:1:1$.

1 Explain why the geneticists involved a large number of parents in this investigation.
A large number of parents was involved so that the results would be reliable.

2 Explain why the geneticists expected a ratio of $1:1:1:1$ in the offspring.
This is the ratio that would be predicted by a genetic diagram from a cross between parents of these genotypes.

Table 26.9 shows the results the geneticists obtained.

Table 26.9 The results of an investigation into the inheritance of human blood groups.

Blood group	Observed (O)	Expected (E)	$(O - E)$	$(O - E)2/E$
A	26			
B	31			
AB	39			
0	24			
				$\chi^2 = \sum \frac{(O - E)^2}{E}$

3 Copy the table and use the prediction of the geneticists to complete the column showing the expected number of children with each blood group (*E*). Explain how you arrived at your answer.

The expected number is found by adding up the total number of offspring (120) and then dividing by 4 (as the offspring are expected in a 1 : 1 : 1 : 1 ratio). The answer is 30% for each blood group.

4 Calculate the values for $(O - E)$, $(O - E)^2$, $\frac{(O - E)^2}{E}$ and $\sum \frac{(O - E)^2}{E}$. Enter these values in your copy of Table 26.9.

Blood group (O)	Observed (O)	Expected (E)	(O – E)	(O – E)²	(O – E)²/E
A	26	30	4	16	0.53
B	31	30	1	1	0.03
AB	39	30	9	81	2.7
O	24	30	6	36	1.2
	120				Σ = 4.46

Note that when (O – E) is calculated the minus sign can be ignored, because in the next stage the result is squared.

The geneticists calculated a value of $\chi^2 = 4.46$. They looked up their calculated value of χ^2 in a probability table. Table 26.10 shows part of this table.

Table 26.10 Values of χ^2 at three probability levels.

Degrees of freedom	Values of χ^2 at each probability level		
	$p = 0.10$	$p = 0.05$	$10 = 0.01$
1	2.71	3.84	6.64
2	4.61	5.99	9.21
3	6.25	7.82	11.34
4	7.78	9.49	13.28

5 How many degrees of freedom were there in the geneticists' data? Explain your answer.
Three degrees of freedom. Degrees of freedom = (number of categories) – 1 and there are four categories.

6 What does the calculated value of χ^2 tell us about the results of this investigation? Explain your answer.
In biology we use a probability of p = 0.05. The critical value of χ^2 is 7.82 for 3 degrees of freedom. The χ^2 value of 4.46 is less than the critical value of 7.82. Therefore we must accept the null hypothesis.

7 State a null hypothesis for this analysis.
The null hypothesis is that this genetic cross gives a 1 : 1 : 1 : 1 ratio of phenotypes in the offspring.

TIP
This example shows how to work out a *t* test manually but there are plenty of online programmes that will calculate the value for you.

TIP
You do not need to recall the formula for the *t* test.

TIP
You may have noticed the similarity between the equations for standard deviation and variance. Standard deviation (see page 490) is often given the symbol *S* in equations and is the square root of the variance.

Using the *t* test

You should choose this test to compare the means of two samples of measured data to see whether they are significantly different.

The *t* test will tell you whether there is a significant difference between the means of two samples, or not.

The *t* test formula is:

$$t = \frac{|\bar{x}_1 - \bar{x}_2|}{\sqrt{\frac{S_1^2}{n_1} + \frac{S_2^2}{n_2}}}$$

where

$\bar{x}_1$ is the mean of sample 1
$\bar{x}_2$ is the mean of sample 2
n_1 is the number of subjects in sample 1
n_2 is the number of subjects in sample 2
S_1^2 is the variance of sample 1, $\frac{\sum(x_1 - \bar{x}_1)^2}{n_1}$
S_2^2 is the variance of sample 2, $\frac{\sum(x_2 - \bar{x}_2)^2}{n_2}$.

Thinking about the example of limpets in Chapter 22 (page 414), first you need to calculate the variance of each sample.

Table 26.11

Sheltered shore				Exposed shore		
x_1	$x_1 - \overline{x}_1$	$(x_1 - x_1)^2$	x_2	$x_2 - \overline{x}_2$	$(x_2 - x_2)^2$
0.44	0.013	0.00017	0.32	0.019	0.000361
0.41	0.043	0.00185	0.32	0.029	0.000841
0.55	0.097	0.00941	0.40	0.053	0.002601
0.43	0.023	0.00053	0.47	0.121	0.014641
0.40	0.053	0.00281	0.32	0.029	0.000841
0.50	0.047	0.00221	0.31	0.039	0.001521
0.39	0.063	0.00397	0.37	0.021	0.000441
0.54	0.087	0.00757	0.30	0.049	0.002401
0.43	0.023	0.00053	0.42	0.071	0.005041
0.39	0.063	0.00397	0.35	0.001	0.000001
0.44	0.013	0.00017	0.30	0.049	0.002401
0.47	0.02	0.00040	0.36	0.011	0.000121
0.47	0.02	0.00040	0.36	0.011	0.000121
0.49	0.037	0.00137	0.31	0.039	0.001521
0.45	0.003	0.00001	0.32	0.029	0.000841
$\overline{x}_1 = 0.453$	$n_1 = 15$	$\Sigma = 0.03537$	$\overline{x}_2 = 0.349$	$n_2 = 15$	$\Sigma = 0.033695$

Now you calculate t.

$$S_1^2 = \frac{\Sigma(x_1 - \overline{x}_1)^2}{n_1}$$

$$= 0.002358$$

$$S_2^2 = \frac{\Sigma(x_2 - \overline{x}_2)^2}{n_2}$$

$$= 0.002246$$

$$\frac{S_1^2}{n_1} = \frac{0.002358}{15} = 1.572 \times 10^{-4}$$

$$\frac{S_2^2}{n_2} = \frac{0.002246}{15} = 1.497 \times 10^{-4}$$

Substituting these values together with the mean values from Table 26.11 into the equation,

$$t = \frac{(0.453 - 0.349)}{\sqrt{1.572 \times 10^{-4} + 1.497 \times 10^{-4}}}$$

$$= \frac{0.104}{\sqrt{3.069 \times 10^{-4}}}$$

$$= \frac{0.104}{0.0175}$$

$$= 5.94 \text{ (2 dp)}$$

You now need to look up this calculated value of t on a table of t values (Table 26.12).

Degrees of freedom $= (n_1 + n_2) - 1 = 29$

Table 26.12 Table of *t* values.

Degrees of freedom	Significance level					
	20% (0.02)	10% (0.01)	5% (0.05)	2% (0.02)	1% (0.01)	0.01% (0.001)
28	1.313	1.701	2.048	2.467	2.763	3.674
29	1.311	1.699	2.043	2.462	2.756	3.659
30	1.310	1.697	2.042	2.457	2.750	3.646

For $p = 0.05$ and 29 degrees of freedom, the critical value of t is 2.043.

Your calculated value of t is 5.94, which is higher than this value.

Therefore, you must reject your null hypothesis. There is a significant difference between the mean height: base diameter ratio of the limpets on the two shores. Can you suggest a hypothesis to explain this?

The limpets on the exposed shore have a lower height: base diameter ratio than those on the sheltered shore, which means they have a greater area of 'foot' to hold them against the rock than the limpets on the sheltered shore of the same height. This means they can grip on to the rock better and are less likely to be washed off by waves. Also, a smaller height and wider base makes them more streamlined when waves hit them, again making it easier for them to stay attached to the rocks.

TIP

It is the difference between results that is significant or non-significant. You should never just write that 'the results are significant'. You must say what it is that is significant and refer to the variables involved.

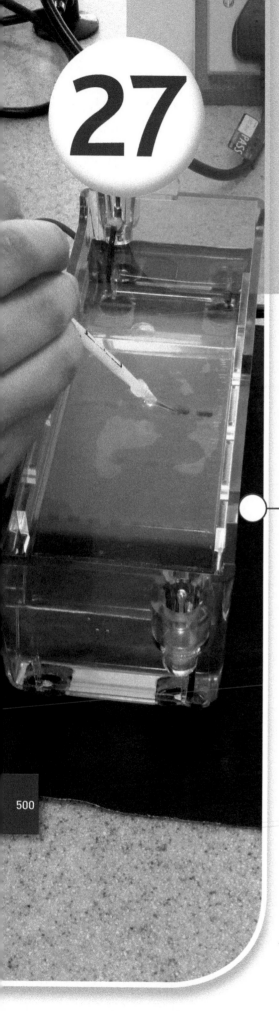

27

Developing practical skills

Alongside developing your knowledge and understanding of biology, you are expected to become familiar with a range of apparatus and competent in a number of practical techniques. You will gain experience of these apparatus and techniques during routine practical work. Your teacher will probably include a number of different practical investigations and activities in the course but there are 12 required practical activities that you must complete. Written papers will assess your knowledge and understanding of the apparatus and techniques in some of these practicals. A list of the required practicals is given at the end of this chapter. But you can demonstrate your competence in these practical techniques during any of the practical work you do.

Investigating a scientific question

In all investigations you control something. This is the independent variable. In the oral rehydration solution investigation in Chapter 2, on enzymes, the independent variable is the concentration of amylase. As a result of these changes, you have to measure something else. This is the dependent variable. In this investigation, the dependent variable is the viscosity of the rice flour solution.

The first step in any investigation is to identify these two variables. The next is to plan how to change the independent variable.

Changing the independent variable

An experimental result that occurs only once could be due to chance. It could also be because the scientist who carried out the investigation did it in a particular way. Experiments only provide **precise** data if they can be repeated by the same experimenter or by someone else to give very similar results. This means that you should give enough detail in describing your method for another scientist to follow the technique exactly as you intended, without the need for any help other than your set of instructions. In this case it is not enough, for example, to say, 'Put the mixture in a water bath'. Another scientist would want to know what temperature you used. 'Put the mixture in a water bath at a constant temperature of 35°C' is much better.

In most experiments, variables other than the one you are planning to change might also affect the results. We call these control variables. You should design your experiment so that these control variables are kept constant. There are several control variables in this experiment.

The incubation time of the enzyme and rice flour solutions, temperature, pH and substrate concentration could all affect the rate of this reaction. If we keep these factors constant, then there is a reasonable chance that any difference in viscosity will be due to the concentration of amylase.

Once you have decided how you are going to change the independent variable and have identified the various control variables whose values you must keep constant, you need to ask another question: is there anything, other than the independent variable, in the way I have set up this experiment that could have produced these results?

If there is, you need a control. A control experiment is one that you set up to eliminate certain possibilities. In this investigation it is possible that the rice flour solution would have become runnier anyway if you just left it. We ought to have a control experiment where everything is the same as in the experimental container but without amylase or with denatured amylase. By setting up this control you eliminate the possibility that something other than amylase reduced the viscosity of the solution.

Measuring the dependent variable

Once you have decided how you will change the independent variable, you should decide how to determine the effect of this on the dependent variable. You need quantitative data and this means taking measurements. These measurements must be made with the appropriate degree of **resolution**.

In the oral rehydration solution investigation, you could measure viscosity. For more resolution you could use a viscometer but it is not very likely that you would have one in your laboratory. There are other, simpler methods, however. You could pour the solution into a burette and see how long it took for the liquid to drain out through the tap, or you could take the time for a marble or other object to fall through a column of the solution.

You also need to bear in mind that if you carry out the experiment once only, you do not know if the results are precise. Carrying it out twice doesn't help you much either. You need to carry it out several times. Only in this way can you distinguish precise data from anomalies.

Making quantitative measurements

Making and recording quantitative measurements is a key practical skill. The measurements you are most likely to make include time, volume, length, mass, temperature and pH. However, if you use instruments such as a colorimeter you will come across some less familiar measurements. A colorimeter measures the optical density of a solution. You may also come across less familiar units such as micrometres (μm) or micrograms (μg), and new units such as millivolts (mV) or unfamiliar combinations of units such as micrograms per square metre per day (μg m^{-2} day^{-1}).

TIP

Quantitative results involve making measurements and recording numbers. Qualitative results, such as those you record if using reagents to identify biological molecules, are descriptive. They simply tell you if a particular substance is present or not, rather than how much is present.

To enable some measurements to be made with more resolution, different apparatus needs to be used. A wall clock could be used to measure time in seconds but a stopwatch would be needed for measurements in milliseconds. Although you will be used to measuring length in millimetres you may not have measured length before in micrometres. You could measure the length of a leaf in millimetres using a **ruler**. But to measure the thickness of the leaf, you might use a piece of equipment called a micrometer and to measure cells or organelles seen with a microscope, a **graticule** would be used. If you are trying to find the biomass of some small animals such as freshwater shrimps, a balance reading to 0.1 g or even 0.01 g would be more suitable than one reading to just the nearest 1 g.

Figure 27.1 Digital balance being used to weigh freshwater shrimps.

You may need more specialised instruments to make some measurements. The pH of the water in a stream might be measured using a **pH meter** (Figure 27.4) whereas a **light meter** might be used to measure the light intensity in a wood. A **respirometer** or a **photosynthometer** measures the volume of gases involved in gas exchange. A **potometer** measures the volume of water take-up by a leafy stem. A **colorimeter** measures how much light can pass through a coloured solution (Figure 27.2).

If you use a colorimeter you will find that it needs to be zeroed, rather like a balance, before making measurements. Zeroing a balance is done with nothing on the balance, but a colorimeter is zeroed using a clear solution, often water, called a **blank**. Samples are placed into a colorimeter in rectangular plastic tubes called cuvettes. Just as it is important not to spill material or solutions onto a balance pan to avoid errors, it is important to avoid spilling drops of liquid on the outside of the cuvette. This is because drops of liquid can cause errors by scattering the light. Knowing and avoiding likely causes of error with each piece of equipment helps to ensure accurate measurements.

TIP

Put the blank back into the colorimeter every so often between measurements and check that it still reads zero. If not, re-zero it. This ensures that all of your measurements remain as accurate as possible.

Figure 27.2 A colorimeter measures the optical density of coloured solutions.

Figure 27.3 Digital callipers being used to measure shells.

Figure 27.4 A pH meter being used to measure pH of a solution.

> **TIP**
>
> Keep the leaves of your leafy stem out of the water while assembling the potometer in the sink. Any water on the leaf surfaces will interfere with the transpiration rate and cause an error in measuring the volume of water taken up by the stem and it is very difficult to dry them without damaging them.

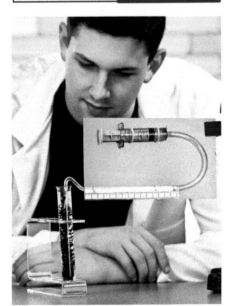

Figure 27.5 A photosynthometer being used to measure the volume of gas produced by plant material during photosynthesis.

Most colorimeters have a method of placing coloured filters in the path of the light. This is because light of a particular wavelength is more suitable for measuring the optical density of solutions of particular colours. For example, red light is most suitable for measuring the optical density of blue Benedict's reagent after it has reacted with glucose. In general, the filter that gives the highest optical density reading for a solution of a certain colour is the best one to use.

Potometers work by measuring how far an air bubble travels along a scale in a certain time. It is important to avoid other air bubbles elsewhere in the apparatus, so they are often assembled underwater in a large sink or bowl. Once they are working without leaking, it is best not to disturb them. Potometers often have a three-way tap and a water reservoir to enable the air bubble to be returned to the start of the tube by letting water back in without having to remove and replace the stem. This enables repeat readings to be made easily.

A respirometer measures the volume of oxygen used by living material during respiration. It works by measuring how far a bead of coloured liquid in a capillary tube travels along a scale in a certain time. The bead of coloured liquid moves because as oxygen is used up the pressure inside the apparatus falls, provided a carbon dioxide absorber is present inside. Leaks at the joints of a respirometer or changes in the room temperature or air pressure can cause errors in the volume measurements.

Once respirometers are working without leaking, it is best not to disturb them. They sometimes have a three-way tap and a syringe to enable the bead of coloured liquid to be pushed back to the start. This enables repeat readings to be made without having to reassemble the apparatus every time.

A photosynthometer measures the volume of gas produced by plant material during photosynthesis and can be used to investigate the effect of variables like temperature and light intensity on the rate of photosynthesis in aquatic plants or algae (Figure 27.5).

Collecting the gas produced by photosynthesising plant material can be done with a simple upturned tube but this does not allow volume measurements. It is better if the gas can be collected in a graduated container or collected and then pulled along a capillary using a syringe. The length of the bubble of air can then be measured. The volume of gas used or produced in respirometers and photosynthometers is calculated from the distance the bead or bubble has moved, in the same way as for a potometer.

> **TIP**
>
> Photosynthesis takes a while to respond to changed environmental factors. A sensible precaution is to wait a while for the plant material to adjust to the new conditions before starting to make measurements.

Using glassware

Measuring volumes accurately using glassware such as measuring cylinders is a key practical skill for both years of the course. This year, you will need to make a dilution series of a glucose solution to produce a calibration curve for a colorimeter. Selecting the appropriate size of measuring cylinder depending on the volume to be measured is crucial. The smaller graduations on the measuring cylinder, the better the resolution of the measurement can be. But obviously the total volume required has to be able to fit into the measuring cylinder selected. Ensuring that the full volume is drained out of the measuring cylinder is also important, as is the need to avoid splashes and spillages, which lead to error.

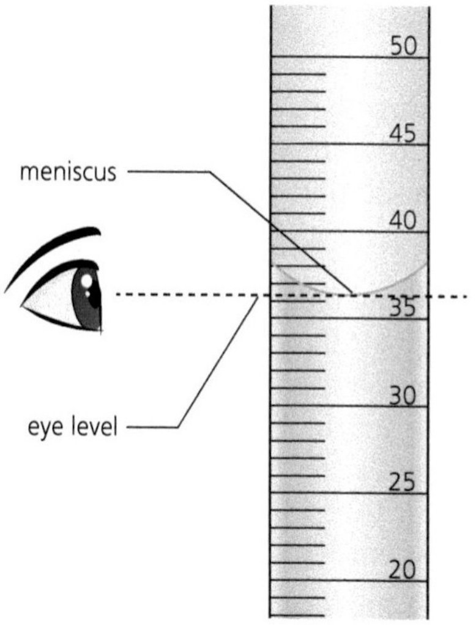

Using a pipette accurately to deliver a fixed volume of reagent to a solution is another practical skill. If you were making a calibration curve to find the concentration of glucose in an unknown sample with a colorimeter, you would need to add exactly the same volume of Benedict's reagent to each glucose dilution. Careful use of a graduated pipette or syringe is vital if the volume of Benedict's reagent is to be properly controlled.

Reading the position of a **meniscus** accurately is also important for accurate volume measurement (Figure 27.6). The correct way to do this is to line up the bottom of the meniscus with the graduation mark, at eye level. You may need to do this to measure the volume of carbon dioxide produced by respiring yeast cells. Fermentation tubes allow the carbon dioxide produced to be collected and measured. This can be done by placing upturned ignition tubes in a yeast culture. These have no graduations, but the volume of gas produced can still be compared qualitatively. Alternatively, specialised pieces of graduated glassware can be used which allow the volume of gas to be measured quantitatively (Figure 27.7).

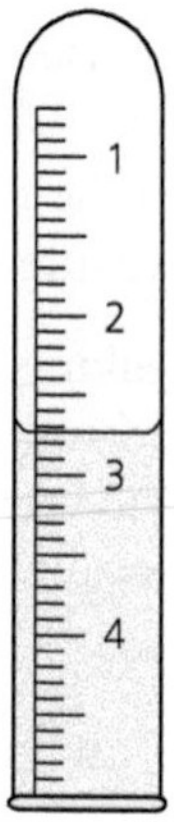

Figure 27.6 How to read a volume correctly from a meniscus.

Figure 27.7 The meniscus formed by collecting carbon dioxide gas in a graduated fermentation tube.

> **TIP**
>
> If you are making up a dilution series, the volume in each of your tubes or beakers should end up the same. Line them up and compare the position of the meniscus in each. They should line up exactly. If they do not, you have made a measuring error.

Using an optical microscope

The optical microscope is a vital tool for biologists because of the small size of many biological structures. Tissues, cells and organelles are all too small to be seen in any detail by eye so their images must be magnified. Optical microscopes work by directing light through a thin layer of biological material supported on a glass slide. This light is then focused through several lenses so an image can be seen through an eyepiece. You can switch from low to high power by rotating a different objective lens into position.

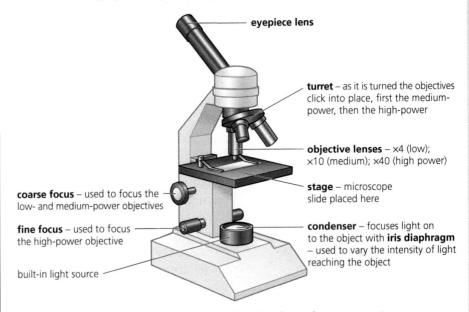

TIP

Even if you are going to look at your material under high power, always start by focusing on low power. It is easier to see what you are looking at using low power and when you change to the high power objective, it should be roughly focused already. This also helps to prevent damage to the lens.

eyepiece lens

turret – as it is turned the objectives click into place, first the medium-power, then the high-power

objective lenses – ×4 (low); ×10 (medium); ×40 (high power)

stage – microscope slide placed here

condenser – focuses light on to the object with **iris diaphragm** – used to vary the intensity of light reaching the object

coarse focus – used to focus the low- and medium-power objectives

fine focus – used to focus the high-power objective

built-in light source

Figure 27.8 A typical optical microscope showing the main components.

Any material you are going to look at using a microscope has to be either transparent or really thin to allow light to pass through it. Some material might already be thin enough, such as the very thin leaves of moss which allow you to see their chloroplasts (Figure 27.9a). Other material such as pea leaf tissue might need to be sliced very thinly using a sharp blade. A drop of a culture of yeast under a cover slip will be spread out into a very thin layer so that individual yeast cells can be seen. Cutting animal tissue into a thin enough slice is much more difficult, so you will probably look at prepared slides of material such as skeletal muscle (Figure 27.9b).

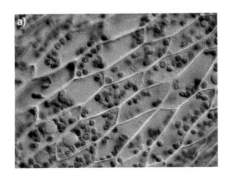

Figure 27.9 a) Chloroplasts visible in the cells of a moss leaf using a light microscope; b) skeletal muscle fibres.

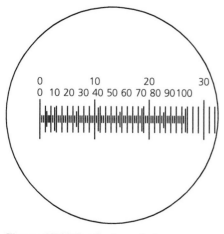

Figure 27.10 Graticule and stage micrometer being used together to calibrate the graticule.

Figure 27.11 Tubes containing homogenised leaf tissue being placed in a centrifuge.

Thin layers of material quickly dry out and shrivel up in the heat of the microscope lamp. This is why a drop or two of water is usually added beneath the cover slip to prevent dehydration damaging the cells.

If you need to make measurements of cells seen using an optical microscope, then a graticule is required. This provides a scale like a ruler in the field of view. It has no fixed units, so it has to be **calibrated** for the objective lens being used. Calibration is done using a stage micrometer, which is a second scale in micrometres engraved on a microscope slide. Using the two together, as in Figure 15.6, it is possible to see how many micrometres each graticule 'unit' is worth. If you continue to use the same objective lens, the graticule can now be used like a ruler.

Graticule A small glass or plastic disc with an engraved scale that is placed into the eyepiece of a microscope where it acts like a ruler superimposed on the image.

Isolating organelles

Isolating chloroplasts for microscopy, or for use in other experiments, requires the use of a centrifuge. Plant material such as chopped spinach leaves is **homogenised** in a cold solution with the same water potential in a blender. It is filtered to remove larger pieces of tissue and the filtrate is then **centrifuged** for several minutes (Figure 27.11). The nuclei are compacted down into a **pellet** at the bottom of the centrifuge tube. The supernatent is carefully poured into a second centrifuge tube and centrifuged at a higher speed for several minutes. This time, the chloroplasts form a pellet.

The **supernatant** can be poured off and the pellet can be re-suspended in an isotonic solution. The chloroplasts can then be observed using an optical microscope. They could also be used to make a chloroplast extract to investigate their dehydrogenase activity.

Making drawings

Making drawings is an important way of recording the results of practical work where the outcome of an investigation is descriptive rather than quantitative. The purpose of biological drawing is to make a clear scientific record of what you have observed rather than an artistic interpretation of the material. This means line drawings in pencil with no shading and no use of colour.

The important things are shape, proportion and scale. The structures drawn should be the same shape as those you observe, the parts of the drawing should be in proportion to one another and the drawing should be large enough to show the details clearly. If you are drawing individual cells such as skeletal muscle cells using an optical microscope, remember that you just draw two or three fibres as examples. If you are drawing tissues, such those in a section of a root with mycorrhizae, you should draw a tissue map (Figure 27.12), rather than attempting to draw all of the individual cells.

Figure 27.12 A section through a plant stem with a tissue map to show how the areas of different tissue should be drawn.

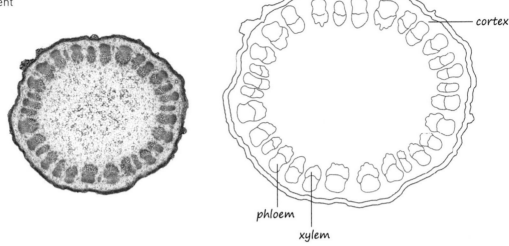

Sometimes you might make drawings from electron micrographs (see Figure 27.13).

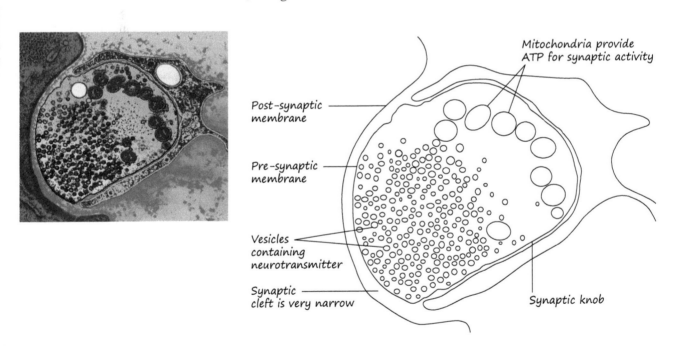

Figure 27.13 An electron micrograph of a synapse together with an annotated drawing of the image.

507

TIP

All drawings should have a title and some indication of scale or magnification. If drawing images seen using a microscope or micrograph, the drawing should show the magnification used to observe the material.

Annotating drawings makes them much more useful. **Annotations** are labels with more information than just the name of the structure. You might include the name of the structure but, equally, annotations can just be descriptive or they might include measurements. Examples of annotation are shown for features of the synapse in Figure 27.13. Remember that, for clarity, you should try to avoid label lines crossing each other or too much of the drawing. Label lines should be straight, drawn in pencil, using a ruler.

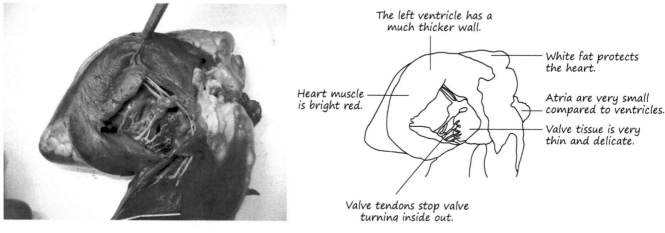

The left ventricle has a much thicker wall.

Heart muscle is bright red.

White fat protects the heart.

Atria are very small compared to ventricles.

Valve tissue is very thin and delicate.

Valve tendons stop valve turning inside out.

Figure 27.14 A sheep heart together with an annotated drawing.

Identifying biological molecules

Being able to identify which biological molecules are present in material from organisms is a useful part of investigating how they work. There are some simple chemical tests that you can use to identify some of the biological molecules that are present in materials or solutions. You can also use Biuret reagent and Benedict's reagent to analyse 'urine' samples from 'patients' with different clinical conditions. Both reagents can be used to produce a **calibration curve** (see Chapter 19, page 349) for a colorimeter to allow the concentration of protein or glucose in an unknown solution to be found. You may also use iodine solution to find out if discs cut from plant leaves and kept in different conditions have been carrying out photosynthesis.

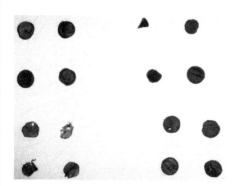

Figure 27.15 Leaf discs tested for starch after being kept in different conditions.

Table 27.1 The four chemical tests for biological molecules and their positive results.

Test for	Reagent	Initial colour	Positive result
Protein	Biuret reagent	Pale blue	Violet solution
Reducing sugar	Benedict's reagent	Blue	Red precipitate
Starch	Iodine solution	Orange	Dark blue solution
Lipid	Ethanol	Colourless	Cloudy white suspension

Biuret reagent and iodine solution are simply added to the sample. A few drops are all that is needed. The colour change showing a positive result becomes apparent immediately if either protein or starch are present. Benedict's reagent must be heated for a few minutes with the sample to show a positive result. A few drops heated to 80 °C for several minutes will give a positive result if reducing sugar such as glucose or maltose is present. The test for lipid is known as the emulsion test. A volume of ethanol equal to the volume of the sample is added and the mixture shaken well. The mixture is then poured carefully into water. If lipid is present, a cloudy white suspension is formed.

Benedict's reagent can also be used to test indirectly for non-reducing sugars. If the sample gives a negative result when tested but a non-reducing sugar such as sucrose might be in the sample, then add a few drops of acid, warm and repeat the test. A positive result the second time around indicates that non-reducing sugar is present.

TIP

You need to learn both the colour of the test reagent and the colour of a positive result.

Separating biological molecules

Chromatography is a way of separating mixtures of biological molecules. It is carried out on paper or on thin layers of solid media such as alumina on glass or plastic plates. A mixture of compounds, such as the photosynthetic pigments isolated from plant leaves, is very carefully pipetted onto an origin line one small drop at a time. In between each drop, the spot is dried. A small but concentrated spot is slowly built up on the origin line with repeated addition of drops and drying between each one. Some 20 or 30 small drops can be applied to the one spot.

When the leaf extracts have been spotted onto the origin line, the plate or paper is held vertically in a small volume of chromatography solvent. The origin line should always be higher than the level of the solvent otherwise the spot dissolves in the solvent rather than moving up the plate or paper.

The solvent slowly moves up the paper or the solid medium on the plate. This may take several hours. As the solvent moves, the different pigments are carried up with it. But they do not all move at the same rate. Those that move faster travel further up the plate or paper so the different pigments separate out into a series of individual spots such as those in Figure 27.17.

Figure 27.16 Spotting solutions onto the origin line of a thin-layer chromatography plate.

Electrophoresis is another separation technique. It separates mixtures of biological compounds on a gel by applying a voltage across the gel. Different compounds move at different rates across the gel depending on their size and charge. Electrophoresis can be used to separate mixtures of proteins or fragments of DNA.

An agarose gel is made by pouring molten agarose into a mould fitted with a plastic comb that forms holes or **wells** in the gel. The gel is allowed to set and the plastic comb is removed carefully.

The gel is placed into an electrophoresis tank. An electrolyte solution is added to the tank so that the gel is submerged. Mixtures of DNA fragments in solution are pipetted carefully into the wells. A power supply is connected to electrodes at each end of the tank so that a voltage is applied across the gel. DNA fragments are negatively charged so they will move towards the positive electrode. This means that the positive wire must be connected to the tank at the end furthest from the wells.

Figure 27.17 Pigments isolated from plant leaves separated by thin-layer chromatography.

Electrolyte A solution that will conduct electricity.

DNA fragments of different sizes will move towards the positive electrode at different rates. This may take several hours. When the separation is complete, the gel is developed using a stain such as methylene blue that will show up the positions of the bands of different sized fragments of DNA. The patterns of bands for different mixtures can then be compared.

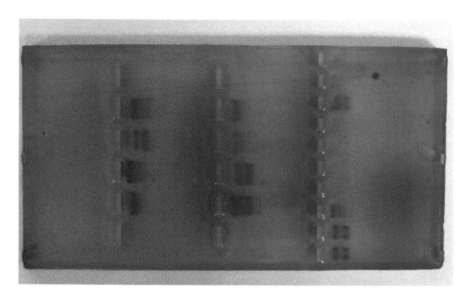

Figure 27.18 A developed electrophoresis gel.

Using living organisms

If you use live animals such as water fleas (*Daphnia*) to observe heart rate or insects such as locusts to look at spiracles, you should take great care to avoid them being harmed. They should only be observed for a short period of time and returned to a larger container as soon as possible. Water fleas should be placed on **cavity slides** so sufficient water can be provided on the slide to protect them from the heat of the microscope lamp and dehydration.

If you use live animals such as maggots or woodlice to investigate animal movement responses you must make sure that they are not harmed during the investigation. Woodlice can be used in choice chambers and maggots can be placed in trays to investigate **taxes** and **kineses**. The effect of factors such as light or humidity on their rate or direction of movement can be discovered.

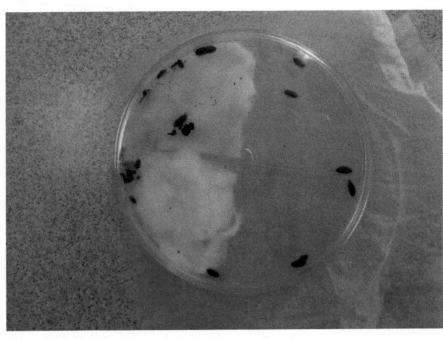

Figure 27.19 Woodlice in a choice-chamber experiment.

If you carry out investigations on other members of your class, for example measuring pulse rate or to measure skin sensitivity or muscle fatigue, you should be very careful to avoid injury. Any exercise should be planned carefully and done in a suitable location to prevent accidents.

Using aseptic techniques to grow microorganisms

Growing bacteria successfully requires sterile conditions so that the culture is not contaminated by other microorganisms that you are not investigating, especially potential human pathogens. It is also important that none of your culture gets into the environment. Appropriate procedures to ensure **aseptic techniques** should be followed.

Aseptic techniques are ways of working that minimise the possibility of contamination or escape of your bacterial culture. Before starting any work with bacteria, the work surface should be sterilised using a suitable disinfectant. Any equipment such as wire loops, pipettes or spreaders that will come into contact with the bacteria should be sterilised before use.

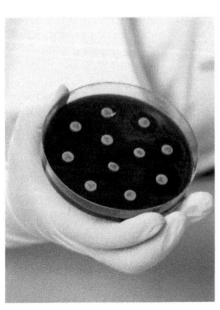

Figure 27.20 Bacteria growing on nutrient agar in a Petri dish, being tested for antibiotic resistance.

It is a sensible precaution to close any nearby windows or doors to prevent draughts that might carry bacteria towards or away from your working area. You should have everything close to hand before starting any transfers of bacteria. Once you begin, you should work quickly but carefully. Bottles or Petri dishes containing bacterial cultures should only be opened for the shortest time possible to reduce the chance of any bacteria getting in or out.

Whenever you open a culture bottle or test tube to remove bacteria, you should immediately move the neck of the container into and out of the Bunsen burner flame while holding the container almost horizontally.

Figure 27.21 Yeast cultures at different stages of population growth.

TurbidityA measure of the cloudiness of a suspension.

This is known as **flaming**. It sterilises the neck of the container and the air above the culture. Before opening the bottle, you should also flame the wire loop you intend using by holding it in the flame until it becomes red hot.

Whenever you open a Petri dish you should hold the lid at an angle above it while you make any transfers as quickly as possible. The lid should then be replaced immediately. Any used pipettes or spreaders should be placed into a suitable disinfectant.

Testing for antibiotic resistance usually involves growing bacteria on nutrient agar (Figure 27.20) and then adding antibiotic on prepared discs of filter paper to the surface of the agar. Where bacteria have been killed by the antibiotic, clear zones are formed around the disc.

You may grow bacteria or yeast in a broth culture to follow their population growth. In both cases, the turbidity of the suspension can be used as a way of measuring the number of cells in the culture. The turbidity is a measure of how cloudy the suspension is. As the population reproduces and grows, it becomes progressively more cloudy. A turbidity meter, or a light sensor or even a colorimeter can be used to measure how much light can get through the suspension at fixed time intervals following inoculation. This would allow a population growth curve to be plotted.

Dissecting animal or plant organs

Observing biological material frequently requires that it be cut open or sectioned so the internal structures are visible. To see the internal features of a heart such as the relative wall thicknesses and the valves requires the ventricles to be cut open. To see the distribution of xylem and phloem in a plant stem, thin cross-sections have to be cut. Sections of gill may need to be removed from a fish such as a herring and the very ends of garlic roots removed for making root-tip squashes to see mitosis.

These activities require the use of sharp instruments such as scissors and scalpels. Handling these confidently but safely is an important practical skill in biology, as is being careful enough to cut and slice the material without damaging it so much that observations are impossible.

Cutting should always be done away from you and your fingers, and on a suitable surface to avoid damage to the work bench. Sharp instruments are better than slightly blunt ones because they require less force to cut and there is therefore less risk of them slipping suddenly.

Using sampling techniques

There are also some outdoor practical skills you need to develop. You should know which sampling techniques to use to estimate the size of a population or to investigate the distribution of organisms. It is much easier to appreciate how the techniques work and their potential problems if you have tried them out for yourself.

The number of individuals in a population cannot normally be counted easily. There are usually too many individuals for this to be practical. Instead, a sample of the population is used to calculate an estimate of the

Motile An organism that moves around; in contrast, non-motile, or sessile, organisms remain in the same place.

population size. How you do this depends on whether the organisms are non-motile, slow moving or motile. In general, animals move around whereas plants stay still. However, there are some animal species such as barnacles that are **sessile** and some that move very slowly such as snails. The population size of plants, sessile and slow-moving animals can all be estimated in the same way, but motile animals require a different approach.

The population size of plants, sessile or slow-moving animals in a given area can be estimated using randomly placed **quadrats**. It is important that the quadrats are placed at **random** to avoid bias, which would result in an over- or under-estimate of the population size. For example, this might happen if you subconsciously positioned your quadrats where there were more of the organisms or if you positioned your quadrats in places that were easier to get at.

A grid is set up across the area and random coordinates are generated using random number tables, dice or the random-number generator on a calculator. Quadrats are then placed at these coordinates.

Figure 27.22 A quadrat being used to sample a population of periwinkles.

The number of individuals in each quadrat are counted. The sample must be **representative** of the population as a whole. You can use a **running mean** to decide when enough quadrats have been used. When the mean number of individuals per quadrat stops changing the sample can be regarded as representative.

The mean number per quadrat is known as the **population density**. This is usually expressed as the number per square metre. The density is then multiplied by the number of square metres in the whole area to give an estimated **population size**.

TIP

The mark-release-recapture method makes some assumptions about the marking and the population and you should be aware of how likely they are to be true when you use this method.

If the population is mobile, the mark-release-recapture method should be used. A sample of the population is caught or trapped and marked in some way so that they can be recognised as having been caught. They are then released and allowed to mix in with the rest of the population. After a suitable time has been allowed for them to mix, another sample is caught or trapped. The technique relies on the fact that the marked individuals will be more diluted in a larger population.

The idea is to see how many marked individuals turn up in the second sample. You can then use the mark-release-recapture equation to calculate the population estimate (see Chapter 22, page 417).

If you want to find out where in a habitat a particular species is found, you will need to set up a **transect**. A transect is a line set out across the habitat along which samples are taken. It is often marked with string or a tape measure. A belt transect is a narrow strip of the habitat, perhaps 0.5 m or 1.0 m wide, alongside the line. Quadrats of the same width are placed in the strip at intervals along the line and the **abundance** of one or more species can be found at each sampling point. Using a belt transect lets you discover if a species is found more in one part of the habitat compared to another.

Figure 27.23 Quadrats placed along a transect can be used to find the distribution of a species in a habitat.

TIP

Using a grid quadrat of a hundred squares reduces the subjectivity of measuring percentage cover. Each square occupied by that species counts as 1% cover.

Figure 27.24 Percentage cover can be used to measure the abundance of a grass species.

Some species are easily counted to find their abundance. But some species are very difficult to count, perhaps because they are very tiny and there are large numbers, or perhaps because it is not clear which individuals are which. In this sort of situation, you can measure abundance by using the techniques of **frequency** or **percentage cover**.

Percentage cover involves making an estimate of the percentage area of the quadrat occupied by the species. If you judge that it occupies about a quarter of the quadrat, the percentage cover is 25%. Obviously this is a little subjective so it is a good idea if the same person in a group always makes the estimate and some practice with this skill helps.

TIP

Frequency and percentage cover used together can tell you different things about the distribution of a species in a habitat.

Another way to measure abundance is by finding the frequency of the species. This is the proportion of samples that contain that species. If the species is found in three out of 10 quadrats, the frequency is 30%. All you have to do to find the frequency is record if the species occurs or not at the sample point and calculate percentage of samples at which the species is found.

Frequency tells you how common a species is, but it does not give information on how much of the species is present. Percentage cover tells you how much space a species occupies in a sample, but does not give any information about how often it may be found in the whole habitat.

These two measures of abundance are related. The more space a species occupies, the more likely it is to be found in more than one sample. Using the two together can give information on how species are distributed in a habitat. A species with a high frequency but a low mean percentage cover occurs as single plants almost everywhere in the habitat. On the other hand, if you find a species with a low frequency but a high mean percentage cover, it would indicate that the species was found in groups but only in certain parts of the habitat.

Using digital technology

Measuring human pulse rate can be done with a pulse sensor linked to a datalogger that can record the data for a specified time period and enable graphs to be produced very easily. When counting your own pulse rate, or that of a partner, it is easy to miscount. A pulse sensor avoids this problem.

The turbidity of a broth culture of microorganisms could be measured at fixed time intervals using a turbidity meter or light meter connected to a datalogger. In the same way, the changes in environmental factors such as light intensity or temperature in an outdoor location could be recorded at fixed intervals during the course of a day. The practical skills you will need to develop are those relating to positioning the sensors, setting up the time intervals for the particular datalogger and downloading the data for analysis, along with presenting and analysing the data and writing conclusions.

Computer programs that model the effects of genetic drift and natural selection are very helpful for illustrating how allele frequencies will change in large and small populations. The advantage of a simulation is that it can model the events over many generations in a short space of time. You will need to choose the conditions that you want to investigate and run the program a number of times while changing the conditions, to see what happens.

Required practical activities in AQA A-level Biology

> **TIP**
>
> Processing and presenting data are important skills. Make sure you know about these and other skills developed in the required practicals.

1 Investigation into the effect of a named variable on the rate of an enzyme controlled reaction (see Chapter 2)

2 Preparation of stained squashes of cells from plant root tips; set-up and use of an optical microscope to identify the stages of mitosis in these squashes and calculation of a mitotic index (see Chapter 5)

3 Production of a dilution series of a solute to produce a calibration curve with which to identify the water potential of plant tissue (see Chapter 3)

4 Investigation into the effect of a named variable on the permeability of cell-surface membranes (see Chapter 3)

5 Dissection of animal or plant respiratory system or mass transport system or of organ within such a system (see Chapter 9)

6 Use of aseptic techniques to investigate the effect of antimicrobial substances on microbial growth (see Chapter 11)

7 Use of chromatography to investigate the pigments isolated from leaves of different plants, e.g. leaves from shade-tolerant or shade-intolerant plants or leaves of different colours (see Chapter 14)

8 Investigation into the effect of a named factor on the rate of dehydrogenase activity in extracts of chloroplasts (see Chapter 14)

9 Investigation into the effect of a named variable on the rate of respiration of cultures of single-celled organisms (see Chapter 14)

10 Investigation into the effect of an environmental variable on the movement of an animal using either a choice chamber or a maze (see Chapter 16)

11 Production of a dilution series of a glucose solution and use of colorimetric techniques to produce a calibration curve with which to identify the concentration of glucose in an unknown 'urine' sample (see Chapter 19)

12 Investigation into the effect of a named environmental factor on the distribution of a given species (see Chapter 22)

28 Exam preparation

Overview

Your exam preparation begins from the moment you enter your first A-level biology lesson. The most successful candidates drive their own learning and do not rely entirely on their teacher or the work they do in lessons. The exams include an element of recall, but simply knowing the facts will not get you a good grade. Most of the marks require a demonstration of understanding and the ability to apply your knowledge in a new context. It is important that you are thoroughly familiar with the topics covered, but also that you have a good biological general knowledge. If something does not make sense, or some information seems to be missing, you should investigate on your own account using the internet and any available books. Scientific magazines and TV programmes should also be used wherever possible to broaden your knowledge base. This will bring you into contact with information, examples and new contexts, which will reinforce your knowledge and understanding of the work on the A-level specification. The BBC News website also has interesting articles in the science section.

A thorough familiarity with the AQA specification is vital. This will tell you precisely what you need to know (and therefore what you do not need to know) and gives details of the assessment process. You will need to be familiar with the format of the exam papers and practical assessment.

If you constantly 'read around' the subject as you go through the course, you will need to spend less time on revision, as many of the basic facts will be hardwired into your brain. Nevertheless, thorough and effective revision will be necessary. If you have a thorough strategy or plan for your revision, you can improve your grades.

Once you get into the exam, the mark you achieve will not depend only on how much you know and understand about the subject. Exam technique plays a big part in the final outcome, as it is essential that you answer precisely what the questions ask and explain yourself clearly. You cannot get marks for material you do not know or understand, but it is essential that you gain every possible mark from the content that you do know.

This chapter outlines some links and tips which will guide you in the right direction. However, different people do learn in different ways and you will need to find out the working practices and revision techniques that work best for you.

Paper 1
Written exam 1 h 30 min
Any AS content, including
relevant practical skills
75 marks
Worth 50% of AS

Short-answer questions (total
65 marks)
Questions on a short
comprehension passage
(10 marks)

+

Paper 2
Written exam 1 h 30 min
Any AS content, including
relevant practical skills
75 marks
Worth 50% of AS

Short-answer questions (total
65 marks)
Structured continuous prose
question (10 marks)

Figure 28.1 The structure of the AS
examination.

The AS examination

The structure of the AS examination is shown in Figure 28.1.

The exam will assess different skills. There are three different assessment objectives, as follows.

A01 Demonstrate knowledge and understanding of scientific ideas, processes, techniques and procedures.

A02 Apply knowledge and understanding of scientific ideas, processes, techniques and procedures:
- in a theoretical context
- in a practical context
- when handling qualitative data
- when handling quantitative data.

A03 Analyse, interpret and evaluate scientific information, ideas and evidence, including in relation to issues, to:
- make judgements and reach conclusions
- develop and refine practical design and procedures.

These skills will be assessed in the two AS papers as shown in Table 28.1.

Table 28.1 Skills to be assessed in the two AS papers.

Assessment objective	Component weightings (approximate %)		Overall weighting (approximate %)
	Paper 1	Paper 2	
A01	47–51	33–37	35–40
A02	35–39	41–45	40–45
A03	13–17	21–25	20–25
Overall weighting of components (approximate %)	50	50	100

A01 questions are the ones that test whether you can understand and recall what you have learned. Notice that this kind of question accounts for less than half the marks overall, so just learning your notes is not enough to get you a reasonable grade in the exam. A02 questions check that you have an in-depth understanding of what you have learned, by presenting you with an example or context that you haven't seen before, and asking you to apply your understanding in a new situation. This means you really need to understand thoroughly what you have learned. These questions account for almost half the marks. Finally A03 questions ask you to evaluate and analyse scientific information or data or relate to practical skills. These are worth up to a quarter of all marks.

In summary, if you hope to achieve a high grade you really need to understand everything inside out, and be good at thinking scientifically.

Table 28.2 presents a list of key words that are used in exam questions, with explanations of what examiners mean when they use each term.

Table 28.2 Key words used in exams.

Describe	This simply means 'tell us about...'. Or 'give an account of'. This is a straightforward test of what you have learned and remember. Alternatively, you might be asked to summarise the trends shown in a graph or table of data that you are presented with.
Explain	This means 'give one or more reasons for'. The best way to make sure that you explain, rather than describe, is to use the word 'because' in your answer. For example, 'This happens because...'.
Suggest	This word is a clue that you are being given an unfamiliar example or context, but you are being asked to apply your AS level understanding in a new or unfamiliar situation. The answer to this question is unlikely to be in your notes, but you should have enough understanding of biology to work out the answer
Evaluate	This means you need to present arguments for and against a point of view or conclusion. A good way to answer questions like this is to think of evidence to support the statement or idea, and then think of evidence why the statement or idea might not be supported. Alternatively, it may be asking you to make a judgement about something, such as an experimental protocol.
Give	Produce an answer from recall or from given information. Don't explain.
Sketch	This means 'draw approximately'. For example, if you are asked to sketch a line on a graph to show what would happen in different circumstances from the line already on the graph, the examiner is only looking to see that you have the right trend and shape on the line you draw. The examiner does not expect mathematical accuracy.

Different types of question

Your exam will consist of several different kinds of question. You should look at past papers and specimen papers so that you recognise the different kinds of question, and understand the best way to answer them.

Structured questions

There are two kinds of structured question. Some require short answers and others require longer answers.

Short structured questions typically score one or two marks. Sometimes only a single word is needed, or a simple calculation.

Longer structured questions typically score three or four marks, although they could be up to six marks. Sometimes you are asked to use information from a table, graph or diagram in your answer, which you must do. Sometimes you are given two command words, such as 'Describe and explain the mechanism that causes forced expiration'. To answer the question well, make sure that you describe and explain. The mark scheme for this question has four marking points:

1 contraction of internal intercostal muscles

2 relaxation of diaphragm muscles/of external intercostal muscles

3 causes decrease in volume of chest/thoracic cavity

4 air pushed down pressure gradient.

Notice how the first two marking points describe the mechanism that causes forced expiration. The second two marking points explain how these actions bring about forced expiration.

Sometimes structured questions involve reading some information at the start of the question. Don't be tempted to skip straight to the questions without reading the information you are given very carefully. Examiners do not give you information to read unless it is necessary to answer the question.

Comprehension questions

In these questions, you will be asked to read a short passage. You should read the passage carefully, then read through the questions that follow. Often, the questions that follow will tell you which lines in the passage the questions are referring to. After you have read all the questions, read the passage again and then answer the questions, one by one. You may be asked to use your own knowledge as well as information from the passage to answer the questions. These questions are testing whether you can relate the biology you have learned in your first year's study to a new situation.

Extended prose answers

These questions typically score five marks or so. These require you to answer clearly and in a well-organised fashion. For example, you may be asked 'describe the differences between active and passive immunity (5 marks)'. The mark scheme may have more than five marking points on it, so you can get full marks by mentioning just five of them. However, be careful that in this example you actually point out the differences. If you simply describe active immunity, and then passive immunity, you are not pointing out the differences. The examiner will not do the work for you. Here is the mark scheme for this question:

1 active involves memory cells, passive does not

2 active involves production of antibody by plasma cells/memory cells

3 passive involves antibody introduced into body from outside/named source

4 active is long term, because an antibody produced in response to antigen

5 passive is short term, because an antibody (given) is broken down

6 active (can) take time to develop/work, passive is fast acting.

Take a look at AQA's website for more sample questions and mark schemes. This will give you a good idea how marks are allocated to sample questions.

Some extended prose questions simply ask you to describe a process, for example: 'When a vaccine is given to a person, it leads to the production of antibodies against a disease-causing organism. Describe how. (5 marks)'. Here, you need to give a clear account of the process, but make sure the points are expressed clearly and in the correct sequence. Good use of scientific terminology is also important.

In this example, the mark scheme has seven marking points and you only need five of them:

1 vaccine contains antigen from pathogen

2 macrophage presents antigen on its surface

3 T cell with complementary receptor protein binds to antigen

4 T cell stimulates B cell

5 (with) complementary receptor on its surface

6 B cell secretes large amounts of antibody

7 B cell divides to form a clone all secreting/producing the same antibody.

Synoptic questions

In answering questions you will be expected to use knowledge and understanding from all the aspects of biology you have studied this year. For example, you may be expected to refer to the primary, secondary or tertiary structure of proteins when answering questions about enzymes and explaining their specificity. Similarly, you may be expected to discuss the specificity of proteins when answering a question about a test that uses monoclonal antibodies.

Answering data questions

Describing a graph

- This means give the trend shown by the graph: don't describe every single point.
- A-level examiners will never give you a simple straight line, so saying 'when *x* goes up, so does *y*' is very unlikely to be the answer required.
- There is usually a change in the gradient of the graph, so make sure you notice this. For example, the answer might be, 'As *x* increases from … to …, *y* increases slowly, but after *x* reaches the value of …, *y* increases much more steeply'.
- Examiners may give you a graph showing a correlation, but this may not be exact. You may be asked to suggest a reason why there is a trend but the correlation is not exact.
- Remember that a correlation between two different factors does not mean that one causes another.

As an example, describe the graphs shown in Figure 28.2.

For example, a description of the first graph would be something like 'As the temperature increases from … to …°C, the rate of sodium ion uptake increases. As temperature increases above …°C, the rate of sodium ion uptake continues to increase, but more slowly'.

If you are then asked to **explain** the data on a graph, you need to give a reason for what you see. Your answer would start 'This happens because…'.

Evaluating experimental designs

- Having a large number of samples or people in an investigation makes the results more **reliable** because any outliers or unusual cases have less of an effect on the overall results.
- We often present results as **percentages** so that we can **compare** the results from two different groups more easily, especially if there are different numbers in the groups.
- A good investigation changes only one variable. For example, if you want to find the effect of temperature on the rate of an enzyme-controlled reaction, what variable do you change and what variables do you keep the same?
- In human studies, we usually can't control all the variables. Therefore, the results might not be completely reliable. This would be the case in a study to find out whether smoking cigarettes increases the chances of developing heart disease.
- It is easier to control all the variables if you do an investigation on animals, but remember that studies on animals might not be applicable to humans because the animals' physiology may be different from ours.

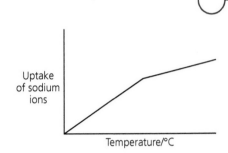

Figure 28.2

- If you are testing a drug, you usually give each person a dose calculated in milligrams per kilogram of body mass, rather than just giving everybody the same dose of the drug. Why?
- Good investigations have an experimental group and a control group. For example, if you are finding the effect of a probiotic yogurt drink on the incidence of colon cancer, you would have your experimental group taking a probiotic yogurt drink every day. What would your control group be doing? And why?
- Remember that the control group is treated in **exactly the same way** as the experimental group, except for the one factor that you are changing.

Sometimes results are given like those in Table 28.3, with standard deviations.

Table 28.3 An investigation to compare loss in body mass over 4 weeks on a low-glycaemic-index (GI) diet and a normal reduced-energy diet.

Group	Mean loss in body mass/kg (± standard deviation)
Normal reduced-energy diet	1.7 ± 0.1
Low GI diet	1.9 ± 0.3

- How would you set up an investigation like this?
- Why did the investigator find a mean loss in body mass for each group?
- Do these data show that the low-GI diet leads to greater weight loss? Explain your answer.
- What does standard deviation show?

Calculations

Make sure you show your working. At AS level, a calculation is often worth two marks, but one mark can be awarded if your method was correct but the answer was wrong.

Exam technique

- Read the question carefully and ensure your answer is accurately targeted to what it asks. If you look at mark schemes you will see that the marking points are very specific and correct but irrelevant points will not be credited.
- Your exams are worth 75 marks and the paper takes 90 minutes. Calculate in advance of the exam how much time you will need to read the questions and then check your answers at the end. Divide the remaining time between the marks available to allocate the time you want to spend on each mark during the exam. Use the number of marks for a question to ensure you have written enough, but not too much. A three-mark question needs three valid points to be made, for example.
- Question writers do not include information that has no use. If you get to the end of the question and there is some information you have not used, alarm bells should ring in your head. If you get stuck on a question, look carefully through the question for 'clues' to the answer.
- Communication must always be clear and detailed. Use appropriate scientific terminology. Avoid vague and general comments which do not include factual information. If you find yourself writing something that you haven't learned during the first year of study of your biology course, then think carefully about whether this is the answer the examiners are looking for.

- Never leave a question blank, especially if it is worth three marks or more. If you find it difficult, leave it and do an easier question, but come back to it at the end of the exam and have a go.
- Don't waste time writing out the question again. If the question is 'Name the monomer from which proteins are made' you just need to write 'amino acids'. Don't waste time with 'The monomers from which proteins are made are amino acids'.
- Make sure you read the whole question before starting to answer. There may be useful information in it, or key words to help you answer the question precisely.
- If you are asked for two things, such as two reasons for something, don't give three. The examiners will only mark the first two answers that you give. If there are more than two answers, and one or more of them is wrong, the examiners will use the wrong answers to cancel out a correct mark earlier in the list.
- Be clear what you are referring to in your answer. Avoid starting the answer with 'It' or 'This'. Make it clear what you are referring to; for example, 'The enzyme…'.
- Write clearly and legibly using a **black pen**. This is because your exam paper will be scanned so it can be marked on a computer. Don't write your answer in the margin or in a space at the bottom of a page. If you need more space, ask for an additional sheet of paper and write the question number clearly before the answer.

Communication skills

Improve these answers, all of which are factually correct but lack clarity or detail.

1 The folded membrane of a mitochondrion is an adaptation for more efficient respiration.

2 The thin wall of the alveolus allows for easier exchange of gases.

3 The reason for the higher blood pressure in the left ventricle is because it needs to pump blood all around the body.

4 The Benedict's test is the test for sugars. The test solution has Benedict's solution added and is then heated. If sugar is present, there is a colour change to brick red.

Revision

To revise effectively, it is useful to have a basic understanding of how memory works. When you receive information, either for the first time or when you are revising something that you have completely forgotten, it enters your working (or short-term) memory. From the point of view of passing exams, working memory is not very useful. It fades very quickly, and if you are going to have any hope of remembering what you've learned, you have got to get it into your long-term memory. Fortunately, this is not difficult. All you really have to do is pay close attention to the information as it comes in, and try to make some sense of it. Information that is meaningless to you will not make it into your long-term memory. It is therefore important that you ask questions of your teacher (or a computer search engine) if you do not understand the information you are given.

However, getting the information into your long-term memory is only half the battle. It is no good having it stored somewhere in your brain if you cannot retrieve it again.

Evidence suggests that the information will stay in your brain for quite a long time, possibly forever, but your ability to access it may fade. Effective revision methods will optimise the retrieval of information, but you will always forget a good proportion of what you have heard or read (see Figure 28.3).

Figure 28.3 Percentage of information retained after different time intervals.

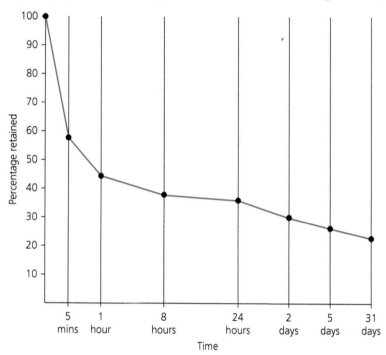

This is not quite as bad as it seems: a lot of what you hear or read is not important. For example, you need to retain the basic ideas, but not the actual sentences used word for word. There are a number of established learning techniques that will greatly increase how much you remember.

- **Repetition**: review new or revised material a week or so after encountering it. This need only be a brief refresh of the memory: you do not have to learn the material all over again. Try to do regular reviews as you go through the course, as this will mean that you already remember much of the material when you start intensive revision.
- **Forming memory cues**: link the information with something you already know. This may be other parts of the course, or personal experiences. The human aspects of biology are particularly suited to this: you have experienced things like breathing, sweating and reflexes, you may know someone with diabetes, etc.
- **Mnemonics**: a mnemonic is a formula or rhyme which assists the memory. A common form is to use the initial letters of a list of words to form a memorable sentence. For example, when memorising the stages of mitosis (prophase, metaphase, anaphase, telophase) you could use the phrase 'penguins march around trees'. When doing this, it is important that the sentence is very memorable: it is better to choose things that are visual and a bit ridiculous! The mnemonic forms a memory cue, but this type of cue is mainly restricted to lists.
- **Translate the information into a different form**: use the information to make a mind map, a picture or a presentation, or explain what you

have learned to someone who knows nothing about A-level biology. If you cannot make someone else understand the information, you may not fully understand it yourself.

- **Do not revise for long without a break**: most peoples' brains cannot take in information continuously for longer than about half an hour without a break. A short break of a few minutes will refresh your brain and allow you to revise more efficiently.
- **Structure your revision**: it has been established that people learn best at the beginning and the end of a session. Start your revision with material that is quite difficult to understand or remember, finish it with similar material (or a review of what you learned at the start) and do the easier content in the middle. This is also a reason for breaks in your session: with breaks, you get more 'beginnings and 'endings'.

The A-level exam

Your A-level exam will consist of three papers.

Paper 1: written exam, 2 hours

- 91 marks, worth 35% of total A-level marks
- Content from sections 1–4, including relevant practical skills. This may involve synoptic questions that make links between these different topics.
- Long- and short-answer questions, 76 marks
- Extended-response questions, 15 marks

Paper 2: written exam, 2 hours

- 91 marks, worth 35% of total A-level marks
- Content from sections 5–8, including relevant practical skills. This may involve synoptic questions that make links between all the topics you have studied over the whole of the A-level course.
- Long- and short-answer questions, 76 marks
- Comprehension question, 15 marks

Paper 3: written exam, 2 hours

- 78 marks, worth 30% of total A-level marks
- Content from sections 1–8, including relevant practical skills
- Structured questions, including practical techniques, 38 marks
- Critical analysis of given experimental data, 15 marks
- One essay from a choice of two titles, 25 marks

Assessment objectives

There are three different assessment objectives describing the skills that **will** be tested in the examinations.

- AO1 (35% of total marks): demonstrate knowledge and understanding of scientific ideas, processes, techniques and procedures. These are the only marks that will benefit from revision shortly before the exam.
- AO2 (35% of total marks): apply knowledge and understanding of scientific ideas, processes, techniques and procedures
  - in a theoretical context
  - in a practical context
  - when handling qualitative data
  - when handling quantitative data.

- AO3 (30% of total marks): analyse, interpret and evaluate scientific information, ideas and evidence, including in relation to issues, to
  - make judgements and reach conclusions
  - develop and refine practical design and procedures.

AO2 and AO3 skills are those you need to develop throughout the course.

The weightings of these assessment objectives are shown in Table 28.4.

Table 28.4 Weightings of A-level assessment objectives.

Assessment objective	Approximate percentage			
	Paper 1	Paper 2	Paper 3	Overall
AO1	44–48	23–27	28–32	30–35
AO2	30–34	52–56	35–39	40–45
AO3	20–24	19–23	31–35	25–30

In addition, the specification states that:

- 10% of the overall assessment of A-level Biology will contain mathematical skills equivalent to level 2 (higher-tier GCSE) or above
- at least 15% of A-level Biology will assess knowledge, skills and understanding in relation to practical work.

These are the 'rules' that examiners must follow when setting examination papers. They represent the **only** thing that you can predict about the series of examination papers you will sit.

The most important thing for you to understand is that, even if you learn your work thoroughly and can recall all the facts, this will only help you to achieve up to 35% of the marks. Therefore, it is vital that you can:

- apply your knowledge and understanding to contexts that you have not been taught
- deal with questions that contain material you have never seen before without panic, because examiners deliberately give you unfamiliar contexts to test your skills
- analyse, interpret and evaluate information you have never seen before.

Your mathematical skills will also be tested. This means you need to go through the mathematical skills chapter in this book, and ensure you understand everything in it. In addition, you need to have a thorough understanding of the practical techniques you have experienced, and be able to apply them in a new situation. This means going through all the practical techniques in this book, and, once again, ensuring you understand the rationale behind all of them. You should be able to evaluate practical designs and methodology as well. You should recognise different kinds of data, and be able to present them in a table or graph, as well as interpret the data. So you must make sure you really understand everything thoroughly. Then check your understanding by doing the questions in this book and any past paper or specimen questions that you can find. Mark your work using the exam board's mark scheme, and make sure you mark your answer very strictly.

Command words

Be very careful to learn what the key command words mean, so that you answer the question in the intended way. Be particularly careful that you know the difference between 'describe' and 'explain'.

Common pitfalls

If you are asked to do a calculation, you will often be told to show your working. This is so that, if you make a mathematical error, you can be credited for your working. Too many students do not show their working, so they fail to gain marks if their answer is wrong.

Avoid starting an answer with 'it'. For example, if the question is 'Give two differences between active transport and facilitated diffusion' and the candidate answers 'It does not use ATP' the examiner does not know what 'it' refers to. If the answer is 'Facilitated diffusion does not use ATP but active transport does' the answer is perfectly clear.

Preparing for the essay question

The essay is worth 25 marks out of a total of 78 on paper 3, so you should allow no more than 40 minutes to answer it. It is important to plan, but don't spend a lot of time on a lengthy plan. Don't spend too long selecting the appropriate topics. A good essay will cover a breadth of relevant examples. Where plant examples are relevant to the title of the essay, remember to include them. Try to choose topics that you studied at different times in your course.

Planning

Suppose you were planning the essay 'The importance of transport in living organisms'. There are different ways you could do this.

One might be to think of different aspects of biology, as shown in Table 28.5.

Table 28.5 Thinking about transport in different areas of biology.

Cells and biochemistry	Physiology
Proteins made and transported by RER to Golgi to vesicle to secretion	Transport of water in plants
	Translocation in the phloem
Transport across membranes	Transport of oxygen by haemoglobin

One method of planning might be to make a spider diagram, as shown in Figure 28.4.

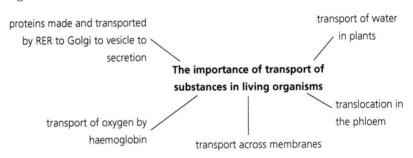

Figure 28.4 A spider diagram outlining transport in different areas of biology.

Whichever method you use, before starting to write you should:

- decide whether each of the topics in your plan is strictly relevant to the essay title and delete those that are not
- decide whether there is a logical order of topics. If there is, write your paragraphs in that order.

Writing

Remember that you need to write as much relevant detail as possible about five topics, in just 35 minutes (assuming you have spent 5 minutes planning), so there is no time to waste.

Do not waste time with an introductory paragraph, saying what you are going to cover. There are no marks for an introduction, so go straight into the first paragraph. 'One way in which transport is important in living organisms is…'. When you have done the first paragraph, it is a good idea to start the second paragraph similarly, as in: 'Another way in which transport in living organisms is important is…'. It doesn't matter if all the paragraphs start similarly, and an advantage of doing this is that it reminds you of the essay title before you write each paragraph, so it helps you to stick strictly to the title. If you include anything that isn't relevant, even if it's really good biology, you will not gain high marks. When you get to the end, just finish at the end of the last biology topic. Don't waste time by writing a final paragraph such as 'So you can see there are many ways in which transport in living organisms is important, and I have talked about some of them…'. This won't gain you any more marks, and you run the risk of being penalised for irrelevance.

> **TIP**
>
> In this example the essay title refers to the **importance** of transport in living organisms. So a student who simply describes transport in living organisms is not answering the essay title. Therefore, if your essay includes transport of water in the xylem, you need to make sure you are saying how this is important and not simply how it happens.

You must use appropriate technical terms in your essay. Avoid terms like 'signals' when you should say 'nerve impulses', 'levels' instead of 'concentrations' and expressions such as 'enzymes being killed' or 'energy being produced'. Write in sentences and make sure you express yourself well.

It is very important that the essay is written in **continuous prose**. This means you should not use bullet points or sub-headings.

Ideally, your essay needs to include some information that is beyond A-level, showing that you have read beyond the specification. If you have answered some of the Stretch and challenge questions in this book you may have some knowledge that goes beyond A-level. Similarly, if you read *New Scientist* magazine, or other science magazines, or regularly check the science items on the BBC News website, you may have some additional information. It is a good idea to flag this up in your essay by writing, 'In a recent *New Scientist* article I read that…'. However, make sure it is relevant. Don't be so keen to show off your additional reading that you include something that does not relate to the essay title. It is not enough to add general background material that anyone would know from watching a natural history programme on TV. Additional material must be A-level standard or higher.

How is it marked?

Examiners are given guidelines for marking the essay.

The examiner reads the essay and annotates it, indicating recall and understanding that is appropriate at A-level, errors, irrelevance, poor use of language and information that goes beyond A-level. Then they read the descriptors for each of the five 'levels' and decide which one fits the essay best. The descriptors for each level represent the middle mark at that level. If the essay doesn't completely fit any of the descriptors, the examiner uses a 'best fit' approach.

Once the examiner has chosen the level, they then decide on which mark to allocate within that level. The mark will be towards the top of that level if it is better than the descriptor, or towards the bottom of that level if it is a little worse than the descriptor.

Table 28.6 An approximate guide to the levels of response.

Mark range	Description
21–25	The biological content is detailed and correct. Several topics are covered, written in clear English using scientific terminology correctly. The topics covered are all relevant to the topic of the question and linked together. To achieve a mark at the top of this band there should be content that is beyond the specification.
16–20	Several relevant topics are covered that are related to each other and the biological content is correct. Most of the topics are covered in detail although there may be one or two topics that are less detailed. Scientific terminology is used correctly. There might be one error or irrelevant topic.
11–15	Most of the topics chosen are relevant to the title but they are not related to each other and not well linked to the title of the essay. Most of the content is correct A-level biology although it may lack detail. Most of the essay uses scientific terminology correctly. There might be more than one irrelevant topic or a small number of errors.
6–10	Only one or two relevant topics are chosen which have some A-level content without much detail. Alternatively, the topics may be more detailed but scientific terminology is not used appropriately. There might be several errors or a few irrelevant topics.
1–5	The response is not well related to the topic of the essay. Biological content may be confined to GCSE level and there may be factual errors as well as irrelevant topics.
0	Nothing relevant included.

Here are some essay titles that you might like to plan.

- The biological importance of proteins
- The importance of water in biology
- Cycles in biology
- The importance of biological polymers
- How the shape of molecules is important to their biological function
- How substances are transferred between living organisms and their environment

Answering the comprehension question

There will be a comprehension question worth 15 marks on paper 2. It will contain a passage for you to read that will relate to your A-level studies but will bring together a number of different topics. It presents these topics in a context that you are unlikely to have seen before. This question is mainly testing AO2 and AO3 skills and not AO1.

The first thing you should do is to read the passage carefully and check that you understand what it is telling you. Then read the questions quickly. This will give you an idea of the parts of the passage you need to focus on especially. Then read the passage again, carefully. Once you have understood the passage, start answering the questions. A line reference is given for each question, so make sure your answer relates to that part of the passage. You are asked to use information from the passage as well as your own knowledge to answer the questions, so make sure you do this.

Answering questions about practical procedures and evaluating data

You will be asked about one or more of the required practical activities that you carried out. When answering, give details that demonstrate you actually carried out these activities. For example, if you are asked how you carried out a Benedict's test for reducing sugar, don't say that you 'warmed' the solution. This is too vague. Give the approximate temperature you used, or say you 'boiled' the solution. If asked to devise a procedure, rather than just saying 'repeat' the investigation, suggest how many times. Say what results you would record, for example, a colour change, time taken for something to happen or a measurement you will make. Then say what you would do with the results, for example plot a graph or carry out a statistical test.

You might find a question that provides information about an investigation carried out by a student. When evaluating such an investigation, consider the student's methodology carefully.

- Did the student keep all the variables constant except the one under investigation? If not, would this affect the validity of the conclusion?
- Did the student use a large enough sample size, carry out sufficient repeats or continue the study for a sufficient length of time?
- Did the student use an appropriate control group to make the investigation valid?

In complete contrast, paper 3 carries a 15-mark question that provides information about an investigation carried out by professional scientists. When answering this question, you must assume that the professional scientists used appropriate methodology, so the above list becomes irrelevant. In this question, you might be asked to justify aspects of the methodology to show your own understanding of scientific investigations. For example:

- if the study was carried out on animals, could the conclusion be applied to humans?
- why was the control group used by the scientists appropriate?
- what was the importance of a double-blind clinical trial?
- can cause and effect be inferred from a correlation between two variables?
- how does the result of a statistical test, or the use of standard errors, help you make a valid conclusion?
- how could the scientists decide an appropriate number of repeats/samples?

Hopefully you will realise by now that success in the exam depends on skills you have acquired over the 2 years of your A-level Biology course, and not simply on revising notes for a few weeks before the exam. From the start of your course, make sure you can carry out the mathematical techniques required and, when you carry out a practical investigation in class, make sure you understand the reasons for everything you are told to do. If you think scientifically throughout the course, you will find it much easier to perform well when you take your written exams at the end of 2 years' study.

We hope you have enjoyed the A-level Biology course and found this book useful. Good luck!

Index

531